International Scientific and Practical Conference on Urgent Problems of Modern Mechanical Engineering

IOP Conference Series: Materials Science and Engineering
Volume 127

Yurga, Russia
17 - 18 December 2015

ISBN: 978-1-5108-2414-0
ISSN: 1757-8981

Printed from e-media with permission by:

Curran Associates, Inc.
57 Morehouse Lane
Red Hook, NY 12571

Some format issues inherent in the e-media version may also appear in this print version.

Printed by Curran Associates, Inc. (2016)

Institute of Physics
Dirac House, Temple Back
Bristol BS1 6BE UK

Phone: 44 1 17 929 7481
Fax: 44 1 17 920 0979

techtracking@iop.org

Additional copies of this publication are available from:

Curran Associates, Inc.
57 Morehouse Lane
Red Hook, NY 12571 USA
Phone: 845-758-0400
Fax: 845-758-2633
Email: curran@proceedings.com
Web: www.proceedings.com

Table of contents

Volume 127

International Scientific and Practical Conference on Urgent Problems of Modern Mechanical Engineering
17–18 December 2015, Yurga, Russia

Accepted papers received: 6 April 2016
Published online: 6 May 2016

IOP Conf. Series: Materials Science and Engineering **127** (2016) 011001 doi:10.1088/1757-899X/127/1/011001

International Scientific and Practical Conference on "Urgent Problems of Modern Mechanical Engineering" took place at Yurga Institute of Technology of National Research Tomsk Polytechnic University on December 17-18, 2015. The Proceedings contain the selected scientific reports submitted to the Conference.

The scientific reports published in these Proceedings have been revised and approved by the editorial team of the Conference. All of the reports exhibit clear, concise, and precise expositions that appeal to a broad international readership interested in mechanical engineering, welding, metallurgy, material science, computer-aided manufacturing. The reports present original ideas or results of general significance supported by clear reasoning and compelling evidence, and employ methods, theories and practices relevant to the research. The authors state clearly the questions and the significance of their research to theory and practice and describe how the research contributes to new knowledge; and provide tables and figures that meaningfully add to the narrative.

Organizing Committee of the Conference thank all the participants of the Conference for their fruitful work and personal contribution to the development of these Conference Proceedings.

Dmitry A. Chinakhov, PhD,
Conference Convenor,
Vice-Chancellor, Yurga Institute of Technology
of National Research Tomsk Polytechnic University

Peer review statement

All papers published in this volume of *IOP Conference Series: Materials Science and Engineering* have been peer reviewed through processes administered by the proceedings Editors. Reviews were conducted by expert referees to the professional and scientific standards expected of a proceedings journal published by IOP Publishing.

IOP Conf. Series: Materials Science and Engineering **127** (2016) 012001 doi:10.1088/1757-899X/127/1/012001

Electric Transport Traction Power Supply System With Distributed Energy Sources

E Y Abramov[1,4], N I Schurov[2,5], M V Rozhkova[3,6]

[1, 2, 3] Novosibirsk State Technical University, 20 Prospekt K. Marksa, Novosibirsk, 630073, Russia

E-mail: [4]ewgeniyabramow@gmail.com, [5]nischurov@mail.ru, [6]marina.rozhkova.63@mail.ru

Abstract. The paper states the problem of traction substation (TSS) leveling of daily-load curve for urban electric transport. The circuit of traction power supply system (TPSS) with distributed autonomous energy source (AES) based on photovoltaic (PV) and energy storage (ES) units is submitted here. The distribution algorithm of power flow for the daily traction load curve leveling is also introduced in this paper. In addition, it illustrates the implemented experiment model of power supply system.

Introduction

The urban electric transport is one of the most important and energy-consuming city agglomeration users. Therefore the arrangement of conditions of sustainable electric transport system development with contemporary high technology solution adoption is necessary for providing high economic growth rate and improving the quality of people's lives. The basis of any transport system is power supply infrastructure, however, electric energy process losses in TSS converter units and traction network are 15-20 % on consumption to perform useful work [1, 2, 3]. Due to Joule-Lenz law the value of resistance losses in TPSS is:

$$\Delta P = \sum_{i=1}^{n}\left(\frac{1}{T}\cdot\int_{0}^{T} R_{TSi}\cdot I_{TSi}^{2}(t)\cdot dt\right)+\sum_{k=1}^{m}\left(\frac{1}{T}\cdot\int_{0}^{T} R_{Grid\ k}\cdot I_{Grid\ k}^{2}(t)\cdot dt\right), \quad (1)$$

whereR_{TS}– resistance of TSS rectifiers and transformers; I_{TS}– TSS current; $R_{Grid\ k}$ – resistance of k-th element of traction network; $I_{Grid\ i}$– current of k-th element of traction network.

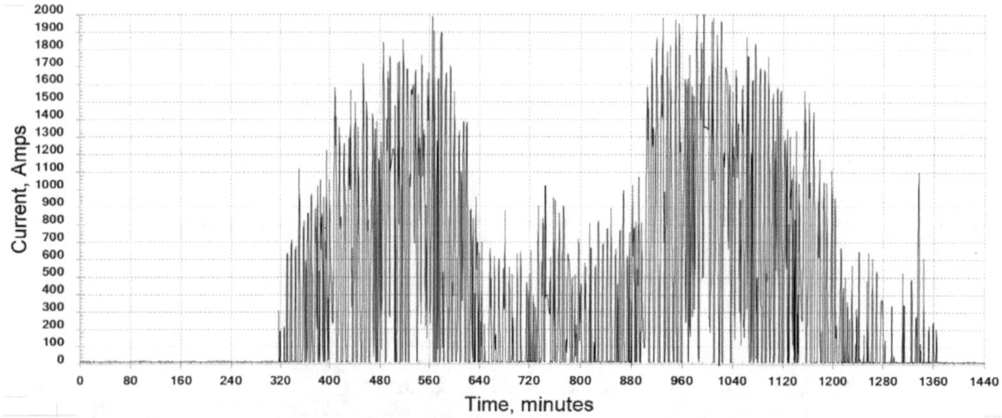

Figure 1. Sample of measured oscillogram TSS current

The researches of instantaneous current and voltage on DC bus made by means of measuring system based on Velleman K 8047/PCS10 in operating TSS of Novosibirsk city ground-based electric transport showed that the functional operation of TSS is an obvious pulse pattern with a power distribution variation by the hour for 24-hour period. In this regard the highest power consumption is due to morning and evening rush hour (figure 1) The load decrease these periods provides sufficient engineering-and-economical effects.

To estimate the energy losses value in TPSS an integrated index of power consumption variation is used as a form factor, given that the index is a ratio of instantaneous value to average and it can be rated as proportional to resistance losses value.

$$K_F = \frac{\sqrt{\frac{1}{T} \cdot \int_0^T P^2(t) \cdot dt}}{\frac{1}{T} \cdot \int_0^T P(t) \cdot dt}. \quad (2)$$

The desired degree of daily traction load curve leveling can be provided by increasing the self-supporting of TPSS. This trend is related to the application of distributed energy sources based on ES and alternative kinds of restricted power sources carrying out the power flow distribution by means of static converters.

The approaches to the integration of distributed TPSS were built in 60^{th}years, however these researches were not evolved due to engineering constraints. New prospects of inventing more effective power systems are emerging at this point and here the number of distributed power supply systems is growing every year [4, 5, 6, 7], in addition, a greater part of generation rating is accounted for solar energy.

A number of researchers throughout the world have extensively been working over the project implementation of power supply systems by using PV and ES systems. But the theory and practice developments on co-utilization of these energy sources in TPSSs were not considered properly.

1. The circuit of TPSS

On basis of research of current distribution through electric circuit branch [1, 8], we can conclude the circuit with distributed connection of PV and ES systems will have the least energy losses proceeding from possible circuit with AES. Besides, such circuit will be more tailored to use PV systems with a large area. Figure 2 presents the circuit of single- side feed with parallel connection of contact wires in which the distributed connection of AES is used.

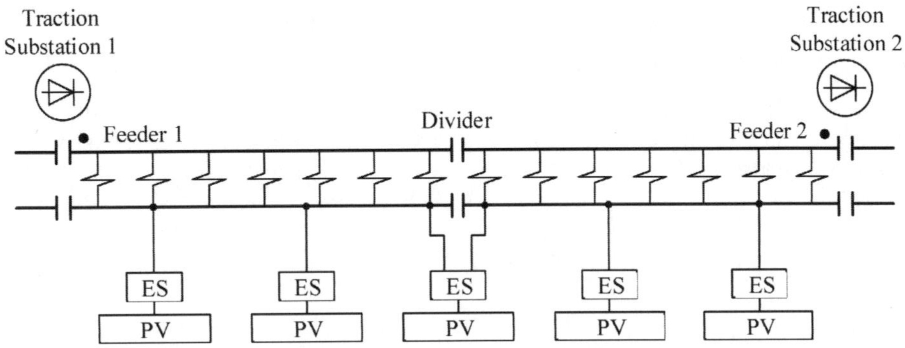

Figure 2. Circuit of TPSS with distributed energy sources

The structure diagram solution is suggested for combining solar batteries with ES, it is shown in figure 3. The diagram concept is based on the integration of electricity storage and primary energy sources as single modules by using power and microprocessing electronic units. [9]

The basic operating mode is the daily traction load curve leveling on account of intelligent power flows distribution between TSS and AESs. The distribution control is provided by means of AES DC/DC converter and controlled rectifier. On basis of represented figure 3diagram the experimental power plant model was implemented into practice.

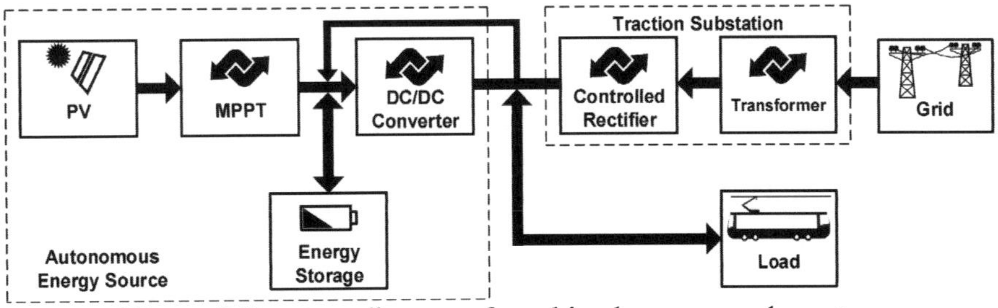

Figure 3. Structure diagram of combined power supply system

2. Control system

Figure 4 shows the operational diagram of power flows distribution in the system.

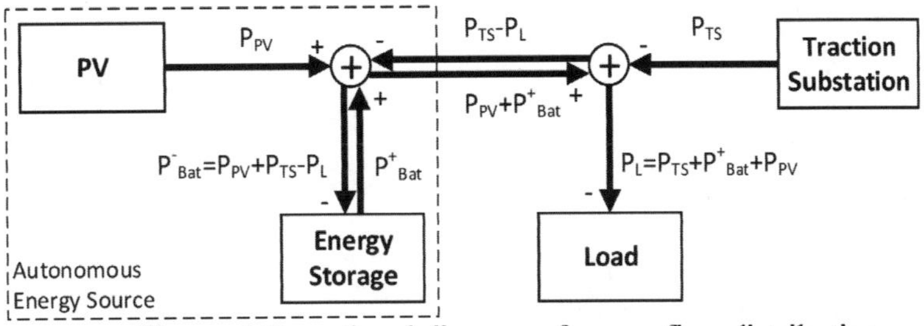

Figure 4. Operational diagram of power flow distribution

The target function is formulated for the basic operating mode:

$$f=\frac{1}{n}\sum_{t=1}^{n}\left[\left(P_{TS}(t)-\left(P^{+}_{Bat}(t)+P_{PV}(t)\right)-\left(P_{T\Pi}(t)-P_{L}(t)\right)\right)-\overline{P_{TS}}\right]^{2}\rightarrow \min, \tag{3}$$

IOP Conf. Series: Materials Science and Engineering **127** (2016) 012001 doi:10.1088/1757-899X/127/1/012001

where $P_{TS}(t)$ – TSS power, $P^{+}_{Bat}(t)$– ES discharge power, $P_{PV}(t)$– PV power, $P_{L}(t)$– load traction power, $\overline{P_{TS}}$ – TSS average power for 24hour period.

The expression (3) introduces the optimization function by minimum criterion dispersion of load TSS. The average value of this load for 24hour period is the mathematical expectation value. Three control actions are emphasized in compliance with the control object structure.

1. Control signal of operating mode circuit. Switching-over from day mode to night one and backwards.

2. PWM signal for MPPT. Adaptive maximum power tracking in output of PVs is carried out on basis of Fuzzy Logic.

3. PWM signal to control load distribution between energy sources. The duty cycle signal is proportional to the capacity credit and it may be used to transmit from AES to the load. This part of power is defined in accordance with a target function. To match the output power the out- of- level reverse control signal is fed to an active rectifier. Thus, the following ratio is correct:

$$P_{TS} = (P_{PV} + P^{+}_{Bat}) \cdot \frac{D_1}{D_2}, \quad (4)$$

whereD_1 – duty cycle of control signal by means of AES DC/DC converter, D_2 – duty cycle of control signal by means of controlled rectifier.

At day mode the charge power is equal to the difference of PV output power and power transmitted to the load due to constraints on maximum charging current rate. Whereas, the PV output power is always supported at maximum level. At night mode the system goes into storage charge from the grid due to an active rectifier, the charge power is defined as:

$$P^{+}_{Bat} = \frac{E_{Bat.MAX} - E_{Bat}}{T_{CH}}, \quad (5)$$

where $E_{Bat.MAX}$ – maximum ES capacity; E_{Bat} – ES state of charge; T_{CH}=1700 c – charge time.

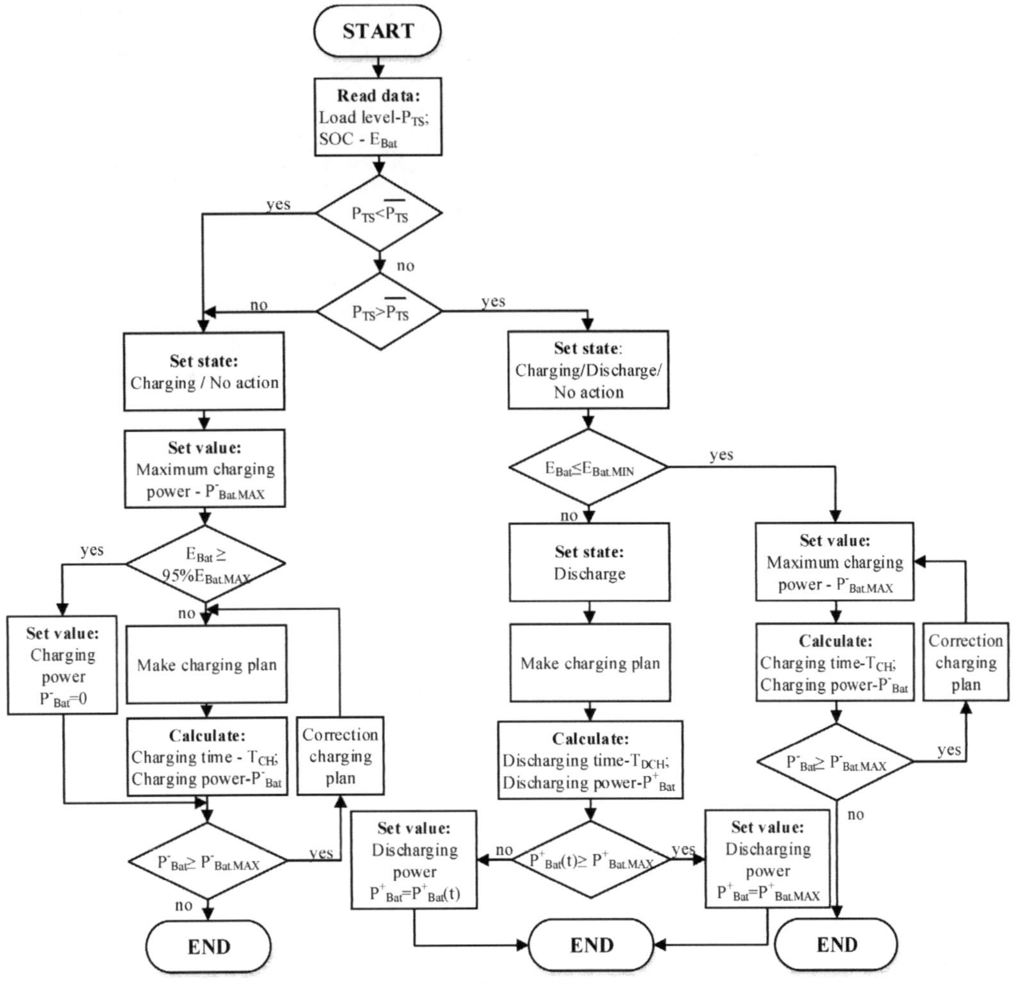

Figure 5. Power flow distribution control algorithm

At day mode the control function is the definition of output power storage value. The solution of the given task and also the charge storage task is presented as algorithm in figure 5.

The load curve leveling principle. The ES charge is carried out at night time hours at zero power consumption (its price is the lowest) as well as at PV power redundancy. The load factoring is provided on account of power generated by PV combined with storage power during the periods of low or zero energy consumption and during the periods of PV power redundancy. This PV energy use is more rational on peak busy hours rather than on its generation period.

3. Experimental results

The experiment model of power generation system based on the integration of PV system, ES and grid is implemented and tested as part of presented solutions. The control program is implemented in Lab View.

To test and demonstrate the operation by means of program-controlled simulator the daily load curve with form factor 1.3 is modeled. The derived oscillograms of currents and voltages on the load are shown in figure 6,a. The oscillogram of PV output power is shown in figure 6,b. These oscillograms demonstrate the experiment model operating results illustrated in figure 6,c and d. As a result of experiment model test on load leveling the form factor of the grid equal to 1.016 is received.

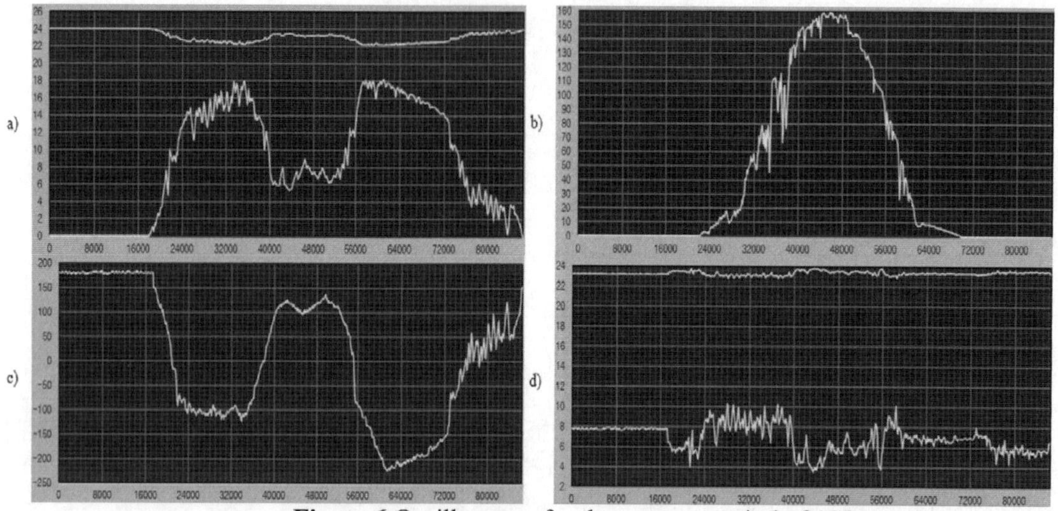

Figure 6. Oscillograms for the average period of 100 s:
a – grid voltage and current without leveling; b – PV output power; c – ES charge/discharge power; d – grid current and voltage when operating at AES

Conclusion

The paper presents the researchers' results on the load leveling of power supply system for urban electric transport by means of PV and ES systems. The distributed connection circuit of these AESs to the traction network for reduction energy losses is proposed here. The experiment model of power generation system and the experiment control program are implemented. The control program is based on the presented algorithm of distributed power flows. Further the program performs the circuit modes switching and tracking PV maximum output power based on Fuzzy Logic. The test on load curve leveling is conducted. At the same time the storage batteries with energy capacity above the mark are used. As a result the form factor of the grid daily-load curve equal to 1.016 is received. Taking into account the fact that the form factor is proportional to resistance losses value then the power loss reduction amounts to as much as 22%.

To improve the program operation it is necessary to calculate the charge time based on statistic data and also update the algorithm by the module of charge and discharge integrator in accordance with the quality management program assessment.

The objective of the future research is a ratio determination between PV power and ES capacity, whereby, the rational rate of leveling traction load curve will be provided. The TPSS with distributed connection circuit of AESs for the light rail transit line in Novosibirsk will be the research object.

References

[1] N I Schurov, E A Spiridonov, A V Larin Modes of traction power supply system in case of electric rolling stock equipped with energy storage Applied Mechanics and Materials **698** (2014) 19-23

[2] V V Biryukov, M V Kalugin, A O Nevolina Ways of decreasing fuel consumption in vehicles Applied Mechanics and Materials **792** (2015) 542 -545

[3] N I Schurov, A A Shtang, M V Yaroslavtsev Application of energy storage devices at the road city transit rolling stock Applied Mechanics and Materials **792** (2015) 536-541

[4] N Hatziargyriou, H Asano, R Iravani, C Marnay Microgrids IEEE Power Energy Magazine (2007) 578-94

[5] N W Lidula, A D Rajapakse Microgrids research: areview of experimental microgrids and test systems Renewable and Sustainable Energy Reviews **15** (2011) 186-202

[6] H Jiayi, J Chuanwen, X Rong A review on distributedenergy resources and MicroGrid Renewable and Sustainable Energy Reviews **12** (2008) 2465-2476
[7] B Kroposki, R Lasseter, T Ise, S Morozumi A look at micro grid technologies and testing projects from around the world, making micro grids work IEEE Power and Energy Magazine (2008) 640-653
[8] N I Schurov, V I Sopov, V V Ivanov, A P Kuznetsov Adaptive systems of traction power supply Russian Electrical Engineering **85** (2014) 729-733
[9] E Y Abramov, A A Stang, S A Enkudinov Transformation of the urban electric transport system when using autonomous energy sourcesAdvanced Materials Research **1040** (2014) 778-783

IOP Conf. Series: Materials Science and Engineering **127** (2016) 012002 doi:10.1088/1757-899X/127/1/012002

The Method of Fundamental Solutions using the Vector Magnetic Dipoles for Calculation of the Magnetic Fields in the Diagnostic Problems Based on Full-Scale Modelling Experiment

Yu A Bakhvalov[1], V V Grechikhin[2],A L Yufanova[3]

[1,2,3] The Platov South-Russia State Polytechnic University (NPI), 132 Prosvesheniyaul., Novocherkassk 346428, Russia

E-mail: vgrech@mail.ru

Abstract. The article describes the calculation of the magnetic fields in the problems diagnostic of technical systems based on the full-scale modeling experiment. Use of gridless fundamental solution method and its variants in combination with grid methods (finite differences and finite elements) are allowed to considerably reduce the dimensionality task of the field calculation and hence to reduce calculation time. When implementing the method are used fictitious magnetic charges. In addition, much attention is given to the calculation accuracy. Error occurs when wrong choice of the distance between the charges.The authors are proposing to use vector magnetic dipoles to improve the accuracy of magnetic fields calculation. Examples of this approacharegiven. The article shows the results of research. They are allowed to recommend the use of this approach in the method of fundamental solutions for the full-scale modeling tests of technical systems.

Introduction

The need to save energy and resources leads to creation of systems with different limit modes and high electromagnetic, mechanical, thermal loads on the materials and structures.All this requires reliable identification of research objects, that is determination of the parameters and characteristics of the materials and constructionsused in the mathematical modeling with sufficient accuracy.However, a number of parameters and characteristics cannot be measured directly.In such cases, the only way to get information is the solution of inverse problems. They allow define the parameters of the subdomains that are inaccessible to measurement, using the results of measurements in available subdomains of the object[1-4]. For the successful implementation of such an approach is necessary mathematical models and methods of physical fields calculation, focused on the implementation of full-scale modeling test systems that provide the definition of the parameters of systems with high precision and speed.

In recent years, wide application received gridless method of fundamental solutions (MFS) and its variants in combination with grid methods (finite differences and finite elements) for calculating fields [5-8]. The calculation of the magnetic field of the electromagnet, moving along the ferromagnetic plate, considered in article. In this article, the calculation of the dipole magnetic field is used in the auxiliary problem.In research presented in [9], the field in the space around the ferromagnet is determined by using a fictitious magnetic charges, the field in the nonlinear medium is calculated by finite element method.This approach is significantly reduces the dimensionality task of calculation of the field and, thereby, reduces calculation time.However, when using dipoles, the field determined with an accuracy that depends on the distance between the charges $+q_i$ and $-q_i$ [10,11].

In this paper, it was proposed to use MFS vector magnetic dipoles to improve the accuracy of magnetic fields calculation in the problems diagnosticof systems based on full-scale modeling experiment.

2. Methodology

For exception of error calculation of the magnetic field by MFS, is usedapproach needed in the calculation of electrostatic fields [12]. This error caused by the introduction of fictitious magnetic dipoles.If charges $+q$ and $-q$ are positioned as shown in figure 1, then value of the potential at an arbitrary point P is determined by expression

$$\varphi_e(P)=\frac{1}{4\pi\varepsilon_0}\left(\frac{q}{r_+}-\frac{q}{r_-}\right)=\frac{q}{4\pi\varepsilon_0}\left(\frac{1}{\left|\vec{r}_-+\vec{l}\right|}-\frac{1}{\left|\vec{r}_-\right|}\right). \quad (1)$$

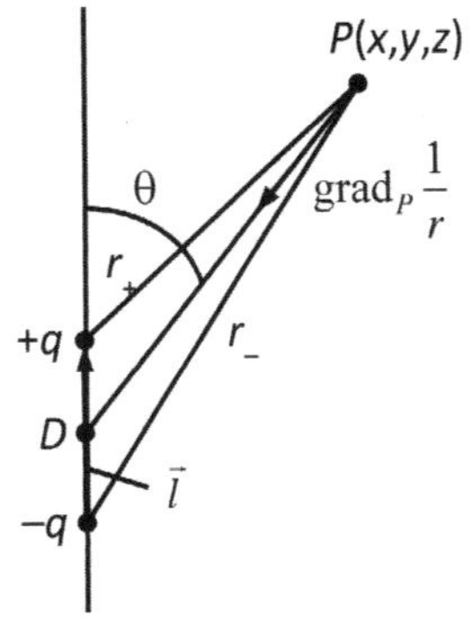

Figure 1. To determine the scalar magnetic potential.

By the definition of the gradient, considering that the quantity $l=\left|\vec{l}\right|$ is sufficiently small compared $\left|\vec{r}\right|$ we obtain

$$\frac{1}{\left|\vec{r}_-+\vec{l}\right|}-\frac{1}{\left|\vec{r}_-\right|}\approx-\vec{l}\,\text{grad}_P\frac{1}{r_{DP}}. \quad (2)$$

Here, the index P is that in terms of the gradient, coordinates of point P are considered as variables. Onthebasisof (1) and (2) weobtain

$$\varphi_e(P)=-\frac{q\vec{l}}{4\pi\varepsilon_0}\text{grad}_P\frac{1}{r_{DP}}. \quad (3)$$

We will reduce the distance l between the charges and to increase q so that $q\vec{l}=\vec{m}_e$ remain constant.

Then, when $l\to 0$ the expression (3) will give the exact value of the potential in the limit

$$\varphi_e(P)=-\frac{1}{4\pi\varepsilon_0}\vec{m}_e\text{grad}_P\frac{1}{r_{DP}}. \quad (4)$$

Vector $\vec{m}_e$ as is $\vec{l}$ directed from a negative to a positive charge. The magnitude $\vec{m}_e$ is called an electric dipole moment. Dipolecharacterized $\vec{m}_e$ iscalledavectorelectricdipole.

Considering the analogy of magnetic and electric fields we will determine magnetic dipole moment of the formula

$$\vec{m}_m = q_m \vec{l}\,,$$

whereq_m – fictitious magnetic charges and the dipole is a vector magnetic dipole$\vec{m}$.

The scalar magnetic potential of such a dipole in a vacuum is defined by analogy with (4)

$$\varphi_m(P) = -\frac{1}{4\pi\mu_0}\vec{m}_m \operatorname{grad}_P \frac{1}{r_{DP}}. \quad (5)$$

Equation (5) forthescalarmagneticpotentialvectordipolecanbeintheform

$$\varphi_m(P) = \frac{1}{4\pi\mu_0}\frac{\vec{m}\vec{e}_r}{r_{DP}^2} = \frac{1}{4\pi\mu_0}\frac{m\cos\theta}{r_{DP}^2}.$$

The potential of vector dipole is zero in the plane normal to the axis of the dipole and extending through its center.

The magnetic field intensity of vector dipole determined by the formulas

$$H_r = -\frac{\partial\varphi}{\partial r} = \frac{1}{2\pi\mu_0}\frac{m\cos\theta}{r_{DP}^3};\; H_\theta = -\frac{1}{r}\frac{\partial\varphi}{\partial\theta} = \frac{1}{4\pi\mu_0}\frac{m\sin\theta}{r_{DP}^3};\; H_\alpha = 0.$$

Figure 2 shows the placement of the normal dipoles to the problem considered in [9].In the figure positive charges dipoles $+q_i$are placed above the planewith $\varphi_m = 0$, the negative charges of the dipoles $-q_i$– under the plane.PointsD_i are located on the planewith $\varphi_m = 0$.Figure 3 shows the placement of vector magnetic dipoles $\vec{m}_i$ for the same problem.

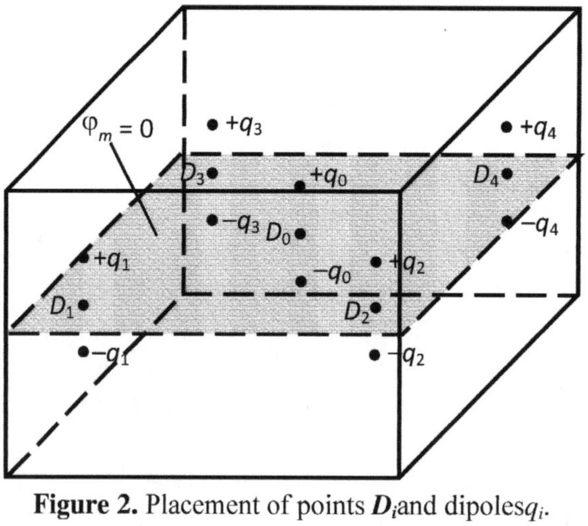

Figure 2. Placement of points $D_i$and dipolesq_i.

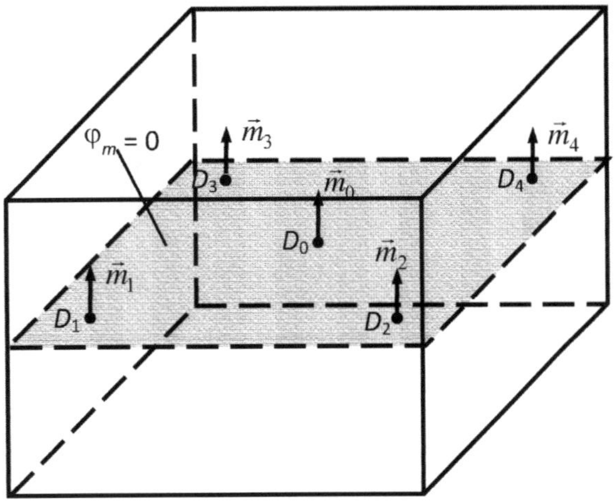

Figure 3. Placement of vector magnetic dipoles $\vec{m}_i$.

3. Results and Discussion

3.1. Solution test problem

Consider the use of vector magnetic dipoles to solve the test problem "the continuous ferromagnetic sphere in a homogeneous magnetic field", using the results obtained in [9].

For the calculation of the magnetic field outside of the sphere in figure 4 is used an ordinary dipole and vector magnetic dipole in figure 5.

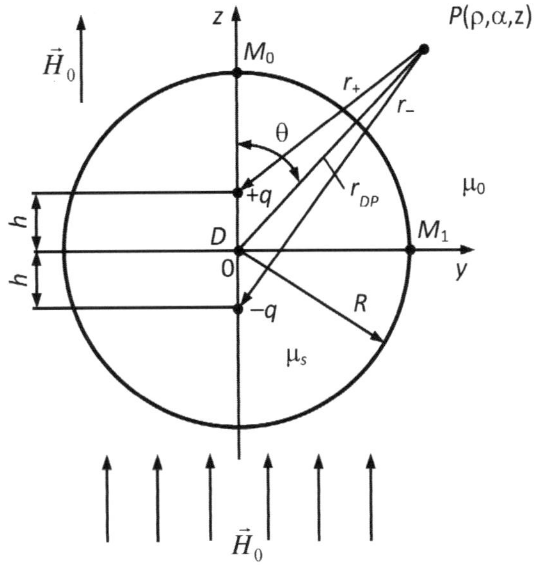

Figure 4. Location of the charges $+q$ and $-q$, used for calculation of the field outside of the sphere.

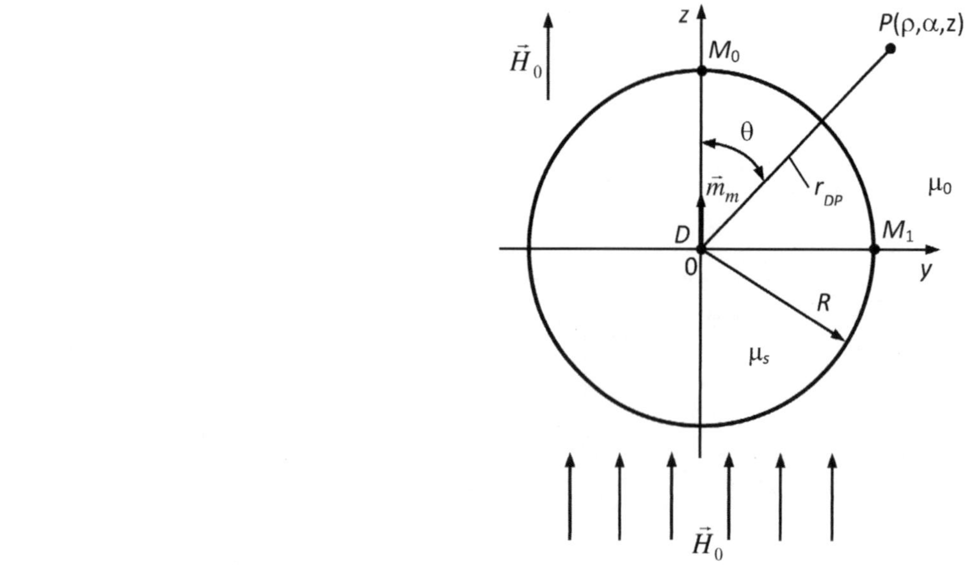

Figure 5. Locationvector magnetic dipole for the calculation of the field outside of the sphere.

Analytic solution of a test problem is presented in [13]. The exact value of the z-component of the magnetic field outside of the sphere at the point M_0 determined by formulas

$$\vec{H}_z^-(M_0) = 2\frac{\mu_s - \mu_0}{\mu_s + 2\mu_0} H_{0z}, \qquad (6)$$

whereH_{0z}–z-component of the intensity external homogeneous known magnetic field; μ_s – sphere magnetic permeability, where $\mu_s >> \mu_0$.

Formulas for determining of the same intensity by MFS (figure 4) has the form

$$\widetilde{H}_z^-(M_0) = q_0\left(\frac{1}{(R-h)^2} - \frac{1}{(R+h)^2}\right). \qquad (7)$$

Suppose $h = R/n$. Substituting$h$intotheequation (7), weobtain

$$\widetilde{H}_z^-(M_0) = q_0 \frac{4n^3}{R^2(n^2-1)^2}.$$

In [9] the test problem is solved by conventional dipoles (figure4) using the boundary conditions at the point M_0. A result we obtain

$$\widetilde{H}_z^-(M_0) = \frac{n^2}{n^2-1} 2\frac{\mu_s - \mu_0}{\mu_s + 2\mu_0} H_{0z}.$$

Estimate relativeerror $\widetilde{H}_z^-(M_0)$

$$\delta\left(\widetilde{H}_z^-(M_0)\right) = \frac{\left|\widetilde{H}_z^- - H_z^-\right|}{\left|H_z^-\right|} 10^2 = \frac{200}{n^2-1}\ \%.$$

Dependence $\delta\left(\widetilde{H}_z^-(M_0)\right)$on $n = R/h$ is presented in the table 1

Table 1.Dependence $\delta\left(\tilde{H}_z^-(M_0)\right)$ on $n = R/h$.

$n = R/h$	2	4	6	8	10
$\delta\left(\tilde{H}_z^-(M_0)\right)$ %	33	6.7	2.9	1.6	1.0

With the same precision were defined other quantities problem: $H_z^+(M_0)$; $\varphi_m^-(M_0)$; $\varphi_m^+(M_0)$.
Consider the solution of a test problem with a vector magnetic dipole (figure 5).
Using the boundary conditions at the point M_0, we obtain the ratios of the following form [9]

$$\begin{cases} \varphi^+(M_0) = \varphi^-(M_0) + \varphi_\delta^-(M_0); \\ \mu_s H_z^+(M_0) = \mu_0 H_z^-(M_0) + \mu_0 H_{0z}, \end{cases} \tag{8}$$

where $\varphi^+(M_0) = -H_z^+ R$; $\varphi_\delta^-(M_0) = -H_{0z} R$; $\varphi^-(M_0) = -m'_m \text{grad}\dfrac{1}{r_{DM_0}} = \dfrac{m'_m}{R^2}$;

$$H_z^-(M_0) = -\frac{\partial\varphi}{\partial r} = \frac{2m'_m}{R^2}; \; m'_m = \frac{m_m}{4\pi\mu_0}.$$

The system (8) are two unknown quantities H_z^+ and m'_m .
Solving the system (8), we obtain

$$\tilde{H}_z^+(M_0) = \frac{3\mu_0}{\mu_s + 2\mu_0} H_{0z};$$

$$m'_s = \frac{3\mu_0}{\mu_s + 2\mu_0} R^3 H_{0z}.$$

From the second equation of (8) can be defined $H_z^-(M_0)$ through $H_{0z}(M_0)$:

$$H_z^-(M_0) = 2\frac{\mu_s - \mu_0}{\mu_s + 2\mu_0} H_{0z}. \tag{9}$$

Comparison (9) and (6) shows that the use of vector magnetic dipoles provides a solution of test problem that identical to the analytical solution.

3.2. Theresultsofexperimentalresearch

We choose to research the magnetic system, consisting of a pair of coils, between which is a part of the active element from a ferromagnetic material and its dimensions match the dimensions of the coil windows(figure6).In the figure: $\vec{\delta}$ – the current density in the coils; V^+ – subdomain occupied by an active element (ferromagnetic with a magnetic permeability μ); V^- – subdomain surrounding the element of space, filled by linear medium with μ_0. The system has the following parameters:
$A_e = B_e = A_c = B_c = 2 \cdot 10^{-3}$ m; $H_e = 1 \cdot 10^{-3}$ m; $H_c = 2 \cdot 10^{-3}$ m; $C_c = 0.5 \cdot 10^{-3}$ m.

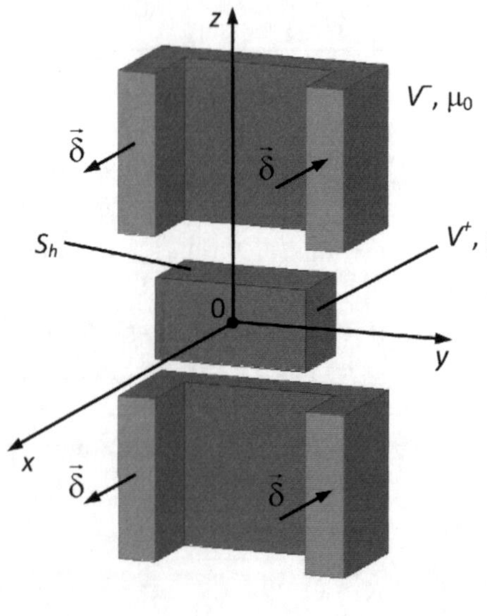

Figure 6. The investigated magnetic system.

In order to estimate accuracy calculation of the magnetic field by MFS using fictitious magnetic dipoles (figure 2) and the vector magnetic dipoles (figure 3), the problem is solved by finite element method (FEM).

The coils and the active element is placed in the center of the parallelepiped with sides $0.5\times0.25\times0.25$ m, it was filled by 87434 tetrahedra.On the edges of the parallelepiped is given vector magnetic potential $\vec{A}=0$. Were determined the values H_0^- when magnetomotive force of each coil $iw_c=10^3$ A in the pointsD_i.Then there were calculated $\varphi_0^-(M_i)$ and determined of the values H_z^+ and φ^+ in the pointsD_i by numerical integration.

A comparison was made received results and calculations performed by the MFS (table 2, 3). In the calculations is taken into account symmetry of the magnetic system.If the number of collocation points $N=5$ (figure 2, 3) then due to the symmetry of their position $q_1=q_2=q_3=q_4$ and unknown $H_z^+(M_i)$, $i=0,1$; $N=13$ –by symmetry positions of the points have four unknowns charge q_0, q_1, q_2, q_3 and unknown $H_z^+(M_i)$, $i=0,1,2,3$.

Table 2.The results of calculations by MFS at the number of collocation points $N=5$.

Numberofpoints(i)	Calculation usingmagneticdipoles	Relative error	Calculation using vector magnetic dipoles	Relative error	FEM calculation
	H_z^+, A/м	$\delta(H_z^+)$, %	H_z^+, A/м	$\delta(H_z^+)$, %	H_z^+, A/м
0	449.2	68.7	445.7	67.4	266.2
1	317.4	49.3	315.0	48.2	212.5

Table 3.The results of calculations by MFS at the number of collocation points $N = 13$.

Numberofpoint s(i)	Calculation usingmagneticdipoles	Relative error	Calculation using vector magnetic dipoles	Relative error	FEM calculation
	H_z^+, A/m	$\delta(H_z^+)$, %	H_z^+, A/m	$\delta(H_z^+)$, %	H_z^+, A/m
0	321.6	20.8	320.2	20.1	266.2
1	247.9	16.6	246.9	16.2	212.5
2	219.3	–7.9	219.5	7.8	238.0
3	286.4	16.2	285.8	15.9	246.4

Comment. The results which are given in the table 2 and 3 are obtained by $\mu = 10^3\mu_0$.

Results of experimental research show that sufficient accuracy for calculating the parameters of the magnetic field for engineering problems can be obtained with a small number of fictitious magnetic charges and vector magnetic dipoles, which is typical for the MFS.The dimension of the problem is significantly reduced when account is taken the symmetry of the magnetic system.

Application in the MFS of vector magnetic dipoles improves the accuracy calculation of the magnetic fields, and in some cases gives the solution that coincides with the analytical solution.

4. Conclusion

The use of the proposed approach provides a significant reduction in calculation time with the required precision of the magnetic field modeling.This allows to effectively solving the problems of design electrotechnical systems and performing the system diagnostics in the process of full-scale modeling experiment. It also allows performing fault analysis and forecasting of systems in real time.

The obtained results allow recommending the use of MFSwith vector dipoles in the calculation of other physical fields (electrical, thermal, etc.).

Acknowledgements

The study results are obtained with the support of the project #1.2690.2014/K Methods to solve inverse problems of complex system diagnostics (in engineering and healthcare industry) based on full-scale and modeling testing, implemented within the scope of a designing stage for the state-given job.

References

[1] N V Korovkin, V L Chechurin Inverse Problems in Electric Circuits and Electromagnetics (2006)
[2] H D Bui Inverse problems in the Mechanic of Materials: An Introduction (1994)
[3] O M Alifanov Inverse Heat Transfer Problems (1994)
[4] Yu A Bakhvalov, N I Gorbatenko and V V Grechikhin Meas Tech **58** (2015) 336-340
[5] C S Chen, A Karageorghis and Y S Smyrlis The Method of Fundamental Solutions – A Meshless Method (2008)
[6] V D Kupradze and M A Aleksidze (1964) 482-126
[7] M A Golberg and C S Chen The method of fundamental solutions for potential, Helmholtz and diffusion problems Boundary Inetgral Methods – Numerical and Mathematical Aspects (1998) 103-176
[8] C S Chen, S Y Reutskiy and V Y Rozov Comput. Assist. Mech. Eng. Sci. **16** (2009) 21-33
[9] Yu A Bakhvalov, V V Grechikhin and A L Yufanova Russian Electromechanics6 (2015)
[10] C J Alves Eng Anal Boundary Elements **33** (2009) 48-61

IOP Conf. Series: Materials Science and Engineering **127** (2016) 012002 doi:10.1088/1757-899X/127/1/012002

[11] Y C Hon and M Li Int J Comput Meth **6** (2009) 181-197
[12] K Simonyi J Theoretisc the Elektrotechnik (1993)
[13] M Stafl Electrodynamics of electrical machines (1967)

Increase Productivity Through Knowledge Management

N A Gavrikova[1], I N Dolgih[2] and E N Dyrina[3]

[1,2,3] Senior Lecturer, Institute of Humanities, Social Sciences & Technologies
Tomsk Polytechnic University, Lenin St. 30, Tomsk, Russia

Abstract Increase in competition level requires companies to improve the efficiency of work force use characterized by labor productivity. Professional knowledge of staff and its experience play the key role in it. The results of Extrusion Line operator's working time analysis are performed in this article. The analysis revealed that the reasons of working time ineffective use connected with inadequate information exchange and knowledge management in the company. Authors suggest the way to solve this problem: the main sources of knowledge in engineering enterprise have been defined, the conditions of success and the stages of knowledge management control have been stated.

Introduction

The key to success in commodity markets is an open economy and competition by State performance. Cost-effective potential is crucial to reduce production costs of goods, while productivity seems the only source of real economic growth and progress. Efficiency characterizes the level of labor resources utilization, taking into account generation, elapsed time and quality of work, as well as the cost of labor per employee. Output shows the social balance and economic efficiency in terms of labor relations based on the principle of economical expenditure of labor.

Competitiveness is not based on sufficiently high and increasing level of work efficiency; it can only take place at the expense of low or declining average wages and/or income of domestic producers. Explanation of this is the fact that without decrease in time per unit of this unit, the producer would reduce the valuation of unit of this time, in other words, the wage for unit of output. Quantitative characteristic of labor efficiency is labor capacity defined as the ratio of output to the number of employees.

Now, in terms of increasingly competitive world market, every country should maintain the competitive productivity level, at least in key sectors, such as mechanical engineering. In this respect, the level of productivity in Russia requires strong raise and appropriate economic policy. In 2012, presidential decree officially defined the challenge of improving productivity in 1.5 times up to 2018.

However, despite the course, adopted by the State, of increasing efficiency in engineering, there has been a slowdown in the growth of this indicator and productivity reduced by 1.7% in engineering according to federal State statistics service (table 1).

In 2014 the plan of measures ensuring productivity was adopted again and it tasked the government to increase productivity more than 5% annually. In 2015, the President marked the challenge of improving productivity again as a priority at the meeting of the Commission on the monitoring of the achievement of the targets of socio-economic development.

Table 1. The ratio of production index and the index of the change in the number of employed persons by economic activity of the Russian Federation

(% to the previous year)		
	2012	2013
Industrial production	104.2	101.3
including:		
manufacture of machinery and equipment	105.3	98.3

The implementation of these tasks requires state enterprises support of the machine-building complex, community initiatives, associations and unions of producers, in order to promote the search and implementation of best practices and methods, sharing of experiences and learning. According to the experts, the average performance of the industry in the year 2014 is only 2.05 million rubles/person/year.

A great role in increasing the productivity of the enterprise staff plays its professional knowledge and experience, while scientific and technological productivity growth factors are often capabilities realization of which depends on a person. Productivity growth in industrial enterprises at the expense of the development of the human factor is reflected in the works of such scholars as Dutse, A.Y. , Ayuba, B. [1] , Yu, X., Dosi, G. , Lei, J., Nuvolari, A. [3], Yus Kelana, B.W., Abu Mansor, N.N., Ayyub Hassan, M [8], etc. In the works of these scientists, it is noted that both a lack of investment in upgrading the material and technical base of production and in terms of innovation, improving education, intellectual capacity, innovation and creativity of the workforce contributes to increased productivity.

1. The evaluation of efficiency level in the use of labor resources for example, the industrial enterprises of mechanical engineering

Traditional tools, which are used in order to collect and analyze information on the effectiveness of the use of labor resources, include working time of monitoring with construction photos of the working day. Based on these tools, we have carried out the study of the structure of time in an Enterprise Engineering (scope of activity «manufacture of electrical machinery and electrical equipment "), an example of the extrusion line operator to overlay the insulation on the electrical cable.

During the execution of the technological process of blending work insulation the following actions perform:

- carries out preparation of raw materials
- makes it possible to configure and launch the production line,
- monitors and, if necessary, adjustment of technology blending modes and winding insulation product in the drum.

Having a set of measurements on the use of working time operator, we have obtained the results presented in table 2.

Following an analysis of the results obtained, it is possible to identify the following issues driving inefficient use of time:

- Lack of structured storage of semi-finished product coming from previous technological operations takes time to search; lack of consistency when placing a semi-finished product causes loss of time associated with its transportation to the production line, unnecessarily increases the load on the transport mechanisms,
- Improper layout of equipment, organization of working places requires the expenditure of time, associated with the movement of the operator increases the time of the changeover.
- Lack of internal regulatory requirements for line speed causes a significant disparity between the formulations that does not correlate with the level of quality of the operation. The speed of the line

is installed a work based on experience within the limited operational documentation on the equipment.

- Lack of structured techniques to conduct control operations, in particular, on the frequency control (it is determined by the operator alone, based on experience and is different for different operators).

Table 2. Structure of the working time of the extrusion line operator to overlay the insulation on the electrical cable.

Activity operator	Function	Type of operation	The time of the operation, %
Preparation of raw materials and materials	Preparation	auxiliary	22
Setting up equipment	Preparation	auxiliary	7
Monitoring/adjustment of the blending process isolation	Cable protection	main	31
Monitoring/adjustment process winding product on the drum	Storage, transport	main	25
Moving operator on the line	-	auxiliary	6
Control operations	Quality control	auxiliary	9

Analyzing the identified problems it may be concluded that their causes lie in the realm of insufficient exchange of information and knowledge management in the enterprise:

- problems with accommodation and finding materials due to insufficient information exchange processes between related technological operations,
- loss of time associated with the movement were due to the lack of information on the process of executing an operation (in particular on the structure of support time) when placing equipment and jobs
- development variation indicators, the amount of time required to perform control operations due to the lack of organization of processes of identification and translation best practices.

On the basis of the foregoing it may be concluded that the company has all the necessary resources to increase productivity, but due to insufficient control of the processes of accumulation and exchange of information, cannot use them.

2. Knowledge management system at the industrial enterprise

By J. Harrington's definition, the information transformed into the ability for effective action is called knowledge.

Knowledge management is a process that aims at creating, sharing and using knowledge to improve the activities and goals of the Organization, with knowledge of the Organization's processes, the specifics of their operation are a kind of skeleton, on which "grow" various types of management information.

The introduction of knowledge management system enables an organization to maintain expertise and provide it for reuse, improve the performance of staff, reduce turnaround time and adaptation period for new specialists.

Knowledge creates a person based on different information resources, it always personally, because man is generated through the prism of their own experience, intellect. Knowledge reflects people's beliefs and values, their skills and professional knowledge as well as their rules, relationships and concepts.

The process of knowledge creation has two levels - empirical and theoretical. Empirical knowledge produced in the course of practical experience, research, observation, experiment. At the theoretical level occurs generalization of results of cognitive activity, abstraction from the object monitoring, theoretical knowledge is a product of the creative activity of thinking. These levels are interconnected and closely intertwined.

Feature operation knowledge in organizations is their diversity. They may exist in two forms - formalized and non-formalized form.

Formalized knowledge (explicit) is a knowledge that can be described, documented, told other people. People can directly send each other formalized knowledge in the form of text, video, audio, software, etc. stored on real physical media (books, paper documents, drawings, diagrams, films, audio and video recordings, electronic files and databases, etc.).

Some types of knowledge exist implicitly-in the minds of employees, this knowledge cannot be seen or documented, and passing them is possible just through personal and direct communication. The implicit knowledge includes the experience, skill, culture of thinking, intuition.

3. Formation and knowledge management

All types of knowledge described above form the knowledge base of the company.

The machine-building enterprise the following sources of knowledge can be identified, potentially forming the basis of efficiency:

Table 3. Sources of information on the types of knowledge.

Internal sources of information	External sources of information
Empirical knowledge	
The results of the analysis of the structure of time expenses for performance of technological operations; Results of a comparative analysis of the personal efficiency of employees; The results of the implementation of the projects of the development and production of new types of products (reasons for success/failure of such projects, complicating factors, etc.); The results of the aggregate data on quality in all types of control: input, stage-by-stage, acceptance testing, causes of defects), as well as complaints of consumers; Dynamics and structure of marketing; The results of internal audits; The results of the appraisal of staff, identified training needs; Proposals for improvement generated by employees based on experience.	The results of marketing researches; Information on the results of external examinations; Information from the media, Governments, consumers, affecting the functioning of the Organization; The knowledge gained in the training organizations and from experts; Information about the features, parameters and features of operation of equipment from suppliers and manufacturers. Information about the features of use, methods of carrying out the entrance control of raw materials; The experience of industry leaders.
Theoretical knowledge	
Internal regulatory documentation: regulations, standards of enterprise status, organizational and regulatory documentation; Design and technological documentation; Technical documentation that defines the requirements for the operation and maintenance, inspection and measurement equipment, engineering networks.	Legislation Normative and technical documentation Special literature Training organization

Knowledge management process can be represented as a diagram, shown in Figure 1.

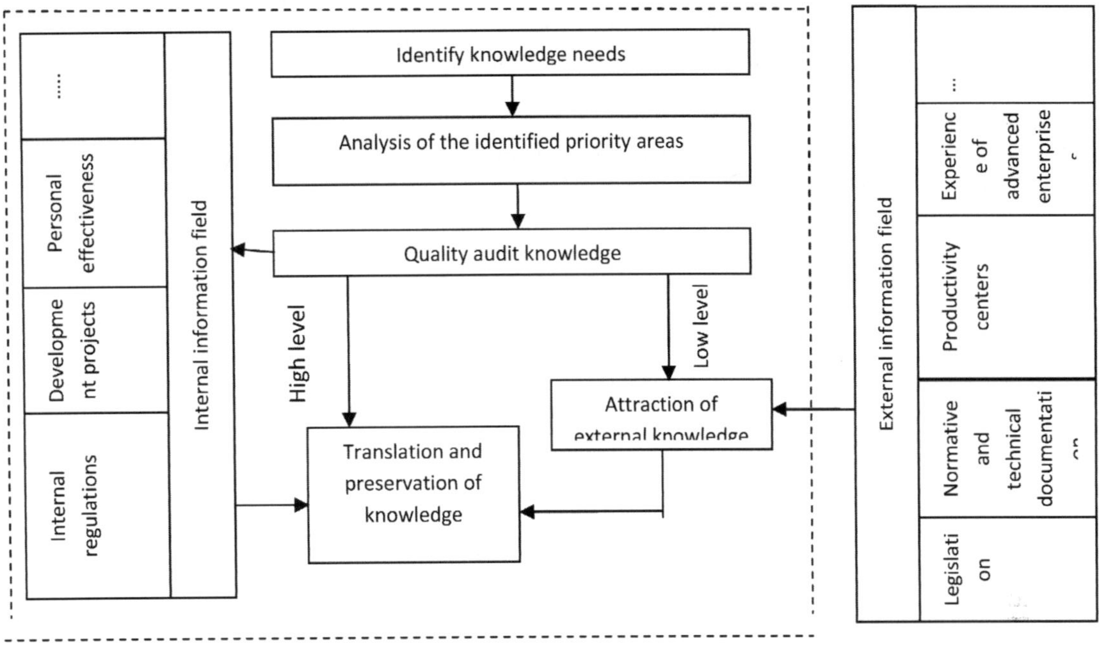

Figure 1. Diagram of knowledge management process

Phase identifying the need for new knowledge involves collection and processing of information on the named source data. For example, an analysis of personal efficiency of employees is carried out to identify staff who consistently demonstrates achievement or even surpassing their goals. This is indicative of the presence of such staff have personal knowledge that the organization needs. Planned to implement projects to successful their realization usually require the formation of new staff knowledge and competency. Performing official duties, the employees of the Organization are often faced with problems whose solution requires additional knowledge. Identifying such needs can be carried out by regular staff surveys, etc.

At the next stage, forming a list of needs in knowledge and operating in an environment of limited resources, the Organization should identify priority needs. Such analysis should take into account the impact of knowledge on the attainment of the objectives of the Organization and the availability of the necessary resources to meet the knowledge needs.

Necessary knowledge can be put at the disposal of the enterprise in the form of lessons learned staff or other sources. However, in order to assess the suitability of the Knowledge Bank for use, it is necessary to conduct an audit of the quality of available knowledge. If the quality level high, the knowledge translation and preservation of existing knowledge.

In case of absence, the source of knowledge (staff) in the Organization, finds the source of knowledge in the environment. The need for knowledge from the environment can be satisfied by the involvement of experts for the execution of works, the training organizations, and recruitment of specialists in other ways.

Translation and preservation of knowledge are to spread knowledge among all staff interested in. Translation and preservation of knowledge can be implemented by:

- Developing a system of training and instruction of personnel that can be received.
- Regular internal master classes and competitions of professional skill.
- Develop a set of internal regulations.
- Development and optimization based on the knowledge needs of corporate information systems.

Factor limiting broadcast knowledge, in addition to the quality of information resources is the competition among workers, in terms of wanting to hide information about the knowledge from colleagues, to achieve higher personal result and gain access to additional resources.

Conclusion

All in all, the exchange and the formation of knowledge exist in any organization, regardless of the will of the leadership. However, only a systematic approach to knowledge management allows organizations ensure the effective functioning of the organization as a whole and manage performance in particular.

It is important to note that the proposed approach to the process of knowledge management will be viable only if certain conditions are met:

- the staff have a desire for new knowledge and willingness to share personal knowledge;
- Defined the format of providing knowledge;
- Developed a clear and fast mechanism for adoption/rejection of knowledge;
- There is no competition for resources among the staff of the organization.

References

[1] A Y Dutse , B Ayuba Human resource capabilities and FDI-related productivity performance: Experience from Nigeria's indigenous manufacturing firms Entrepreneurship **1** (2013) 774-784

[2] Soltan Hassan and Sherif Mostafa Lean and Agile Performance Framework for Manufacturing Enterprises Materials Industrial and Manufacturing Engineering **2** (2015) 476-484

[3] Yu X G Dosi, J Lei, A Nuvolari Institutional change and productivity growth in China's manufacturing: The microeconomics of knowledge accumulation and "creative restructuring" Industrial and Corporate Change **3** (2015) 565-602

[4] Iuga Virginia and Kifor Claudiu Succesful Lean Manufacturing Implementation: Internal Key Influencing Factors ACTA Universitatis Cibiniensis **1** (2015) 121-126

[5] Anoop C Nair Ramalingam and Ashvini Ravi Knowledge Creation Within the Automobile Industry (2015)

[6] Fabio Lotti Oliva Knowledge management barriers, practices and maturity model Journal of Knowledge Management **6** (2014) 1053 – 1074

[7] Lu Zhang Lean Production "with Chinese Characteristics" A Case Study of China's Automobile Industry International Journal of Sociology **2** (2015) 152-170

[8] B W Yus Kelana, N N Abu Mansor, M Ayyub Hassan Does sustainability practices of human resources as a new approach able to increase the workers productivity in the SME sector through human resources policy support Advanced Science Letters **5** (2015) 1501-1504

[9] M Balle Lean attitude considering attitude in lean production Manufacturing Engineer (2014)

[10] Waldemiro Francisco Sorte Junior Standardisation for incremental innovation: a case study in the Brazilian automobile industry J Lean Enterprise Research **1** (2014) 81-93

IOP Conf. Series: Materials Science and Engineering **127** (2016) 012004 doi:10.1088/1757-899X/127/1/012004

Pulse-Width Control in Ladder Structure Four-Phase Rectifier for AC-Electromotive

V V Ivanov [1], S V Myatez[2], E G Langeman[3], N I Schurov[4]

[1,2,3,4]Novosibirsk State Technical University, Novosibirsk, 630073, Russia

E-mail: ivanov.etk@yandex.ru

Abstract. Based on these studies the ways of power factor of the single-phase rectifiers operating in a single-phase AC network improving are suggested. The ladder four-phase rectifier is offered as a technical mean using a pulse-width method of controlling the rectified voltage. The pulse-width control efficiency as a way of the power factor rectifier with a ladder structure for AC electromotive improving is evaluated.

Introduction

Electromotive with the one-phase bridge AC rectifier in contrast to others has a gradual acceleration in a traction mode but also has a not optimal efficiency factors, e.g. power factor. In a normal electromotive's operation mode its power factor doesn't exceed the meaning 0.8. This phenomenon is explained by the forced delay in the used four-phase transducer with the bridge rectifier structure with two commutation circuits (small and big) which cause the move of the rectified voltage electrical center along the current in the input and output circuits.

Low power factor value shows that there are high reactive power consumption and supply current form distortion. In this case electric-traction network and electromotive electrical equipment is loaded by the particular reactive current which causes the increased energy consumption while traction mode.

1. Methods of the power factor's increasing

The factors which influence on the power factor should be evaluated for the problem of power factor increasing in the first rectifying zone:

$$\chi = \frac{P}{\sqrt{P^2 + Q^2 + T^2}} = K_m \cdot K_d \qquad (1)$$

where P, Q and T – active, reactive power and the power of distortion correspondingly, K_m – move coefficient, K_d – distortion coefficient.

As a technical mean of the energy effectiveness factor increasing improved four-zone rectifier with so called semiconductor distortion factorswitch ladder structure [1] was proposed by the authors.

By means of the special scheme solution topology general commutation process duration is reduced by half in the rectifier for the majority of the operation zones. This allows increasing the power factor by 4-5%.

The increasing of the AC working freight electromotive energy effectiveness by the zone rectifier control algorithm improvement is the effective solution [2].

As far as creation of the rectifier opening delay by a firing delay angle in the zone rectifier couldn't be refused, for the power coefficient value increasing in the first zone the distortion fundamental component compensation should be made.

This principle is known as a sector controlling. For common SCR-semiconductor switches forced commutation is attended with the commutation key out circuits' injection which is not cost-effective. However power semiconductors perfection allows using the sector controlling principle with new element base.

Sector controlling principle increases the transformation energy efficiency by means of reakting power, so $K_m \leq 1.0$. Sector controlling effect is carefully described in source [3].

For the additional power factor increasing it is necessary to increase distortion factor value. The distortion factor is the ratio of the effective range of the primary winding first current harmonic to the effective range of the primary current.

For reducing the influence of higher harmonic component on the power factor was offered to use the pulse-width controlling of the rectifier first zone voltage.

The pulse-width controlling is a controlling of the average on-load voltage value by means of the pulse duty factor and control key.

The pulse-width controlling is realize as well as the sector controlling with the help of the semiconductors switches substituting with fully controlled power semiconductors.

2. Current curves' harmonic analysis

It is known, that any periodic function with a finite number of discontinuities of the first kind and a finite number of highs and lows for the period, can be arranged in the trigonometric series (Fourier series):

$$f(\omega t) = A_0 + A_{1m} \cdot \sin(\omega t + \phi_1) + A_{2m} \cdot \sin(2\omega t + \phi_2) + \ldots = \\ = \sum_{k=0}^{\infty} A_{km} \cdot \sin(k\omega t + \phi_k) \quad (2)$$

where $k = 0 - A_{km} = A_0;\ \phi_k = \phi_0 = \pi/2$.

The first member A_0 of the series is called constant component, the second $A_{1m} \cdot \sin(\omega t + \phi_1)$ - primary or first harmonic. The remaining members of the series are called higher harmonics.

If the expression (2) discloses the amount of each harmonic of Sines, it takes the form:

$$f(\omega t) = A_0 + B_{1m} \cdot \sin \omega t + B_{2m} \cdot \sin 2\omega t + \ldots + B_{km} \cdot \sin k\omega t + \ldots + \\ C_{1m} \cdot \cos \omega t + C_{2m} \cdot \cos 2\omega t + \ldots + C_{km} \cdot \cos k\omega t + \ldots = \\ = A_0 + \sum_{k=1}^{\infty} B_{km} \cdot \sin k\omega t + \sum_{k=1}^{\infty} C_{km} \cdot \cos k\omega t \quad (3)$$

where $B_{km} = A_{km} \cos \phi_k; C_{km} = A_{km} \sin \phi_k$.

When the $f(\omega t)$ functions number is analyticaly set, then coefficients can be calculated using the formulae:

$$A_0 = \frac{1}{2\pi} \int_0^{2\pi} f(\omega t) d(\omega t); \quad (4)$$

$$B_{km} = \frac{1}{\pi} \int_0^{2\pi} f(\omega t) \sin(k\omega t) d(\omega t); \quad (5)$$

IOP Conf. Series: Materials Science and Engineering 127 (2016) 012004 doi:10.1088/1757-899X/127/1/012004

$$C_{km} = \frac{1}{\pi}\int_0^{2\pi} f(\omega t)\cos(k\omega t)d(\omega t). \quad (6)$$

then you can move to the form:

$$A_{km} = \sqrt{B^2{}_{km} + C^2{}_{km}};\ \phi_k = \mathrm{arctg}(C_{km}/B_{km}). \quad (7)$$

Distortion factor is defined as the ratio of the effective value of the main harmonics to the current value of the entire function:

$$K_И = A_1/A. \quad (8)$$

3. The pulse-width controlling modulation

The rectifier model was made by the authors in analytical modeling environment MathCad for the current curves' harmonic analyses while the pulse-width voltage. For the analitycal model with pulse-width controlling unipolar signal front sweep was used.

Consider a program for spectral modeling method of regulating the sector:

Diskretytime setting:

$$N = 5000;\ \ Q = 0; \quad (9)$$

$$d = 2\cdot\frac{p}{N};\ n = Q...N; \quad (10)$$

$$O_n = n\cdot d. \quad (11)$$

Sweep signal:

$$r = \frac{-1}{p}\cdot a\tan(\sin(A\cdot O_n - \frac{p}{2})) + \frac{1}{2}; \quad (12)$$

Control signal setting:

$$m = 0.8\sin(O_n); \quad (13)$$

$$i_n = \mathrm{if}(il_n > 0, il_n, -(il_n)). \quad (14)$$

Definition of the modulator output signal and converter current:

$$m_n = \mathrm{if}(i_n < S_n, 100, 0); \quad (15)$$

$$i_n = \mathrm{if}(n > 2500, -(m_n), m_n). \quad (16)$$

Picture 1 shows the sweep wave form – r, control signal – m, line current – i(t) and the first harmonic current $i_1(t)$.

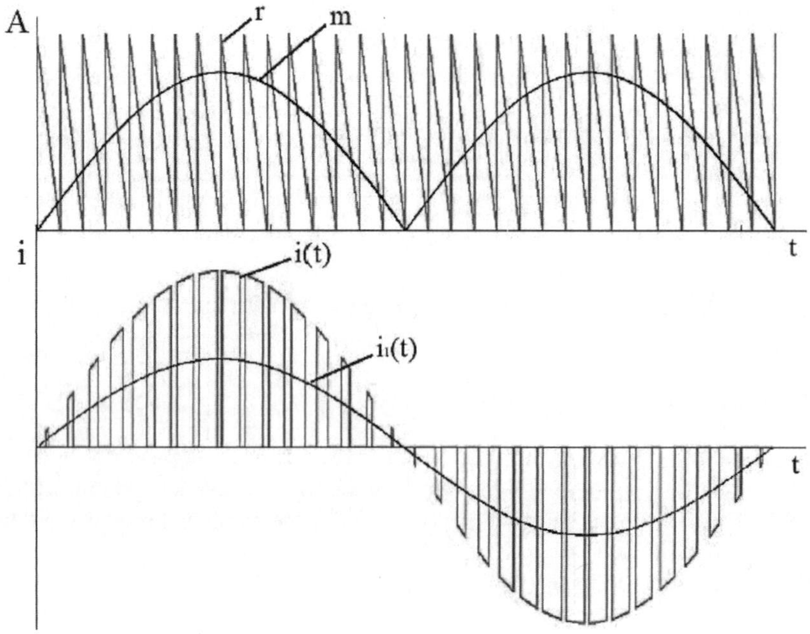

Figure 1.The rectifier model with the sector controlling in the first controlling zone

The control signal m amplitude changing causes the fill-factor value D changes the, which is defined as:

$$D = \sum_{n} m_n . \tag{17}$$

To determine the range of output current it is necessary to calculate the coefficients of a Fourier series [3]:

$$k = 1...100; \tag{18}$$

$$A_k = d \cdot (\sum_{n} \frac{i_n \cdot \cos(k \cdot n \cdot d)}{\pi}); \tag{19}$$

$$B_k = d \cdot (\sum_{n} \frac{i_n \cdot \sin(k \cdot n \cdot d)}{\pi}); \tag{20}$$

$$C_k = \sqrt{A^2_k + B^2_k} . \tag{21}$$

With help of this model output current spectrum of the rectifier was defined (Fig. 2).

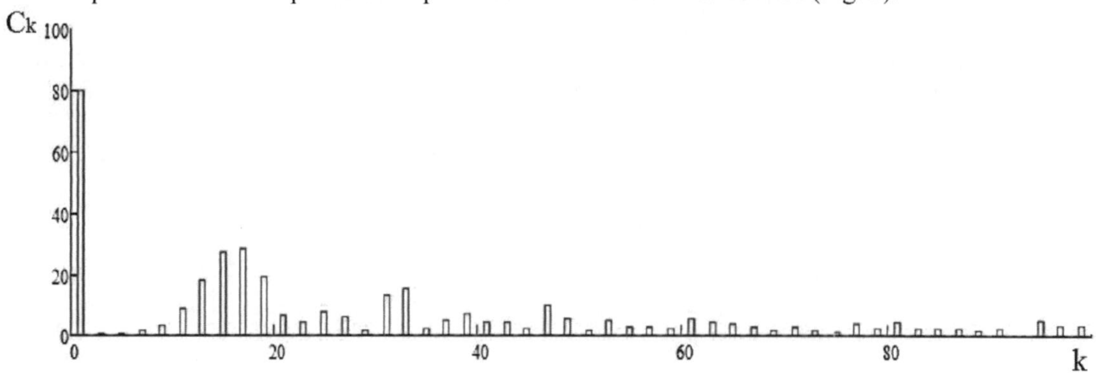

Figure 2. The output current spectrum of the pulse-width controlling rectifier

The result of the harmonic analysis is the dependence of distortion from the fill factor. The comparison of the phase, sectoral and pulse-width voltage regulation methods is presented in Figure 3. The graph shows that the principles of sectoral and pulse-width controling methods are significantly more effective phase control method. Values of the 2 curve are above the curve 1 22% on average, but the usage of the pulse-width controling method increases the power factor on the first 10% of the regualtionrange, changing the duty factor (D) from 0 to 1.

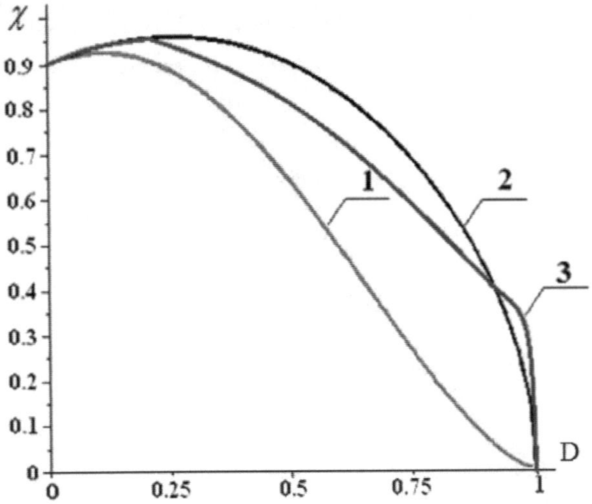

Figure 3. The differences between the power factor of the ladder structure rectifier 1 – phase controlling 2 – sector controlling 3 - pulse-width controlling on the first zone

4. Conclusion

As the analytical modeling shows by means of thepulse-width controlling principle the rectifier energy indicator increases by 10% on the first controlling range. Authors offer to combine sector and pulse-width methods of the voltage controlling for obtaining better energy indicator.

Reference

[1] S A Evdokimov, L G Evdokimova One-phase AC-DC rectifier, Patent of Russia (2009) 2398344

[2] G S Zinovev Bases of the power electronics Novosibirsk NSTU (2003)

[3] S V Myatezh, N I Schurov, M M Dzhaborov Sector controlling in ladder structure four-phase rectifier for AC-electromotive Elektrotehnika (2014)

[4] A V Plaks, Sistemy upravleniya elektricheskikh podvizhnym sostavom. Ucheb Dlya vuzov zh.-d. transporta Moscow: Marshrut (2005)

[5] B G Yuzhakov Elektricheskii privod i preobrazovateli podvizhnogo sostava Moscow: Uchebno-metodicheskii Tsentrpo Obrazovaniyuna Zh-d Transporte (2007)

[6] B A Tushkanov, N G Pushkarev and L A Pozdnyakova Rukovodstvo po ekspluatatsii Moscow: Transport (1992)

[7] Yu M Kulinich Elektronnayai preobrazovatel'naya tekhnika: Uchebn posobie Khabarovsk: Far Eastern State Transport Univ (2008)

IOP Conf. Series: Materials Science and Engineering **127** (2016) 012005 doi:10.1088/1757-899X/127/1/012005

Effect of Distributed Photovoltaic Generation on the Voltage Magnitude in a Self-Contained Power Supply System

B V Lukutin[1], E B Shandarova[1], A F Makarova[1] and I B Shvartsman [2]

[1]National Research Tomsk Polytechnic University, 30 Lenin Avenue,Tomsk, 634050, Russia
[2]Tallinn University of Technology, Ehitajate tee 5, Tallinn ,19086, Estonia

E-mail: shandarovaelena@mail.ru

Abstract. A promising way to increase the technical and economic characteristics of stand-alone power supply systems is to incorporate renewable energy installations in their structure. This saves fuel and extends the operational life of diesel power stations. The most common option is a hybrid system with photovoltaic power stations incorporated into the local network of the diesel power station. This paper deals with the dependence of the deflection voltage and power losses in the electric power transmission line on the graphs of electrical loads, the parameters of elements of the power supply system, connection points and the capacity of distributed photovoltaic power stations. Research has been carried out on the common low-voltage power supply systems of the radial type (0.4 kV) with an installed capacity of up to 100 kW. The studies have been conducted by simulating the operating modes of hybrid power systems of various configurations. As a result of these studies recommendations to reduce losses and voltage variations in the network by selecting the power and photovoltaic power connection points have been put forward.

Introduction

Consumers' electrification is implemented by connecting them to the central power grid or by creating their own power station. In areas with a high density of electrical loads a centralized energysupply has both technical and economic advantages. It is more expedient to provide electricity to remote sparsely populated areas that have low power consumption from their own autonomous power stations.
Russia can serve as an example of the parallel development of centralized and decentralized energy. The centralized power system covers only 30% of Russian territory, mainly in its European part and southern regions of Siberia. For the rest of the territory autonomous power sources are used to generate power, the vast majority of which (98%) is generated by diesel power stations (DPS). The average fixed capacity of DPS is around 340 kW.

The remoteness of decentralized electrification facilities, the harsh climate and complex transport logistics account for the major share of the tariff for electricity - costs for the purchase and delivery of fuel. This part of the operating cost amounts to 50-60% of the produced electricity cost. The tariff on DPS electricity in certain regions of Siberia reaches 100 rubles / kWh and can be higher.

One method to improve the technical and economic characteristics of the autonomous power supply systems (APSS) is to use renewable energy units in their structure [1,2,3,4].

The most universal of natural energy sources are wind energy and solar radiation [5,6]. The reduced cost of photovoltaic panels combined with their ease of installation and operation has sparked interest in photo diesel autonomous power in Russia.

The integration of the photoelectric station (PES) into the autonomous power supply system with DPS requires solutions to many problems related to the unbalance of load graphs of the electrification object and solar radiation during the day, and the power ratio of the diesel station and photovoltaic units and the autonomous power supply system configuration among others.

A study of autonomous power systems with renewable sources including photovoltaic power stations is deemed to be essential. The works [7,8,9,10] have all contributed to the development of thisproposition.

However, the complexity and diversity of the research problems as well as the intent to improve the technical and economic characteristics of the photo diesel energy sector require new research.

Research objectives

In previous studies, the issues of quality and the loss of electricity in an autonomous power system with photovoltaic distributed generation were not reflected. Of particular interest is the stability of the voltage and losses in power transmission lines depending on the power ratio of DPS and PES, daily schedules of insolation and electricity consumption, photoelectric stations' connection points to the autonomous electrical network and the type of inverter PES.

Object of study

Research was carried out on the geographical conditions in the middle latitudes of Russia. We considered an autonomous power supply system of the radial type 0.4 kV characteristic of that used for the electrification of small settlements.

Figure 1 shows the computational equivalent circuit. The letter C on the circuit designates the source DPS, equivalent to the diesel power station, and the inductive x_{ln} and active resistance r_{ln} of the circuit sections up to the consumers' connection points to the total power throughput in a complex form $\dot{S}_n$. Consumers are connected via the sections of lead-in lines with the parametersx_{nn},r_{nn}. Photoelectric stations are equivalent to the power source $\dot{J}_{ph}$ that can be connected to various points of the network the voltage of which is indicated as $\dot{U}_n$.

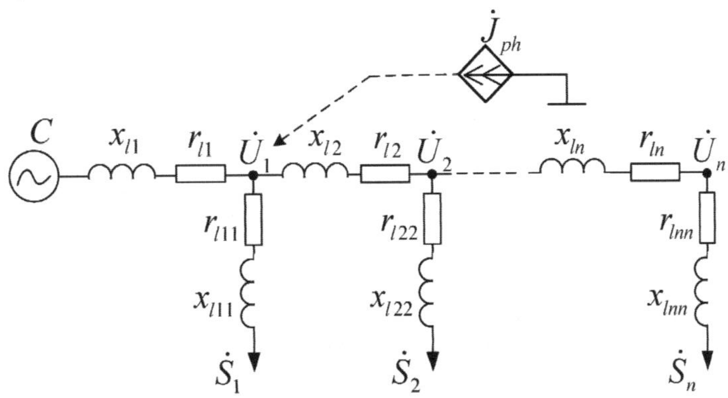

Figure 1. Computational equivalent circuit.

Photo panels transmit DC generated power via the network invertor. Modern invertors with MPPT functions in their output characteristics can be equated to the source of current [9]. DPS with control of excitation of synchronic generators is equated to the voltage source. The State Standard specifies a limit value toany deviation of not more than 10%. Thus, the line length will depend upon known factors: the specific linear resistance wires and power loads.

Taking into account the need for a large number of calculations related to the daily changes in insolation and electricity as well as an analysis of circuit configurations with different connecting points of PES with varying power, in this environment Matlab modelling is used as a method of study.

Results of the Study

Initial data for modelling are the daily schedules of electricity and insolation. Consumers of electricity in decentralized areas are mostly small towns with a population of up to several hundred people. The volume of electricity consumption in settlements is determined mainly by the communal household load as well as by small enterprises processing local natural resources and agricultural products.

For modelling we used typical daily schedules for the communal household load for conditions in Russia shown in Figure 2 and Figure3, where P, Q – active and reactive load power; w_s – total energy of the solar radiation on a horizontal surface within 24 hours at the latitude 55-56° (in summer).

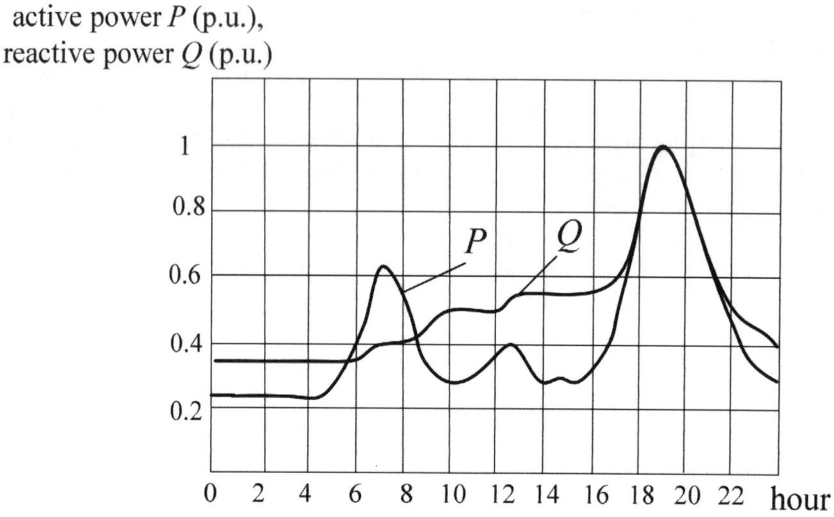

Figure 2. Winter consumer diurnal load graph.

For a seasonal factor of the summer load peak P_{maxs} against the winter $P_{maxw}$the value 0.7 is recommended. In the graphs of power consumption in named units it is convenient to use the maximum coefficient $k_{max}=P_{max}/P_{av}$ for a typical graph. Average power consumption P_{av} is determined by reporting data on monthly electricity generation by DES of the settlement.

The distribution of electricity consumers' power connected to the points of the power line along its entire length is important to the goal of this study. The main consumers are private houses and social facilities: schools, libraries, clubs, kindergartens, etc. For small and medium-sized towns with a population of 50-1000 people the established norms of the annual electricity consumption are 760-855 kWh / person (in the absence of stationary electric stoves). If we assume that the number of family members living in the house ranges from two to five people, the annual rate of power consumption will be 1710-4275 kWh. The average power consumption of the house is within 0.2 ÷ 0.5 kW. At peak hours it increases in proportion to the maximal coefficient. Social facilities are more powerful electrical energy consumers. The maximum amount of power used by a school or a club for the considered settlements is 7-10 kW.

This paper investigates the power supply system for summer season conditions which is expedient as winter insolation in central Russia (55-56° N) is 15 times less than for summer. It defines the creation of PES that will only be effective for the summer season. In winter, with an increase in power consumption and a decrease in insolation, the role of PES is significantly reduced and it has virtually no effect on the electrical mode of the power supply system. Full power of the electricity transmission line in terms of voltage loss does not exceed 30 kVA.

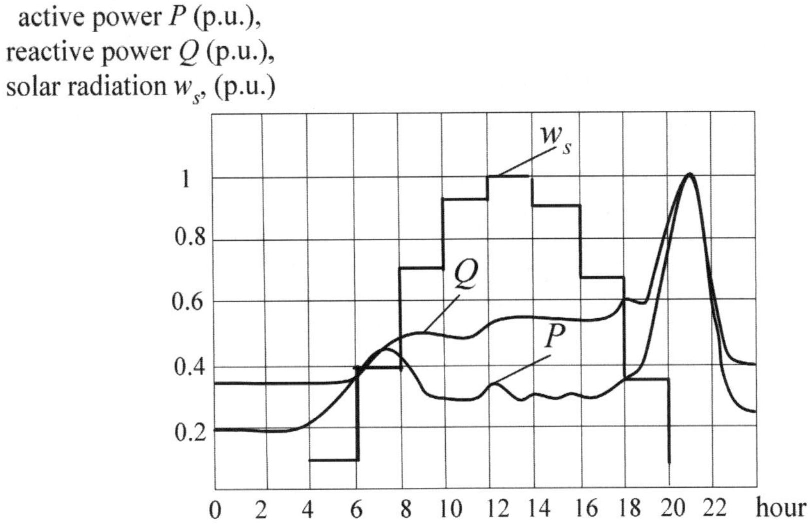

Figure 3. Summer consumer diurnal load graph and insolation.

Comparing a graph of the relative magnitudes of the total solar radiation w_s on a horizontal surface within 24 hours and a graph of power loads (Figure 3) shows that a morning load peak corresponds to only 40-50% of possible generation in relation to the midday solar insolation. The evening load peak corresponds to a reduction in insolation of 10% or less.

Thus, the question of a rational installed capacity of networked PES in the autonomous power supply system without an energy storage system arises. Similar schemes with power invertors are also spreading across the autonomous power supply systems due to their simplicity and low-cost implementation [11,12].

It is known that the network invertor can operate in parallel only with a sufficiently powerful electrical supply network. In autonomous objects such a power system is created by a diesel power station. Leading manufacturers of "solar invertors" do not recommend reducing the current capacity of DESin photo-diesel engines to less than 40-50% relative to the power of total generation. PES does not practically add its power during the morning and evening peak loads.

Thus, when taking into consideration the geographical conditions, the nature of electrical consumption of the object and restrictions on the mode work of the network invertor, it is recommended thatthe available electric power of the PES be limitedduring the summer period to a level of no more than half of the average daily consumed power without taking load peaks into consideration. In the course of modelling this recommendation is depictedby the corresponding current value $\dot{J}_{ph}$.

PESs with energy storage have more opportunityto match the graphs of electricity consumption and insolation. The power of storage batteries can be redistributed at specific times,for example, participating in ensuring the evening peak.

The matter of the connection point of the PES to an autonomous electrical system is critical. The PES is often connected in parallel to the DPS on the general switchgear. This practice is more economical on the switchgear and relay protection. Such a connection does not cause the electric mode of the distributed electrical network to change. Rather, it is the value of the transmitted current and resistance of the wires that determine the loss of power and voltage in the power line.

Connection of the current source, equal to PES, to other points on the electrical transmission line leads to redistribution of the currents in the electrical circuit diagram and, accordingly, to changes in levels of power loss and voltage at the connecting points of the loads.

In the nominal work mode of the DPS the voltage at the end of a simulated power line is reduced by 9.75% relative to the nominal. The amount of lost electricity is 8% of the transmitted power.

When an additional power source $\dot{J}_{ph}$ is connectedpower loads can be reallocatedbetween DPS and PES. The result is a change in the current valueat various parts of the network whichreduces power losses in the conductors and provides greater voltage stability at the connection points of consumers of electric power.

The modelling results showed that using power fromthe PES with the energy storage system at the rate of one half of the nominal load of the power supply system will reducevoltage loss at the end of the line by up to 2-3%. This is achieved by connecting the PES at a distance from DES that isnot less than half of the total length of the electrical transmission line. By sodoing the loss value is reduced to lessthan 1%. Under substantially uneven load power of the distributing power transmission lines, it is advisable to connect the PESat the connection point of the most powerful electro-consumer.

At low power supply system loads in the summer period when, consequently, thereis a need to reduce the PES power, the voltage stability over the transmission line is significantly higher and the level of energy loss is significantly lower. Implementation of the recommendations at the connection point of the PES improves the parameters of the electric mode butthese changes are not as noticeable as in the rated load of the system.

Conclusions

1. The effectiveness of using the PES inautonomous power systems without energy storage systems is significantly limited by the operating conditions of the network invertors.

2. For the geographical areas in the northern mid-latitudes the big difference between thevalues of insolation in summer and winter determines expediency of the maximum use of PES in the summer, which is contraryto reducing the summer electricity consumption.

3.To stabilize the voltage in the power system and reduce electricity losses it is advisable to use PES with energy storage systems and connect them closer to the end of the distribution line or to the connection point of the most powerful consumer. The generation power of the PES is limited to halfof the maximum capacity of electricity consumption.

References

[1] R Passey, T Spooner, I MacGill, M Watt and K Syngellakis The potential impacts of grid-connected distributed generation and how to address them: A review of technical and non-technical factors Energy Policy **39** (2011) 6280-90

[2] Y M Atwa, E F El-Saadany, M A Salama and R Seethapathy Optimal renewable resources mix for distribution system energy loss minimization IEEE Transactions on Power Systems **25** (2010) 360-370

[3] P Georgilakis, N D Sand Hatziargyriou IEEE Transactions on Power Systems 28 (2013) 20-28

[4] D H Popović, J A Greatbanks, M Begović and A Pregelj Placement of distributed generators and reclosers for distribution network security and reliability JEPE 27 (2005) 398-408

[5] Méndez Quezada V H, Rivier Abbad J and Gómez San T Román Assessment of energy distribution losses for increasing penetration of distributed generation IEEE Transactions on Power Systems 21 (2006) 533-540

[6] C Wang and M H Nehrir Analytical approaches for optimal placement of distributed generation sources in power systems IEEE Transactions on Power Systems19 (2004) 68-76

[7] C L Masters J Power Engineering Journal 16 (2002) 5-12

[8] M H Moradi, S Reza Tousi, M MandAbedini JEPE 56 (2014) 117-126

[9] T V Kabanova, E O Kuleshova, Y N Isaev, N V Shishka and Y V Shishkovskaya MEACS, 69 (2014) 86903

[10] R S Al Abri, E F El-Saadany and Y M Atwa Optimal placement and sizing method to improve the voltage stability margin in a distribution system using distributed generation IEEE Transactions on Power Systems 28 (2013) 326-334

[11] I I Perpinias, N P Papanikolaou and E C Tatakis Fault ride through concept in low voltage distributed photovoltaic generators for various dispersion and penetration scenarios Sustainable Energy Technologies and Assessments 12 (2015) 15-25
[12] A Hoke, R Butler, J Hambrick and B Kroposki Steady-state analysis of maximum photovoltaic penetration levels on typical distribution feeders IEEE Transactions on Sustainable Energy 4 (2013) 350-357

IOP Conf. Series: Materials Science and Engineering **127** (2016) 012006 doi:10.1088/1757-899X/127/1/012006

The Structure and Mechanical Properties of Bridge Steel Weldings With Glass-Steel Liners

V N Muzalev[1], **B S Semukhin**[1,2], **V I Danilov**[2,3]

[1]Tomsk State Architecture and Building University
2, Solyanaya sq., Tomsk, 634003, Russia
[2]Institute of Strength Physics and Materials Science SB RAS
2/4, Academichesky pr., Tomsk, 634055, Russia
[3]Yurga Institute of Technology National Research Tomsk Polytechnic University
26, Leningradskaya st., Yurga, Kemerovo region, 652055, Russia

E-mail: dvi@ispms.tsc.ru

Abstract. A new technology is developed for welding multi-span bridge constructions. The mechanical properties and structure of the low-carbon bridge steel welds have been studied. The welding parameters and application of steel-glass liners provide for long-term service of steel constructions in conformity with the welding industry specifications.

Introduction

Most critical steel constructions in Russia contain a multitude of non-detachable junction parts produced by the method of electric arc welding. This has led to a widespread use of field welding of all-climate bridge constructions, e.g. highway span bridges, foot bridges, including overpasses, viaducts, trestle bridges, etc. [see, e.g. STP 005-97 'Field welding technology of steel bridge constructions']. Field welding technologies are widely used in highway span bridge construction, which provides savings in materials, time and manpower. In particular, welding of roll stock having thickness up to 16 mm has been generally performed applying copper and glass-copper liners without cutting the wedges of the weld. Considering the extreme climatic conditions of the Far North and Siberia, fast heat removal from the copper liner frequently appears unwarranted. Since copper is expensive material, more cost-effective materials for welding liners are needed.

More specifically, new affordable liner material possessing low thermal conduction is required. This liner should provide a means for the preservation of the mechanical properties and uniform structure of the welded material in the area immediately adjacent to the weld. The researchers in [1] maintain that due to the insufficient heat removal from existing glass-copper liners, large-sized crystallites would form in the welding material at right angles to the welding pool, while small-sized crystallites aligned parallel to the glass-copper liner are much less numerous. Such structure is known to create imperfections in the welding junctions and often results in defects (cracks) in the butt joints of orthotropic plates near the longitudinal joints. These defects frequently occur during the longitudinal sliding of bridge spans in the process of bridge construction. For example, the longitudinal sliding of 30-m bridge spans along three-span pathway is known to cause crack initiation in 50% butt joints, although the stress level in the plates due to the longitudinal sliding did not exceed

150 MPa. The given paper is concerned with the glass-steel liner as an alternative material for the welding of bridge joints. The main purpose of the paper is to understand whether the glass-steel liner has better structural and mechanical properties allowing for less welding defects and cracks in the orthotropic bridge plates.

2. Research methods

The studies were performed for the samples of 12 mm thick model welds. The welds were produced by automatic welding on the horizontal plane of the low-carbon bridge steel (0.12-0.18% C, 0.4-0.7% Si, 0.4-0.7% Mn, 0.6-0.9% Cr, 0.3-0.6% Ni, 0.2-0.4% Cu, <0.035% P, <0.04% S, <0.08% As, <0.008% N) plates with glass-steel liners. There were also 8-mm gaps between the plates; no cutting of the weld edges was performed. First 50-mm long tack welds were manufactured at a distance of 200 mm from one another, using the E50 UONI-13/55 electrodes (GOST 9467-75). In so doing, manual adjustments were made between the welding operations. Next, the gap between the plates was filled with metal nibs prepared from the Sv-10NMA wire 2 mm in diameter. No chemical agents were used for initiating crystallization of the weld. Single pass welding was performed using the welding wire Sv-10NMA 3 mm in diameter; the flux AN-47 was also employed which had been subjected to calcination at 350-400^{0}C. The welding station was equipped with a tractor ADF-1202 and a rectifier VDU-1202. The welding conditions were as follows: arc welding voltage and current 36 V and 800 A, respectively; welding productivity 18 m/h; electrode wire departure 12 mm.

The model weld sample had dimensions 500×400 mm^2; its centerline weld joint was parallel to the long side of the sample. The following test samples were cut from the model weld samples in accordance with the GOST 6996–66.

a) The flat samples were tested by uniaxial static tension. These had dimensions 400×20×6 mm^3; the weld was normal to the extension axis.

b) The samples subjected to free bend testing had a rectangular cross-section. The sample length was 170 mm; the sample had the same thickness as the plate, i.e. 12 mm; its width was equal to 1.5 plate thickness, i.e. 18 mm^3. The weld was right in the middle of the sample; its bead was cut off.

c) The V-notch samples had the weld right in the middle and normal to the longitudinal axis. These were intended for impact testing in accordance with the GOST 9454–08.

The main mechanical properties of the welds were determined for the weld samples, which were tested in an Instron 1185 testing machine; the crosshead rate was 1 mm/min. The gauge length of the sample was 300 mm and the deformation rate was $5.5 \cdot 10^{-5}$ s^{-1}.

The samples were also tested in three-point slow bending in an Instron 1185 testing machine. The test fixture for three-point bends consisted of two cylindrical supports 25 mm in diameter. The separation between the supports was 110 mm, which determined the gauge length of the test sample. The test sample was arranged so that the weld root was at the bottom, i.e. in the area subjected to tension. The cylindrical bending mandrel was 25 mm in diameter; its rate was 5 mm/min. The test was stopped as soon as a crack appeared on the sample surface or the sample bended to 90°.

The value KCU^{-60} was determined on an impact pendulum-type testing machine having maximum impact energy 300 J.

The macrostructure analysis of the cross-section of the weld metal was performed with the aid of deep etching method using nitrous-hydrofluoric acid solution in the 1:1 ratio. Metallographic sections were prepared from the weld metal and adjacent zones by polishing and etching in 4% alcohol solution of nitrous acid. The metallographic sections were used for examination of the microstructure of the weld material in the microscope'Neophot-21'.

3. Discussion of Results

The microstructure of the weld material was examined using the metallographic cross-sections prepared from the samples tested by uniaxial static tension. The entire cross-section of the weld with the exception of the cap beads is presented in Fig. 1a. Evidently, the width of the weld zone exceeds the thickness of the welded plates. A plausible explanation for this fact is too large volume of the welding bath coupled with too low rate of heat removal.

The weld had a coarse structure typical for most cast metals. The large dendrites were aligned horizontally with respect to the weld axis, which suggested that the heat removal had occurred mostly to the welded metal, which caused an insignificant growth of crystallites in the vicinity of glass liner (Fig. 1 a, b). Thus the macrostructure of the weld material had all the unfavorable features reported in [1]. Hence, the effectiveness of the technology in question must be assessed in terms of specifications of structural steels for bridge construction.

The main mechanical properties of the weld material were determined for the weld samples tested in tension [2]. The deformation curves were plotted for the test samples of as-received low-carbon bridge steel. These had a sharp yield point and a yield plateau. The physical yield strength, ultimate strength and plasticity were determined for the test samples; the values obtained were $\sigma_{т}$ = 368.9±6.8 MPa; σ_B = 537.3±2.9 MPa and δ = 17.6±1%, respectively (cf. the values given in the GOST19282-73, i.e. $\sigma_{0.2} \geq 345$ MPa; $\sigma_B \geq 490$ MPa and $\delta \geq 21\%$ [3]). The latter value was probably due to the deformation behavior of the tensile sample. Thus the fracture probably occurred in the sample bulk, while the material of the weld and that of the adjacent zone remained practically non-deformed. The plasticity was determined for the entire test sample; hence a low value δ.

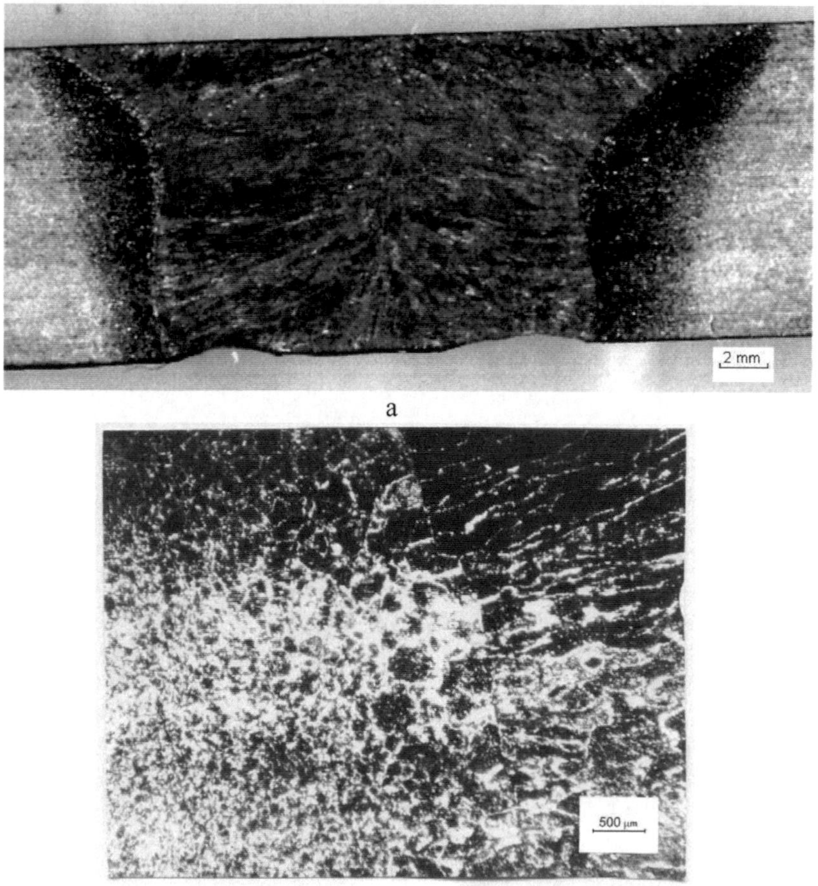

a

b

Figure 1. The macrostructure of the welding with glass-steel liner

The samples were tested in bending to an angle ≤90°. No cracks were observed for either the weld and heat-affected zone (HAZ) material or the bulk material. Therefore, all the samples were also tested in static bending.

The impact tests were carried on at the temperature –60 °C. The resultant value KCV^{-60} = 101.1±19.2 J/cm^2 was much higher than the value given in the GOST 19282–73 for the investigated steel, i.e. 34 J/cm^2. All the test samples were found to undergo ductile-brittle fracture, with fibriform fracture surface showing large cleavage facets. The chemical composition of the V-notch zone of the weld differed from that of the base metal, since it had been altered due to the welding filler wire. The same zone had also greater thickness relative to that of the welded plates. Thus, the results of mechanical tests suggested that the weld joints with the glass-steel liner met all the requirements of the GOST specifications. It has been found that the macro-structure of the weld joints with glass liners is far from ideal. Evidently, the problem invites further investigation [4].

The base metal area immediately adjacent to the weld metal can be clearly seen in the macro-section; a portion of the macro-section is also illustrated in Fig. 1 (a, b, respectively). A well-defined difference apparently exists between the structure of the weld metal and that of the base metal on going from the fusion border to the original structure of the base metal (to the right and left sides of Fig. 1, respectively). The microstructure of the weld metal in zone I was examined (in what follows the zones are numbered in accordance with the classification in [4, 5]). The columnar crystals observed in the macro-section (Fig. 2) were defined as elongated ferrite plates with finely dispersed ferrite-cementite mixture occurring in between. In the course of material crystallization significant hydrogen-related stratification would probably occur.

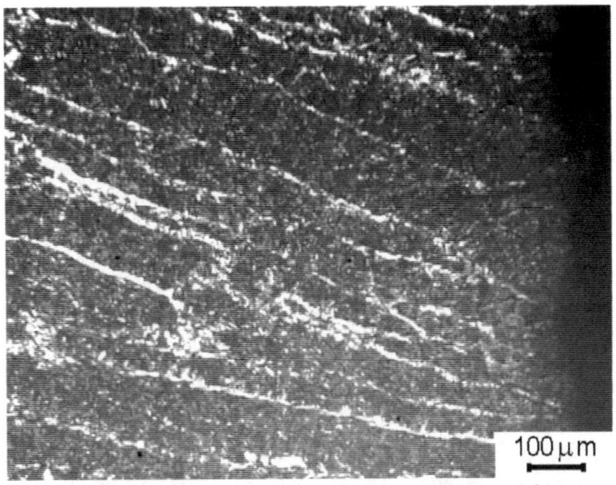

Figure 2. The microstructure of the weld metal in zone I

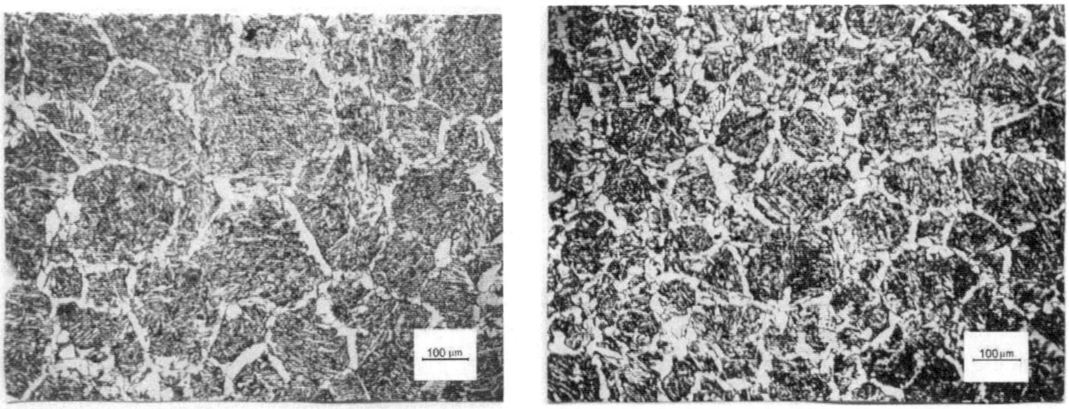

a b

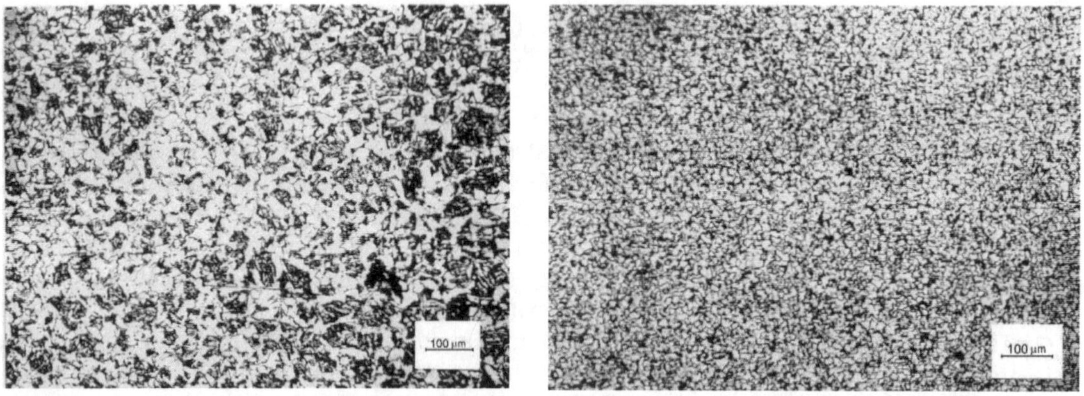

c d

Figure 3. The microstructure of the weld material in zones V – VIII

The metal structure of the heat-affected zone 4.5 mm wide was distinguished by a dramatic inhomogeneity. Thus a narrow zone occurred in the vicinity of the fusion boundary where the temperature exceeded A_4 point (Fig. 1 a). The metal within this zone had an austenitic structure with medium-sized old austenitic grains elongated along the temperature gradient. This zone evidently corresponded to zones II, III and IV. The material of these zones has been subjected to a variety of thermal cycles, which was responsible for different types of discontinuities, e.g. transformation of austenite to δ-ferrite, local melting of austenite grain boundaries, formation of segregation areas and nonmetallic sulfide inclusions. The material of the melted zone would undergo crystallization, generally forming primary and secondary gain boundaries, depending on the cooling rate. In this case, ferrites were found to form at the primary austenitic grain boundaries and carbide particles, within the grains. The above transformations occurring within the HAZ might have had a drastic effect on the properties of the weld [4]. However, this investigation has provided no support for this view.

The material of the zone 1.25 mm in width has a coarse grain microstructure. According to the welding specification, the given area is a HAZ or zone V (Fig. 3 a) in which intensive austenite formation would occur. The microstructure of material of the latter zone was distinguished by the occurrence of Widmanstatten ferrite plates ~20 µm wide, which surrounded the primary austenitic grains having average size ~300 µm, which contained misoriented thin ferrite plates. Besides, an insignificant amount of gas pores having size 20…30 µm also occurred in the same zone.

Another zone is located on going from the fusion border down to a depth ≈1.9 mm (Fig. 3 b). The material within the latter zone (VI) heated in the course of welding to the temperature exceeding A_3 point; however, the temperature was not sufficiently high to favor intensive growth of austenite. Thus the austenite formation also occurred in the base metal; however, the austenite grains had average size ≤150 µm. The austenite grains were also surrounded by Widmanstatten ferrite plates, which frequently broke at the grain boundaries, forming a set of polyhedral grains. The structure of the primary austenite grain was similar to that described above, although a more disperse one. Locally, diffusion transformation of the kind $\gamma \rightarrow \alpha$ took place with resultant formation of perlite eutectoids. The gas pores also occurred in the given zone; however, these were smaller than those of zone V.

4. Conclusion

The studies were carried on for the glass-steel liner welds. The mechanical properties of the welds from the low-carbon bridge steel were examined. The results suggest that the weld properties are in conformity with the GOST welding standards as well as the specifications of welding production. It has been found that the structure of the weld metal was far from perfect. Moreover, the width of the heat-affected zone was too large, which might cause problems by assembling and service of items. The structural defects might have emerged in the weld due to insufficient heat removal to the glass-steel liner, with the main energy flux passing through the base metal. To address the situation, the energy flux might be reduced. In point of fact, the characteristic of the metal that is being welded and the joint type dictate the welding parameters. The methods for testing the weld material specified by the STP 005-97 are not sufficiently advanced to certify the weld joint. In particular, notch toughness is typically measured by the V-notch sample to provide information on the fracture toughness of the weld material. However, the experimental evidence obtained in this study suggests that the properties of the heat-affected zone are one of the significant factors to consider when evaluating the soundness of the weld joint.

Acknowledgement

The work was performed owing to the financial backing of the RFFI and in the frame of the RFFI grant N 14-08-00299; it was instrumentally supported by the 'Nanotech'.

References

[1] V S Ageyev, S V Morozov, M P Shurygin, G I Nikonov, S A Shelkunov on the cause of crack initiation in the welds of orthotropic deck Transport Building 11 (2002) 18-20

[2] E A Zernin, I R Sabirov, M A Kuznetsov The estimation of efficiency f welded joints Mining Informational and Analytical bulletin J Scientific and Technical Journal Issue 3 (2010) 318-322

[3] V G Sorokin, V G Sorokin, M A Gevrasyev Steels and Alloys Manual M Intermet Engineering. (2001) 608

[4] L S Livshits, A N Khakimov Physical metallurgy welding and thermal treatment of welds M Machine Building (1989) 336

[5] D P Il'yaschenko, D A Chinakhov, V I Danilov, GV Shlyakhova, YM Gotovschik Increasing Strength and Operational Reliability of Fixed Joints of Tubes by MMA Welding IOP Conf Series Materials Science and Engineering **91** (2015)

IOP Conf. Series: Materials Science and Engineering **127** (2016) 012007 doi:10.1088/1757-899X/127/1/012007

Investigation of Input Signal Curve Effect on Formed Pulse of Hydraulic-Powered Pulse Machine

M V Novoseltseva[1], I A Masson[2], E N Pashkov[3]

[1,2,3]Department of Theoretical and Applied Mechanics,
Institute of Natural Resources National Research Tomsk Polytechnic University
Tomsk, Russia

E-mail: novmv@tpu.ru, iam2@tpu.ru

Abstract. Well drilling machines should have as high efficiency factor as it is possible. This work proposes factors that are affected by change of input signal pulse curve. A series of runs are conducted on mathematical model of hydraulic-powered pulse machine. From this experiment, interrelations between input pulse curve and construction parameters are found. Results of conducted experiment are obtained with the help of the mathematical model, which is created in Simulink Matlab. Keywords – mathematical modelling; impact machine; output signal amplitude; input signal curve.

Introduction

At the present day, special emphasis is placed on studying the output signal curve of drilling machines [1]. Key performance indicators of existing impact machines influence the overall impact machine efficiency; the most essential indicator is impact energy. Comparison of impact energy is always used to rate different impact machines and determine their possible applications in borehole drilling [5,7]. Impact energy value is dependent on machine type and borehole diameter [2]. Other key performance indicators are amplitude and oscillation period [4]. As well as impact energy, they have an effect on overall drilling efficiency [6]. This studying is devoted to find the interrelation between oscillation period and amplitude and to find applicable pulse curve in order to increase overall efficiency of drilling process. A series of runs are conducted on mathematical model of hydraulic-powered pulse machine, it allows investigating the effect of input signal on output characteristics of formed pulse.

In general, mathematical models describe the basic properties of an object, process or system, its parameters, internal and external communications with the help of logical-mathematical structures [10]. MATLAB software is used to conduct the experiments. Software offers a variety of methods to find solution for Cauchy problem for differential equations: ordinary differential equation (ODE) solvers, special-purpose programs for model investigation of controlled systems, SIMULINK as analog simulation tool [3, 8]. Scheme, which generated in SIMULINK, consists of units that simulate process, its properties and construction parameters.

It is possible to choose one of the following input signal by changing variable in - sine, sawtooth or square. Output frequency can be measured in hertz (Hz) or in radians per second (rad/s); it is also possible to adjust the amplitude of the input signal by changing the plunger stroke length [9]. Sample amplitude of 1 mm is applied for model calculation. Signal curve is not affected by decreasing or increasing of this parameter.

Results & Discussion

Mathematical model is tested for following system parameters as depicted in Table 1.

Table 1.Impact machine parameters

	Parameter	Value
c	Spring stiffness	$18600 \frac{N}{m}$;
m	Active mass	$100\,kg$
Cv	High-pressure hose volume elasticity	$3,441 \bullet 10^{-9} \frac{m^3}{Pa}$
k_f	Coefficient of friction	0,2

Fig.1-3 show the result of mathematical modelling for different type of input signal curve.

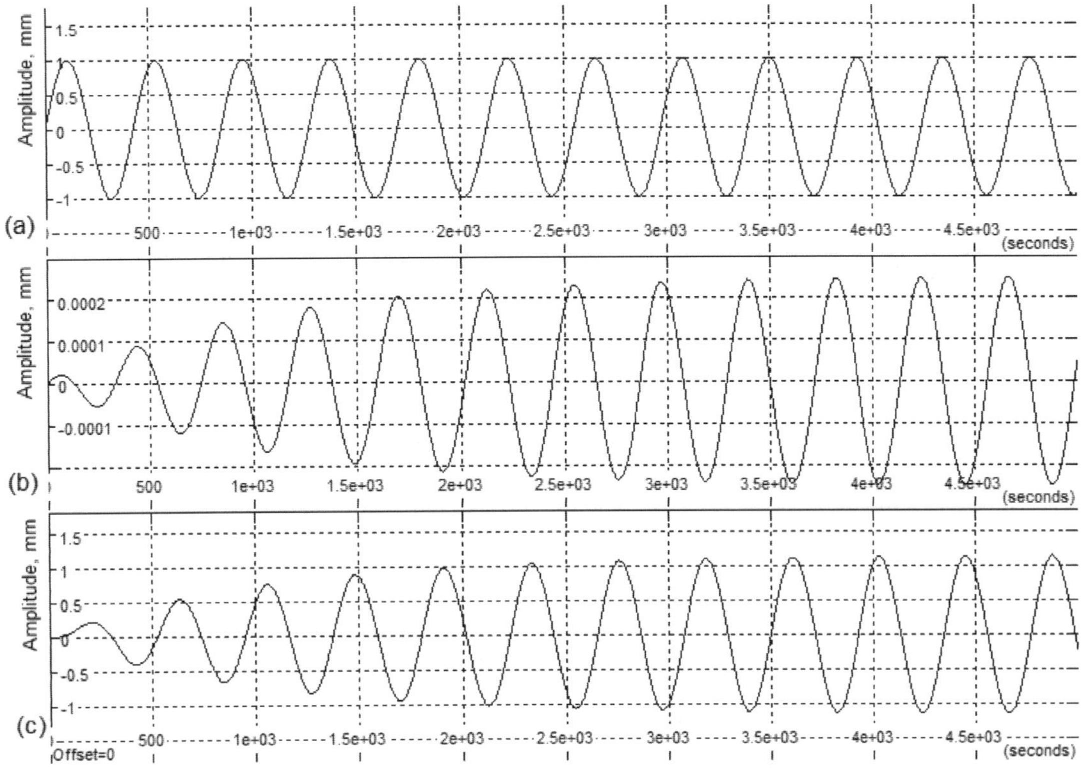

Fig.1. Dependence graphs of active mass displacement (b) and formed pulse (c) for sine input signal (a) on time interval, for which pulse acts

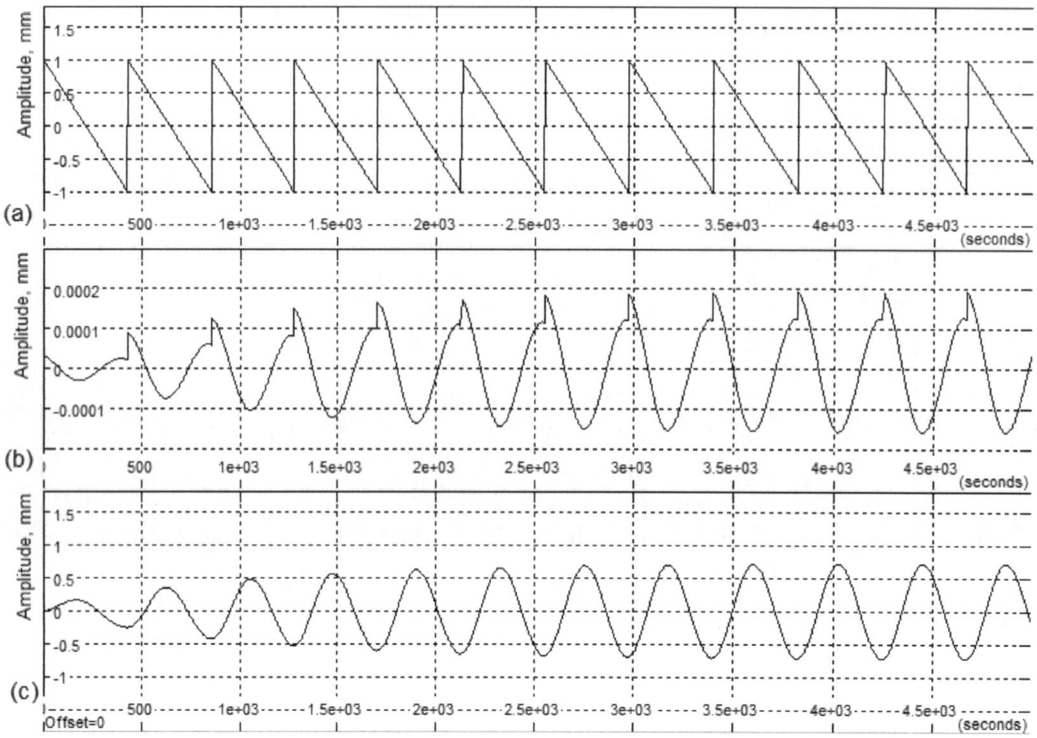

Fig.2. Dependence curves of active mass displacement (b) and formed pulse (c) for sawtooth input signal (a) on time interval, for which pulse acts

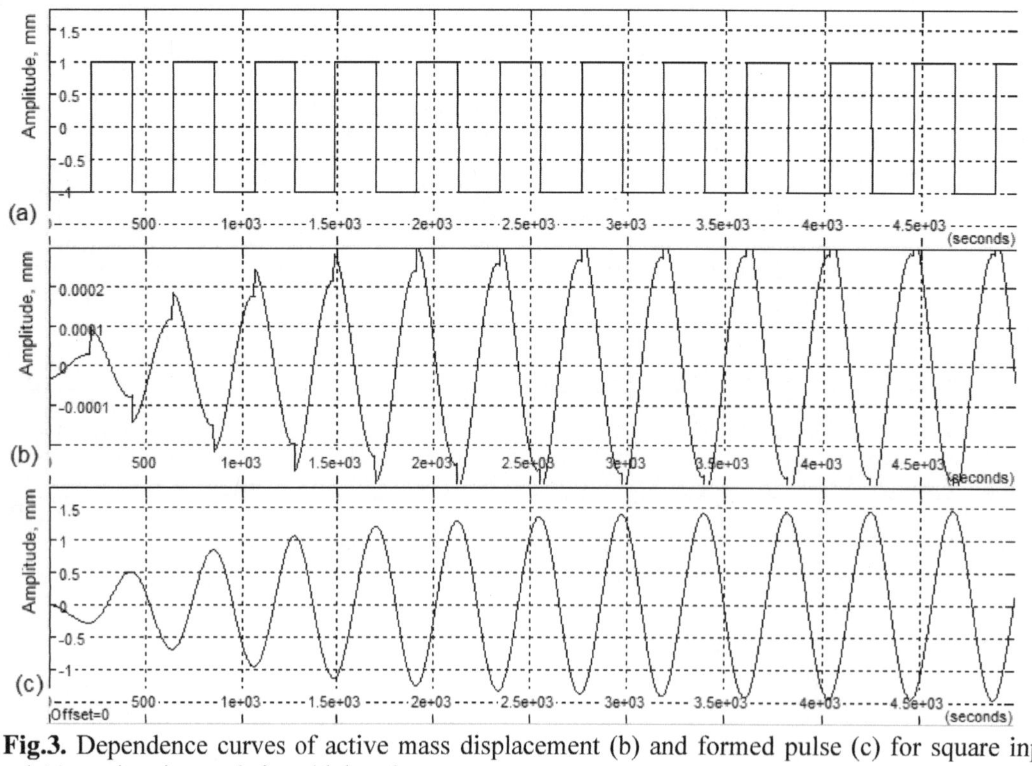

Fig.3. Dependence curves of active mass displacement (b) and formed pulse (c) for square input signal (a) on time interval, for which pulse acts

Formed pulse curve is sine for three experiment runs; it comes out from oscillation concept of machine operation. Output signal becomes stable after eight oscillation periods for all types of input signal curve and then reaches its maximum amplitude. Maximum amplitude is reached by modelling with usage of square input signal. It is 1.3 times higher than amplitude for sine input signal curve and 2 times higher than amplitude for sawtooth input signal curve.

Fig.4-5 show the result of mathematical modelling. Spring stiffness varies from 18600 N/m to 76000 N/m, while active mass value remains constant.

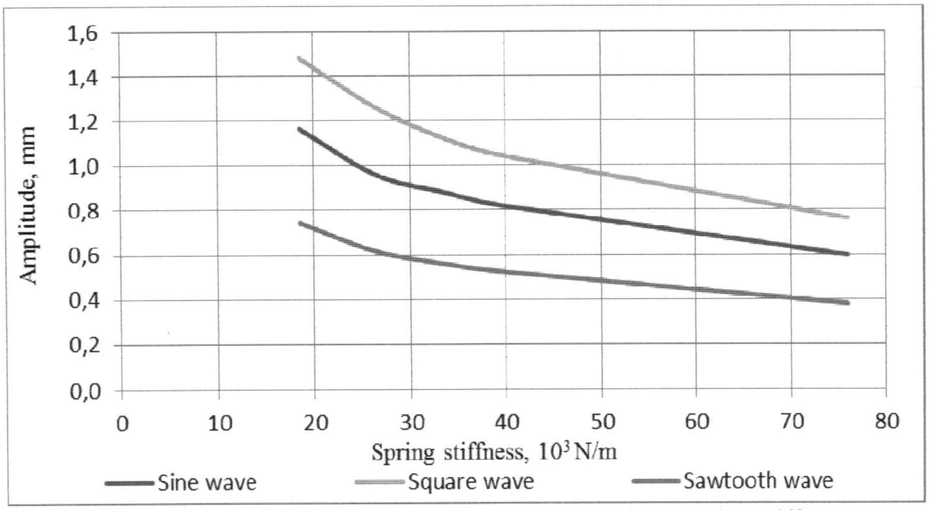

Fig.4. Dependence curve of formed pulse amplitude on spring stiffness

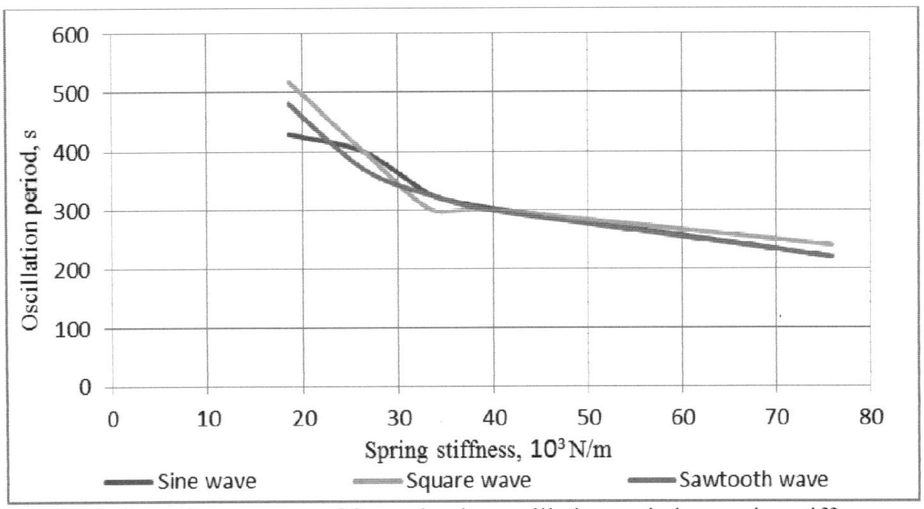

Fig.5. Dependence curve of formed pulse oscillation period on spring stiffness

Formed pulse oscillation period decreases with the increase of spring stiffness for all types of input signal curve. Oscillation period does not depend on type of input signal curve. Small difference for several values of spring stiffness is visible due to the measurements uncertainty. Formed pulse amplitude also decreases with the increase of spring stiffness for all types of input signal curve. Amplitude interrelation remains constant for all values of spring stiffness for different types of input curves.

Fig.6-7 show the result of mathematical modelling. Active mass varies from 100 to 800 kg, while spring stiffness remains constant.

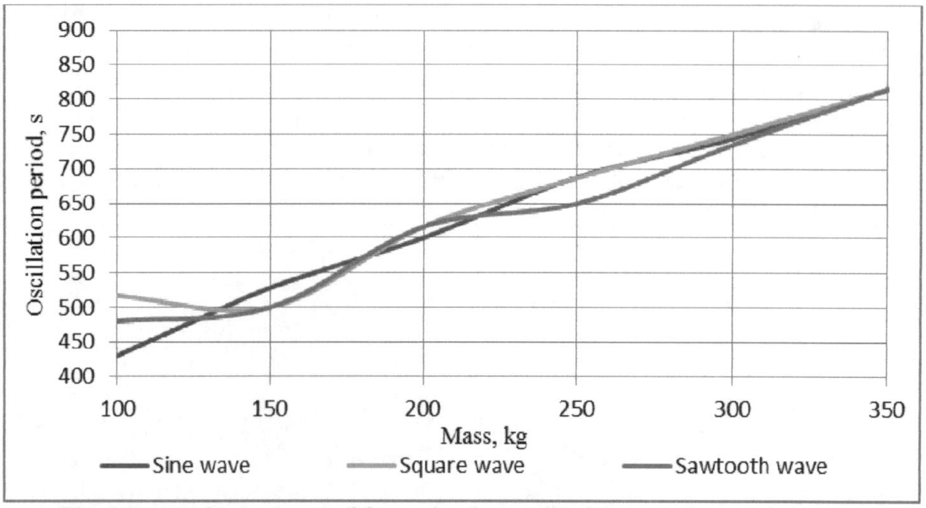

Fig.6. Dependence curve of formed pulse oscillation period on active mass

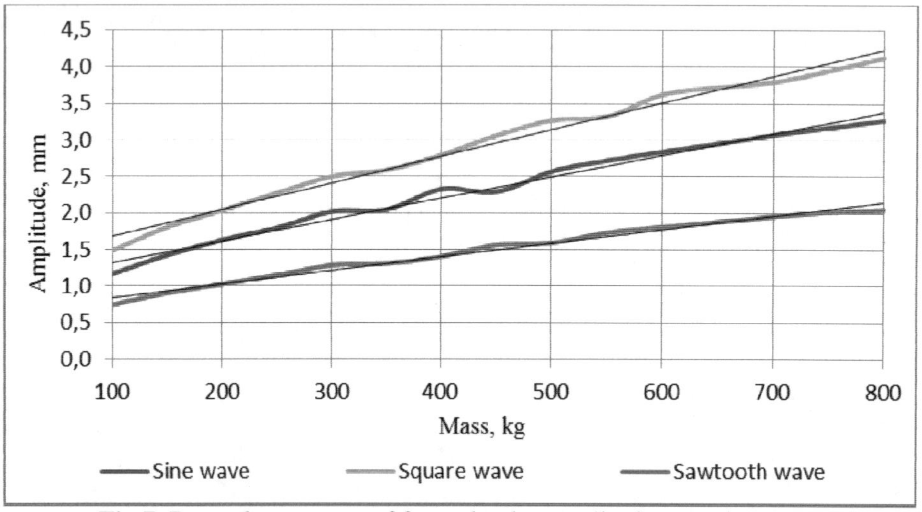

Fig.7. Dependence curve of formed pulse amplitude on active mass

Formed pulse oscillation period increases two times, when active mass is increased from 100 kg to 350 kg. Oscillation period does not depend on type of input signal curve. Small difference for several values of active mass is visible due to the measurements uncertainty. Formed pulse amplitude also increases with the increase of active mass for all types of input signal curve. Amplitude interrelation remains constant for all values of spring stiffness for different types of input curves. Amplitude increases two times, when active mass is increased from 100 kg to 500 kg.

Conclusion

Results of conducted experiment show that input signal curve has an effect on amplitude, while oscillation period value is not dependent on curve. Oscillation period is affected by construction parameters, such as spring stiffness and active mass. Regardless of construction parameters, amplitude is also affected by input pulse curve. Square input pulse curve allows reaching the maximum possible value of amplitude. It is 1.3 times higher than amplitude for sine input signal curve and 2 times higher than amplitude for sawtooth input signal curve. Further studying will be on finding interrelationship between construction parameters, which have an effect on formed pulse, including parameters that are not covered in this article.

References

[1] O O Angatkina, P Y Krauinsh, V N Deryusheva Influence of the Shaper Design of Pneumohydraulic Impact Device on the Form and Duration of the Impact Impulse Applied Mechanics and Materials **756** (2015) 394-401

[2] V F Gorbunov, A G Lazutkin, L S Ushakov, Hydraulic power unit in mining industry Novosibirsk Nauka (1986) 197

[3] E N Pashkov, N V Martyushev, I A Masson, Vessel ellipticity and eccentricity effect on automatic balancing accuracy IOP Conference Series Materials Science and Engineering **66** (2014) 1-5

[4] M V Novoseltseva Investigation of the Influence of Hydroimpulsive Mechanism Design Parameters on the Formed Impulse J Advanced Materials Research **1040** (2014) 682-685

[5] A M Ashavsky, A Y Volpert, V S Sheinbaum Power pulse systems Moscow Mashinostroenie (1978) 200

[6] X Wei, C Y Wang, H L Yuan, Z H Xie Study of Fracture Mechanism of Rock Cutting Key Engineering Materials **250** (2003) 200-208

[7] K K Tupitsyn Surveillance of dynamically balanced machines Novosibirsk Nauka (1974) 157

[8] E N Pashkov, G R Ziyakaev, M V Novoseltseva Differential equations of processes for the hydropuls power mechanism of drill machines Applied Mechanics and Materials **379** (2013) 91-94

[9] L S Ushakov, Y E Kotylev, V A Kravchenko Impact hydraulic units Moscow Mashinostroenie (2000) 416

[10] A N Glazov, I V Kuznetsov Distributing System of Percussion Mechanism in Compressed-air Rock Drills J Mining research and information journal **4** (2013) 515-517

Analysis of the Formation of Multi-Layer Carbon Nanotubes in the Process of Mechanical Activation of the Pyrolysis Products of Vegetable Raw Materials

V P Reva[1], A E Filatenkov[2], V U Yagofarov[3], D A Gulevskii[4], V G Kuryavyi[5], Yu N Mansurov[6]

[1,2,3,4,6]Far Eastern Federal University, Vladivostok, Russia
[5]Institute of Chemistry, Far Eastern Branch, Russian Academy of Sciences, Vladivostok, Russia

E-mail: festurvp@mail.ru

Abstract. The carbon nanotubes are formed by pyrolytic and mechanochemical technology. Amorphous carbon is produced at 950°C and then subjected to mechanochemical treatment in a planetary mill for 1—46 h.Analysis ofinfluence of duration of mechanical activation of amorphous carbon on the morphology of moldable multilayer carbon nanotubes. It is demonstrated that prolonged mechanical activation of carbon composite in a vario-planetary mill promotes to formation of aggregates and amorphous carbon and to loss of thermal stability of nanotubeswith furtherconduct ofvacuum annealing.

Introduction

Material generatedbased on carbon nanotubeshave unique propertiesand were therefore findwide application as components of radio electronics, modifiers of structural materials, additives to lubricants, varnishes and paints, hydrogen accumulators of fuel cells of electric cars, high-efficiency adsorbents, gas diffusion layers of fuel cells, etc. Use of carbon nanotubes in fine chemical synthesis, biology, and medicine is promisingas well asfor producing composites with improved mechanical and electrophysical properties, including for imparting antistatic and conducting properties to polymers [1].

Properties of multilayer carbon nanotubes (MCNT) formed in the process of mechanical activation of amorphous carbon of plant origin in a vario-planetary mill for 1-27 h were studied earlier [2-4]. However, the problem of maximally possible MCNT yield under fixed conditions of mechanical activation of amorphous carbon remained unclarified. This study was carried out to determine the influence of duration of mechanical activation of amorphous carbon on the nanotube content in the end product.

The carbon nanotubes were produced by employing pyrolytic and mechanochemical technologies. Rusty (brown) peat moss *(Sphagnum fuscum)* and Magellan's (midway) peat moss *(Sphagnum magellanicum),* shoots of corn of the variety Katerina SV, thorny (spiny) bamboo *(Bambusablumeanaschultes),* cotton plant of the variety Priozernyi (Lake region)-4, and stems (stalks) of okra *(Abelmoschusexculentus)* of the variety Zelenyibarkhat(Green velvet) were used as the feed material. The plant material was initially dried and sieved to remove excess moisture and foreign matters, and disintegrated to 100-150 μm particles. Modifications of carbon having amorphous structure were obtained at 950° C. The mineral matters were removed from the carbon by chemical washing with a mixture (1:1) of 25% HNO_3 and HCl solutions at 100°C. The treatment was carried out in an analytical autoclave from Wiegand International GmbH (Germany) for 50 min. Next, the amorphous carbon was removed from the mixture of the acid solutions on a Keramtech (Czech Republic) filter and washed with distilled water in an Elamasonic S 30 (Germany) ultrasonic washer for 30 min up to neutral pH (pH 7.0). Thereafter, it was dewatered in a Sigma Laborzetrifugen (Germany) centrifuge and then dried in a Binder drying oven for 60 min at 125-130°C.

Further, the amorphous carbon was treated in a Fritsch (Germany) Pulverisette-4 vario-planetary mill. The mechanoreactor of the vario-planetary mill consisted of a leak-proof corrosion-resistant steel container with a VK-6 (WC-6) hard alloy insert: the pulverizing bodies were VK-6 hard alloy balls 16 mm in diameter. The mill operation conditions were: rotation speed of main disk 400 rpm and of planetary pinions 800 rpm, intensity (ratio of weight of original materials to weight of pulverizing balls) 1:50.

The specific surface area of the carbon materials was studied on a Sorbtometr-M-series [ZAO (CJSC) KATAKON, Novosibirsk] specific surface area analyzer and the specific surface area was determined by thermal desorption of nitrogen. The X-ray energodispersion microanalysis was performed on an EVO-50XVP (Carl Zeiss) scanning electron microscope in conjunction with an INCA Energy-350 (England) X-ray energodispersion spectrometer. The structure of the MCNT was studied on a Hitachi S5500 (Japan) high-resolution scanning electron microscope with an attachment for transmission microscopy.

Table 1.The chemical composition of the amorphous carbon obtained by pyrolytic treatment of brown sphagnum peat moss.

Spectrum No.	Content*, wt. %							
	0	C	W	Ni	Cu	Co	Fe	Ti
1	0.10	99.5	0.020	0.003	0.011	0.004	0.011	0.002
2	0.12	99.3	0.010	0.001	0.015	0.005	0.015	0.005
3	0.14	99.4	0.009	0.001	0.018	0.003	0.017	0.004

Note.*K, Na, Ca, Mg, Zn,Mn, Mo are also presentin the analysis samples in traces.

a b

IOP Conf. Series: Materials Science and Engineering **127** (2016) 012008 doi:10.1088/1757-899X/127/1/012008

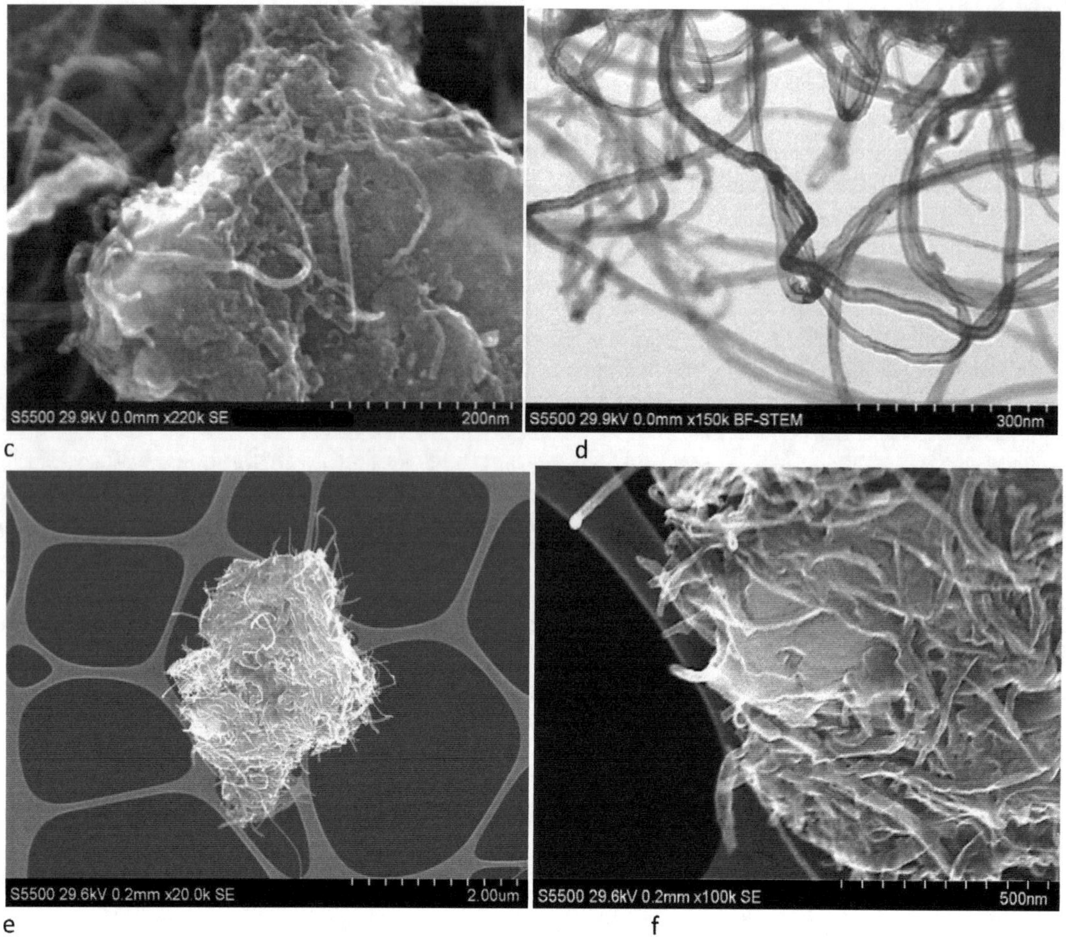

Figure1.Change in structure of carbon from brown sphagnum peat moss *(ad)* and corn (e and *f*) after mechanical activation for: *a* - 6 h; b, *c* - 8 h; *d* -10 h;*e,f*- 16 h; *a, b, c, e, f* -SEM image; *d* - TEM image.

The carbon composites were prepared for vacuum annealing using coal toluene (OAO Bagleikoks, Ukraine). The carbon mass (0.4 g) was mixed with 50 ml of toluene and filtered on a finely disperse biologicalfilter and then dried in a drying oven at 60°C. The amorphous carbon was removed by three-stage vacuum annealing at 220-870°C in a laboratory vacuum furnace with a System VII-series graphite electrode from Contorr Vacuum Industries (USA). The temperature was 220°C in the first stage, 550°C in the second, and 870°C in the third stage.

At the initial stage of the study, carbon modifications with amorphous structure were obtained from the plant material at the pyrolysis temperature of 950°C. The chemical composition of the amorphous carbon obtained by pyrolytic treatment of brown (rusty) sphagnum peat moss is given in the Table 1. In all cases, the morphology of the amorphous carbon is represented by the original structure typical for a specific type of plant material(figure 1).

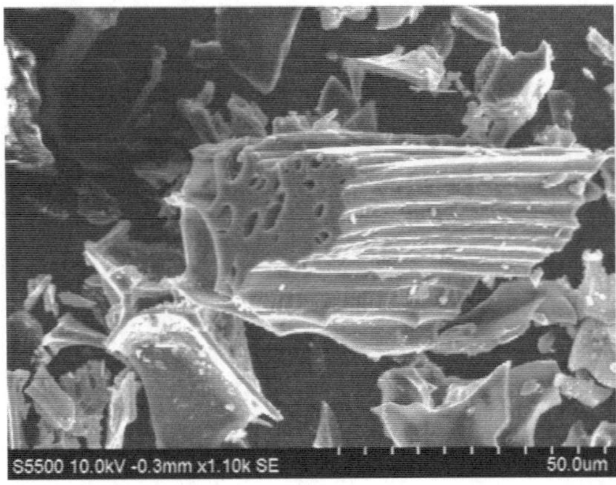

a

b

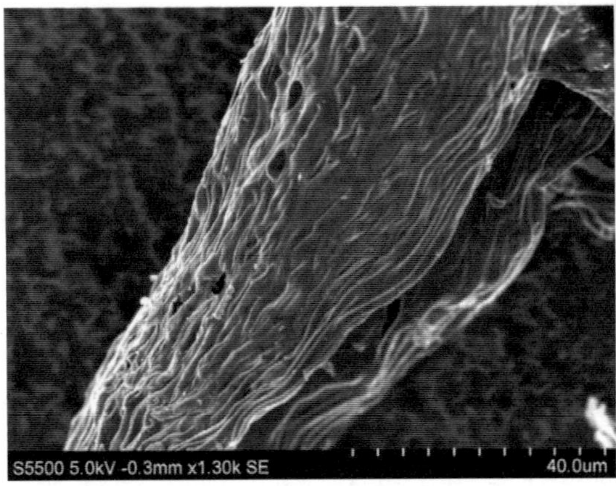

c

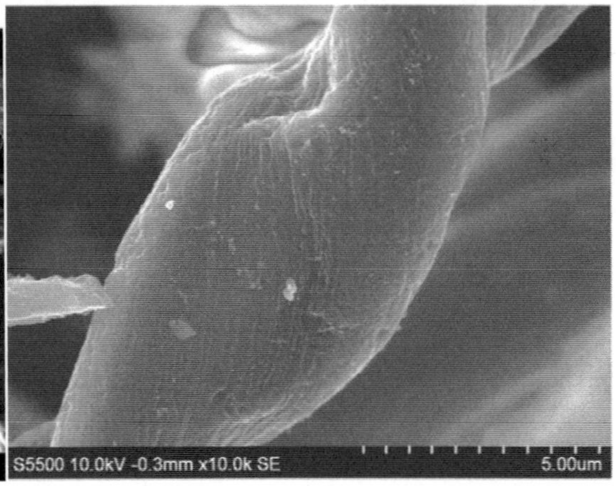

d

e

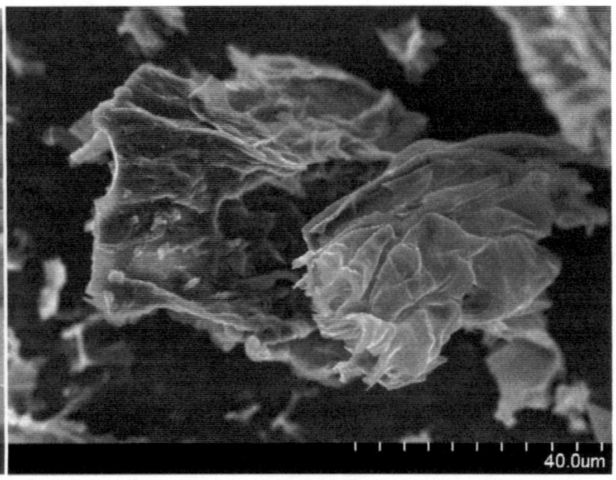

f

Figure 2.Structure (SEM) amorphous carbon after pyrolysis at 950 ° C: a - bamboo; b - corn; c- sphagnum brown; d - cotton; e - okra stalks; f - мagellan's peat moss

To ascertain the sequence of MCNT formation, the amorphous carbon was submitted to mechanical activation for 1 to 46 h. The change in the structure of the amorphous carbon in the course of its mechanical treatment in a vario-planetary mill is shown in Figure1. As can be seen from Figure 1a, in the first 1-6 h, formation of MCNT is not discernible by electron microscopic methods (the amorphous carbon retains its lamellar form). After 8 h of treatment (see figure2 b), begins the process of formation of a nanofiber structure, which is realized in the mass of the carbon particle (see figure2 c), whereupon carbon nanotubes with diameters of 10-20 nm are formed. After 10 h of mechanical activation, the whole volume of the treated material consists of carbon nanotubes with diameters ranging from 10 to 70 nm (figure2 d).

Note that some particles up to 3 μm in size survive up to 16 h of mechanical treatment (figure2e,f) when amorphous carbon is produced by pyrolysis of corn and Magellan's peat moss and up to 27 h when amorphous carbon from cotton plant, bamboo, and okra are used.

Further extension of mechanical treatment time to 27 h leads only to increased defects in the carbon nanotubes with partial formation of "enclosednanocones" and "bamboo" type of structures. The MCNTs produced by mechanical activation of amorphous carbon have a fairly large specific surface area (S_{sp}= 400-510 m^2/g and a low ash content (~ 1.5 wt. %).

It was shown earlier that the quantity of nanotubes formed upon mechanical activation of amorphous carbon depends linearly on the time of mechanical treatment (1-27 h) of carbon composite in a vario-planetary mill [2-4].To determine the nanotube content in the carbon composite, we carried out in this work three-stage vacuum annealing of the latter, which facilitates removal of amorphous carbon. It is well known that carbon nanotubes retain thermal stability during high-temperature vacuum annealing up to 1900°C and even above and that no change in morphology of the nanotubes occurs after thermovacuum treatment, which allows one to effectively purify CNTs and get a product with a purity of no less than 99.9 wt. % [5].

As will be seen from Table 2, a substantial quantity of CNT is formed in just 4 h of mechanical activation of the amorphous carbon, although electron microscopic methods do not show them up until 8 h of mechanical activation. The reason for this is CNT formation inside the amorphous carbon particles. The yield of nanotubes reaches the maximum for all the studied amorphous carbon modifications when the duration of mechanical activation is 36 h. In this case, the MCNT content is found to be maximum (79.48 wt. %) after vacuum annealing of carbon composite obtained by mechanical activation of rusty (brown) sphagnum peat moss pyrolysis products, which is twice as much as the yield of CNTs obtained by similar treatment of another sphagnum moss, namely, Magellan's peat moss.

Table 2.Substantial quantity of CNT from 4 - 46h.

Initial plant matter	Mechanical-activation time, h						
	4	7	10	16	27	36	46
Brown sphagnum moss	8.28	18.98	43.71	52.67	74.87	79.48	11.49
Corn	6.42	11.52	24.45	34.06	42.75	50.0	8.55
Magellan's peat moss	13.42	23.98	25.4	27.54	33.42	37.26	8.26
Cotton	4.14	8.45	21.84	24.25	26.55	28.72	7.12
Spiny bamboo	1.95	4.12	8.82	12.14	15.91	18.31	5.22
Okra stalks	2.16	6.28	7.31	8.5	10.77	12.42	6.21

The results of scanning electron microscopic study of CNTs formed upon 36- and 48-h of mechanical activation of the amorphous carbon produced from rusty (brown) sphagnum peat moss are shown in Figure 3. In this case, the carbon composites were not submitted to vacuum annealing. As can be seen, the carbon nanotubes formed in 36 h of mechanical activation are segregated from each other, and formation of agglomerates and aggregates in the carbon material mass is not observed.

IOP Conf. Series: Materials Science and Engineering **127** (2016) 012008 doi:10.1088/1757-899X/127/1/012008

a b

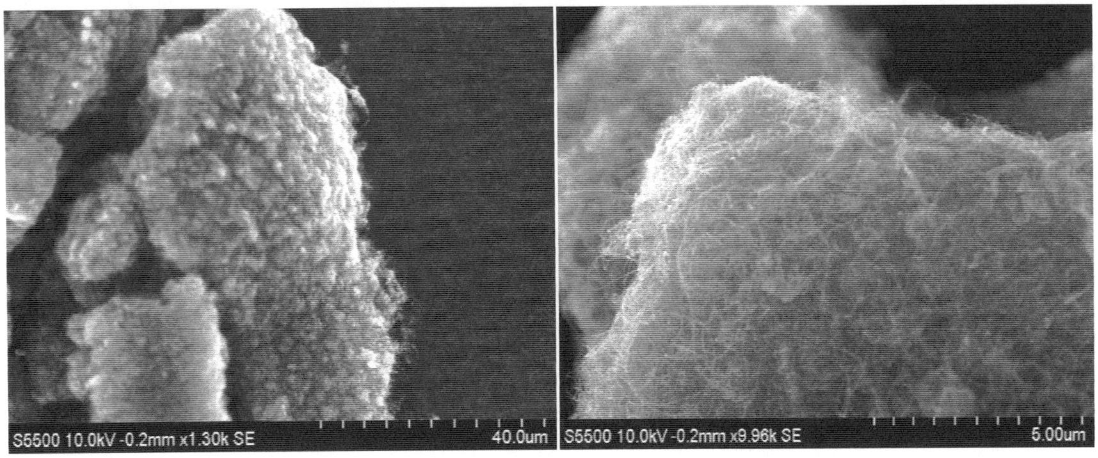

c d

Figure3.Structure (SEM) of carbon nanotubes obtained by mechanical activation for:*a, b* - 36 h; *c,d*- 46 h.

Extension of mechanical activation time to 46 h leads to formation of 20-100 nm tomentoseaggregates (nanocomposites) consisting of carbon nanotubes and amorphous carbon. The aggregates are formed apparently due to electrostatic interaction of MCNTs. If this carbon nanocomposite is submitted to vacuum annealing for removing amorphous carbon, the nanotube content in the annealing products decreases several folds (see table 2).

Conclusions

Thus, the maximum carbon nanotube yield is observed after 36 h of mechanical activation of amorphous carbon obtained by pyrolysis of plant materials. Longer mechanical activation produces MCNT + amorphous carbon nanocomposite aggregates, which subsequently leads upon vacuum annealing to loss of thermal stability of the carbon nanotubes constituting the aggregate. This fact has to be taken into consideration while optimizing the technology of production and purification of carbon nanotubes formed from plant materials treatment products and while using MCNTs as modifiers of composites submitted to thermal action.

References

[1] A L Martinez-Hernandez, Carlos Velasco-Santos, V M Castano Carbon Nanotubes Composites: Processing Grafting and Mechanical and Thermal Properties Current Nanoscience **6** (2010) 12-39

[2] D V Onishchenko, V P Reva Use of multilayer carbon nanotubes made from plant materials as sorption- based hydrogen storage containers Chemistry and Technology of Fuels and Oils **4** (2013) 281-286

[3] D V Onishchenko, V P Reva, V V Chakov Promising nanocomposites based on renewable plant resources Metallurgist **9** (2013) 679-683

[4] D V Onishchenko, V P Reva, B A Voronov Adsorption activity of carbon nanotubes formed from brown sphagnum moss in J mechanical activation Journal of Analytical Chemistry **5** (2014) 403-407

[5] W Huang, Y Wang, carbon nanotubes by vacuum high-temperature annealing, Carbon **13** (2003) 2585-2590

IOP Conf. Series: Materials Science and Engineering **127** (2016) 012009 doi:10.1088/1757-899X/127/1/012009

Physical and Mathematical Modeling of Heat Transfer in Intumescent Thermal Protective Coatings Under Radiative Heating

V G Zverev[1], V D Goldin[2], and A V Teploukhov[3]

[1] Tomsk State University, Tomsk, Russia
[3] Moscow Institute of Thermal Technology, Moscow, Russia

E-mail: zverev@niipmm.tsu.ru

Abstract. Radiative heating of a metal plate protected by an intumescent coating layer has been studied experimentally and theoretically. Special aspects of physical modeling of intumescent coating heating on a test bench for radiative heating are considered. Heat resistance testing of metal structures is justified. A conjugate mathematical model of radiative-conductive heat transfer in an intumesced material layer is proposed. The experimental and calculation data on the metal plate temperature beneath the coating are shown to agree well.

Introduction

Different-purpose structures operating under extreme conditions require an improvement of their heat resistance to fires of different origin, intensity and duration. The notion of heat resistance implies the ability of structures to operate properly during fire [1]. The heat resistance of metal structures is defined by the time of achieving the ultimate temperature T_{fr} at which material strength decreases. For example, T_{fr} for steel is equal to 500 °C and T_{fr} for aluminum is 250 °C.

A solution to this problem is the application of thermal protective materials [1]. Intumescent coatings have a range of advantages over conventional materials and are currently one of the most efficient means for passive thermal protection of structures.

Intumescent coatings operate through an increase of the protective layer thickness during heating and formation of char that has low porosity and thermal conductivity. The char serves as a barrier for heat transfer and reduces the heat flux to the substrate [2]. The development and testing of intumescent thermal protective coatings on structures requires that tests reproduce the full-scale fire conditions and theoretical investigation methods be developed to determine optimal parameters of fire protection systems.

In view of the fact that it is very difficult and expensive to perform full-scale experiments, a large part of investigations can be carried out using test benches. In this case objects for investigation are materials and structural elements, and modeled thermal loading conditions are equivalent to those in real fires. This approach has certain advantages as it allows reproducing results, reduces material and time costs, and provides more information. It is widely applied in industry [3] to improve the heat resistance of structures [3] and hence it can be naturally used to study, test and optimize the application of fire protection materials.

This paper is aimed to study experimentally and theoretically heat transfer in intumescent coatings under radiative heat load, to model fire resistance tests of metal structures using a test bench for radiative heating, and to describe mathematically the heating of intumescent coatings with regard to radiation transfer in the char layer.

1. Analysis of thermal loading on a test bench for radiative heating.

Thermal radiation is one of the hazardous factors in fire [1]. To analyze the heat resistance of materials and structures, it is important to model thermal loading conditions of fire. Modeling is usually performed with the use of various setups operating on the principle of fuel combustion, electric or sun heating. Most of them model a narrow heat load range or do not allow repeated reproduction of real fire parameters.

Test benches for radiative heating offer more opportunities for creating the space-time laws of thermal loading [3]. They use tubular incandescent halogen lamps as the radiator. The lamps are simple to use and can be easily mounted into heating panels of various shape and size. Figure 1a shows the principal scheme of a test bench for radiative heating [3].

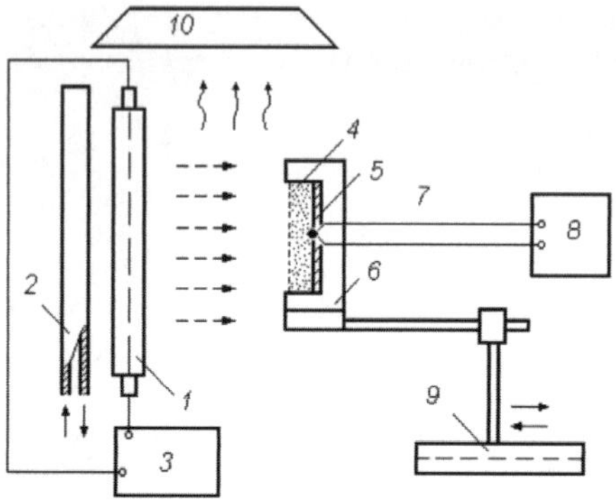

(*a*)

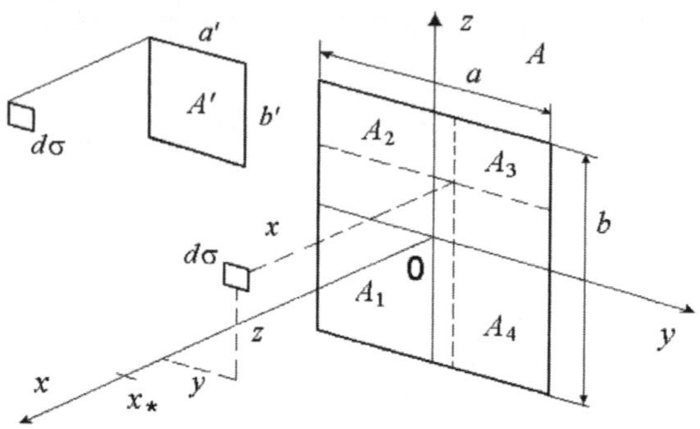

(*b*)

Fig. 1. Schematic of a test bench for radiative heating [3] (*a*) and schematic of calculating the radiative heat flux (*b*). *1* – halogen lamp panel KG 220-2000, *2* – water-cooled reflector, *3* – heating power control system, *4* – coating, *5* – substrate, *6* – heat insulation, 7 – thermocouples, *8* – recorders, *9* – device for moving the sample, *10* – ventilation.

The variation of the spatial field of radiative heat load can be estimated in calculations and measurements using heat flux sensors. Figure 1b schematically illustrates the calculation of the radiative heat flux field in the approximation of a continuously emitting panel A of dimension $(a \times b)$, where A_1–A_4 are the parts of panel A, A' is the plane $(a' \times b')$, $d\sigma$ is the unit surface area, and x_* is the fixed cross section along x.

The unit surface area $d\sigma(x, y, z)$ is under radiation flux of density

$$q_r(x,y,z) = q_{r0}\varphi_{d\sigma\to A}(x,y,z), \qquad q_{r0} = \varepsilon_r \sigma T_r^4 \tag{1}$$

where T_r, ε_r is the radiation temperature, emissivity factor of the radiator, and σ is the Stefan–Boltzmann constant. The angular coefficient $\varphi_{d\sigma\to A}$ between the unit surface area $d\sigma$ and parallel plane A' spaced distance x is equal to [4]:

$$\varphi_{d\sigma\to A'} = \varphi(x,a',b') = \frac{1}{2\pi}\left\{\frac{a'}{\sqrt{a'^2+x^2}}\,arctg\frac{b'}{\sqrt{a'^2+x^2}} + \frac{b'}{\sqrt{b'^2+x^2}}\,arctg\frac{a'}{\sqrt{b'^2+x^2}}\right\} \tag{2}$$

For the unit surface area $d\sigma(x, y, z)$ we have:

$$\varphi_{d\sigma\to A} = \varphi_{d\sigma\to A_1} + \varphi_{d\sigma\to A_2} + \varphi_{d\sigma\to A_3} + \varphi_{d\sigma\to A_4} =$$

$$= \varphi(x,\frac{a}{2}+y,\frac{b}{2}+z) + \varphi(x,\frac{a}{2}+y,\frac{b}{2}-z) + \varphi(x,\frac{a}{2}-y,\frac{b}{2}-z) + \varphi(x,\frac{a}{2}-y,\frac{b}{2}+z). \tag{3}$$

Dependences (3) for the points lying on the x axis ($y = 0$, $z = 0$) and y axis ($z = 0$, $x = x_*$) for a laboratory emitting panel with dimensions $a = b = 0.25$, 0.5 m and industrial panel with dimensions $a = b = 1$ m are given in Fig. 2.

Analysis of curve *1* in Fig. 2a shows that radiative heat load decreases significantly with distance from the panel along the x axis. For example, for distances $0.1 < x < 0.2$ m for a laboratory panel it comprises 65% and 33% of the initial value. An increase in the panel dimensions up to $a = b = 0.5$ m (curve *1'*) reduces this dependence down to 88% and 66%, and at $a = b = 1$ m (curve *1''*) down to 97% and 88%, respectively. This imposes certain requirements on the performance of tests on intumescent coatings which are related to the coating thickness growth during heating.

For example, the surface of the coating of initial thickness 2 mm located at the distance $x_* = 0.15$ m from the emitting panel would have the coordinate $x_* \approx 0.11$ m at a 20-fold expansion of the material. This increases the heat load up to 32% in the first case ($a = b = 0.25$ m) and up to 12% in the second ($a = b = 0.5$ m).

Thus, to preserve a given load level on the coating surface, larger panels or a special device for moving the sample for small laboratory panels should be used [1].

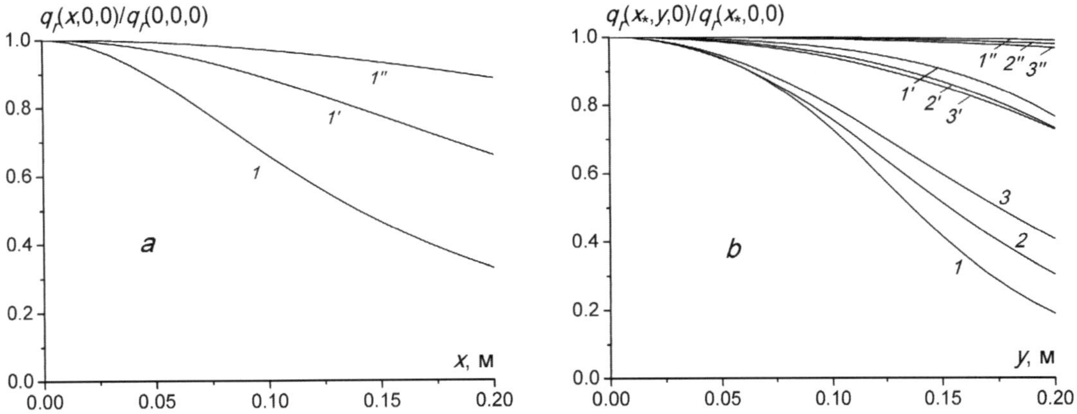

Fig. 2. Density variation of the radiative heat flux: (*a*) – in the x coordinate (symmetry axis), (*b*) – in the y coordinate. $a = b = 0.25$ m; *1–3* – $x_* = 0.1$; 0.15; 0.2 m. $a = b = 0.5$ m; *1'–3'*, the same x_*. $a = b = 1$ m; *1''–3''*, the same x_*

The radiative flux in each section along x (Fig. 2b) decreases with y increase. The decrease is less pronounced for large x (curves *2*, *3*), and at $y < 0.1$ m all curves almost coincide. According to calculations, the irregularity degree of the radiative flux field within the range $-50 \le y \le 50$ mm is less than 5%, which corresponds to the data from [3]. This means that the required sample dimensions must not exceed 100×100 mm if we model uniform radiative loading of the surface for a laboratory panel of dimensions $a = b = 0.25$ m.

2. Modeling of fire resistance test of metal structures.

Let us consider in detail the performance of fire resistance tests of metal structures. Metal structures have different cross section shapes. Owing to high thermal conductivity of metal there is no temperature gradient in the bulk of metal structure, and the equation for its heating in fire has the form:

$$\rho_m c_m V_m \frac{dT}{dt} = \{\alpha_f (T_f(t) - T) + \varepsilon_{fw}\sigma(T_f^4(t) - T^4)\} S_f, \quad T(0) = T_0 , \tag{4}$$

$$T_f(t) - T_0 = 345 \cdot \lg(0.133t + 1)$$

where T, $T_f(t)$ is the temperature of metal structure and "standard fire" [1], t is the time, ρ_m, c_m, V_m is the density, specific heat capacity and metal structure volume, S_f is the heated surface area, $\varepsilon_{fw.}=(1/\varepsilon_f+1/\varepsilon_w-1)^{-1}$ is the reduced emissivity factor of the "flame – metal structure surface" system, and α_f is the convective heat transfer coefficient. The subscript m is the metal, f is the flame, and w is the metal structure surface.

Let us denote the reduced metal thickness through $\delta_{red}=V_m/S_f$, for elongated metal structures $\delta_{red}\approx S_{sec}/P_f$, where S_{sec} is the cross section area, P_f is the heated area of the metal structure perimeter. Then, Eq. (4) takes the form:

$$\rho_m c_m \delta_{red}\frac{dT}{dt}=\{\alpha_f(T_f(t)-T)+\varepsilon_{fw}\sigma(T_f^4(t)-T^4)\},$$

which coincides with the equation for heating a plane metal plate of thickness δ_{red} with thermally insulated back surface. As a result, the full-scale fire resistance testing of metal structures on a test bench for radiative heating can be reduced to testing of a plane plate with reduced thickness δ_{red}.

Heat resistance tests can also be conducted directly on metal structure fragments. Let us consider it using an I-beam as an example (Fig. 3). Since thermal loading is single-sided, only a part of the surface is subjected to radiative heating. The reduced metal thickness under test bench conditions is $\delta_{stand}=V_m/S_{inc}$, where S_{inc} is the irradiated surface. Thus, δ_{stand} exceeds (S_f/S_{inc}) times the δ_{red} value determined by the cross section geometry. Consequently, the bench test results correspond to full-scale test results on fire resistance of a similar section with other dimensions, for which $\delta_{red}=\delta_{stand}$.

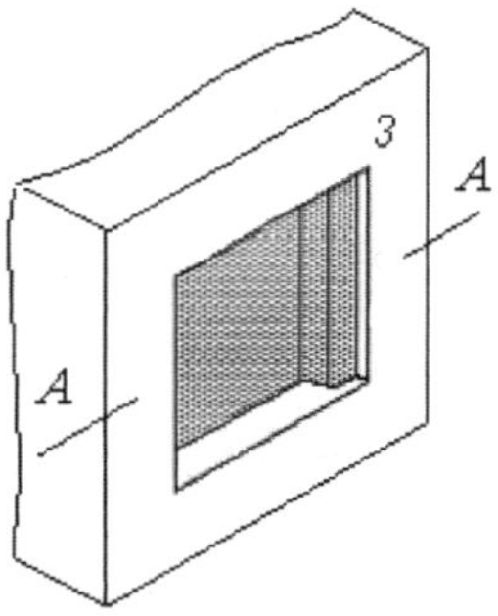

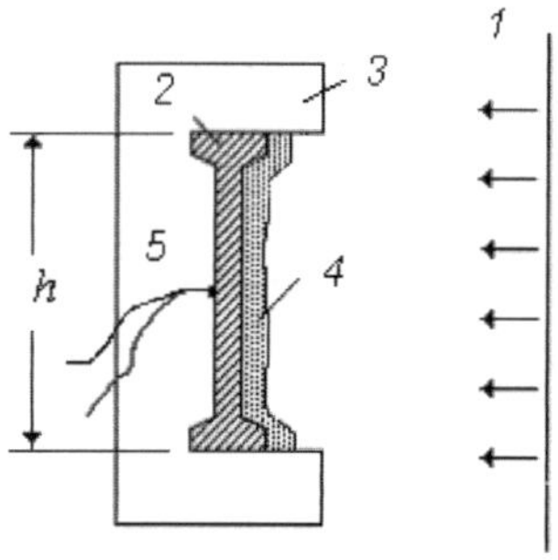

(*a*) (*b*)

Fig. 3. Schematic layout of a metal structure fragment in a test bench for radiative heating.
a – front view, *b* – section *A-A*. *1* – emitting panel, *2* – I-beam, *3* – insulation, *4* – coating, *5* – thermocouple, *h* – cross section height of I-beam.

3. Heat transfer in intumescent coatings under radiative heating.

The mathematical models and approaches used to study heat transfer processes in intumescent coatings are reviewed elsewhere [5]. Of practical interest are theoretical models closed by integral characteristics of intumescent coatings which can easily be obtained within a simplest laboratory experiment. These characteristics can be the mass loss and volume change determined under isothermal heating conditions. The model of this type is proposed in [6]. In order for mathematical modeling to be more accurate, it is necessary to account for additional heat transfer mechanisms among which is the radiation transfer in the coating layer.

The intumescent coating will be considered as a porous two-phase reacting medium that contains a solid (char) and gaseous phase [6]. Heating causes a mass loss in the material. We assume that its expansion is irreversible and one-dimensional. In view of high porosity we will consider radiation transfer in the char layer, neglecting its heterogeneous burnout and shrinkage. Without going into details of the intumescence and thermal destruction mechanisms, we determine the local mass fraction and expansion ratio of an elementary intumescent coating volume through $\overline{m}=m/m_0$ and $\theta=V/V_0$.

In the Lagrangian coordinate system s related to solid material the laws of conservation of mass and energy for the intumescent coating read [6]:

$$\rho_0 \frac{\partial \overline{m}}{\partial t} + \frac{\partial G_g}{\partial s} = 0, \qquad s \in (0, h_0), \quad x = h_b + \int_0^s \overline{\theta} ds, \tag{5}$$

$$(\rho c_p)_{ef} \overline{\theta} \frac{\partial T}{\partial t} + (G c_p)_g \frac{\partial T}{\partial s} = \frac{\partial}{\partial s}\left(\frac{\lambda_{ef}}{\overline{\theta}} \frac{\partial T}{\partial s} - q_r \right) + Q \rho_0 \frac{\partial \overline{m}}{\partial t}, \tag{6}$$

$$\frac{\partial}{\partial s}\left(\frac{1}{3(k+\beta)\overline{\theta}} \frac{\partial U_r}{\partial s} \right) + k\overline{\theta}(4\sigma T^4 - U_r) = 0, \qquad q_r = -\frac{1}{3(k+\beta)\overline{\theta}} \frac{\partial U_r}{\partial s}, \tag{7}$$

$$\frac{\partial \overline{m}}{\partial t} = -R_s(T, \overline{m}), \tag{8}$$

$$(\rho c_p)_{ef} = (\rho \varphi c_p)_g + (\rho \varphi c_p)_s, \quad \lambda_{ef} = (\lambda \varphi)_g + (\lambda \varphi)_s, \quad \varphi_g + \varphi_s = 1, \quad \rho_g / \rho_{g0} = T_0 / T. \tag{9}$$

The thermal state of an opaque substrate is described by the heat conduction equation

$$(\rho c)_b \frac{\partial T}{\partial t} = \frac{\partial}{\partial x}\left(\lambda_b \frac{\partial T}{\partial x} \right), \qquad x \in (0, h_b). \tag{10}$$

The boundary conditions for Eqs. (6), (7), (10) have the form:

$$x = 0: \qquad \lambda_b \frac{\partial T}{\partial x} = 0, \tag{11}$$

$$x = h_b, \; s = 0: \quad T\big|_{h_b} = T\big|_{s=0}, \quad \lambda_b \frac{\partial T}{\partial x}\bigg|_{h_b} = \frac{\lambda_{ef}}{\overline{\theta}} \frac{\partial T}{\partial s}\bigg|_{s=0} - q_r, \quad G = 0, \tag{12}$$

$$-\varepsilon_b U_r - 2(2 - \varepsilon_b) q_r + 4\varepsilon_b \sigma T^4 = 0,$$

$$s = h_0: \qquad \frac{\lambda_{ef}}{\overline{\theta}} \frac{\partial T}{\partial x} = \alpha_f (T_f(t) - T), \tag{13}$$

$$\varepsilon_f U_r - 2(2 - \varepsilon_f) q_r - 4\varepsilon_f \sigma T_f^4 = 0.$$

The initial conditions correspond to the thermal equilibrium between the substrate – coating system and environment:

$$t = 0: \qquad T = T_0, \quad \overline{m} = 1, \quad \overline{\theta} = 1, \quad s = x - h_b. \tag{14}$$

Here, ρ, c_p, λ are the density, heat capacity and thermal conductivity, G_g is the filtration gas mass flux, s, x are the Lagrangian ($s = 0$ for substrate – coating) and Euler ($x = 0$ for substrate back surface) coordinates, Q, R_s are the thermal effect and rate of thermal coating destruction, φ is the volume fraction of phase, ε, σ are the emissivity factor and Stefan–Boltzmann constant, U_r is the quantity proportional to the bulk radiation density, q_r is the radiative heat flux, k, β are the absorption and scattering coefficients, α is the heat transfer coefficient, and h_b, h_0 are the substrate and coating thickness (initial). The subscripts s is the solid material, g is the gaseous phase, f is the environment, r is the radiation, b is the substrate, ef is the effective quantities, and 0 are the initial values.

Equation (5) is the mass conservation law for intumescent coating, Eq. (6) is the energy equation, Eq. (7) is the radiation transfer equation in the diffusion approximation [4, 7], Eq. (8) is the equation of thermal destruction of material, Eq. (9) are the effective thermophysical characteristics of porous medium of the coating [1]. Condition (11) corresponds to perfect heat insulation of the substrate, Eq. (12) corresponds to the thermal conjugation of the substrate – coating interface and to radiation reflection, Eq. (13) to convective heating of the coating surface and to specifying the radiative heat load. The model is closed by determining the parameters $\overline{m}$, $\overline{\theta}$ that integrally characterize material expansion [6].

The system of equations was solved numerically using Petukhov's difference method of the 4th order of accuracy with respect to coordinate and 1st order with respect to time [8]. The comparison of calculation results and experimental data for temperature is illustrated in Fig. 4a for a two-phase system heated on the test bench: steel – 2 mm, heat insulation (glass fiber cloth) – 40 mm; and in Fig. 4b for a three-layer system with a 2-mm thick layer of coating SGK-1 [6, 9]. The incident radiative heat flux was 59 kW/m^2. As one can see, there is a good agreement with experiment. Calculations revealed a strong dependence of the obtained results on the coefficients k, β affecting the radiation energy transfer mechanism within the volume.

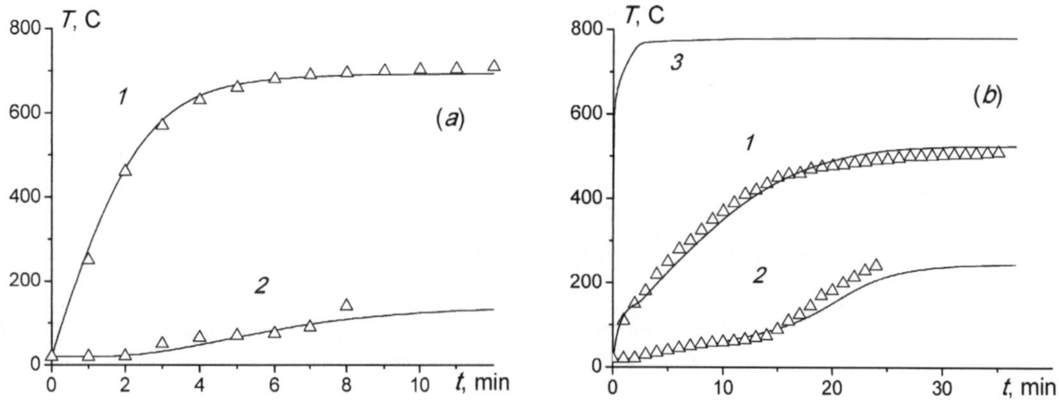

Fig. 4. Heating of a metal plate unprotected (a) and protected (b) by coating SGK-1. Calculation: *1* – metal, *2* – heat insulation (back surface), *3* – coating surface, Δ – experiment. $h_0 = 2$ mm, $h_b = 2$ mm

Conclusion

1. The possibility of fire resistance testing of metal structures with intumescent coatings on a test bench for radiative heating has been shown experimentally and theoretically. The spatial distribution of radiative heat load has been studied, and optimal dimensions of the studied samples have been determined.

2. A conjugate mathematical model of intumescent coating heating with regard to radiation transfer in the char layer has been developed.

3. The thermal regime of a metal plate with coating SGK-1 has been studied on the test bench for radiative heating. Calculation results are shown to be in a good agreement with experimental data, which bears witness to the adequacy of the proposed model for describing heat transfer processes in the char layer.

The work was supported by Project No. 9.1024.2014/k of the RF Ministry of Education and Science.

References

[1] V L Strakhov, A M Krutov, N F Davydkin Ognezashchita Stroitelnykh Konstruktsiy Information and Publishing Centre "TIMR" (2000) 433 p

[2] G J Griffin The modeling of heat transfer across intumescent polymer coatingsJ Journal of Fire Science **3** (2010) 249-277

[3] G B Sinyarev Obshchie printsypy i organizatsiya stendovykh ispytaniy General principles and organization of the thermal test bench Trudy MGTU **392** (1982) 3-16

[4] R Siegel, J R Howell Thermal Radiation Heat Transfer Taylor Francis. (2002) 868

[5] I S Reshetnikov, A V Antonov, N A Khalturinskii Mathematical description of combustion of intumescent polymer systems Combustion Explosion and Shock Waves **6** (1997) 669- 684

[6] V G Zverev, V V Nesmelov, A F Tsimbalyuk, V D Gol'din Modeling heat and mass transfer in intumescent fire-retardant coating. Combustion Explosion and Shock Waves **2** (1998) 198-205

[7] V G Zverev, V D Gol'din, V A Nazarenko Radiation-Conduction Heat Transfer in Fibrous Heat-Resistant Insulation under Thermal Effect High Temperature **1** (2008) 108-114

[8] I V Petukhov Chislennyy raschet dvumernykh techeniy v pogranichnom sloe Numerical methods for solving differential and integral equations (1954) 309-325
[9] Sostav ognezashchitnyy vspenivayushchiysya SGK-1 Spets energotekhnika Publ (2002) 36

Modeling of the Process of Welding Aerosol Formation Taking Place During Mining Equipment Fabrication

V M Grishagin[1], A V Filonov[2], S V Kiselev

Yurga Institute of Tehnology (branch) of Tomsk Polytechnic University
652055 Russia, Yurga, ul. Dostoevskogo, 6

E-mail: [1]znaesh007@yandex.ru, [2]grishagin.v_@list.ru

Abstract. In the paper the authors formulate the thermodynamic model of welding aerosol formation. The thermodynamic parameters of chemical compounds and aerosol phases are calculated. The authors develop a program for numerical calculation of various elements emission under varied parameters changing the welding conditions.

1.Introduction

At the present moment practically in all branches of industry various kinds of welding including elctroslag, contact, electron beam, plasma-arc welding, have become widely spread. Application of the given methods of metal working leads to formation of welding aerosol (WA) in the air of manufacturing areas. WA is a combination of the finest particles formed as a result of condensation of molten metal vapor, electrode coating, ingredients of the flux cored wire or fluxes. WA composition depends upon the components of welded and welding materials. Mainly, WA consists of iron and its oxides, as well as manganese, chrome, nickel, aluminum, cuprum, zinc, fluorine, silicon, nitrogen and other. By the nature of its formation the WA is a condensation aerosol and presents a disperse system where the dispersion phase is presented by fine particles of solids and the dispersion medium is the mixture of gases. So it is necessary to distinguish the solid component of the welding aerosol (SCWA) and the gas component of the welding aerosol (GCWA) in the WA composition [1].

Arc welding processes take place in the open type reactors (Fig. 1) where, effected by high temperature and infrared radiation, the components of coatings, fluxes and metals are intensively evaporated and then, in the process of oxidation and condensation outside the arc, they form submicron aerodispersion particles; thermochemical transformations of fluorides and fluorisilicates, reduction of carbon dioxide or incomplete oxidation of organic binders of coating and plasticizers with formation of HF and SiF_4, CO; nitrogen is oxidized ($N_2 + O_2 + hv = 2NO$); oxygen of the air turns into ozone ($O_2 + hv \Leftarrow O^* + O^*$; $O_2 + O^* \Leftarrow O_3$). When completing the works associated with application of calcium carbide which, as a rule, is contaminated with calcium phosphide, phosphine is formed (PH_3). In case of welding (cutting) parts, units, constructions or materials covered with conserving agents, paint or (and) technological contaminators some other substances may also be released into the gaseous phase, such as: chlorine hydride (HCl), chlorine (Cl_2), phosgene ($COCl_2$), carbon tetrachloride (CCl_4), trichloroethylene ($CCl_2{=}CHCl$), sulphur dioxide and trioxide (SO_2, SO_3), hydrogen sulphide (H_2S), phosphorus (P_4), phosphoric anhydride (P_2O_5), phosphine (PH_3) and other toxic compounds.

It is known that 80-90% of the chemical composition of the SCWA is determined by the content of the welding electrode [1]. In the process of covered-electrode welding 1-3% of electrode material evaporates [3]. The rate of evaporation of the molten mass and the composition of WA are determined by the welding conditions and the welding method, composition of electrode coating, that of the base and filler metal. In the process of welding we usually observe not melting of pure metals but that of alloys, then, vapor will represent a mixture of gaseous components of the alloy. The lower the evaporation heat of the substance and the higher the heating temperature, the more intensive the evaporation process is. Thus, the greater part of the vapor will consist of components with lower evaporation heat. That means that separate components of multicomponent systems are evaporated selectively, for example, from the iron and manganese melts it is manganese to be evaporated first.

When covered-electrode welding metal sequentially passes first the stage of a drop and then the stage of a weld pool. When covered-electrode welding the same stages are also observed for the slag phase. The drop phase is characterized by higher degree of overheating and the specific surface exceeding the specific surface at the stage of the weld pool. Due to the temperature gradient effected by arc pressure and plasma flows convective mass transfer of the evaporated substances is completed from the area of the arc discharge. In the area of lower temperatures the vapor phase condenses and some products of condensation (metals and oxides with the lowest oxidation degree) are subjected to further oxidation. In the works of M. Kobayashi and others [4, 5] we conclude that WA results mainly from electrode metal evaporation (coating and core) in the area of the arc at the stage of the drop and the role of the melt at the stage of the weld pool is less significant. The formed vapor effected by arc pressure and plasma flows travels to the outside environment with lower temperature where it is condensed.

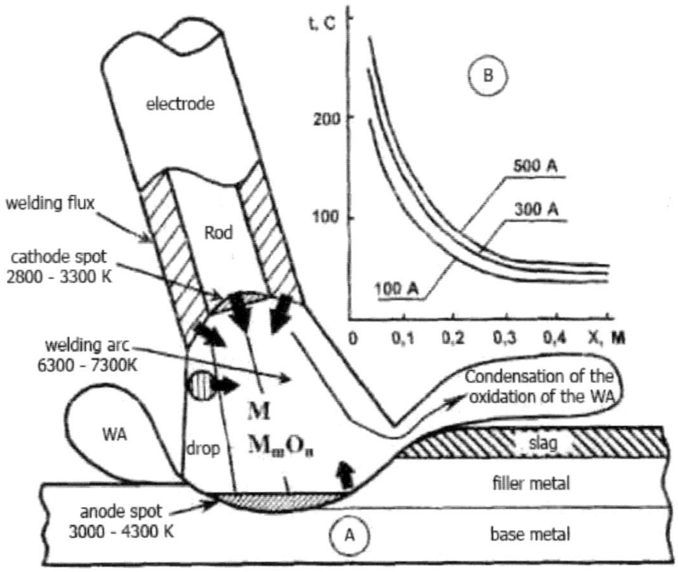

Figure 1 - Mechanism of WA formation: a – basic diagram of heat-mass exchange; b – radial temperature variation

In their works R. Heile and D. Hill [6] describe two possible mechanisms of WA formation. According to the first mechanism, components of electrode and base metals are melt and evaporated and the produced vapor is condensed after it leaves the high-temperature area of the arc and oxidizes. The second mechanism involves formation of volatile oxides under the action of the oxygen on the surface of molten electrode (for example, SiO). The oxygen can be found in the arc atmosphere when welding in the gas mixture (92% Ar + 6 % CO_2 + 2 % O_2). Formation of such oxides may lead to more intense formation of WA. The degree of involvement of each of the described above mechanisms changes according to the welding method and the shielding gas composition, with reduction of oxidizing potential of the shielding gas the second mechanism plays a less significant role in WA formation.

According to the data of modern physical and chemical research (X-ray diffraction, spectral and other methods of analysis) the composition of the solid component of the welding aerosol (SCWA) formed in the process of covered-electrode welding includes basic (Mn, Fe, Si, K, Na, Ca, Mg, Ti, Al, Cr, Ni, F) and impurity (As, Cu, Rb, Sn, Zr, Sr, Nb, Mo, Ag, Sb, Ba, Pb) as well as complex mixtures of metals, simple and composite oxides of metals, spinels $MnFe_2O_4$, $CaFe_2O_4$, $(Fe,Mn)O{\cdot}Fe_2O_3$, $K_2Cr_2O_7$, $Na_2Cr_2O_7$, Fe_3O_4 and other, ftorides (NaF, KF, K_3FeF_6, K_2SiF_6, CaF_2 and other), silicates ($CaSiO_3$, -Si-O-Si-O-Si-, $Fe_2[SiO_4]$, $Mn_2[SiO_4]$ and other). Application of special steels with increased content of manganese (11…14.5 %) and other alloying components for mining equipment production (35L, 45L, 110G13L, 35HGSL, 30HGSA) increases the toxicity of the welding aerosol. The information about gross emissions of welding aerosol and its chemical composition is necessary to estimate the hygiene and sanitary work conditions effecting the people working in welding production. As a rule, these data is obtained with application of spectrophotometric method of analysis which is characterized by long time and complexity of experiment and, thus, it is important to be able to predict the working conditions of the welders without completing special experiments, by creating a thermodynamic model of welding aerosol formation.

2. Results and discussion

In the given paper we chose 30HGSA, 14HG2ASD and 14G2AS steels and welding wire Sv-08G2S and Sv-08GSMT to estimate the welding aerosol composition and concentration as these grades of steel and types of welding wire are most frequently applied for welding mining equipment. The process was completed in CO_2 atmosphere. We estimated the mass concentrations of three elements (Fe, Mn, Si) which occur in the composition of SCWA produced when welding with the given wires, and the intensity of the given elements emission in the process of welding. The model allowing calculating the composition of SCWA was suggested by the authors [1,2]. The model is based on the equation which makes possible taking into consideration the equilibrium evaporation of the elements from the molten metal of the weld pool and nonequilibrium (explosion-like) transition of the components into the aerosol.

$$C_i^a = \alpha_i C_i^p + (1-\alpha_i) C_i^n , \qquad (1)$$

where C_i^a - concentration of the i-th element in SCWA,

C_i^p - concentration of the i-th element in the molten metal (in the composition of wire),

C_i^n - concentration of the i-th element in the saturated vapor over the molten,

α_i - coefficient of nonequilibrium formation of the aerosol characterizing the amount of non-fraction component in the process of SCWA formation [2].

On the base of the given equation and on the base of the results of experimental measurements of SCWA amount and composition, also basing on the equations for determining composition and concentration of elements in vapor according to works [7,8], in paper [2] coefficients α for three basic elements (Fe, Mn, Si) under various arc power were estimated. Basing on the obtained coefficients in [2] the authors suggest to solve the problem of SCWA determining for wires with different composition and for the whole range of arc power. In [1] the authors provide numerical results of SCWA composition and the results of calculations of element concentration in the saturated vapor. Proceeding from the meaning of the introduced coefficient α, its values, in our opinion, must range from zero to one, but the authors obtain negative values of the coefficient according to experimental results and calculations.

According to [7,8] concentration of elements in the saturated vapor over the molten can be obtained from the correlation of vapor pressure under the certain temperature of the molten. The mole composition of the saturated vapor is found from the correlation

$$M_i^n = P_i^{om} \cdot [M]_i / (\sum_i P_i^{om} \cdot [M]_i), \qquad (2)$$

where $P_i^{om} = P_i^y / P_{Fe}^y$, relative pressure of vapor of the i-th element is considered as relation of vapor pressure of the given element over this element melt to vapor pressure of iron over the iron melt,

$[M]_i$ - element concentration in the melt (wire).

Vapor pressure was estimated from Clausius-Clapeyron relation

$$P_i^y = \exp\left(-\frac{\Delta H_i}{RT} + C_i\right), \qquad (3)$$

where $\Delta H_i, C_i$ - heat of element evaporation and constant of integration, accordingly, which are taken from the table data or found from the data on vapor pressure of the elements under the given temperature.

Molar concentration is transformed into the mass one according to the equation

$$C_i^n = M_i^n \cdot A_i / (\sum_i M_i^n \cdot A_i), \qquad (4)$$

where A_i - atomic mass of the given element.

The values of element concentrations in the vapor over the melt found by us do not correspond to the results obtained by Levchenko [2] which we relate to significant disagreement of reference data applied by different authors.

In [2] the author also suggests applying approximation of α coefficient dependence upon arc power to obtain the formula for SCWA elements concentration influenced by the varied welding conditions. To take into consideration the effect of arc power upon elements concentration in the vapor under equilibrium evaporation we need to consider the influence of power upon the temperature of welding wire drops. This influence can be taken into consideration if we put down and solve the heat balance equation which is a nontrivial task which requires considering the welding method, welding conditions, environment, welding productivity and wire cross-section, so temperature change can be obtained only by estimation. In general, to find the SCWA composition we need to apply a thermophysical model which takes into consideration formation of chemical compounds in the process of welding and the welding conditions.

In simplified form by analogy with the model in [2], we suggest changing the expression (1) in such a way not to break the physical meaning of the required coefficients

$$C_i^a = \beta\alpha_i C_i^p + (1-\beta)C_i^n, \qquad (5)$$

where β - coefficient, characterizing the amount of nonequilibrium aerosol formation which is similar in meaning to coefficient α in works [1,2], depends upon the welding conditions and parameters and it is the same for all required elements. The values of the coefficient are within the range from 0 to 1.

α_i - non-fraction coefficient of aerosol composition under its non-equilibrium formation, actually is a characteristics of concentration change of each element in relation to the melt composition. The coefficient cannot be negative.

Finding the SCWA composition according to the expression above is a complicated problem as it depends upon two unknown coefficients which, in their turn, depend upon the welding conditions and, thus, are hard to obtain experimentally. In this regard we have to introduce some suppositions concerning the dependence of coefficient β upon arc power. Basing on the results of SCWA composition measurements in work [1] and calculation of the coefficient, and also understanding that small input power leads to the growth of nonequlibrium component of welding aerosol, we took the changes of coefficient β from 0.7 to 0.5 under changing the input power from 4.5 kW to 13 kW. Coefficients α_i were calculated under such supposition on the base of measuremt results of SCWA composition in work [1]. The coefficients obtained for Sv-08G2S wire and concentration of elements in SCWA from work [1] and the calculated concentration of elements in the saturated vapor under equilibrium evaporation for the capacity of 13 kW are provided in Table 1.

Table 1 – Results of calculations

Element	Mo	Ni	Cr	Mn	Si	Fe
C_i^n,%	0	12.72	0.13	23.83	5.47	57.85
C_i^a,%	-	-	-	15.66	7.54	76.78
α_i	-	-	-	1.82	13.31	1.1

Proceeding from the linearity of coefficients dependence upon capacity in formula (5) we obtained the formula for concentration of elements in SCWA depending upon the wire composition and input power (similar to [1,2], coefficient 0.68 used in these works was not used by the authors of the given paper as its physical meaning is unknown and its application was not reasoned by the authors).

Calculation of saturated vapor composition and that of element concentrations in SCWA according to (5) was programmed and the program allows obtaining the dependences of element concentration in SCWA upon the capacity and the wire composition inserted by the user. In the process of program checking we applied two models of calculation of the saturated vapor composition over the melt described in [7]. As a result we chose the model described in the first part (equations (2)-(4)).

Also, we programmed the algorithm of calculation of intensity and specific emission of SCWA components on the base of equations taken from work [2]

$$V_i = 0.167 \cdot 10^{-4} \cdot I \cdot U \cdot (\beta_0 - K \cdot I \cdot U) \cdot C_i^a, \text{ g/min} \quad (6)$$

where $I \cdot U$ - welding arc power, kW;

β_a, K - coefficients of linear approximation of specific rate of SCWA formation dependence upon the arc power.

$$G_i = V_i / G_p, g / kg; \quad (7)$$

where G_p - productivity of welding material diffusion, kg/min.

Results of calculating the SCWA composition and intensity of emission for three metals (Mn, Si, Fe) are shown in figures 2-7. The changes in concentration of the given elements correlate with experimental data in [2].

The dramatic impact upon the cooling rate of metal vapor and, thus, upon quantity distribution of particles in the welding aerosol, is produced by temperature distribution and the flow rate in the arc [9].

Thus, the results prove that intensity of manganese, iron and silicon emission in SCWA when welding the considered steels applied for mining equipment production is maximal under arc power from 8 to 12 kW which allows making recommendations about choosing the welding conditions. Besides, taking into consideration the peculiarity of SCWA mixing under two physical phenomena taking place in the process of welding, we obtained a valid expression for SCWA composition allowing forecasting total emission and intensity of SCWA emission [10]. At the same time, due to two coefficients

IOP Conf. Series: Materials Science and Engineering **127** (2016) 012010 doi:10.1088/1757-899X/127/1/012010

in the model depending upon welding conditions, their experimental estimation is complicated. For more accurate solution of this problem we need to develop a thermophysical model of the welding process taking into consideration formation of chemical compounds and phases. This model could describe nonequilibrium formation of welding aerosol and its amount in the emitted aerosol. We think that for correct predictive modeling we also need to consider the equation of temperature dependence of melt drops upon the applied power. To obtain such dependence we need to derive an equation of heat balance which can be solved with application of a number of experimentally obtained values.

The dramatic impact upon the cooling rate of metal vapor and, thus, upon quantity distribution of particles in the welding aerosol, is produced by temperature distribution and the flow rate in the arc [9].

Thus, the results prove that intensity of manganese, iron and silicon emission in SCWA when welding the considered steels applied for mining equipment production is maximal under arc power from 8 to 12 kW which allows making recommendations about choosing the welding conditions. Besides, taking into consideration the peculiarity of SCWA mixing under two physical phenomena taking place in the process of welding, we obtained a valid expression for SCWA composition allowing forecasting total emission and intensity of SCWA emission [10]. At the same time, due to two coefficients in the model depending upon welding conditions, their experimental estimation is complicated. For more accurate solution of this problem we need to develop a thermophysical model of the welding process taking into consideration formation of chemical compounds and phases. This model could describe nonequilibrium formation of welding aerosol and its amount in the emitted aerosol. We think that for correct predictive modeling we also need to consider the equation of temperature dependence of melt drops upon the applied power. To obtain such dependence we need to derive an equation of heat balance which can be solved with application of a number of experimentally obtained values.

The opportunities for detailed analysis of particles formation and distribution of particles according to their dimensions are provided by digital modeling [11]. Application of digital modeling as contrasted to most welding experiments may allow establishing a clear cause-and-effect link.

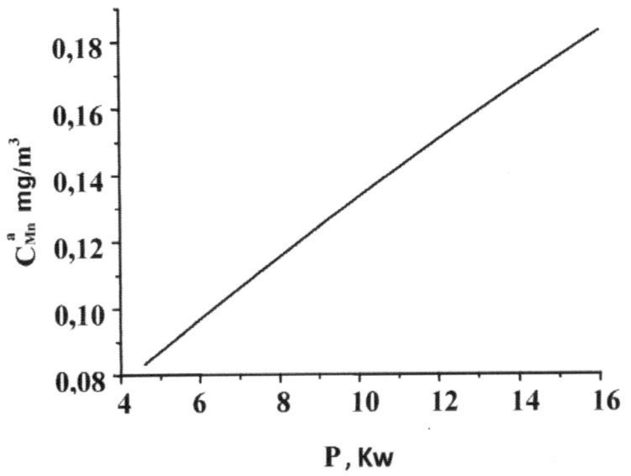

Figure 2 – Dependence of manganese concentration in SCWA upon arc power

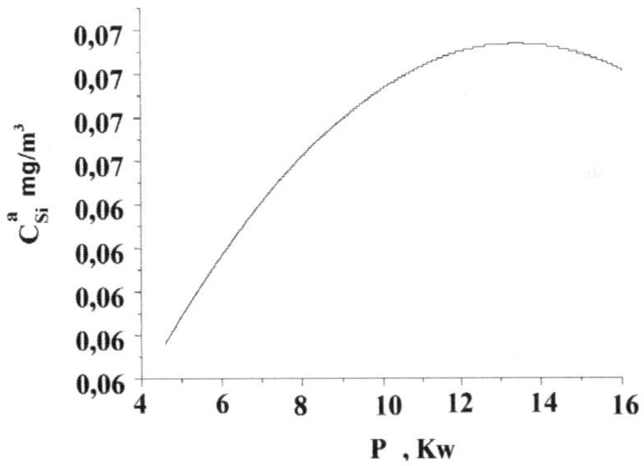

Figure 3 – Dependence of silicon concentration in SCWA upon arc power

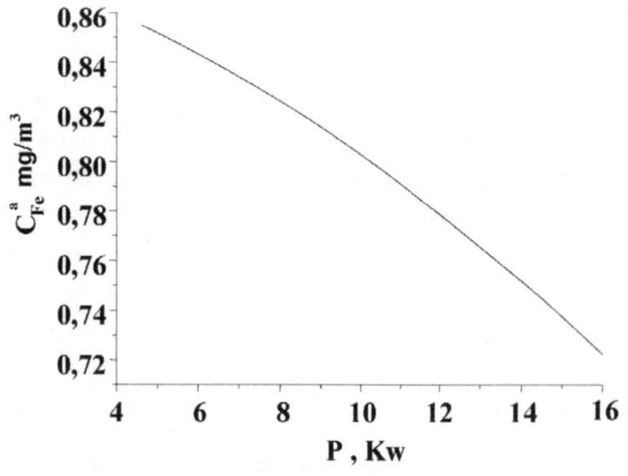

Figure 4 – Dependence of iron concentration in SCWA upon arc power

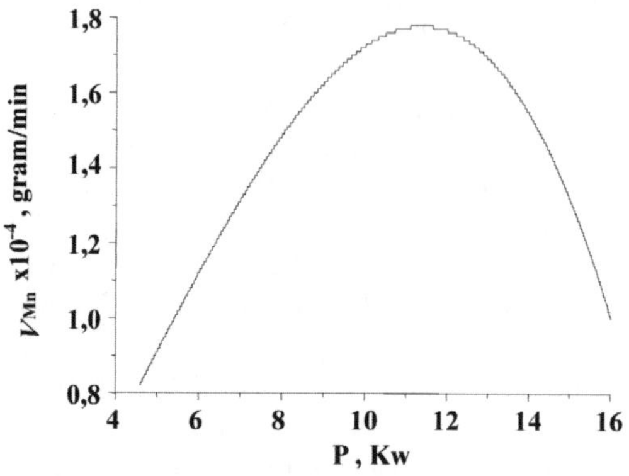

Figure 5 – Intensity of manganese emission in SCWA upon arc power

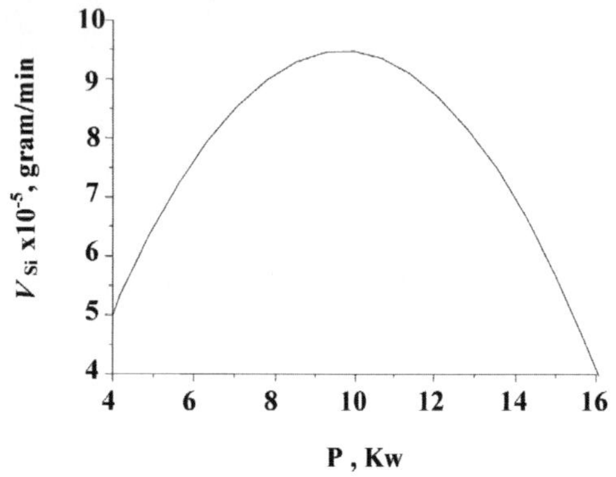

Figure 6 – Intensity of silicon emission in SCWA upon arc power

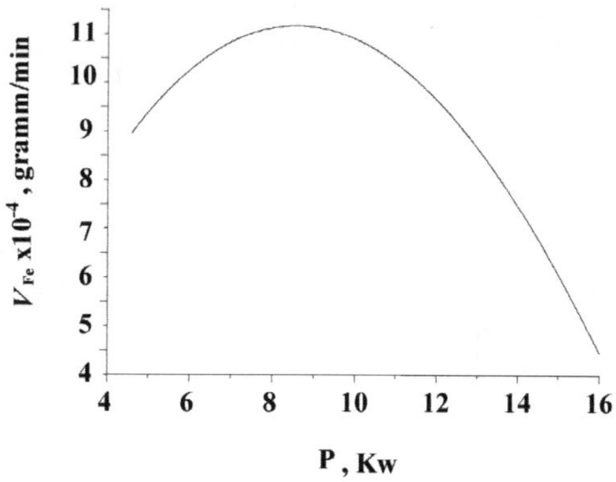

Figure 7 – Intensity of iron emission in SCWA upon arc power

3. Conclusion

The developed model allows calculating the amount of various elements emitting changing due to alteration of parameters effecting the welding conditions. In contrast to experiments we can separately consider the intensity and type of current, type of electrode metal transition, etc. to estimate separate phenomena, such as coarse-droplet transfer with short circuit of the arc space and overheating of electrode end resulting from it, without additionally effecting data variableness.

The given results will lay the foundation for developing a complex model of welding aerosol formation allowing considering the welding method, welding conditions and environment.

References

[1] Pokhodnya, I K, Yavdoshin, I R, Paltsevich, A P, Shvachko, V I, and Kotelchuk A S Arc welding metallurgy. interaction of metal with gases Kiev Naukova dumka 2004

[2] Levchenko O G mathematical modeling of chemical composition and level of welding aerosol emission under arc welding Welding production. 2001 №7 P 25-27

[3] Jenkins N, Met A, Moreton J etal. Welding Fume, vol 2. Published by the Welding Inst. Abington

Hall Abington. Cambridge CB16AL 1981 P 211-506
[4] Kobayashi M, Maki S, Hashimoto Y and Suga T Some considerations about the formation mechanisms of welding fume Weld. World 1978 16 N 11/12 P 238-245
[5] Kobayashi M, Maki S Ohe 1 Factors affecting the amount of fumes generated by manual metal arc welding V IIW Doc II-E-211 76. P 22
[6] Heile R. F, Hill D C Particulate fume generation in arc welding processes Weld. J. 1975 N 7 P 210
[7] Buki A A Modeling physical and chemical processes of arc welding Mashinostroyeniye 1991 P 287
[8] Pokhodnya I K Arc welding metallurgy. Processes in the arc and electrode melting Under the editorship of Kiev Naukova dumka 1990 P 224
[9] Roze C Conception of formation and reduction of smoke emissions released under gas-shielded welding of metals with considerations to new variants of the process. Perm. Works of international scientific technical conference devoted to 125th anniversary of consumer electric arc welding invention by N G Slavyanov Welding and control 2013 P 322–343
[10] Grishagin V M, Safronova A V Solid components of welding fumes as a synthetic composites filler *Applied Mechanics and Materials Trans Tech Publications Vol. 682 2014 P 369-374*
[11] Tashiro S. et al.: Numerical analysis of fume formation mechanism in arc welding. J, Phys. D: Appl. Phys., Vol. 43 2010

Dissemination of Weak Waves in Granular Materials Under Short-Term Impulse Loads

A P Bobriakov[1], V P Kosykh[2], A F Revuzhenko[3], A A Kazantsev[4], N V Ivushkina[4]

[1,2,3]Institute of Mining of Siberian branch of the Russian Academy of Sciences, Russia, Novosibirsk, Red avenue, 54
[4]Yugra Technological Institute branch of Tomsk polytechnic University, Russia, Kemerovo region, Yurga, Leningradskaya str. 26

E-mail: bobriakov@ngs.ru

Abstract. The results of experiments with dry high-silica sand are presented. Multiple point impacts have been revealed to improve waveguide properties of the material because a conductive channel, containing force "chains", is formed there. Quasistatic alternating shears condition the change in the particle packing; destroy "chains", and deteriorate the channel conductivity. Further multiple impulse loads lead to restoration of the "chains" and conductivity of the channel.

Introduction

One of the main properties of powder, granular and granulated materials is multiplicity of their forms of equilibrium, i.e. similar boundary conditions are met by diverse disseminations of inner stresses. It is possible because conditions on the contacts between particles are in form of inequalities. That is why the behavior of material is generally nonlinear and nonholonomic, i. d. the response of material to outward loads can essentially depend on the whole antecedent history of its deformation.

Let us consider this property on the example of wave dissemination in granular materials. The experiments were carried out in dry high-silica sand; the average grain was ~0.3 mm.

1. Methods of experiments

The first difficulty arising when carrying out experiments is to get quite homogenous samples of granular materials. Here, to avoid data scattering samples are to be almost identical in all experiments. To get homogenous packing we used a movable case of a uniform shear machine [1]. As the material is filled up in the case initial particle packing has a number of defects, vacancies, arches, dislocations etc. If the sample is subjected to the cyclic alternating shear, dry friction among the particles is replaced by the pseudo-viscous one and a lot of defects are eliminated. Quite homogenous closed packing is formed; its properties are stable in all the experiments.

The uniform shear machine (Figure 1) consists of a case with three movable and one fixed side walls and a square bottom with the side of 250 mm, transforming into a parallelogram under the shear. Under the cyclic shear parameters of the packing get stabilized within the first dozens of cycles. For the purpose of experiments the samples were deformed within 30 cycles with the range of shear angle

±1.5 degree. As the state close to the stationary one had been attained, cyclic deformation was stopped.

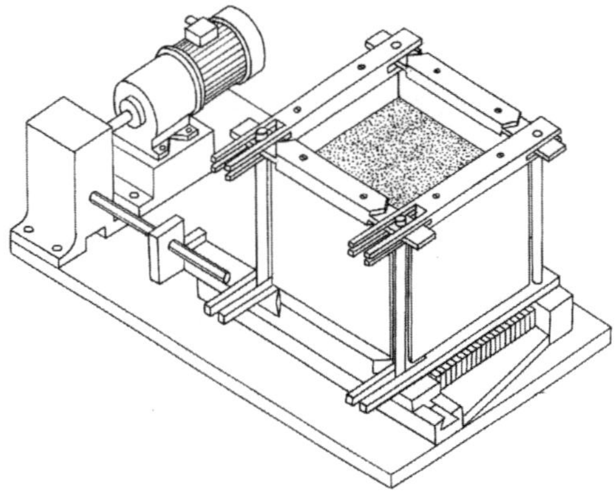

Figure 1. The overview of the uniform shear case

The single point impact of the physical pendulum calibrated against energy and located on the vibroplatform, which was on the fixed wall of the case and submerged into the granular material generated the wave in the material (not shown in Figure 1). The length of the pendulum was 300 mm; a steel ball was 35 g. Impact energy (E) was calculated according to the deflection angle from the equilibrium and amounted to 0.175; 0.350; 1.40; 2.45; 3.50; 4.55 $J \cdot 10^{-3}$ in experiments.

Impulses of acceleration were measured by two accelerometers. Accelerometer 1 was stationary fixed on the vibroplatform (it generated the initial reference impulse). Accelerometer 2 was placed in the granular material in front of the vibroplatform and at a particular distance from it. The axis of sensor sensibility was oriented towards the dissemination of longitudinal waves in the sample (in the direction of impact). Impulses of acceleration were registered by digital oscillograph Agilent 1000 Series. The oscillograph triggering was performed from the front of the reference impulse. Processing of signals involved assessment of the change in their parameters as soon as they had passed through the granular material with respect to the distance length and impact energy. The mean speed of wave dissemination was calculated according to the time of pulse delay.

2. The results of experiments

Figure 2 *a* outlines the impulse acceleration registered by accelerometer 1, a and Figure 2 *b* impulses registered by accelerometer 2, the number of impacts was growing (numbers of impacts are given under oscillograms).

While passing from the reference impulse (Figure 2 *a*) to the first impulse one can see in Figure 2 *b*, that its amplitude is decreased by 1200 times and its duration is grown by 8 folds as compared with the initiating pulse. Comparing curves 1 – 19 one can conclude that impulses are transformed, as well as their forms, amplitudes and duration (the time of impulse passing Δt through the granular material (delay time) is shown along the x-coordinate axis).

To explain the effect let us consider the results of the study on micro-mechanism of granulated material deformation. Paper [2] presents photoelastic analysis of stresses and deformations in conditions of uniform shear in photosensitive broken glass simulating a granular material. Papers [2, 3] point at formation of force chains with maximal contact stresses along them in the material. New additional force contacts are generated if outward load is growing due to micro-rotations and displacement of particles. Therefore, the load-carrying capacity of the structure gets improved. As the result of further increase in loads the generated structure gets destroyed, the resistance of material is deteriorated; and stresses are relieved. The results of study on the sliding behavior under the edge shear of a crack filled with granular material are presented in papers [4, 5] in view of the mechanism of force chains. Advancing and transformation of acoustic waves in high-silica sand are studied in [6, 7, 8].

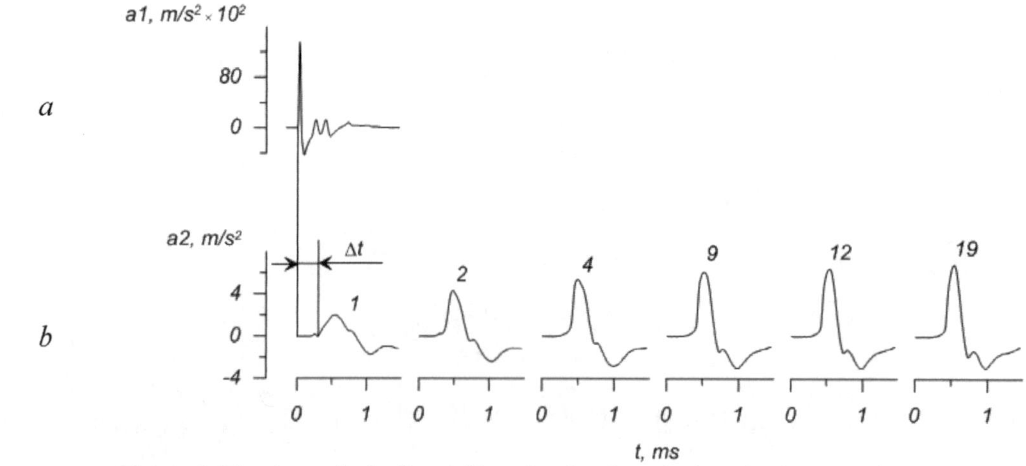

Figure 2. The change in the form of impulse signals under iterative impacts of constant energy E=4.55·10^{-3} J

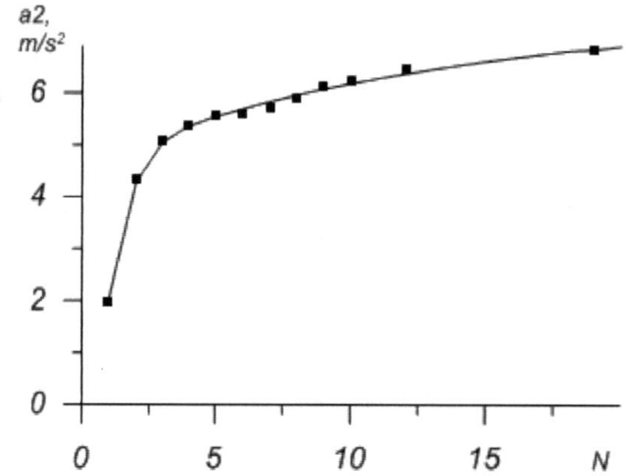

Figure 3. The change in the signal amplitude at the end of the distance L=45 mm in conditions of growing impacts

In the case under consideration each impact conditions a compression wave in the material towards accelerometer 2 from the vibroplatform. Deformations are accumulated with each impact. That is why we have all reasons to assume that particles approach each other and generate stable contact combinations, aligning towards the wave dissemination. As the number of impacts increases, contacts of this kind are multiplied. Here, the conductivity of the material rises too. Energy is transferred both through harder force chains formed under deformation and weakly loaded contacts with more relative slippage of particles. Therefore, a conducting channel is formed in the material under iterative impact loads. Increased pulse duration and reduced amplitude in the unformed conducting channel are possible because of losses in weak contacts. High-frequency components of impulse oscillations are subjected to the strongest influence. In [9] it is revealed that weak contacts as compared with the strong ones have a demodulating property – an ability to weaken or absorb high-frequency components of the signal spectrum. The form of signals and a simple rule «the broader the impulse is, the narrower is the spectrum» enables drawing a conclusion that weak contacts have a dominant influence on signal transfer in the unprepared granular material. As one can see in Figure 2, further impact improves the conductivity of the channel. New force chains formed of weakly loaded contacts support passing through high frequencies, becoming apparent in increased amplitude *a* 2 and reduced pulse duration τ 2 to a constant value.

Figure 3 shows changing amplitude of the signal at the end of the distance according to the number of impacts *N*, as one can see, as soon as 19 impacts had been done, the amplitude was increased nearly by 3.5 times. The delay Δt was quite the same and amounted to 192 μs, being in line with the mean speed of wave dissemination *V*=234 m/s, as the number was changed. After 20 – 25 impacts the state of the channel is distinguished by insignificant further change in its conductivity.

Let us consider the effect of impact energy on basic parameters of wave dissemination in the formed channel. Figure 4 *a* outlines the curve of changing signal delay with respect to the impact energy, which shows the delay Δt is reduced 0.39 to 0.28 ms, being in line with increasing speed of wave dissemination *V* 115 to 160 m/s as the energy grows. Pulse duration $\tau 2$, shown in Figure 4 b grows slightly and gets constant as the impact energy is increased. The growth in pulse duration alongside with the increase in energy points at deteriorated material conductivity.

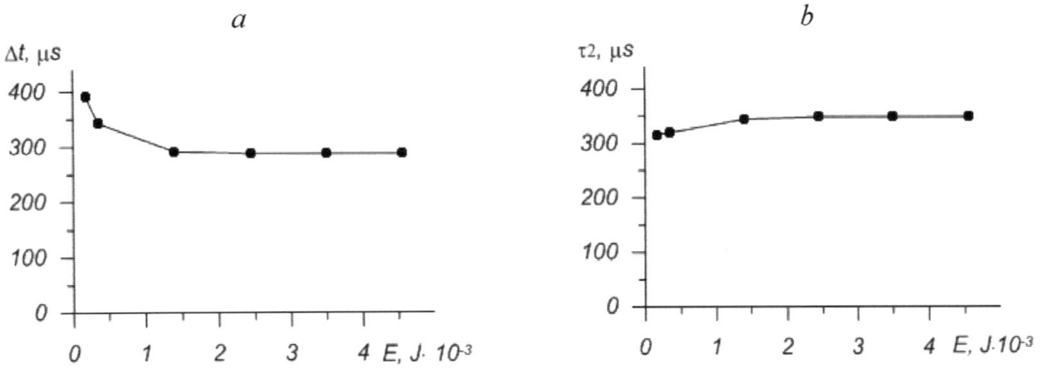

Figure 4. Time of pulse advancing (*a*) and duration (b) at the end of the distance in respect to the impact energy

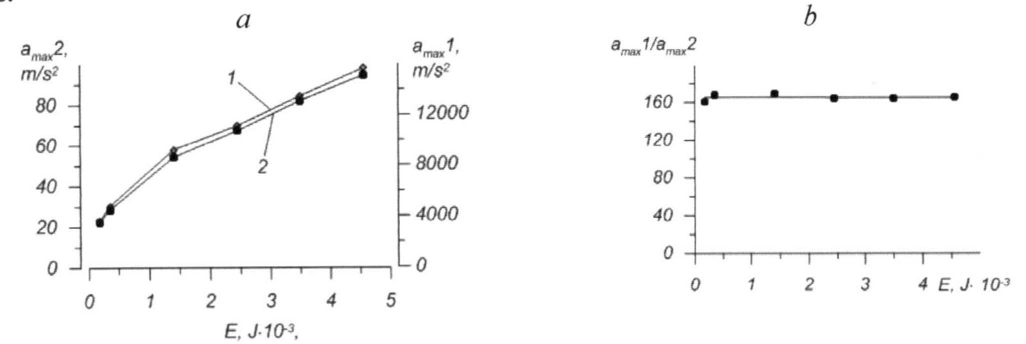

Figure 5. The change in maximal amplitudes of impulses (*a*) and their correlations (b) according to the impact energy on the distance *L*=45 mm

Let us analyze impact transfer through granular material in conditions of fixed length *L*=45 mm. Figure 5*a* shows maximal amplitudes of acceleration *a*1 and *a*2 (curves 1 and 2, respectively) according to impact energy, and Figure 5 b outlines its correlation. One can see amplitudes of acceleration change proportionally, and their correlation is stable in the range under consideration and amounts to 164; that is in line with wave attenuation by 99.4%. The situation is different if the distance between sensors is changed (Figure 6). The dependence on impact energy is becoming apparent slowly. The higher it is, the longer is the distance covered by the wave, and the stronger is attenuation. This dependence agrees with one given in [10].

This dependence can be accounted for the flowing force along the force chains including separate particles in the conductive channel. The process of their formation is highly influenced by the amplitude of iterative impulse impact. Repetitively applied short-term loads with a small amplitude is the reason for activation of unloaded and weakly loaded contacts, which enlarge the number of force chains in the channel; whereas big amplitudes destroy this structure and build a new one with fewer contacts and lower conductivity, consequently.

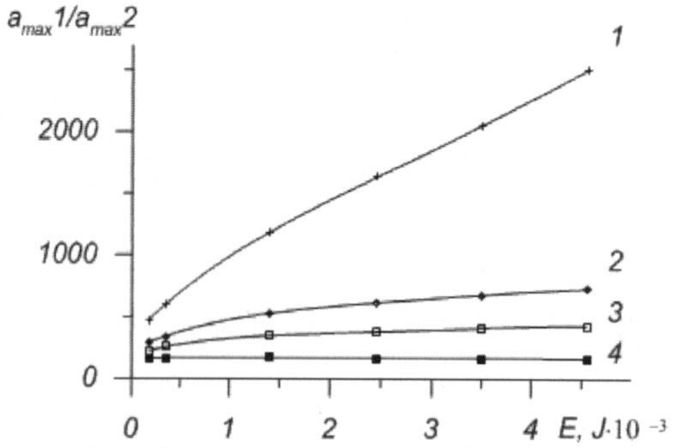

Figure 6. Reference signal attenuation after passing through different distances L, mm: *1* - 180; *2* - 100; *3* - 80; *4* - 45 with respect to impact energy

The property of channel conductivity is also significant in this respect and it comes to light when passing from one impact energy to the other. The experiments have revealed: if the channel is formed by weak short-term loads, the further effect of high loads alongside with the growing number of impacts results in decrease of the amplitude $a2$ (change of higher to lower conductivity in the channel); provided that the channel is the result of high loads, the further effect of weak loads alongside with their growing number leads to gradual amplitude rising (change of lower to higher conductivity in the channel).

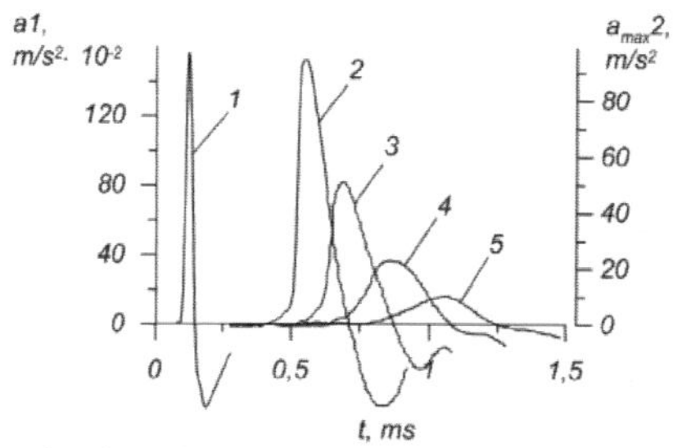

Figure 7. Impulses of acceleration for four distances L, mm: *2* – 45; *3* – 65; *4* – 100; *5* – 150 and reference impulse *1*

Figure 7 presents the reference impulse 1, measured by accelerometer 1 and impulses of acceleration, measured by accelerometer 2 on various distances L. The velocity curve of wave passing through the channel formed by impacts with energy E=4.55·10^{-3} J with respect to the distance length L is shown in Figure 8*a*, and the curve of maximal wave attenuation – in Figure 8 *b* in the same experimental conditions. One can see the speed of wave passing was doubled as the distance was increased. It is possible because acoustic properties of the material get improved due to the effect of low-amplitude deformations arising on the long length, the deformations result from signal attenuation as the distance gets longer (Figure 8*b*). The presented below results are obtained for dry high-silica sand with porosity

$P = (V_{por}/V_{total})\cdot 100\% = 33.6\%$,

where V_{por} – volume of pores in the total volume of the material V_{total}.

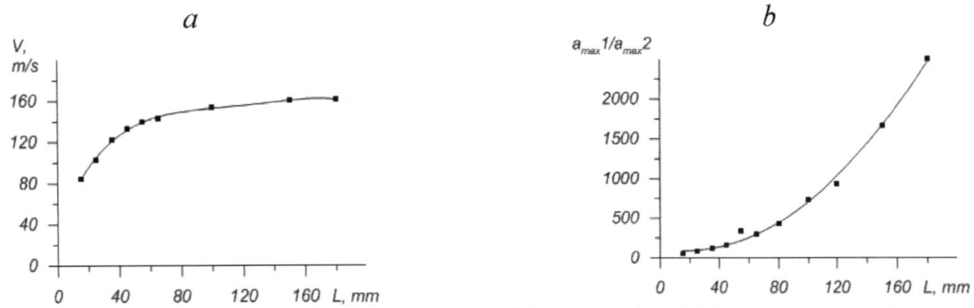

Figure 8. The speed of longitudinal wave passing (*a*) and its attenuation (*b*) in respect to the distance length under constant impact energy E=4.55·10^{-3} J

Let us consider the resistance of formed force chains to external actions. The disturbance was obtained via deforming the sample by alternating cyclic shear with the angle range α=±1.5 degrees.

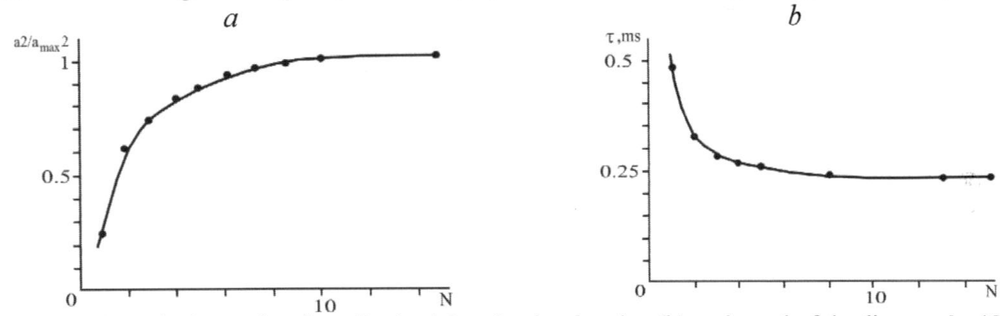

Figure 9. Dimensionless ratio of amplitudes (*a*) and pulse duration (b) at the end of the distance L=45 mm under constant impact energy E=3.5·10^{-3} J according to the number of impacts

The small angle shear conditions slight expanding of the material. The maximal change in the volume was 0.2%, but the porosity got almost restored when the case came back in position α=0. One can think there is no significant change in material properties after a cycle of shear. However, the experiments demonstrated acoustic parameters of sand were rather sensitive to changes arising in particle packing. Similar results concerning other experiments are presented in papers [11, 12, 13]. After being disturbed prevailing conductors made of force chains with strong contacts were destroyed in the conductive channel. The experiments have revealed while the wave is passing through the new structural material amplitude $a2$ goes down sharply. This decline approximates to 70% for the first impact as if compared with the previous maximal amplitude $a_{max}2$. The prior conducting state of the packing can be obtained rather simply via affecting it by some iterative impulses. Figure 9*a* outlines the diagram presenting how the channel conductivity is being restored – dependences of dimensionless ratio of the amplitudes at the end of the distance L=45 mm to the serial number of the constant energy impact E=3.5·10^{-3} J. The relation equal to the one presented in the diagram is in line with completely restored conductivity.

The Figure shows that maximal prior conductivity of the channel has been restored after 15 sequential impacts. In view of experiments N does not depend or is weakly dependent on the impact energy in the range under consideration. Enormously widened spectrum of frequencies to be passed as each impact improved the conductivity of the channel, resulted in double reduction of pulse duration, which amounted to the range $\tau2$ = 520÷260 μs (Figure 9*b*) in this experiment.

3. Conclusions

We have considered low intensive acoustic waves passing through dry high-silica sand under weak impacts. Multiple short-term loads applied up to 20 times are the reason for formation of a channel distinguished by high conductivity of disturbances, advancing from particle to particle along the force chains. The formed chains are irresistible to shear deformations of the sample in general, but can get restored after multiple impacts.

References

[1] A P Bobriakov, A F Revuzhenko, E I Shemyakin Uniform shear of granular material Dilatancy J Journal of mining sciences **5** (1982) 23–29

[2] H G Allersma Photo-elastic stress analysis and strains in simple shear Deform and Failure Granular Mater Rotterdam (1982) 345–353

[3] A Drescher, G de Josselin de Jong Photoelastic verification of a mechanical model for the flow of a granular material J Mech Phys Solids **5** (1972) 337–351

[4] G G Kocharyan, V K Markov, A A Ostapchuk, D V Pavlov Mesomechanics and resistance to the shear along the crack with the filler J Physical Mesomechanics **5** (2013) 5–15

[5] G G Kocharyan, A A Ostapchuk Acoustic emission under various modes of inter-block movements J Journal of mining sciences **1** (2015) 3–13

[6] N A Vilchincka The wave of sand repacking and acoustic emission FEAS USSR **3** (1982) 568–572

[7] Vilchinska N A 2007 Force chains in granular media and ultrasound impulse propagation in sand specimen under load E.J. Technical Acoustics 20

[8] Kolesnikov Yu I, Mednykh D A 2004 On some special features of acoustic waves dissemination in wet sand J. Physical Mesomechanics 7 no 1 pp 69–74

[9] Zaitsev V Yu, Gurbatov S N, Pronchatov-Rubtsov N V 2009 Non-linear acoustic phenomena in structurally inhomogeneous materials

[10] Kulikov V A, Sibiryakov E B 2003 Dissemination of strong waves in inhomogeneous granular materials J: Dynamics of continuous medium. Acoustic of inhomogeneous materials 121 pp 103–110

[11] Goldin S V, Kolesnikov S V, Polozov S V 1999 Dissemination of acoustic waves in soils in conditions of changing shear stress (up to destroying) J. Physical Mesomechanics 2 no 6 pp. 105–113

[12] Kolesnikov S V, Goldin S V, Polozov S V 2002 The change in acoustic properties of wet soils under their shearing (results of laboratory experiments Dynamics of continuous medium: Collected papers 121 pp 97–102

[13] Kolesnikov Yu I, Goldin S V, Polozov S V 1999 Laboratory study on acoustic properties of soils in conditions of destructive shearing Geodynamics and stress state in the Earth's interior: Proceedings of international conference pp 195–199

Influence of the Geometry of Beveled Edges on the Stress-Strain State of Hydraulic Cylinders

G B Buyalich[1,2], A V Anuchin[1], K P Serikov[1]

[1] Yugra Technological Institute branch of Tomsk polytechnic University, Russia, Kemerovo region, Yurga, Leningradskaya str. 26
[2] T.F. Gorbachev Kuzbass State Technical University, Russia, Kemerovo, Vesennyaya str.28

E-mail: gdb@kuzstu.ru, avanuchin@tpu.ru

Abstract. The studies were carried out to determine the influence of forms obtained when preparing edges for welding a cylinder for hydraulic legs; the maximum stresses were defined at the location of weld roots, depending on variable parameters. The stress-strain states were calculated using finite element method.

1. Introduction

Underground mining methods are mainly employed to extract expensive coking coal. Coal is extracted mechanically from long wall faces. Workloads when using longwall mining techniques have been steadily increasing over the last 15 years, while the number of long walls is decreasing. During a given period, the number of longwall faces has been reducing from 170 to 75 [1]. The reliability and durability of longwall mining equipment is of vital importance in this context.

The main basic equipment for longwall mining consists of hydraulically powered supports having a large number of sections. Hydraulically powered supports provide mechanization of roof control and movement of equipment. The main element used to support the roof is a hydraulic leg (Figure 1). The malfunction of this element causes the halt of the whole process of mining, and the malfunction of hydraulic legs in numbers leads to emergency shutdown situation in the face. Research works towards improving the performance of hydraulic legs are relevant.

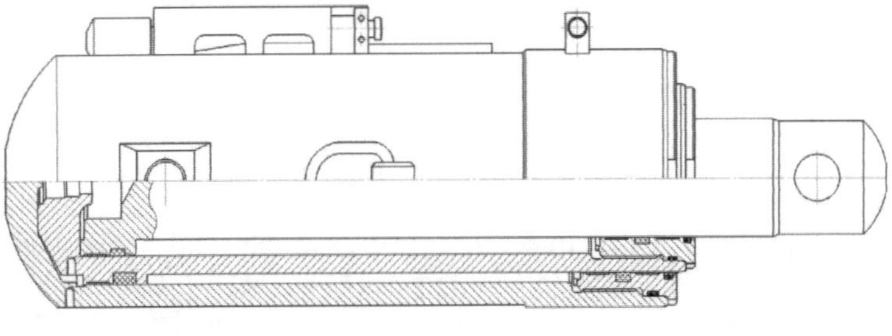

Figure 1. A Hydraulic Leg of Powered Roof Support.

A hydraulic leg consists of a hydraulic cylinder, piston rods of the first and the second stages. Its sealing capacity depends on the design of its main elements [2], forming radial clearances in the working cylinders, and the operation of cup seals in these clearances [3]. The hydraulic cylinder and the first-stage piston rod are welded components consisting of a cylinder and a bottom. Service capabilities of these components depend on the soundness and quality of welds. Metal welding processes are being studied quite extensively [4-8], while the influence of joint configurations prepared for welding is not presented enough in researches at the moment.

2. Influence of the geometry of beveled edges on the stress-strain state

Analyzing the designs of two-stage hydraulic legs with an inside diameter of 220 mm manufactured by various companies, we find a variety of welding edge joints (Figure 2). They can be divided into four types: square groove edge joints (Variant 1); U-groove edge joints (Variant 2); special V-groove edge joints with an opening angle of 36° (Variant 3); special U- groove edge joints with an opening angle of 40 ° (Variant 4).

Using a square groove edge joint is characterized by a smaller amount of weld metal required to fill in the joint, and this, in turn, reduces heat-affected zones. In case of the square groove edge joints, the hydraulic cylinder bottom has a functional step, which is necessary to build a precise gap between parts prepared for welding. The main limitation for using this type of weld joints is lack of fusion at the location of the step, which forms the stress concentrator and increases the risk of occurrence of cold cracks, causing the rapid development of cracks under alternating loads with further structural damage to hydraulic leg cylinders.

The design of U- groove edge joints makes possible to ensure penetration in the zone of the functional step, but due to the improper joint dimensions to allow enough access of the electrode into the joint, there is a risk of lack of fusion in the weld root, that can also cause the appearance of cracks. It can be assumed that the third and the fourth variants have been derived after considering the limitations of the first and the second variants. These variants allow more assess of the electrode into the joint and ensure appropriate fusion in the region of the functional step, avoiding the appearance of stress concentrating point. The limitation for using the latter groove edge joints is large amounts of weld metal needed and additional manufacturing costs.

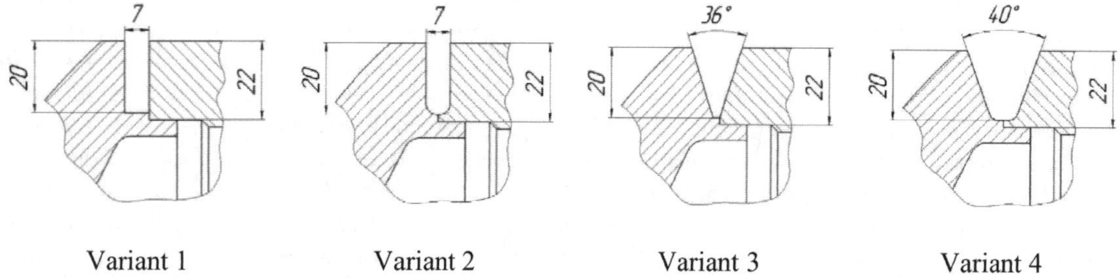

Figure 2. Configuration of edges beveled for welding bottoms and cylinders:
a) square groove edge joint;
b) U-groove edge joint;
c) V-groove edge joint with an opening angle of 36°;
d) special U- groove edge joint with an opening angle of 40°.

We analyzed the influence of prepared for welding edge joints on the stress-strain state using the SolidWorks Simulation package. The cylinder designs of hydraulic legs M138 were tested. The forms of bottom and cylinder cutting edges are given in Figure 3. The inside diameter D is 220 mm and the overall length L is 750 mm and remain unchanged. To construct a model we used a stop piece, which is a part of tooling used during the real-life tests of hydraulic legs for their strength.

A three-dimensional model was constructed by turning contours of the working cylinder, the bottom and the weld joint around the longitudinal axis of symmetry to a central angle $\alpha = 90°$ (Figure3). According to the researches performed earlier [9], the model with a central angle α of 90° is the most relevant in terms of the smallest error and the maximum computing performance. The designed weld had regions of penetration between the welded workpieces. Steel 30KhGSA was used for making the cylinder and the bottom, and wire SV-08G2S was used for welding. Table 1 shows the main properties of the materials used.

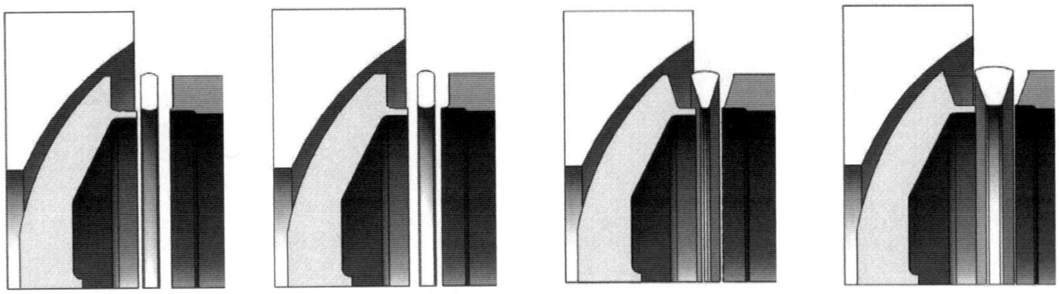

Variant 1 Variant 2 Variant 3 Variant 4

Figure 3. Solid model of the cylinder with a technological stop piece.

Table 1. Properties of Material used for Model

Material	Components	Module of Elasticity E, GPa	Ultimate Tensile Strength σ_T, MPa	Poisson's Ration μ
Steel 30KhGSA	Bottom, cylinder	205	736	0.3
Wire Sv-08G2S	Welding joint	220	415	0.275

The boundary condition "Symmetry" was applied to the surfaces of the simulated model to compensate for the effects of omitted parts in the structure. In setting the conditions of interaction between parts used in the assembly, the No Penetration contact condition with the Surface-to-Surface Contact option was applied. The weld joint between the tube and the bottom of the cylinder was simulated separately, and allowed for penetration zones of the parts welded. For the first variant, the functional step penetration was not taken into account. For other variants, the value of penetration is determined according to formula 1:

$$h_p = C - 0.5b \tag{1}$$

Where C is the height of the functional step, b is the gap between the parts to be welded. According to our calculations, the value of penetration was 2 mm.

For adding loads on the cylinder model, the pressure load equal to the test pressure of working fluid was applied to the inside surface of the cylinder and the bottom. According to the program and procedure of testing this hydraulic leg for strength, the test pressure $P_{test} = 60$ MPa.

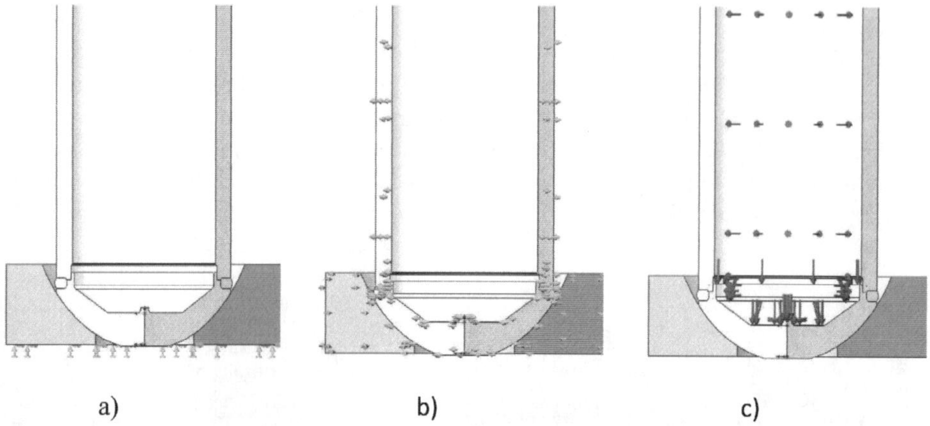

Figure 4. Boundary conditions in the cylinder model simulation:
a) Fixed Geometry; b) Symmentry; c) Load.

The type of mesh used is 'Solid Mesh' with the 'Curvature-based Mesh' option. Previously conducted numerical experiments to determine the von Mises stresses [9-10] showed that the optimal element sizes in applying Finite Element Analysis are as follows: 4 mm for cylinders and cylinder bottoms, and 0.4 mm relating to weld. A decrease in the element size increases the inaccuracy of research and the time for calculation. To reduce the calculation time we used the function "Mesh Control Property Manager" to change a mesh density at the regions where the components are in contact.

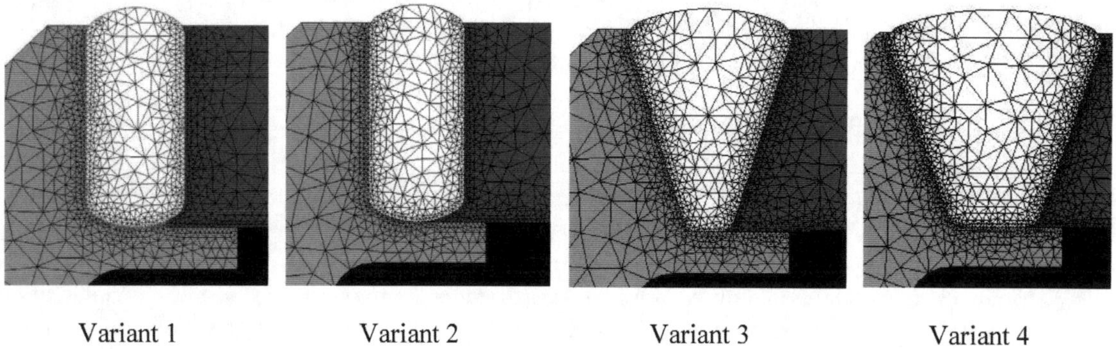

Figure 5. Mesh density with different element sizes at the region of weldment.

The results of numerical experiments performed to determine the von Mises stresses are given in Figure 6. As it can be seen from the figures, the maximum stresses occur at the root of a weld joint over the boundary of penetration of the connected cylinder components. The value of stress changes depending on a type of beveled edge joints. Thus, the value of stress is 634 MPa with the square groove edge joint; the value of stress is 558 MPa with the U-groove edge joint; the value of stress is 675 MPa with the special V-groove edge joint and an opening angle of 36°; the value of stress is 405 MPa with the special U- groove edge joint and an opening angle of 40°. The first three variants have zones where the stresses exceed the ultimate tensile strength of the weld material by 52%, 34% and 62% respectively. In the case of the fourth variant, the value of stress also approaches to the ultimate tensile strength of the weld material, but the the region of stress is small in comparison with the region of penetration.

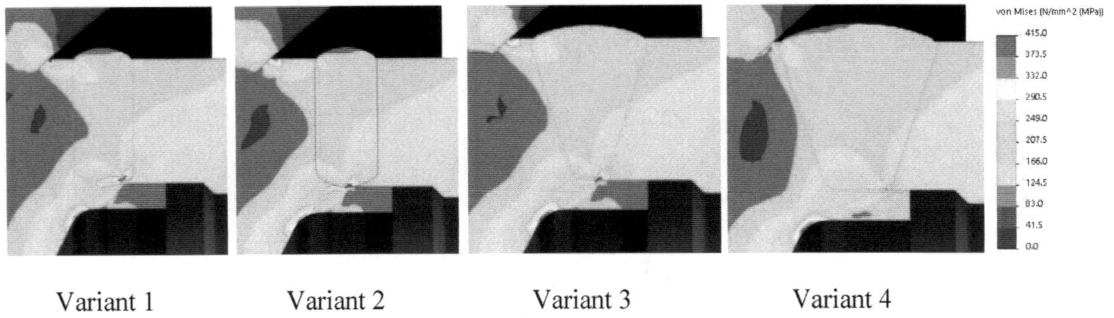

Figure 6. Von Mises stresses defined for different varients of a weld.

3. Additional research

The occurrence of such regions is caused by lack of fusion at the location of the functional step. These regions can be called as stress concentrators that are formed due to the distinct features of different joint edges beveled for welding.

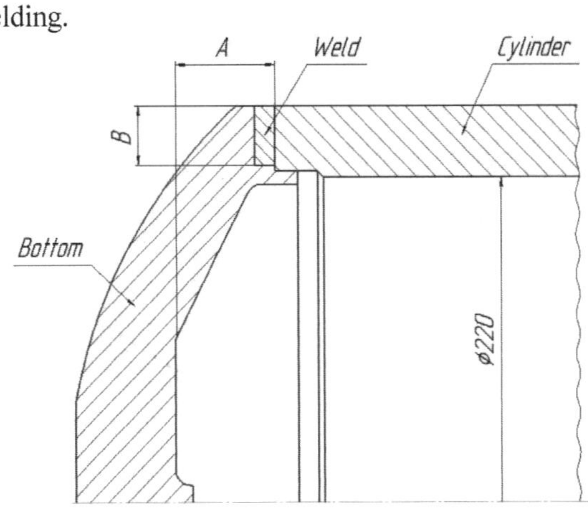

Figure 7. Variable parameters A and B.

In order to determine the change in maximum stresses at the region of the weld root, we carried out additional experiments with introducing variable parameters (Figure 7): A is the distance between the inside surface of the bottom and the cylinder; B is the height of weld. The value of parameter A changes from 31.5 mm to 71.5 mm with an increment of 10 mm. The value of parameter B varies within a range of 20 mm to 12 mm with an increment of 2 mm. For researches in the context of the parameter A, a model was built by means of changing the bottom and the cylinder when dimensions and forms of beveled edges are constant. The modeled weld having penetration regions remained unchanged for all four variants, i.e. the parameter B was constant and equal to 20 mm. In studies relative to the parameter B, the parameter A remained constant and was 31.5 mm.

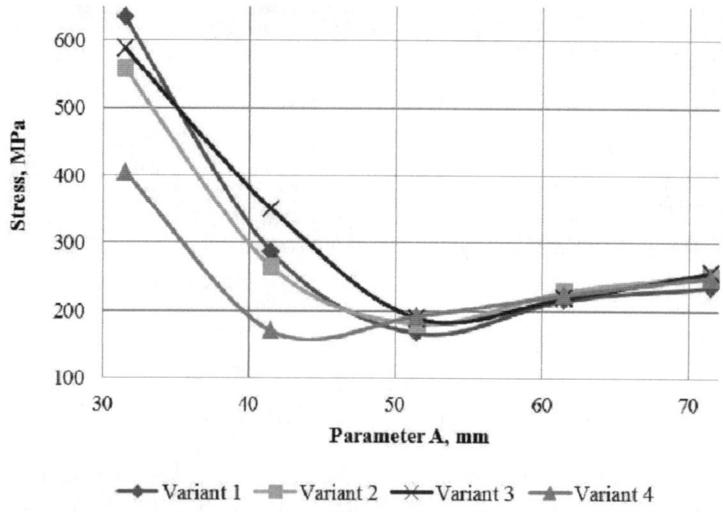

Figure 8. Changing of stresses in the weld root relative to parameter A when B = 20 mm, for four variants.

Results of new studies are presented in Figure 9. From these dependencies relating to changes of the parameter A it is seen that when the weld shifts away from the cylinder bottom, a sharp reduction in stresses occurrs in the root zone of the weld, for all four models. The minimum stress in a given zone is obtained with the square groove edges (Figure 2, a) used to joint the bottom and the cylinder within a parameter range of 50 mm to 55 mm. A further increase in the parameter results in a growth of stresses. The parameter A can not be increased indefinitely, as the overall size of the bottom increases greatly, and this structure without changing the first - stage piston rod is impossible to apply.

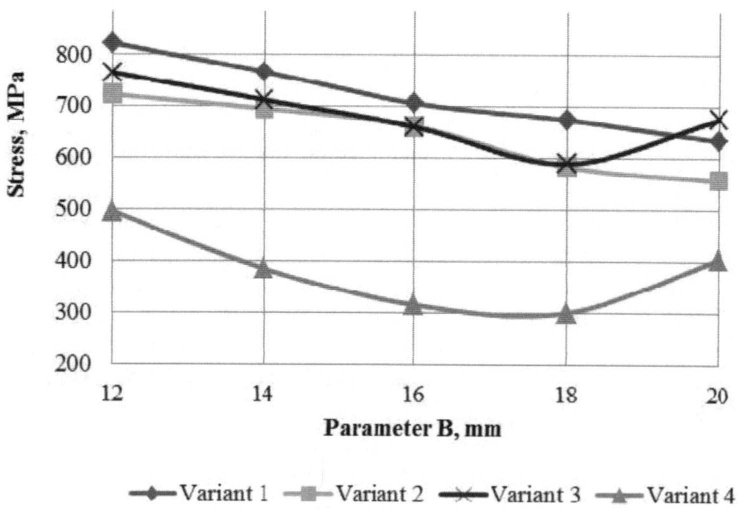

Figure 9. Changing of stresses in the weld root relative to parameter B when A = 31,5 mm, for four variants.

As it can be seen from the graph, changing a depth of the weld (parameter B) results in a reduction of stress in the region of the weld root only for the third and the fourth variants (models with the special V-groove and U- groove edge joints and opening angles of 36 ° and 40 ° respectively (Figure 2, c, d)). The values of stress for these variants are 300 MPa and 590 MPa when the parameter B = 18 mm. Relative to other variants of beveling edges, stresses increase linearly and the depth cannot be reduced. For the fourth variant, a decrease in depth of the weld causes the reduction of stresses that positively affects the entire structure as a whole.

4. Conclusion

The analysis results have proved that all variants of cylinders under study have stress concentrations in the region of the root of welds. These stresses exceed an ultimate tensile strength of materials used for welding by 34-62%. The value of stress directly depends on the geometry of beveled edges and the value of penetration of the functional step. The larger the dimension of opening, the deeper the fusion extends into the functional step. Even full penetration does not ensure optimal stresses in the zone of the weld root, what is obvious when analyzing the fourth variant. The stress can be reduced by changing the design of the bottom with decreasing the depth of weld.

References

[1] Tarazanov I 2014 Results of the coal industry of Russia in January-September 2014. *Ugol (Coal)* **No.12** pp 57-71

[2] Buyalich G D Buyalich K G Voyevodin V V 2015 Radial deformations of working cylinder of hydraulic legs depending on their extension. *IOP Conference Series: Materials Science and Engineering* **Vol. 91** № 1

[3] Buyalich G D Buyalich K G 2014 Modeling of Hydraulic Power Cylinder Seal Assembly Operation, *Mining 2014 : Taishan Academic Forum – Project on Mine Disaster Prevention and Control: Chinese Coal in the Century: Mining, Green and Safety, China, Qingdao, 17–20 October 2014*. Amsterdam, Paris, Beijing. Atlantis Press, pp. 167–170

[4] Chinakhov D A 2011 Study of Thermal Cycle and Cooling Rate of Steel 30ХГСА Single-Pass Weld Joints *J. Applied Mechanics and Materials* **Vol. 52-54** pp 442–447

[5] Chinakhov D A 2013 Simulation of Active Shielding Gas Impact on Heat Distribution in the Weld Zone, *J. Applied Mechanics and Materials* **Vol.762** pp 717-721

[6] Malushin N N Valuev D V Valueva A V Serikbol A 2011 Testing of metallurgical equipment on contact strength and wear resistance *J. Applied Mechanics and Materials* **Vol.682** pp 53-57

[7] Malushin N N Valuev D V 2013 Measuring stress condition in hard faced work rolls for cold rolling *J. Applied Mechanics and Materials* **Vol. 379** pp 71-76

[8] Malushin N N Osetkovsky V L Valuev D V Valueva A V Serikbol A 2014 Selection criteria parameter thermal cycle during surfacing heat resistance high hardness parts of metallurgical equipment *J. Applied Mechanics and Materials* **Vol. 682** pp 69-74

[9] Buyalich G D Anuchin A V Dronov A A 2015 The Numerical Analysis of Accuracy of Hydraulic Leg Cylinder in Modeling Using Solid Works Simulation *J. Applied Mechanics and Materials* **Vol. 770** pp 456-460

[10] Buyalich G D Buyalich K G Voyevodin V V 2015 Radial deformations of working cylinder of hydraulic legs depending on their extension *IOP Conference Series: Materials Science and Engineering* **Vol. 91**

IOP Conf. Series: Materials Science and Engineering **127** (2016) 012013 doi:10.1088/1757-899X/127/1/012013

Study of Gasdynamic Effect Upon the Weld Geometry When Concumable Electrode Welding

D A Chinakhov, E G Grigorieva, E I Mayorova

Yurga Institute of Technology of National Research Tomsk Polytechnic University, 652055, Leningradskaya st., 26, Yurga, Kemerovskaya region, Russia

e-mail: chinakhov@tpu.ru

Abstract. The paper considers the ways of weld geometry controlling when consumable electrode welding under single-jet and double-jet gas shielding. The authors provide comparative results of experimental studies on the effects of shielding gas supply upon the weld geometry in weld joints produced from construction carbon steel 45. It has been established that gas-dynamic effect of the shielding gas has a significant impact upon shaping and weld geometry when consumable electrode welding under double-jet gas shielding.

1. Introduction

During the recent decades much work has been done to ensure high service reliability, safety and efficiency of engineering products. Technological advance imposes new requirements on production methods and welding practices. One of the most wide-spread welding techniques is shielded arc welding. This welding technique has a number of advantages [1]: the opportunity of human observation, opportunity of welding items with varying thickness, application of robotic technologies, etc. The basic technological parameters of consumable electrode gas shielded welding are intensity of arc current, arc voltage, density and type of arc current, cross-section area (diameter) of the electrode, welding rate. During gas-shielded welding gas is applied to protect the welding area . The gas is supplied into the welding area in the firm of a jet. The type of gas applied in the process determines the physic, metallurgic and operational characteristics of the welding method [2].

Controlling weld geometry is one of the most important tasks when consumable electrode welding as the geometrical dimension of the weld determines the efficiency of the metallic construction. Complementary parameters: length of the welding wire, gas composition and flow rate, method of gas shielding, electrode and article position when welding, etc. [2].

There are various ways of controlling weld geometry and improving the operational properties of welded metallic structures but, at the same time, none of them can be considered as a universal approach to problem solution. Many scientists study the opportunities of controlling weld and HAZ shaping when consumable electrode welding. Many techniques and devices contributing to controlling weld joints shaping [2]: changing welding conditions, current pulse pile-up, programming welding conditions, application of magnetic fields over the weld pool, formation of mechanical impulses and electrode vibration, changing electrode geometry, adding various chemical elements into the wire composition, changing shielding gas composition, etc.

Authors [3-12 and other] studied how the changes of shielding gas flow effect the process of consumable electrode welding and note that with increase of gas flow rate the quality of the weld shape and protection of the weld area improve. Increasing the stiffness of the gas jet is especially important when the welding works are completed in open areas [6, 8]. In works [2, 4] the authors show that the electrode metal drop, beside the basic forces, is significantly effected by the pressing force of the shielding gas jet.

Application of double-jet gas shielding [12] as contrasted to traditional (single-jet) one ensures reliable protection of the weld pool, refinement of the weld metal structure, graded transition from the weld to the basic metal, improvement of mechanical properties of the weld joints, decrease of chemical inhomogeneity of the weld due to more rapid gas-dynamic mixing of the molten metal in the weld pool. In other words, it allows controlling the operational properties of the weld joints.

Aim of the work: determine the influence of gas-dynamics of single-jet and double-jet gas shielding upon the weld geometry when consumable electrode welding in the gas mixture of 82% Ar + 18% CO_2.

2. Methodology

To estimate the effect of shielding gas jet upon the weld pool surface we completed research with the physical model of the weld pool presented as a pan with liquid. Molten liquid metal of the weld pool was simulated with glycerin (identical viscosity). For the ease of liquid motion observation convenience the researchers added aluminum powder to the glycerin. The process of simulation of shielding gas flow from the weld nozzle and its effect upon the glycerin surface was recorded with high-speed camera "VideoSprint". The experiment was carried out under the following conditions: gas flow rate Q = 40 l/min, welding rate V = 25 cm/min, length of the welding wire L = 12 mm (Fig. 1).

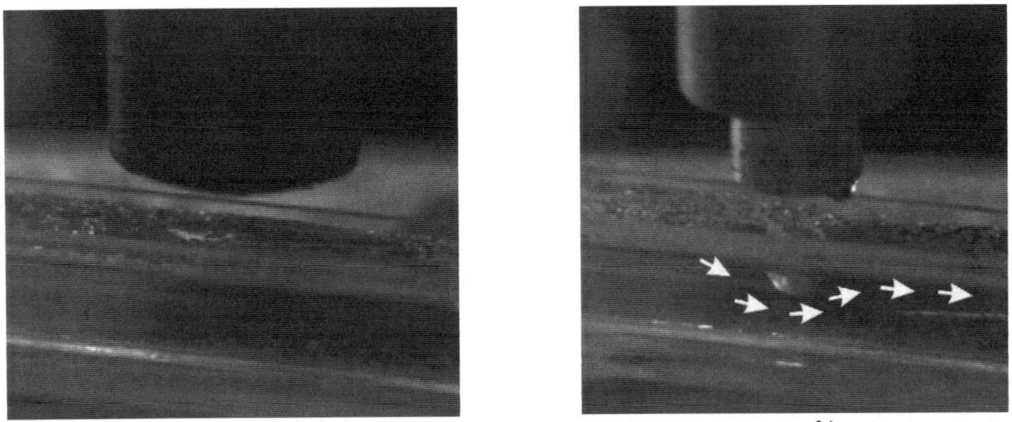

a) b)

Figure 1. Effect of shielding gas jet upon the surface of glycerin:
a) traditional single-jet gas shielding; b) double-jet gas shielding

The experiment showed that under the traditional single-jet gas shielding the surface of the glycerin slightly trembled and the powder particles in the surface layer of glycerin did not practically move (Fig. 1, a). Under the double-jet gas shielding glycerin surface buckled forming a hollow 4-5 mm in diameter and powder particles in the upper glycerin layer were moving from the front area, over the bottom of the hollow, rounding the gas jet, to the back area of the simulated weld pool (Fig. 1, b). This way, significant effect of the shielding gas jet upon the weld pool surface and possibility of weld geometry controlling when consumable electrode welding under controlled gas-dynamics of the shielding gas has been established.

To determine the effect of the gas-dynamics of single-jet and double-jet gas shielding upon the weld geometry when consumable electrode welding in the gas mixture of 82% Ar + 18% CO_2 an experiment was completed on the real object of research. In the course of the experiment we completed bead welding over a plate of steel 45 10 mm thick with welding wire Sv-08G2S, 12 mm in diameter in the gas mixture of 82% Ar + 18% CO_2. Bead welding was completed with traditional (single-jet) and double-jet gas shielding. Welding conditions: I=195…200 A, length of the welding wire L= 10 mm, gas flow rate was changed from 5 to 25 l/min at the interval of 5 l/min (Table 1), arc voltage U = 25…26 V, welding rate V = 5 mm/s. Power supplies Shtorm-Lorch V 50 AC/DC, welding unit VD – 1500.

Table 1

Planning matrix of the experiment

The controlled parameter	Experiment №									
	Single-jet gas shielding					Double-jet gas shielding				
	1	2	3	4	5	6	7	8	9	10
Shielding gas flow rate Q, l/min	5	10	15	20	25	5	10	15	20	25

3. Results and Discussion

The geometrical parameters of the welds of obtained welded samples were studied (e – weld width, g – weld reinforcement, h – penetration) (Table 2).

Table 2

Geometry of welds in relation to gas flow rate

Shielding gas flow rate Q, l/min	Single-jet gas shielding			Double-jet gas shielding		
	Weld width e, mm	Penetration h, mm	Weld reinforcement g, mm	Weld width e, mm	Penetration h, mm	Weld reinforcement g, mm
5	10	3	2.8	10	3	2.7
10	10	3	2.9	11	3	2.7
15	10	2.5	2.6	11	2	2.6
20	10	2.5	2.8	12	2	2.6
25	10	2.5	2.7	12	1.8	2.5

According to the results of experimental study the geometrical parameters vary significantly under similar welding conditions and gas flow rate but different gas shielding methods. Figure 2 shows that geometry of the bead produced with application of double-jet welding nozzle is flatter. It contributes to significant increase of weld joints strength especially under variable load.

a) b)

Figure 2. Experimental samples welded under gas shielding Q = 15 l/min
a) single-jet gas shielding b) double-jet gas shielding

The composition of the shielding gas has a significant effect on the manufacturing characteristics of the welding process. When welding in the gas mixture of 82% Ar + 18% CO_2 stable transfer of electrode metal, minute spattering and fine weld formation are observed (Fig. 2).

All geometrical dimensions of steel 45 weld joints (E, h, g) under the conditions of the given experiment are effected by gas shielding method and shielding gas flow rate. Other significant controlled parameters (welding current, voltage, welding rate, stick-out distance of electrode wire) were not changed under the conditions of the given experiment.

As gas flow rate and correspondingly, gas consumption increase we observe decrease of penetration both for single-jet and for double-jet gas shielding. Weld reinforcement changes insignificantly within the whole range of shielding gas flow rate changes. Weld width for traditional gas shielding does not change with alteration of shielding gas flow rate. In the case of double-jet gas shielding weld width increases with the growth of shielding gas flow rate which is explained by high density of the shielding gas jet and its pressure over the weld pool metal surface.

When welding with double-jet gas shielding we observe smooth transition from the weld joint to the base metal improving efficiency of weld joints. Under traditional welding noticeable bulge of the weld joint is observed which is a negative factor.

Analysis of penetration, width and weld reinforcement changes showed that controlled parameter of the welding conditions Q significantly effects formation of the weld joint.

The diagrams showing the dependences of penetration, width and weld reinforcement changes upon shielding gas flow rate are presented in Fig. 3, 4 and 5.

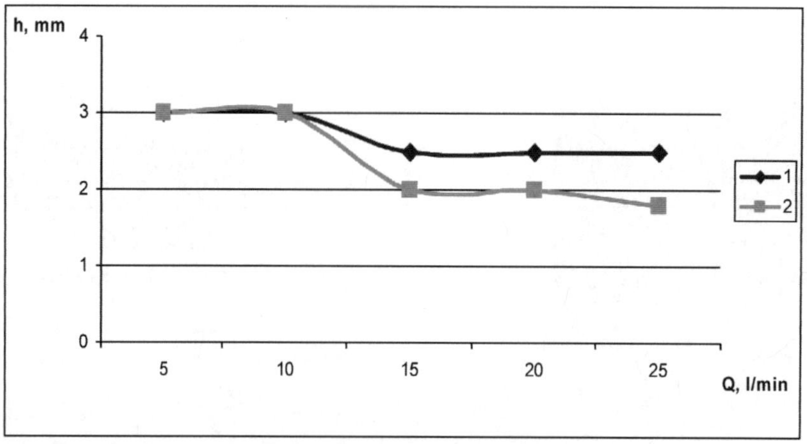

Figure 3. The diagram of penetration changes dependence upon shielding gas flow rate:
1 – single-jet gas shielding 2- double-jet gas shielding

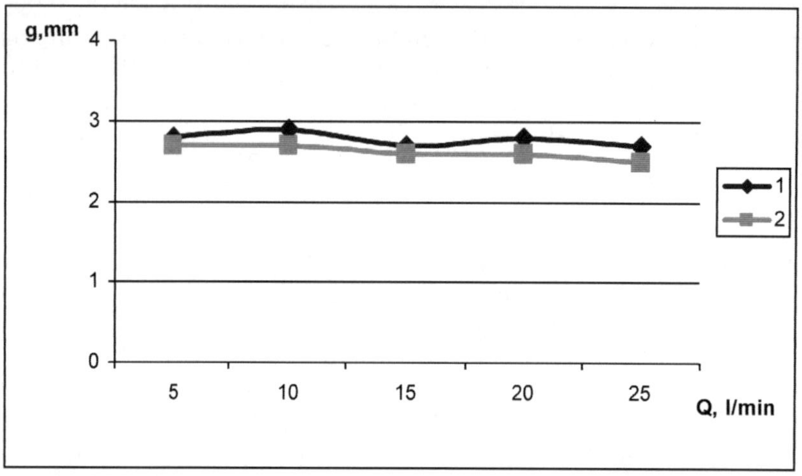

Figure 4. The diagram of weld reinforcement changes dependence upon shielding gas flow rate:
1 – single-jet gas shielding 2- double-jet gas shielding

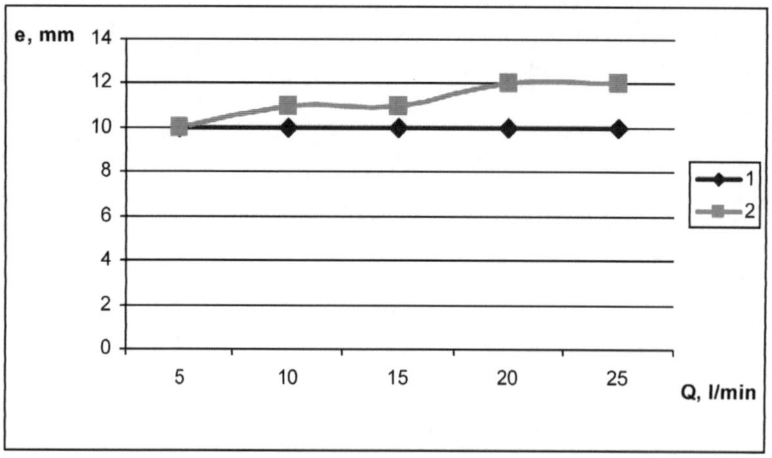

Figure 5. The diagram of weld width changes dependence upon shielding gas flow rate:
1 – single-jet gas shielding 2- double-jet gas shielding

Conclusion

It has been established that gas-dynamic effect of the shielding gas has a significant impact upon formation and geometry of the weld when consumable electrode welding under double-jet gas shielding. As the gas-dynamic effect (gas consumption and gas flow rate) on the weld pool surface and processes in the weld area increase we also observe increase of the weld width, decrease of penetration and insignificant alteration of the weld reinforcement. Gas-dynamic control of formation and geometry of the weld joint is of great practical importance and allows improving maintenance reliability of the welded construction without extra charges.

The reported study was funded by RFBR according to the research project No 16-38-00194 mol_a.

References

[1] Kononenko V.Ya. Consumable welding in shielding gas atmosphere. Reference book for students. K.: Publ.centre "Nika-Print", 2007. 266 p.
[2] Potapievsky A.G. Consumable welding of steels in shielding gas atmosphere. Technologies of the future: monograph. A.G. Potapievsky, Yu.N. Sarayev, D.A. Chinakhov; Yurga Institute of Technology. Tomsk: TPU publishers, 2012. 208 p.
[3] Novozhilov M.N. The principles of arc welding in shielding gases. M.: Mashinostroyeniye. 1979. 231 p.
[4] Chinakhov D.A. Calculation of Gas-dynamic Impact of the Active Shielding Gas on the Electrode Metal Drop in Gas Jet Shielded Welding. Applied Mechanics and Materials. Vol. 379 (2013). Pp. 188-194.
[5] Stolbov V.I. Weld pool: Monograph. Tolyatti: TSU. 2007. 147 p.
[6] Fedorenko G.A., Ivanova I.V., Sinyakov K.A. Improvement of techniques and procedures of welding in shielding gases in the wind. Welding production. 1996. No 1. P. 6–13.
[7] Gribovsky G., Kravchuk B., Lenivkin V.A. Effect of double-layer annular flow of shielding gases upon consumable welding process. Welding production. 1996. No 4. P. 6-8.
[8] Popravka D.L., Khvorostov N.Ye. Arc welding in shielding gases in open areas. M.: Mashinostroyeniye, 1979. 64 p.
[9] Tarasov N.M. Detachment of electrode metal drop by brief gas flow. Machine welding. 1986. No 7. P. 10-13.
[10] Ostrovsky O.Ye., Novikov O.M. A new method of arc welding with pulse supply of shielding gases. Welding production. 1994. No 11. P. 10-12.
[11] Sheiko P.P., Zhernosekov A.M., Shevchuk S.A. Technological characteristics of consumable welding low-alloy steels with alternating supply of shielding gases. Machine welding. 1997. No 8. P. 32-36.
[12] Chinakhov D.A. Gas Dynamic Control of Properties of Welded Joints from High Strength Alloyed Steels. China Welding. Vol. 23 No.3 (2014) pp. 27-31.

Simulation of Stress-Strain State of Shovel Rotary Support Kingpin

A A Khoreshok[1,a], P V Buyankin[2,b], A V Vorobiev[3,c], A A Dronov[4,d]

[1,2]Department of Mining Machines and Complexes/Mining Institute,
T.F. Gorbachev KuzSTU, Russia, 650000, Kemerovo, Vesennyaya street 28,
[1,3,4]Department of Mining Equipment, Yurga Institute of Technology, TPU affiliate
Russia, 652055, Yurga, Leningradskaya street 26

E-mail: [a] haa.omit@kuzstu.ru, [b] pv.buyankin@gmail.com, [c] vorob@tpu.ru,
[d] aadronov@tpu.ru

Abstract. The article presents the sequence of computational simulation of stress-strain state of shovel's rotary support. Computation results are analyzed, the kingpin is specified as the most loaded element, maximum stress zones are identified. Kingpin design modification such as enhancement of fillet curvature radius to 25 mm and displacement of eyebolt holes on the diameter of 165 mm are proposed, thus diminishing impact of stress concentrators and improving reliability of the rotary support.

1. Introduction

The shovel's rotary support (or RS) is one of the critical components determining the safe and reliable operation of the machine as a whole. Therefore, this element requires special attention both in design and in manufacture and operation. However, there is a lack of studies on rotary support loadings.

In domestic engineering estimation of rotary support loadings is performed by mathematical calculations of distributed vertical loads on the roller circle and pull in the kingpin. These approaches are static; wherein the impact of operational loadings is accounted by a dynamic coefficient Cdyn, taken equal to two, and by preset safety factors of components.

In general, these methods yield good results, but can not calculate some specific features of loading, such as impacts of maximum cutting force and excavating pressure. In some cases, this leads to calculation accuracy degradation and, as a consequence, diminishing of bearing capacity and reliability of components which adversely affect machines' operations. As for excessive metal consumption, it increases the cost of the machine and hurts its price competitiveness.

2. Description of design and sequence of stress-strain state simulation of the rotary support kingpin

RS layout scheme of EKG-10 shovel (the most common) is a structure with a unilinear roller circle where cylindrical rollers are arranged between upper and lower rails. The kingpin serves to withstand the loadings and avoid tipping if the loss of balance. During operation, the rollers contact with the rails linearly or rectangularly; the kingpin withstands the stresses arising periodically [1]. These impacts will inevitably lead to a wear and tear of conjugate elements and their failure (Figure 1).

Figure 1. EKG-10 shovel's kingpin fracture under operational loadings

Computational model of operational loading impact on stress-strain state of EKG-10 shovel's RS is developed with SolidWorks Simulation. SolidWorks Simulation is widely used for analysis of stress-strain state of structures, including mining machines [2-11]. MathCad software complex was used to obtain quantitative estimation of set loads.

General algorithm of SolidWorks Simulation for solving a static problem in linear formulation is as follows [1, 4].

1. A geometric model of the structure is developed (Figure 2).
2. A finite element grid is constructed, that is, sampling of the volume occupied by a part or assembly is performed. For a solid body the area is divided into tetrahedrons with edges approximated by linear or parabolic coordination functions.
3. Relationship of transformation of grid points translocations to Global Coordinates are determined. Local axial translocations are degrees of freedom for solid finite elements.
4. Finite element stiffness matrices are computed. Coordinates of the points, elastic modules and Poisson's ratios are used in the formulas for computation of finite element stiffness matrix components.
5. The resulting stiffness matrices transform from Local to Global coordinates, which in turn, are connected into a single Global stiffness matrix.
6. Set boundary (static and kinematic) conditions are converted to loads and points translocations, expressed in Global coordinates.
7. The resulting linear system of equations is solved; dislocations in the finite element points are determined. Direct and iterative methods can be used; calculation of strains at the points is performed.
8. Calculation of the equivalent tensions in base of a selected strength criterion is performed from parameters of the stress-strain state and material strength.

To reduce the volume of calculations some of the details were simplified (Figure 2) (the threaded underside of the kingpin nut, the turntable spigot), which effects were offset by the corresponding boundary conditions.

The turntable frame was simulated as an upper rail and a spigot. The sub-frame was simulated as a cast and a bottom rail. Side planes of the components were connected by means of a condition "Rigid connector", which provides stable bonds of selected surfaces and constant distances between them. This condition describes operation of the structure correctly, since the upper rail is fixed to the turntable; the bottom rail and the kingpin cast are fixed to the sub-frame. A kinematic boundary condition "Fixed" was applied to side plane of the bottom rail. This type of fixing limits linear displacements along three coordinate axes taken to zero.

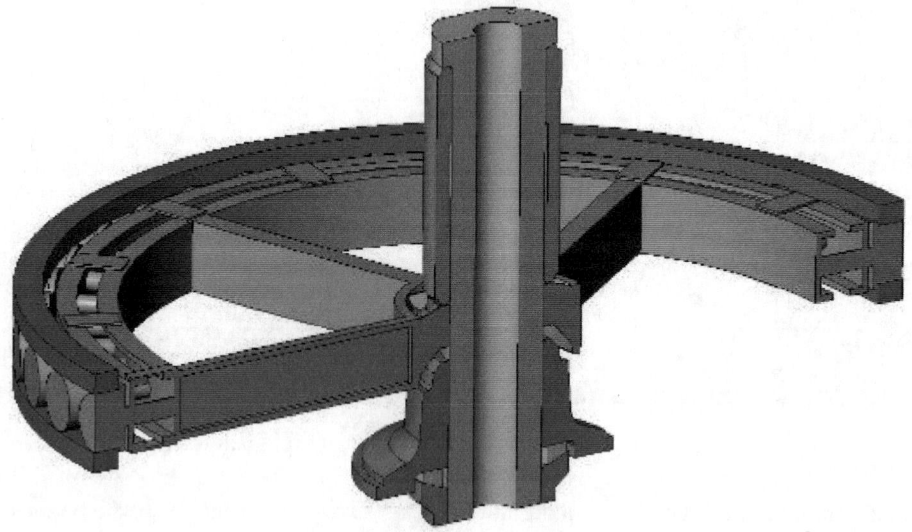

Figure 2. Geometrical model of RS of EKG-10 shovel (section)

Since a geometric model, material properties and imposed loads are symmetrical relative to the vertical plane, a boundary condition "Symmetry" was used. The geometrical model was cut axially; a kinematic boundary condition, inhibiting normal displacements to the cutting plane, was applied to the surfaces of the section. This reduced dimension of the problem; which, in turn, enhanced grid density and accuracy of finite element calculations.

Because of complexity to define boundary conditions of load application on RS elements, boundary forces based on weights of the arm and rock-filled bucket (G=250 kN) and on cutting force and pressure (P_{01}= 330 kN и P_{02}=100 kN) were used in the computing. It just is a specific feature of loading, when maximum cutting force and pressure resulting from rock mass excavation are taken into account. The points of force application were adopted in accordance with Figure 3.

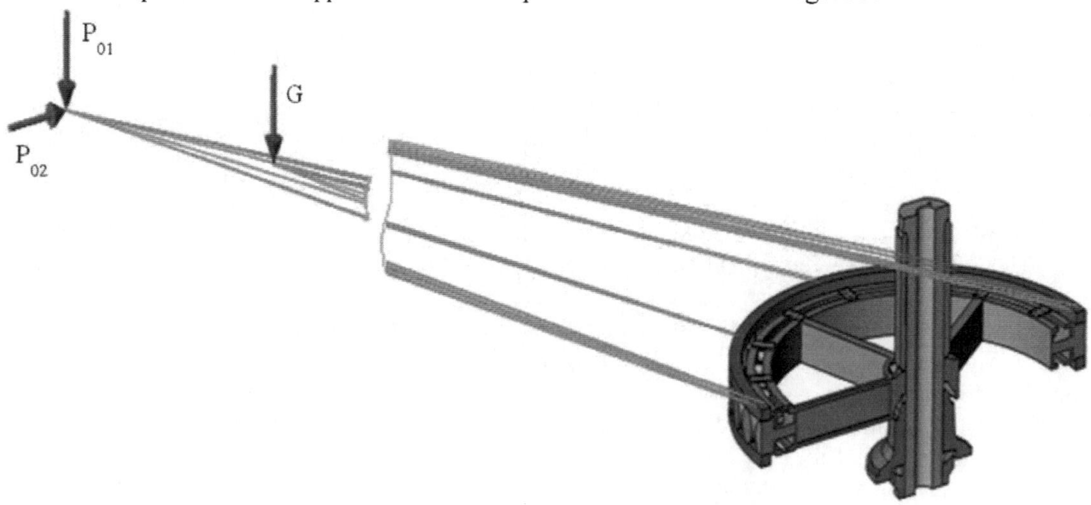

Figure 3. Scheme of rotary support loading

Forces were transferred from the points of application on the upper rail face using the boundary condition "Remote Loading" with the option "Rigid Connector".

Since the calculation was performed for the assembly, conditions of contact interaction of contacting faces were described. The condition "No Penetration", that eliminates interference of components but admits clearances, was used with the option "Surface to Surface". These contact conditions maximize the accuracy of contact problem's solution, if smooth curvilinear interacting edges, but it requires the greatest computational resources.

The grid with parabolic finite elements (FE) in the form of tetrahedrons was used for geometric model digitizing. Parabolic finite elements provide a better grid description of model geometry and enhance accuracy of computing because of more points, if compared with linear FE. Grid parameters: FE size - from 10 to 50 mm; ratio of element size enhancing - 1.6; automatic grid tightening was not used. For a detailed study of the kingpin the grid was tightened. Maximum size of FE was diminished to 30 mm. A powerful program «FFEPlus», which uses advanced matrix reordering, was used for the computation.

3. Results and Discussion.

We have got an equivalent stress distribution pattern, by Mises criterion, in RS elements of the EKG-10 shovel when exposed to operational loads (Figure 4) [4].

Figure 4. Equivalent stress distribution pattern by Mises criterion in RS elements of the EKG-10 shovel

It was found that the kingpin is one of the most loaded RS elements. Maximum of stress is located in the fillet zone, and the additional stress concentrator is the eyebolt hole (Figure 5a). This figure shows the zone where stresses exceed the kingpin material's yield strength (for Steel 40 - 335 MPa by State Standard 1050-88). This confirms that service loads are destructive and may cause damage to the kingpins. Also, the stress distribution pattern shows the most loaded zones of the upper rail in contact places with seven front rollers.

In order to avoid stress concentration in the kingpin, displacement of eyebolt holes closer to the kingpin axis has been proposed. Initially, the holes were arranged on a diameter of 230 mm; it has been proposed to diminish the distance to 165 mm, with the same depth of the hole. Fillet radius has been enhanced from 10 to 25 mm (Figure 5b). These improvements do not result in significant changes in structure of other (conjugate) elements of the RS, but will significantly diminish maximum stresses (Figure 5b).

IOP Conf. Series: Materials Science and Engineering **127** (2016) 012014 doi:10.1088/1757-899X/127/1/012014

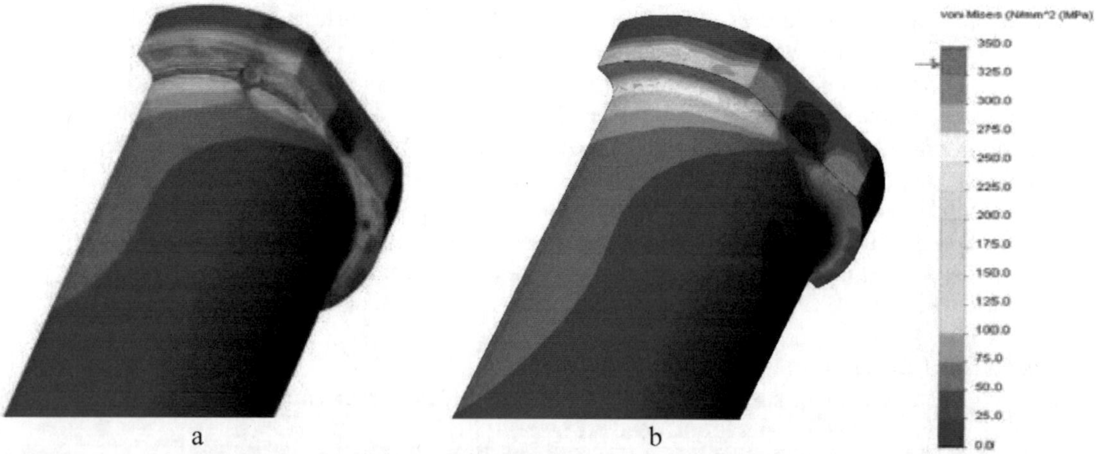

Figure 5. Equivalent stress distribution pattern by Mises criterion in kingpin: a - maximum stress concentration and exceeding of steel 40 yield strength; b - stress concentration after modification

4. Conclusion

The resulting pattern of stress distribution in the RS elements shows how technical condition of the component changes during operation up to the wear and fracture of individual elements. The proposed model of stress-strain state enables to develop both various designs of individual elements and layout diagrams of the RS by simulating maximum loads required for detection of stress concentration, for example, kingpin design modification such as enhancement of fillet curvature radius to 25 mm and displacement of eyebolt holes on the diameter of 165 mm. These improvements can diminish stress concentrations when subjected to operation loadings.

References

[1] Buyankin P V 2014 Computing of loadings in shovel's SRG *Herald of Kuzbass State. Tech. Univ.*, №2. pp. 19-21.

[2] Aksenov V V, Beglyakov V Yu, Sadovets V Yu 2015 Development of math model of geokhod bladed working body interaction with geo-environment *IOP Conference Series: Materials Science and Engineering.* vol. 91, p. 012085

[3] Prokopenko S A, Sushko A V, Kurzina I A 2015 New design of cutters for coal mining machines *IOP Conference Series: Materials Science and Engineering.* vol. 91, p. 012058

[4] Horeshok A A, Bogomolov I D, Buyankin P V, Vorobyev A V 2014 Evaluation of operational loads on supporting-rotating gear of shovels *Mining Equipment and Electromechanics.* №6. p. 43-46.

[5] Khoreshok A A, Bogomolov I D, Buyankin P V, Vorobev A V 2015 Effect of operating stresses on elements of shovel swing bearings *Applied Mechanics and Materials.* vol. 770. pp. 445-448.

[6] Khoreshok A A, Mametyev L E, Borisov A Yu, Vorobiev A V 2014 Stress-deformed state knots fastening of a disk tool on the drums of roadheaders *Mining 2014. Taishan academic forum - project on mine disaster prevention and control. Chinese coal in the XXI century: Mining, green and safety. – Qingdao, China, October 17-20, 2014, Atlantis press, Amsterdam-Paris-Beijing.* pp. 177-183.

[7] Chernukhin R V, Dronov A A, Blashchuk M Y 2015 Application of analytic hierarchy when choosing layout schemes for geokhod pumping station *IOP Conference Series: Materials Science and Engineering.* vol. 91. № 1. p. 012086.

[8] Aksenov V V, Walter A V, Gordeyev A A , Kosovets A V 2015 Classification of geokhod units and systems based on product cost analysis and estimation for prototype model production *IOP Conference Series: Materials Science and Engineering.* vol. 91. № 1. p. 012088.

[9] Efremenkov A B, Timofeev V Y 2012 Determination of necessary forces for geokhod movement *IFOST Proceedings - 2012 7th International Forum on Strategic Technology,* September 18-32, 2012, Tomsk. – p. 1 – 4.
[10] Chinakhov D A, Agrenich E P 2008 Computer simulation of thermo-mechanical processes at fusion welding of alloyed steels *Materials Science Forum.* Vol. 575-578 PART 2 pp. 833-836.
[11] Prokopenko S A, Ludzish V S, Kurzina I A, Sushko A V 2015 Results of industry testing of multiple use rock-cutting picks *Gornyi Zhurnal.* № 5 pp. 67-70.

On One Class of Dual Problems of Mechanics of Deformable Solids

A F Revuzhenko[1], O A Mikenina[1], A A Kazantsev[2], T S Kust[2]

[1] - Institute of Mining of Siberian branch of the Russian Academy of Sciences, Russia, Novosibirsk, Red avenue, 54
[2] - Yugra Technological Institute branch of Tomsk polytechnic University, Russia, Kemerovo region, Yurga, Leningradskaya str. 26

E-mail: revuzhenko@yandex.ru

Abstract. Inelastic body's plane deformation is described by two vector fields: vector stress potential (gradient of Airy stress function) and vector displacement field. Conditions for possibility of proceeding to the dual problem, when variables change the roles, are described: stress potential is interpreted as displacement field and vice versa. Both a perfectly plastic body model and its dual model of perfectly solidifying matter are considered.

Introduction

Mathematical models of solids' deformation are based on two general concepts: displacement vector and stress tensor. Vector displacement field yields kinematic characteristics of stress; and stress tensor yields power and dynamic ones. In other words, kinematics and dynamics of the matter are described qualitatively by different mathematical objects. In the first case we have to use a first-rank tensor (vector), in the second case - a second-rank tensor (tensor itself). Accordingly there is an asymmetry in the description. It is possible, however, to find a class of problems, when both power and kinematic descriptions are symmetrical and can be described with the aid of two vector fields. Moreover this class of problems is quite extended. The main idea of this work is that, since we are dealing with objects of identical rank, then after solving the equation, their roles can be interchanged. That is, the stress vector field can be regarded as the kinematic field, and vice versa. Thus we get the solution of a new (dual or conjugate) problem.Mathematical models of solids' deformation are based on two general concepts: displacement vector and stress tensor. Vector displacement field yields kinematic characteristics of stress; and stress tensor yields power and dynamic ones. In other words, kinematics and dynamics of the matter are described qualitatively by different mathematical objects. In the first case we have to use a first-rank tensor (vector), in the second case - a second-rank tensor (tensor itself). Accordingly there is an asymmetry in the description. It is possible, however, to find a class of problems, when both power and kinematic descriptions are symmetrical and can be described with the aid of two vector fields. Moreover this class of problems is quite extended. The main idea of this work is that, since we are dealing with objects of identical rank, then after solving the equation, their roles can be interchanged. That is, the stress vector field can be regarded as the kinematic field, and vice versa. Thus we get the solution of a new (dual or conjugate) problem.

1. Formulation of primal and dual problems.

Let us assume that deformation is plane, and inertial forces and weight can be neglected. Then we have equations for a wide range of deformable continua:

$$\frac{\partial\sigma_{11}}{\partial x_1}+\frac{\partial\sigma_{12}}{\partial x_2}=0,\ \frac{\partial\sigma_{12}}{\partial x_1}+\frac{\partial\sigma_{22}}{\partial x_2}=0, \tag{1}$$

$$\begin{aligned}&\frac{\partial u_1}{\partial x_1}=a_{11}\sigma_{11}+a_{12}\sigma_{22}+a_{13}\sigma_{12}\ ,\\&\frac{\partial u_2}{\partial x_2}=a_{21}\sigma_{11}+a_{22}\sigma_{22}+a_{23}\sigma_{12}\ ,\\&\frac{\partial u_1}{\partial x_2}+\frac{\partial u_2}{\partial x_1}=a_{31}\sigma_{11}+a_{32}\sigma_{22}+a_{33}\sigma_{12}\ ,\end{aligned} \tag{2}$$

where 0x1x2 – Cartesian coordinate system, σ_{ij} and u_i - components of stress, displacement or their velocities, $i,j=1,2,$ coefficients $a_{11},\dots$ are set. Notations of models (1), (2) are classical. However, in many respects they are "abnormal". [1] For example, five first-order differential equations are reduced to one fourth-order equation only. A natural notation can be obtained by inserting the vector field instead of the stress field $\{p_1,p_2\}$- vector potential of the stress field:

$$\sigma_{11}=\frac{\partial p_2}{\partial x_2},\sigma_{22}=\frac{\partial p_1}{\partial x_1},\sigma_{12}=\frac{\partial p_2}{\partial x_1}=\frac{\partial p_1}{\partial x_2}, \tag{3}$$

$$\begin{aligned}&\frac{\partial p_1}{\partial x_2}-\frac{\partial p_2}{\partial x_1}=0,\ \frac{\partial u_1}{\partial x_1}=a_{11}\frac{\partial p_2}{\partial x_2}+a_{12}\frac{\partial p_1}{\partial x_1}-a_{13}\frac{\partial p_1}{\partial x_2},\\&\frac{\partial u_2}{\partial x_2}=a_{21}\frac{\partial p_2}{\partial x_2}+a_{22}\frac{\partial p_1}{\partial x_1}-a_{23}\frac{\partial p_1}{\partial x_2},\\&\frac{\partial u_1}{\partial x_2}+\frac{\partial u_2}{\partial x_1}=a_{31}\frac{\partial p_2}{\partial x_2}+a_{32}\frac{\partial p_1}{\partial x_1}-a_{33}\frac{\partial p_1}{\partial x_2}.\end{aligned} \tag{4}$$

The system is noted in two vector fields $\overline{p}=\{p_1,p_2\}$ and $\overline{u}=\{u_1,u_2\}$. Suppose that some its solution is built up. As mentioned, the main idea is to interchange roles of vectors $\overline{p}$ and $\overline{u}$: vector $\overline{u}$ correlates with the stress field, and vector $\overline{p}$- with the displacement field of the dual problem. Clearly, this can only be done when $\overline{p}$ and $\overline{u}$ are included in the system in some symmetrical manner. Thus, it is necessary that one of the displacement equations would have the structure of the system's first equation (4). Therefore it is necessary to consider the case of incompressible continuum:

$$\begin{aligned}&\frac{\partial p_1}{\partial x_2}-\frac{\partial p_2}{\partial x_1}=0,\frac{\partial u_1}{\partial x_1}+\frac{\partial u_2}{\partial x_2}=0,\\&\frac{\partial u_1}{\partial x_1}-\frac{\partial u_2}{\partial x_2}=c_1\frac{\partial p_2}{\partial x_2}-c_2\frac{\partial p_1}{\partial x_1}-c_3\frac{\partial p_1}{\partial x_2},\\&\frac{\partial u_1}{\partial x_2}+\frac{\partial u_2}{\partial x_1}=d_1\frac{\partial p_2}{\partial x_2}+d_2\frac{\partial p_1}{\partial x_1}-d_3\frac{\partial p_1}{\partial x_2},\end{aligned} \tag{5}$$

where coefficients $c_1,\dots,d_3$ are set.

Let us examine the boundary conditions. Suppose $\overline{n}$ is an external boundary normal, s is its natural parameter, α is an angle between the axis 0x1 and a normal $\overline{n}$. Stress vector components on the area with the normal $\overline{n}$ are equal to:

$$\sigma_{n1} = \sigma_{11}\cos\alpha + \sigma_{12}\sin\alpha,$$
$$\sigma_{n2} = \sigma_{12}\cos\alpha + \sigma_{22}\cos\alpha \tag{6}$$

Substitution of (3) into (6) gives the following result:

$$\sigma_{n1} = \frac{\partial p_2}{\partial s}, \sigma_{n2} = -\frac{\partial p_1}{\partial s}.$$

Consequently, accurate within a constant, setting of the vector border $\{p_1,p_2\}$ is equivalent to setting of the stress vector $\{\sigma_{n1}, \sigma_{n2}\}$. Here, kinematic boundary conditions remain unchanged:

$u_1 = u_1(s)$, $u_2 = u_2(s)$.

Also more complex conditions, such as dry friction, Winkler elastic foundation and others, are possible. They all are reduced to inter-component relations p_1, p_2, u_1, u_2 , set on the border.

Let us now turn to the dual problem. Relevant variables will be denoted by ˜ "tilde".

$$\tilde{p}_1 = -u_2,\ \tilde{p}_2 = u_1,\ \tilde{u}_1 = -p_2,\ \tilde{u}_2 = p_1,$$
$$\tilde{\sigma}_{11} = \frac{\partial \tilde{p}_2}{\partial x_2}, \tilde{\sigma}_{22} = \frac{\partial \tilde{p}_1}{\partial x_1}, \tilde{\sigma}_{12} = -\frac{\partial \tilde{p}_2}{\partial x_1} = -\frac{\partial \tilde{p}_1}{\partial x_2}. \tag{7}$$

System of equations (5) does not change by such a substitution and takes the following form:

$$\frac{\partial \tilde{u}_2}{\partial x_2} + \frac{\partial \tilde{u}_1}{\partial x_1} = 0\ , \frac{\partial \tilde{p}_2}{\partial x_1} - \frac{\partial \tilde{p}_1}{\partial x_2} = 0\ ,$$
$$-2\tilde{\sigma}_{12} = -c_1\frac{\partial \tilde{u}_1}{\partial x_2} - c_2\frac{\partial \tilde{u}_2}{\partial x_1} - c_3\frac{\partial \tilde{u}_2}{\partial x_2}\ , \tag{8}$$
$$\tilde{\sigma}_{11} - \tilde{\sigma}_{22} = -d_1\frac{\partial \tilde{u}_1}{\partial x_2} + d_2\frac{\partial \tilde{u}_2}{\partial x_1} - d_3\frac{\partial \tilde{u}_2}{\partial x_2}\ .$$

Here one new factor arises. The original system (1), (2) is invariant with reference to rotation of the body as a rigid solid. This is a necessary correctness condition of any problem on solid deformation. Therefore, system (5) will be invariant. However, after the change of roles, invariant variables $\frac{1}{2}\left(\frac{\partial \tilde{u}_2}{\partial x_1} - \frac{\partial \tilde{u}_1}{\partial x_2}\right)$ may not occur with reference to a new turn. The dual problem would be invariant if the system does not depend on $\{\tilde{u}_1,\tilde{u}_2\}$ vector's rotor, that is, on the following combination of derivatives $\frac{\partial \tilde{u}_2}{\partial x_1} - \frac{\partial \tilde{u}_1}{\partial x_2} = \frac{\partial p_1}{\partial x_1} + \frac{\partial p_2}{\partial x_2} = \sigma_{22} + \sigma_{11}$.

Hence it follows that $c_1 = c_2 = c$, $d_1 = - d_2 = d$.

Thus, the direct problem is reduced to solving of the equations

$$\frac{\partial p_1}{\partial x_2} - \frac{\partial p_2}{\partial x_1} = 0\ , \frac{\partial u_1}{\partial x_1} + \frac{\partial u_2}{\partial x_2} = 0\ ,$$
$$\frac{\partial u_1}{\partial x_1} - \frac{\partial u_2}{\partial x_2} = c\cdot\left(\frac{\partial p_2}{\partial x_2} - \frac{\partial p_1}{\partial x_1}\right) + c_3\frac{\partial p_1}{\partial x_2},$$
$$\frac{\partial u_1}{\partial x_2} + \frac{\partial u_2}{\partial x_1} = d\cdot\left(\frac{\partial p_2}{\partial x_2} - \frac{\partial p_1}{\partial x_1}\right) - d_3\frac{\partial p_1}{\partial x_2}.$$

The equations of the dual problem take the form

$$\frac{\partial \tilde{u}_2}{\partial x_2}+\frac{\partial \tilde{u}_1}{\partial x_1}=0\,,\frac{\partial \tilde{p}_2}{\partial x_1}-\frac{\partial \tilde{p}_1}{\partial x_2}=0\,,$$

$$\frac{\partial \tilde{p}_2}{\partial x_1}+\frac{\partial \tilde{p}_1}{\partial x_2}=c\cdot\left(-\frac{\partial \tilde{u}_1}{\partial x_2}-\frac{\partial \tilde{u}_2}{\partial x_1}\right)+c_3\frac{\partial \tilde{u}_2}{\partial x_2},$$

$$\frac{\partial \tilde{p}_2}{\partial x_2}-\frac{\partial \tilde{p}_1}{\partial x_1}=d\cdot\left(-\frac{\partial \tilde{u}_1}{\partial x_2}-\frac{\partial \tilde{u}_2}{\partial x_1}\right)-d_3\frac{\partial \tilde{u}_2}{\partial x_2}.$$

The limitation, obtained above, is acceptable. It means that in the original problem principal stresses' maximum displacement and orientation should not depend on hydrostatic compression.

2. Model of ideal plasticity.

This model and its generalizations nave been studied completely and are widely used for engineering problems [2,3]. The ideal plasticity's closed equation system can be written down as follows:

$$\frac{\partial p_1}{\partial x_2}-\frac{\partial p_2}{\partial x_1}=0\,,\frac{\partial u_1}{\partial x_1}+\frac{\partial u_2}{\partial x_2}=0\,,$$

$$\left(\frac{\partial p_2}{\partial x_2}-\frac{\partial p_1}{\partial x_1}\right)^2+\left(\frac{\partial p_1}{\partial x_2}+\frac{\partial p_2}{\partial x_1}\right)^2=4\tau_s^2\,, \tag{9}$$

$$\frac{\dfrac{\partial u_1}{\partial x_2}+\dfrac{\partial u_2}{\partial x_1}}{\dfrac{\partial u_1}{\partial x_1}-\dfrac{\partial u_2}{\partial x_2}}=-\frac{\dfrac{\partial p_2}{\partial x_1}+\dfrac{\partial p_1}{\partial x_2}}{\dfrac{\partial p_2}{\partial x_2}-\dfrac{\partial p_1}{\partial x_1}}.$$

where u_1 and u_2 are components of velocity vector, $\boldsymbol{\tau_s}$ is yield stress.

Let us consider the dual problem.

$$\frac{\partial \tilde{u}_2}{\partial x_2}+\frac{\partial \tilde{u}_1}{\partial x_1}=0\,,\frac{\partial \tilde{p}_2}{\partial x_1}-\frac{\partial \tilde{p}_1}{\partial x_2}=0\,,$$

$$\left(-\frac{\partial \tilde{u}_1}{\partial x_2}-\frac{\partial \tilde{u}_2}{\partial x_1}\right)^2+\left(\frac{\partial \tilde{u}_2}{\partial x_2}-\frac{\partial \tilde{u}_1}{\partial x_1}\right)^2=4\gamma^2\,, \tag{10}$$

$$\frac{\tilde{\sigma}_{11}-\tilde{\sigma}_{22}}{2\tilde{\sigma}_{12}}=\frac{\dfrac{\partial \tilde{u}_1}{\partial x_1}-\dfrac{\partial \tilde{u}_2}{\partial x_2}}{\dfrac{\partial \tilde{u}_1}{\partial x_2}+\dfrac{\partial \tilde{u}_2}{\partial x_1}}.$$

Here, the constant τs is denoted by γ. Thus, in the dual problem, conditions of incompressibility, coaxiality and constancy 2γ of deformation field displacement velocity are met.

Some similarities (9), (10) can be seen in a simple example. Let us take (9) a uniform stress distribution and a linear velocity distribution, i.e. consider a case of perfectly plastic body affine deformation:

$$\sigma_{11}=\frac{\partial p_2}{\partial x_2}=\tau_s,\sigma_{22}=\frac{\partial p_1}{\partial x_1}=-\tau_s,\sigma_{12}=\frac{\partial p_2}{\partial x_1}=\frac{\partial p_1}{\partial x_2}=0.$$

From (9) it follows that

$$p_2 = \tau_s x_2,\ p_1 = -\tau_s x_1,\ u_1 = \gamma \cdot x_1 - \Omega \cdot x_2,\ u_2 = \Omega \cdot x_1 - \gamma \cdot x_2. \tag{11}$$

Here 2γ is deformation velocity, which (for a perfectly plastic body) is set with border conditions, Ω is a rate of rotation. In [4] it is shown that kinematics (11) is realized under complex loading, when the body is continuously rotated at a rate of Ω in the direction of expansion and contraction.

Such deformations occur in cases of tidal deformation when the body rotates continuously towards the disturbing body.

In addition, study of such strains is of interest for pressure processing of inelastic materials [5,6]. In [4] it is shown that if $\Omega > \gamma$, elliptic area self-converting corresponds to kinematics (11).

Wherein, on the ellipse border, velocity is directed tangentially to the border; and its value is subjected to Kepler's Law: the radius vector, connecting the ellipse center and the boundary point, encloses equal areas for equal time. Let us proceed to the dual problem. Substituting (7) into (11) gives the following result: $\tilde{u}_1 = -\tau_s x_2,\ \tilde{u}_2 = \tau_s x_1,\ \tilde{p}_1 = -\Omega \cdot x_1 + \gamma \cdot x_2,\ \tilde{p}_2 = \gamma \cdot x_1 - \Omega \cdot x_2.$

$$\tilde{\sigma}_{11} = \frac{\partial \tilde{p}_2}{\partial x_2} = -\Omega, \tilde{\sigma}_{22} = \frac{\partial \tilde{p}_1}{\partial x_1} = -\Omega, \tilde{\sigma}_{12} = -\frac{\partial \tilde{p}_2}{\partial x_1} = -\frac{\partial \tilde{p}_1}{\partial x_2} = -\gamma.$$

Thus, rotation uncertainty Ω in the original problem corresponds to additive hydrostatic compression uncertainty in the dual task; shear uncertainty 2γ in a perfectly plastic body corresponds to shear stress uncertainty in a perfectly solidifying body. Conditions of incompressibility and coaxiality are retained in both problems. More trivial examples can be constructed, using exact solutions, outlined in [7] and [8].

3. Summary

The concept of the vector potential was introduced in a number of works under different names. First it appears in [9], after independently (and for other reasons) in [1,10]. A more general case is examined in [11]. Issues of duality are considered in [9,11,12,13]. Despite the fundamental character of these works, potential possibilities of this field of research are far from being exhausted. Thus, condition of shear velocity constancy (the second equation in (10)) is the key in problems of optimal blasting fragmentation of rocks [14,15]. In [16,12] condition of maximum shear velocity constancy has been studied in the framework of the theory of ideal solidified substance with use of duality with reference to equations of ideal plasticity. In [17,18], this model has been studied in connection with problems of brittle fracture. These articles also contain additional lists of scientific publications.

4. Conclusions

1. Description of inelastic body plane deformation can be reduced to a system of four differential equations coupling components of vector potential stress and displacement vector field.
2. Under certain conditions it is possible to proceed to the dual problem when roles of the noted vector fields are mutually changed.
3. The model of perfectly solidifying body is dual in relation to the model of perfect plasticity.

References

[1] A F Revuzhenko On deformation of granular continuum. part 2 Study of plane model J of mining sciences **5** (1981) 3–13
[2] A Y Ishlinskii, D D Ivlev. Mathematical theory of plasticity (2001)
[3] Y N Radaev On one achievable bottom estimate of three-dimensional Coulomb-Tresk stress invariant by systems of "two-dimensional" shear stresses Mechanics of limit state **4** (2012) 3–16
[4] A F Revuzhenko Tidal waves and directional transport of the Earth mass (2013)
[5] E I Shemyakin, A P Bobryakov, A F Revuzhenko Method of producing composite workpieces and device for its implementation Bulletin 47
[6] Y A Afinogenov, A P Biryukov, V M Zhigalkin, A F Revuzhenko, E I Shemyakin Device for

transverse rolling Bulletin 30
[7] A Nadai Ductility and fracture of solids (1969)
[8] A F Revuzhenko Mechanics of Granular Continuum (2006)
[9] R Hill 1956 About conjugate pairs of plane problems of elasticity theory J Mechanics **6** 71–79
[10] S V Lavrikov, O A Mikenina, A F Revuzhenko Description of elastic bodies' plane strain using vector of internal forces and non-Archimedean mathematical analysis. Bulletin of I Yakovlev CSPU Series: J Mechanics of limit state **1** (2009) 160–171
[11] D D Ivlev On theory of differential correspondences in continuum mechanics (2002)
[12] M A Artemov, D D Ivlev On theory of perfect solidifying continuum J DAN USSR 5 vol **355** (1997) 623-625
[13] V Prager 1956 About conjugate conditions of plane strain J Mechanics **6** 87-90
[14] V M Kuznetsov, E N Sher Principle of uniform blasting fragmentation of Solids J DAN USSR **2** (1976) 226
[15] Sher E N, Chernikov A G About shear method of uniform blasting rock fragmentation J of mining sciences **6** (1989) 20-23
[16] D D Ivlev On the ory of perfectly solidifying continuum J DAN USSR 4 vol **130** (1960) 742–745
[17] E I Shemyakin About brittle fracture of solids J Izvestiya RAN Mechanics of solids **2** (1997) 145–150
[18] E I Shemyakin On invariants of tension and strain state in mathematical models of continuum mechanics RAS **5** 632–634

Methodical Approach to Developing a Decision Support System for Well Interventions Planning

V A Silich[1], A O Savelev[2], A N Isaev[3]

[1] Professor at Tomsk Polytechnic University, Information system department
[2] Assistant at Control system optimization department of Tomsk Polytechnic University
[3] Head of Quality management Center of Tomsk Polytechnic University

e-mail: sava@tpu.ru

Abstract. The paper contains aspects of developing a decision support systems aimed for well interventions planning within the process of oil production engineering. The specific approach described by authors is based on system analysis methods and object model for system design. Declared number of problem-decision principles as follows: the principle of consolidated information area, the principle of integrated control, the principle of development process transparency. Also observed a set of models (class model, object model, attribute interdependence model, component model, coordination model) specified for designing decision support system for well intervention planning.

1. Introduction

The development of control systems for information support of decision-making process in the field of oil production engineering is a complex and actual task. The complexity of oil production engineering process is caused by incompleteness and ambiguity of the data about the exploration field [1].

A stabilization of crude oil output volume and equalization oil extraction ratio belong to most important and complicated tasks in the oil engineering area. One of the ways to solve these tasks is to optimize the process and methods of well interventions selection (a proper combination of geological, technological and technical measures) [2-7].

2. Methodical approach to developing a Decision Support System (DSS) for well interventions planning

The process of well interventions planning is a complex, multicriterion task. It is also characterized by poor formalizability of processed data and atypicality of system components. All these specific aspects induce to develop a special methodical approach to developing a DSS that provide and effective well interventions planning process [8].

The following requirements to the methodical approach were determined according to aspects referred above:

- versatility and high level of approach generality;
- operational flexibility and simplicity of customization (according to specific tasks);
- clearness of models;
- openness and high integrity level of development tools;

– availability of development tools integration with applications that use other approaches.

According to requirements determined a new Methodical approach to developing DSS for well interventions was discovered. The Methodical approach possesses:

– methodical principles of system design (based on determined requirements list of well interventions planning actor), that improves an efficiency of well interventions DSS development;
– a system-based well interventions planning sequence;
– a set of interrelated well interventions planning DSS (WIP DSS) object-oriented models, that decreases total time of WIP DSS development and increases economical efficiency of well interventions planning decision-making process.

3. A system-based well interventions planning sequence

To build a system-based well interventions planning sequence the problem domain was analyzed including following basic aspects: methodical and regulatory documents, documentary standards of oil industry and procedural regulations of single oil companies. As a result following stages of well interventions planning sequence were determined (fig. 1) [9]:

1. A detection of oil producing well with underused capacity.
2. Prediction of well capacity after providing technological process optimization and well interventions.
3. Decision-making of well interventions.
4. Well interventions providing.
5. An efficiency analysis of well interventions provided.

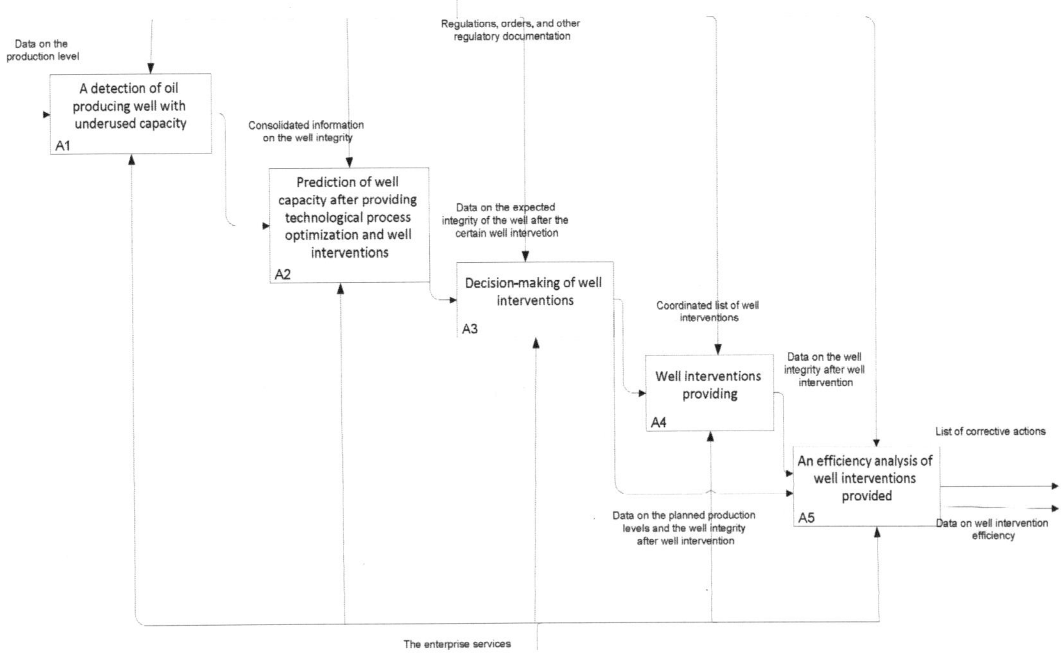

Figure 1. Well interventions decision-making process

4. Principles of well interventions planning DSS design

4.1 The principle of consolidated information area

It is supposed to implement a consolidated information area to increase the efficiency of information exchange among organizational units within an oil company and to integrate up-to-date data about current state of field production. The consolidated information area should be considered as a unity of data bases (DB) and data repository (including procedures of their usage, management and support), telecommunication systems and tools that perform according to common rules and procedures.

The implementation of the consolidated information area comes to solve following tasks:

1. To ensure the unity of data storage structure.
2. To ensure the standardization of data exchange processes among different informational infrastructure elements of oil-field.

The first task could be solved by usage of datamarts that can be considered as a local subject-oriented repository. Datamarts are connected to global data repository and provide services to definite groups of well interventions planning actors.

To solve the second task a set of common rules, procedures and integrated technologies of data transmission should be developed and implemented within IT-infrastructure of oil-field. In this case a unified standard of data transmission (from data-source to global data repository and backwards) can be considered as a main instrument for problem solving. To determine a unified standard of data transmission a WITSML standard is proposed to be used within the new methodical approach to developing a WIP DSS described in the paper.

4.2 The principle of integrated control

Assume that WIP DSS and its global data repository are represented by unity of separate software components that are located on different geographically distributed servers we have to specify monitor system for net infrastructure and information system monitoring. The specified monitor system assure availability and operability of elements of the DSS designed.

The WIP DSS' net infrastructure monitoring subsystem should exhibit following set of properties:

- the subsystem should provide monitoring of all net infrastructure objects;
- all indications read by basic collecting systems should be transmitted through the global data repository that unites net infrastructure representation;
- monitoring should be provided in a real-time mode (24 hours a day, 7 days per week);
- process of collecting current net infrastructure state indications should not have any sufficient implications for the net operating.

4.3 The principle of development process transparency

The structure of DSS should be represented by set of components (modules). The formal specification of modules should include definitions of attributes and functions used by the module. When possible "responsibility" of every component (functions operated by the module) should cover decision of very specific tasks, in order to minimize possible changes.

An OMSD technology (Object Model for System Design) is proposed to be used as a main instrument of the principle implementation. The technology is supposed to form and use process-models of designing systems that contain problems and resolve them.

5. A set of WIP DSS models

A set of WIP DSS model, developed on the basis of OMSD-technology, is represented by unity of models, that belong to one of five determined model types:

$$M = \langle M^C, \{M^O\}, \{M^A\}, M^S, \{M^K\} \rangle, \quad (1)$$

wherein M^C – class model, M^O – object model, M^A – attribute interdependence model, M^S – systems' components model, M^K – coordination model.

5.1 Class model

Class model includes variety of classes (e.g. fig. 2) that are used for system's components specification and for description of relation among classes. A class includes variety of class attributes and class methods. An attribute is specified by triplet "name-type-set of values". A variety of methods comprises an imperative subset of access methods to value of attributes. Different classes can be related to each other through sub-classing, that allows specifying a new class on the bases of already existing class. A sub-classes' variety of methods and attributes includes appropriate varieties of parent class. In this case overlaping attributes of sub-class narrow the range of definition of overlaping attributes of parent class. And at the same time overlaping sub-class methods redeclare overlaped parent-class methods.

Hydraulic fracturing
Well test : object; Well: object; Proppant: object; Fluid volume: real; Projected production rate: real;
Set_value (Well test, object) Set_value (Well, object) Set_value (Proppant, object) Set_value (Fluid volume, %)

Figure 2. Representation of class "Hydraulic fracturing"

5.2 Object model

An object model is formed by the class model. An object is specified in the following way:

$$o_k = \langle n(o_k), c(o_k), \{d_k(a_m)/cf(d_k(a_m))\}\rangle, \qquad (2)$$

wherein $n(o_k)$ – name of the object; $c(o_k)$ – indicator of the class, that was taken as a base for object implementation;

$d_k(a_m) \in D(a_m) | a_m \in A(c(o_k))$ – value of attribute; $cf(d_k(a_m)) \in [0,1]$ – factor of attribute value warranty (from 0 – total uncertainty to 1 – absolute reliability).

A variety of one class implementation variants is represented as a Multiobject (e.g. table 1). A Multiobject is a set of objects of one class that are specified according to a particular characteristic. A Multiobject is used to present a plenty of system states in different moments in time. The type of characteristic that drive the multiobject to be created can be of three kinds: time-divisional, three-dimensional or group.

An object model is used to solve tasks of pattern detection that is based on: comparison of objects that form a multiobject; selection of an optimal variant of system's component implementation. The last can be done by forming alternative variants of the component, evaluation of alternatives and optimal variant choice.

Table 1. A fragment of the Multiobject «Well interventions economic efficiency»

Economic efficiency	Well 7058, Date: 22.01.01	Well 3451, Date: 07.02.01	Well 3277, Date: 10.02.01	Well 3263, Date: 12.02.01
Production rate before Hydraulic fracturing, ton\day	10	26	18	7
Production rate after Hydraulic fracturing, ton\day	75	83	96	62
....				

5.3 Attribute interdependence model

An attribute interdependence model is used for representation of complex attributes dependencies with alternative output paths. It allows combining different methods of attribute dependencies formation and implementing various strategies of decision-making.

A network of attributes' functional dependences is represented by variety of class attributes that are linked by functional dependence relation and specify any component of a system:

$$T^A(c) = \langle A(c), R^{fd}\rangle \qquad (3)$$

The network is a nonstrict hierarchy that can possess some roots (network sink). Basic attributes are considered as a network source.

For evident representation of complex attributes dependencies with alternative output paths was developed a different model. The model allows combining various methods of attribute dependencies formation and implementing various strategies of decision-making.

The specified attributes dependence model helps to solve following tasks:

1. Problem of interpretation. In this case values of basic attributes are initially stated and calculation is provided to find target attributes values that correspond to initial data.
2. Admissible decision search. The task is used to define alternatives (well candidate selection). In this case the target attribute is a planned pay-back period of an intervention. The calculation of a planned pay-back period is driven by Pressure Transient Test data and physical characteristics of geological horizon.

5.4 Components model

A components model (e.g. fig. 3) is used to compose a hierarchy of subsystems. A set of subsystems S = $\{s_i\}$ integrates following components: global system, considered as a top level "Black Box" model; subsystems of a designed system and environment.

A subsystem can be represented as a tuple:

$$s_i=<n(s_i),c(s_i),M^O(c(s_i)),M^A(c(s_i)),\{M^L(c(s_i))\}>, \quad (4)$$

wherein $n(s_i)$ – name of a subsystem; $c(s_i)$ – class, that is referred by subsystem (it is sub-classed from basic class "Subsystem"); $M^O(c(s_i))$ – objects model, associated with the subsystem; $M^A(c(s_i))$ – attribute interdependence model, that specifies the subsystem; $M^L(c(s_i))$ – coordination model associated with the subsystem.

A subsystem tree represents the structure of the system designed. The subsystem tree is formed by sequential subsystems decomposition. The tree poses a stratified hierarchy, where one system is considered on different levels of the tree through various abstraction scales. A higher-level system can be considered as a root of the subsystem tree. As well it includes environment systems relevant to system designed.

Elements can be also considered as components besides subsystems. In this case elements are considered as entities that take part in subsystems functioning (for example, final products, performers, subjects of activity, mechanisms), or integrated properties, that are specified by set of characteristics (for example, technological parameters, economical results, technical specification, etc.).

The element $e_i \in E$, as far as subsystem, is associated with the name, class and also can be associated with object model and attribute interdependence model:

$$e_i=<n(e_i), c(e_i), M^O(c(e_i)),M^A(c(e_i))>, \quad (5)$$

wherein n(ei) – name of the element; c(ei) – class of the element (inherited from basic class "Element"; $M^O(c(e_i))$ – object's model, associated with the element; $M^A(c(e_i))$ – attribute interdependence model, that specify the element.

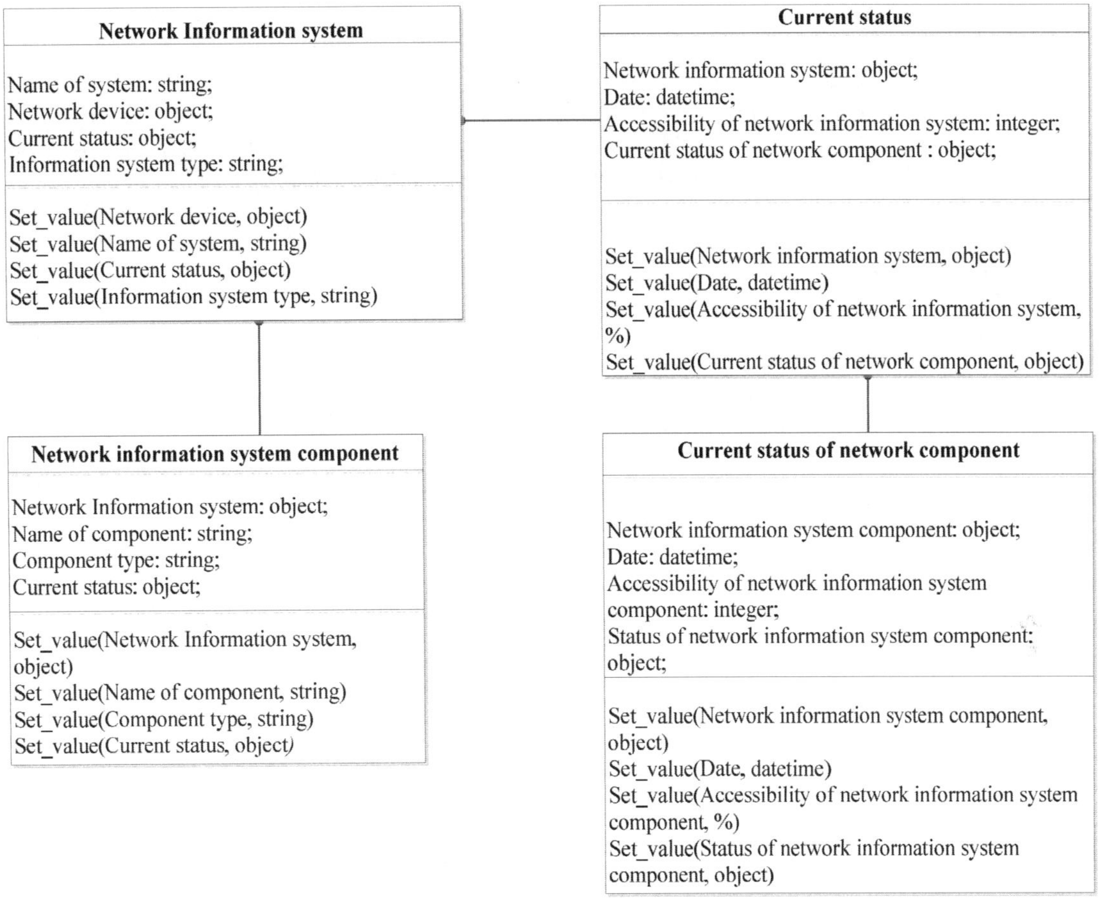

Figure 3. A fragment of the components model of network infrastructure monitoring system

5.5 Coordination model

A coordination model is built to explore interdependences among attributes of subsystems of the same tree-level. It is also used to solve tasks of integration (when it is necessary to specify attribute values of the parent system through interlevel dependences description) and coordination tasks (in this case parent system attribute values are considered as restrictions of descendant attributes admissible values). A coordination model associated with any subsystem s_0 of a complex system includes an interdependences network of subsystems attributes, its descendant components and methods to work with the coordination model:

$$M^K(s_0)=\langle N^K(s_0),P^K\rangle \quad (6)$$

A set of methods P^K includes following groups: methods of model formation (knowledge elicitation); methods of problem-solving (among them integration methods and coordination methods).

6. Conclusion

The presented methodical approach to developing a DSS for well interventions planning makes it possible to:

- increase the efficiency of DSS for well interventions planning development process;
- decrease the total time of DSS for well interventions planning development process;
- increase the economic efficiency of the planning process through the increasing of decision-making operational efficiency.

References

[1] Chen Z, Bunger A P, Zhang and Jeffrey R G 2009 Cohesive zone finite element-based modeling

of hydraulic fractures *Acta Mechanica Solida Sinia* **22** 443-52.

[2] Zhang G M, Liu H, Zhang J, Wu H A and Wang W W 2010 Three dimensional finite element simulation and parametric study for horizontal well hydraulic fracture *Journal of Petroleum Science and Engineering* **72** 310-17.

[3] Adachi J, Siebrits E, Peirce A and Desroches J 2006 Computer simulation of hydraulic fractures *International Journal of Rock Mechanics and Mining Sciences* **44(5)** 739-57.

[4] Gu H R , Siebrits E and Sabourov A 2008 Hydraulic-fracture modeling with bedding plane interfacial slip *SPE Paper 117445, SPE Eastern Regional/AAPG Eastern Section Joint Meeting.*

[5] Jeffrey R G , Settari A, Mills K W, Zhang X and Detournay E 2001 Hydraulic fracturing to induce caving: fracture model development and comparison to field data *Rock Mechanics in the National Interest* **1-2** 251-59.

[6] Zhang X, Detournay E and Jeffrey R 2002 Propagation of a penny-shaped hydraulic fracture parallel to the free-surface of an elastic half-space *International Journal of Fracture* **115(2)** 125-58.

[7] Lecampion B, Detournay E 2007 An implicit algorithm for the propagation of a hydraulic fracture with a fluid lag *Computer Methods in Applied Mechanics and Engineering* **196(49-52)** 4863-80.

[8] Silich V A, Savelev A O and Cherkashin A Yu 2016 Algorithm of the Alternatives Generation in the Design of the Geological and Engineering Operations *Key Engineering Materials* **685** 902-6.

[9] Silich V A and Savelev A O 2016 Development of the Decision-Making Algorithm on the Geological and Engineering Operations Based on the Infrastructure of the Digital Oil Field *Key Engineering Materials* **685** 907-11.

Research on Geometric Errors of Intermediate Unit Shell of a Geokhod

A V Walter[1], I F Borovikov[2], R V Chernukhin[1], S S Nozirzoda[1]

[1] Yurga Institute of Technology, TPU, 26, Leningradskaya Str., 652050, Yurga, Russia
[2] Bauman Moscow State Technical University, 5/1, ul. Baumanskaya 2-ya, 105005, Moscow

E-mail: avwalter@rambler.ru

Abstract. This article includes the research results on production errors of intermediate unit shell of a geokhod prototype model. There has been a problem stated concerning the accuracy of geokhod shells; constructive and technological particularities of a intermediate unit have been specified. A short summary has been conducted of the types of approach to the determination of shell errors of large equipment. The research was performed on the basis of data received by means of coordinate control of a gekhod prototype model in a production environment. The captured data have been researched by means of statistical methods and analyzed with the purpose of unveiling the factors affecting the errors. It was showed in the article that the errors can be partially explained by the production errors of body sectors and the errors of their relative assembling position. It was demonstrated that at least a part of errors is conditioned by reasons which are not taken into account in the process of a pure geometric description of the intermediate unit manufacturing process.

Introduction

The intermediate unit is one of the most responsible assembly units of a gekhod. It takes up high loads from the transmission as well as from the geological environment [1]. From the technological point of view, the intermediate unit is the most precise large-sized assembly unit of a gekhod characterized by a high unit man-hours and significant labor costs for mechanical handling and assembly [2]. As for construction, the intermediate unit consists of two coupled solids of revolution — a body and an outer body (Figure 1). Each of the bodies consists of four sectors which are tridimensional welded assemblies divisible into shell rings, flanges, stiffening elements and transmission interworking elements.

In the work process, the outer body of the intermediate unit is driven around the body by the transmission; and its shell interacts with the geological environment [3]. In the process of gekhod exploitation, the intermediate unit shell goes a long distance against the surrounding massive material; this leads to its significant wear which can result in structural and rigidity failure of the whole construction as well as in an abnormal performance of responsible elements of the power system assembled inside the geokhod body [4]. This is a serious problem because the increase of wear-resistance of the shell by means of known technological methods analogous to the ones described in [5], [6] is hindered due to a big extension of shell surface.

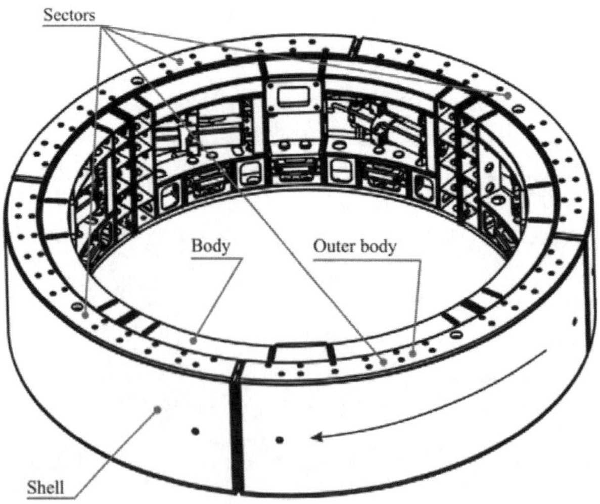

Figure 1. General view of the intermediate unit.

Wear rate depends on many factors; among these is the distribution pattern of bearing pressures on the shell surface. The rate, the character and the permanence of bearing pressures immediately depends on the shell form. The distribution of bearing pressures affects the forces that influence the transmission [7] as well as conditions energy loss connected with the resistance to traction [8].

When researching these issues, it is necessary to consider not the designed shell form but its actual form conditioned by the errors that appear in the process of its manufacture. As follows from model [9], geometric errors of geokhod shells can reach large values and depend on the particularities of the body assembly realization. Similar models based on the consideration of allowances of the form of coupled surfaces are represented in work [10], [11]. In work [12], it was mentioned that the accuracy is also affected by the intermediate sequence of elements in ring-type products. The listed models do not take into account the factors that are able to influence shell accuracy because they are based solely on the geometric approach. At the same time, the manufacturing practice of such products shows the necessity to count some hardly simulated factors, such as welding deformations [13], as well as to count a complex character of mutual influence of various factors on the combined error. Currently implemented project of launching into manufacture of a geokhod model FYURA.612322.401 allows to estimate the rate and the character of actual geometric errors of the intermediate unit shell taking into account a real production environment.

Thus, we should set the following task: to determine the values of geometric errors of the intermediate unit shell of a geokhod prototype model as well as to unveil their character.

2. Research methods

In order to solve this task, we researched on the outer body of the intermediate unit of a geokhod prototype model FYURA.612322.401. The research was conducted by means of coordinate measurement using a mobile coordinate-measuring machine (CMM) FARO Arm Edge 9. The body of the intermediate unit was mounted onto the control plate against the frontal flange. The CMM was mounted onto the plate next to the body. Due to large dimensions of the researched object, the measurement was conducted in four different CMM positions. In order to bring all coordinates received in various positions into a single coordinates system, each CMM position was bound by means of calibration cones. As a result of measurement, a data set has been formed including the coordinates of the points uniformly spread over the shell surfaces (Figure 2).

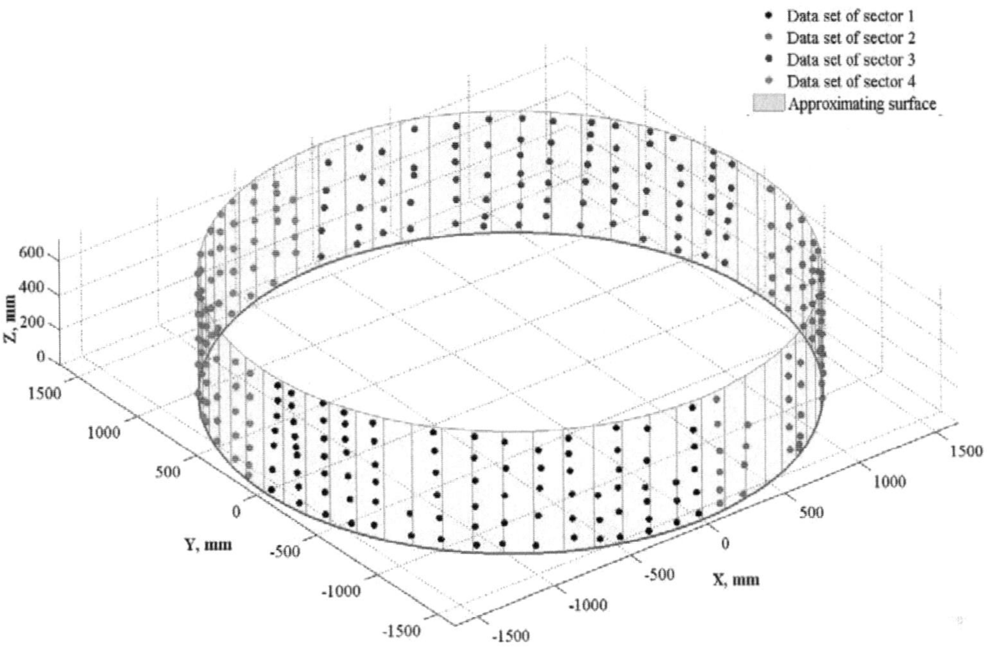

Figure 2. Measurement data sets of the intermediate unit shell as well as approximation cylinder.

For the analysis and the definition of error values of the position of sectors, a specially developed software was used. The coordinates data of the points were imported into the program and automatically split into four data sets according to each of the sectors. Further, based on the data sets, there were regression cylindrical models created and researched. The surface models were defined on the basis of the following regression equation:

$$\sqrt{A^2+B^2+C^2}-r+\varepsilon_i=0;\ a_x^2+a_y^2+a_z^2=1;$$
$$A=-a_y z_i-a_z(y_0-y_i);\ B=a_z(x_0-x_i)+a_x z_i;\ C=a_x(y_0-y_i)-a_y(x_0-x_i); \quad (1)$$

x_i, y_i, z_i are the coordinates of approximated points; a_x, a_y, a_z are the coordinates of the axis vector of the approximation cylinder; x_0, y_0 are the coordinates of the point through which goes the axis of the approximation cylinder; r is the approximation cylinder radius; ε_i are the residuals of the regression model.

The regression coefficients in equation (1) have been defined according to the method of Gauss [14]. There were five models of cylindrical surface created: one was based on the whole measurement data set, and four were based on data sets for each of the sectors. The research of the model based on the whole data set allows to define the values and the character of shell deviations in general. The research of models based on each of the sectors allows to estimate the value of deviations connected with the manufacturing errors of shell rings and the error of their relative position, as well as to define the rate and the character of errors not connected with above mentioned factors. The models were researched by means of calculation of general geometric characteristics, definition of typical statistic characteristics in the residual model rows as well as by means of correlation and regression analysis of residual rows in the sequence described in work [15].

After the coefficients of equation (1) were calculated, the coordinates of the points in the data set were transferred from the coordinates system XYZ created in the process of measurement into the coordinates system UVW connected with the axis of the corresponding cylindrical surface model (axis W of the new coordinates system must match the rotational axis of the cylinder. The transformations were performed using homogeneous coordinates and a generalized mapping grid:

$$\begin{bmatrix} u_1 & \dots & u_n \\ v_1 & \dots & v_n \\ w_1 & \dots & w_n \\ h_1 & \dots & h_n \end{bmatrix} = \begin{bmatrix} u_x & v_x & a_x & x_0 \\ u_y & v_y & a_y & y_0 \\ u_z & v_z & a_z & 0 \\ 0 & 0 & 0 & 1 \end{bmatrix} \begin{bmatrix} x_1 & \dots & x_n \\ y_1 & \dots & y_n \\ z_1 & \dots & z_n \\ 1 & \dots & 1 \end{bmatrix} \tag{2}$$

u_i, v_i, w_i, h_i are homogeneous coordinates of the points in the coordinates system connected with the axis of the approximation cylindrical surface; x_i, y_i, z_i are the coordinates of the points in the coordinates system created in the measurement process; u_x, u_y, u_z and v_x, v_y, v_z are the coordinates of directional vectors of the axles U and V respectively.

The vectors of axles U and V were set following the condition that the data set for each of the sectors would get into one octant of the coordinates system UVW. Moreover, vectors and theirs components are interconnected in the following way:

$$u_z = 0;\ u_x^2 + u_y^2 = 1;\ \mathbf{v} = \mathbf{u} \times \mathbf{w};\ \mathbf{w} = \mathbf{a} = \begin{bmatrix} a_x & a_y & a_z \end{bmatrix}^T \tag{3}$$

Following equations were used in order to define the deviances of the position of the sector against the mutual axis of the cylinder:

$$\cos\alpha = \mathbf{w}_\Sigma \cdot \mathbf{w}_S;\ \cos\beta = \mathbf{n}_{inc} \cdot \mathbf{v}_\Sigma;\ \mathbf{n}_{inc} = \frac{1}{|\mathbf{w}_\Sigma \times \mathbf{w}_S|}\mathbf{w}_\Sigma \times \mathbf{w}_S;\ \operatorname{tg}\gamma = \frac{x_0}{y_0};\ d = \sqrt{x_0^2 + y_0^2} \tag{4}$$

α – sector canting angle, β – sector canting surface angle; γ – direction angle of the dislocation of the sector axis in the reference plane; d – dislocation of the sector axis in the reference plane; $\mathbf{v}_\Sigma$, $\mathbf{w}_\Sigma$ – directing vectors of axles V and W of the coordinates system connected with the axis of the mutual cylinder; $\mathbf{w}_S$ – directing vector of axis W of the coordinates system connected with the axis of a separate sector; $\mathbf{n}_{inc}$ – normal vector of the plane of the largest bend of a sector (see Figure 3).

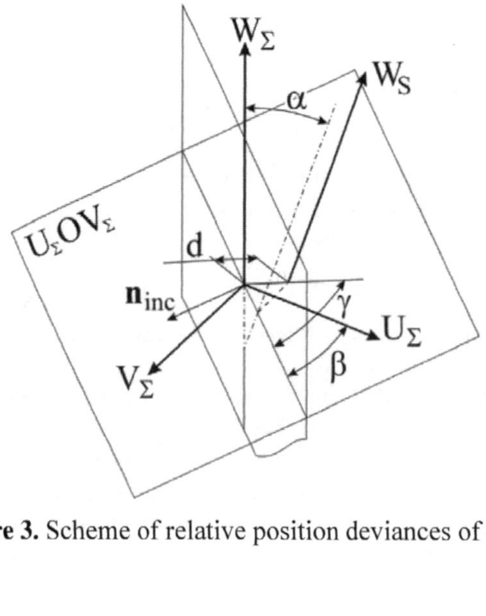

Figure 3. Scheme of relative position deviances of axles.

3. Results and discussions

The approximation results of point sets of the shell surface in general and for separate sector are represented in Table 1. The analysis of error values shows that the sectors of the body significantly differ by their actual dimensions, however, the errors are situated within defined allowances. The error value is close to the limit (10 mm per radius). The sectors are bent with different angles, in different directions as well as dislocated to different distances against the mutual axis.

Table 1. Parameters of model surfaces.

Parameter	The whole shell	Sector 1	Sector 2	Sector 3	Sector 4
Radius of the approximation cylinder, mm	**1,598.23**	**1,601.93**	**1,601.44**	**1,597.13**	**1,602.06**
Number of measured points	**311**	**90**	**64**	**83**	**74**
Standard deviation, mm	**1.72**	**0.92**	**0.94**	**0.70**	**0.68**
Maximal deviation, mm	5.26	2.42	2.11	2.23	1.92
Minimal deviation, mm	-4.22	-1.82	-2.29	-1.55	-1.63
Sector canting angle – α, deg	0	0.326	0.551	0.422	0.431
Sector canting plane angle – β, deg	0	157.272	-155.484	-51.954	55.645
Angle of dislocation direction of the sector axis in the reference plane – γ, deg	0	-131.666	24.363	-135.94	118.77
Dislocation of the sector axis in the reference plane – d, mm	0	2.17	3.82	3.96	1.28

The dislocations reach significant values comparable with allowances for shell manufacturing. Moreover, the fact is interesting that facing sectors are dislocated unidirectionally which is mostly obvious in the case of sectors 1 and 3. This effect is apparently connected with the particularities of sector position alignment in the process of assembly: the sectors are positioned not by the general base, but by their relative position. In order to define the possibility of explanation of the errors of intermediates unit shell with the manufacturing errors and relative position of sectors, the dependence curves were built of the radius of measured points against the axis of the mutual cylinders along the angular coordinate θ (Figure 4) and the height coordinate W (Figure 5).

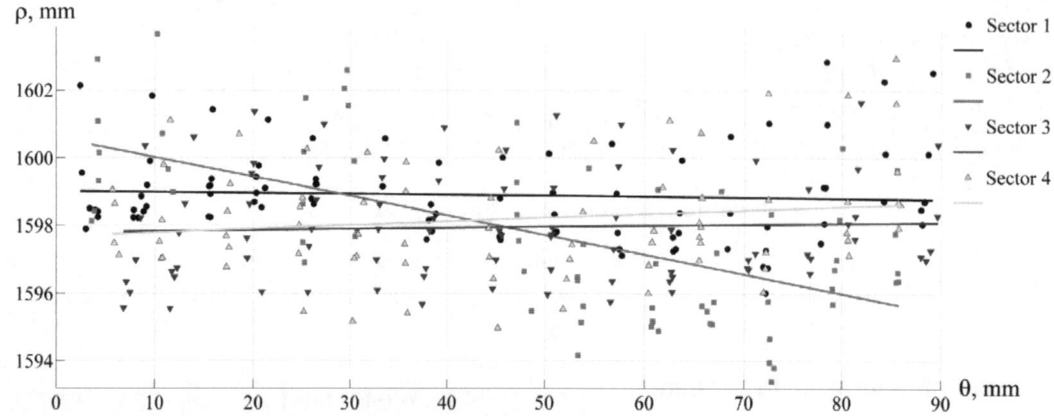

Figure 4. Dependence of the radius of measured points against the axis of the mutual cylinder along the angular coordinate θ.

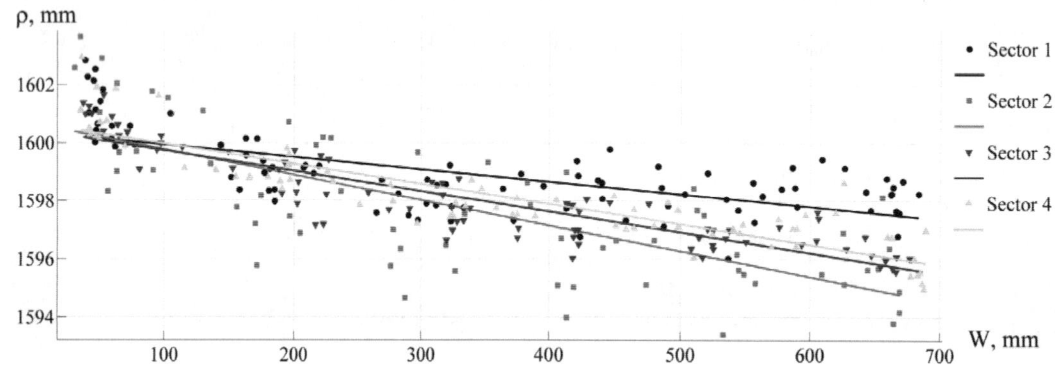

Figure 5. Dependence of the radius of measured points against the axis of the mutual cylinder along the height coordinate W.

As follows from the curves in Figure 4, for all sectors except sector 2, the dependence of the radius on the angular coordinate is hardly recognizable. At the same time, as follows from Figure 5, there is a marked dependence of the radius on height W.

The analysis of residual rows of sector models (Figure 6, Figure 7) shows the lack of an expressed trend. This gives evidence that at least a part of shell errors can be explained by the manufacturing errors of sectors and the inaccuracy of their relative position in the process of assembly.

At the same time, the results of the statistical model analysis represented in Table 2 show that, apart from the above stated reasons, the shell errors can be affected by other factors not counted in the model and provoking systematic errors. This can be verified by the presence of auto-correlation in the most residual rows as well as by the presence of a significant correlation in some rows.

The residuals follow the normal distribution law and are random according to the criteria of turning points test; thus, sector models adequately enough describe the character of geometric errors of the shell and explain about 50 per cent of the cumulative error value.

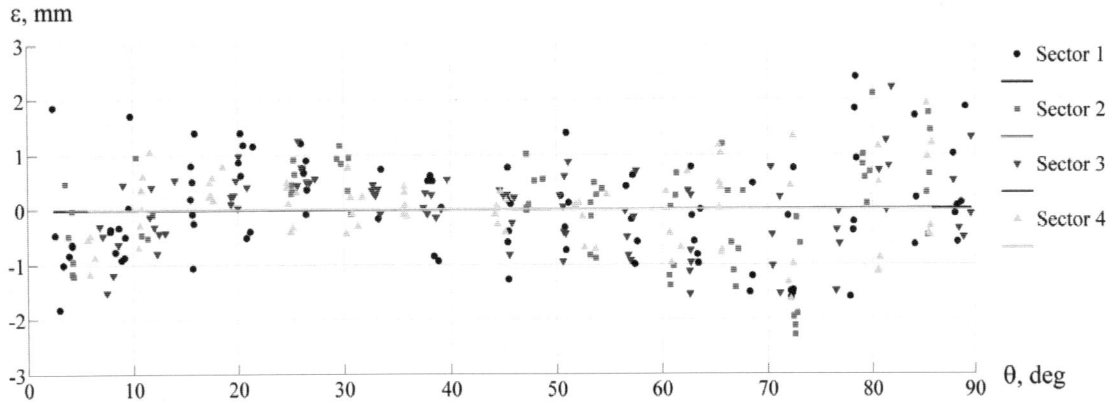

Figure 6. Residual rows of sector models along the angular coordinate θ.

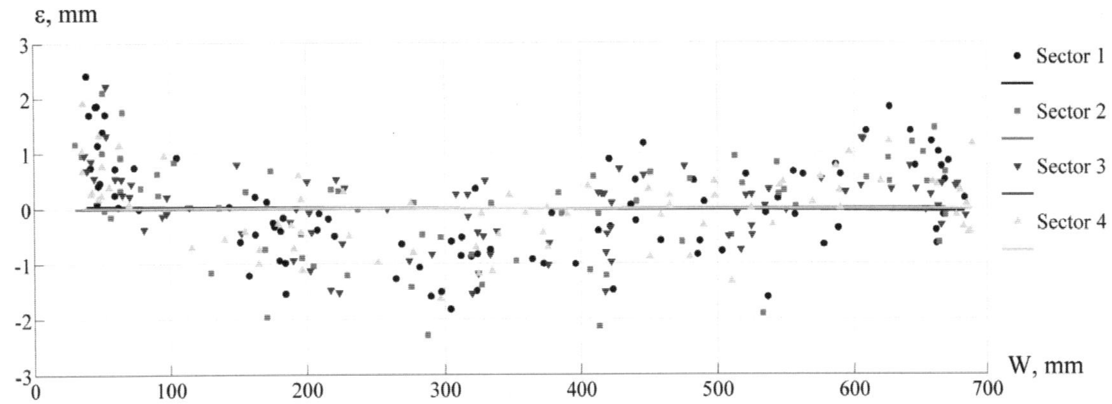

Figure 7. Residual rows of sector models along the height coordinate W.

Table 2. Results of statistical analysis of sector models.

Qualitative characteristic	Sector 1	Sector 2	Sector 3	Sector 4
Lack of significant correlation of residuals along θ	Yes	No	Yes	Yes
Lack of significant correlation of residuals along W	No	No	No	No
Normality of residual rows according to Jarque-Bera test	Yes	Yes	Yes	Yes
Randomness of residuals according to turning points test along θ	Yes	Yes	Yes	Yes
Randomness of residuals according to turning points test along W	Yes	Yes	Yes	Yes
Lack of auto-correlation according to Durbin-Watson test along θ	No	No	No	Yes
Lack of auto-correlation according to Durbin-Watson test along W	No	Yes	No	No

4. Conclusion

- The coordinates control of the outer body shell of the intermediate unit of a geokhod prototype model showed that the error values are close to the limit which can negatively affect the quality of products in serial production.
- A significant part of cumulative error of the shell can be explained by the production errors of body sectors and the error of their relative position in the process of assembly.
- The statistical analysis of sector models suggests that the manufacturing errors of body sectors and the errors of their relative position in the process of assembly are not the only significant factors in the formation of shell errors of the intermediate unit.

References

[1] A B Efremenkov and V Y Timofeev Determination of necessary forces for geohod movement (2012) 1–4

[2] V V Aksenov, A V Walter, A A Gordeyev, A V Kosovets Classification of geokhod units and systems based on product cost analysis and estimation for a prototype model production IOP Conference Series: Materials Science and Engineering (2015)

[3] A B Efremenkov, Forming the subterranean space by means of a new tool in 6th International Forum on Strategic Technology Proceedings **1** (2011) 348–350

[4] R V Chernukhin, A A Dronov, M Y Blashchuk The application of the analytic hierarchy process when choosing layout schemes for a geokhod pumping station, IOP Conference Series: Materials Science Engineering **91** (2015)

[5] V A Klimenov, Z G Kovalevskaya, V Y Ul'yanitskii, K V Zaitsev, and V. Y. Borozna, Effect of ultrasound treatment of the substrate on the formation of the coating in detonation spraying, Welding International **10** (2010) 803–807

[6] Z G Kovalevskaya, V A Klimenov, K V Zaitsev Interfacial Adhesion between Thermal Spray Coating and Substrate Achieved by Ultrasonic Finishing Applied Mechanics and Materials **682** (2014) 459–463

[7] M. Y. Blashchuk, A. A. Kazantsev, and R. V. Chernukhin Capacity Calculation of Hydraulic Motors in Geokhod Systems for Justification of Energy-Power Block Parameters Applied Mechanics and Materials **682** (2014) 418–425

[8] V V Aksenov, A A Khoreshok, V Y Beglyakov Justification of creation of an external propulsor for multipurpose shield-type heading machine – GEO-WALKER Applied Mechanics and Materials **379** (2013) 20–23

[9] A V Walter, V V Aksenov, Determining deviations in geometry of the geokhod shells Applied Mechanics and Materials **770** (2015) 439–444

[10] S Lowth, D A Axinte, An assessment of variation conscious precision fixturing methodologies for the control of circularity within large multi-segment annular assemblies Precision Engineering **38** (2014) 379–390

[11] T Hussain, Z Yang, A A Popov, S McWilliam Straight-build assembly optimization A method to minimize stage-by-stage eccentricity error in the assembly of axisymmetric rigid J components Journal of Manufacturing Science and Engineering Transactions of the ASME, **133** (2011) 3

[12] M Memon, T Hussain, Z A L I Memon Mehran University Research Journal of Engineering Technology **4** (2012) 743–754

[13] Z Chen, Z Chen, R A Shenoi, Ocean Engineering **106** (2015) 271–280

[14] B Y Lemeshko, S B Lemeshko, M S Nikulin, N Saaidia, Modeling statistic distributions for nonparametric goodness-of-fit criteria for testing complex hypotheses with respect to the inverse Gaussian law Automation and Remote Control **7** (2010) 1358–1373

A V Walter Modeling of Track Overlapping Effect on Layer Integrity in the Process of Laser Sintering of Polymer Suspension in International Forum on Strategic Technology **2** (2012) 125–128

IOP Conf. Series: Materials Science and Engineering **127** (2016) 012018 doi:10.1088/1757-899X/127/1/012018

Integrated Model to Assess Cloud Deployment Effectiveness When Developing an IT-strategy

S Razumnikov[a] and D Prankevich

National Research Tomsk Polytechnic University Affiliate, 652050, ul. Leningradskaya 26, Yurga, Russia

E-mail: [a]demolove7@inbox.ru

Abstract. Developing an IT-strategy of cloud deployment is a complex issue since even the stage of its formation necessitates revealing what applications will be the best possible to meet the requirements of a company business-strategy, evaluate reliability and safety of cloud providers and analyze staff satisfaction. A system of criteria, as well an integrated model to assess cloud deployment effectiveness is offered. The model makes it possible to identify what applications being at the disposal of a company, as well as new tools to be deployed are reliable and safe enough for implementation in the cloud environment. The data on practical use of the procedure to assess cloud deployment effectiveness by a provider of telecommunication services is presented. The model was used to calculate values of integral indexes of services to be assessed, then, ones, meeting the criteria and answering the business-strategy of a company, were selected.

1. Introduction

Looking for an approach to IT-infrastructure which is able to provide various scaling a growing number of companies want to use cloud-based technologies having huge potential for gain in performance without prejudice to efficiency [1-4]. Nevertheless, to gain this advantage and have the maximal return on investment companies have to consider diverse difficulties and peculiarities of deploying a cloud-based IT, specific for particular circumstances [5, 6]. Deployment of cloud technologies necessitates a strategic plan to set a correct IT goal and further its achievement, control and adjust the process, ensuring the best results, synchronize IT and business performance, make IT functioning efficient and transparent for business to succeed [7].

The process of managerial decision-making concentrated on deployment of cloud-based technologies is to be organized in line with assessment of their economic efficiency and utilization risks. The specifics of strategic decision-making consist in incompleteness and inaccuracy of reference information, typical for conditions that kind of decisions is made in. As the consequence, an expert has to describe circumstances relying both on quantity and quality characteristics. An expert's knowledge is a determining factor when selecting a cloud service. Moreover, he/she has to develop a single methodology combining quality and quantity assessments of possible alternatives [7].

This problem can be resolved provided that a multi-criteria approach and fuzzy sets theory are applied for decision-making, enabling simulation of a smooth change in object properties, as well as revealing unknown functional dependencies expressed as attribute-based relations.

2. Procedure of calculating criteria and integral index "Effectiveness of a cloud-based service"

The problem of insufficient comprehensive methodological base and tools to support decision making, which rely on the processes of efficiency assessment and risks in conditions of uncertain decision making environment is currently the urgent one in the sphere of cloud-based technologies integration. This issue is relevant for businesses of all branches and levels [6]. To solve a problem of this kind a system approach is to be applied, as well as method of system analysis. This approach can be reduced to giving a more precise definition of the problem and structuring it in a number of tasks soluble by economic and mathematical methods, revealing criteria for their decision, and making purposes more detailed.

The following procedure based on evaluation of 6 group criteria is offered to assess effectiveness of cloud-based IT-services:

1. Efficiency for business (*Eb*)
2. Financial benefit (*Fb*).
3 Technological priority (*Tp*)
4. Work reliability and information safety (*Is*) [8]
5. Risk degree of cloud service utilization (*Dr*)
6. Psychological factor influence (*Pf*)

The integral index is calculated by hybrid additive and multiplicative formula (1):

$$K_{ecs} = Is \cdot (a_1 \cdot Eb + a_2 \cdot Fb + a_3 \cdot Tp + a_4 \cdot Dr + a_5 \cdot Pf), \tag{1}$$

where K_{ecs} – integral index "Effectiveness of a cloud-based service";
Eb – value of criterion "Efficiency for business";
Fb – value of criterion "Financial benefit";
Tp – value of criterion "Technological priority";
Is – value of criterion "Work reliability and information safety";
Dr – value of criterion "Degree of risk";
Pf – value of criterion "Psychological factor".
a_1, a_2, a_3, a_4, a_5 – weight coefficients.

The above criteria have ranks (weight coefficients) to ensure compliancy. Calculating the coefficients an expert has to take into account the range of criteria scale and average statistical numerical scores of a criterion. Research revealed the distinction between weights determined by an expert and those ones resulted from his/her activities. The weights of the most essential criteria are usually underestimated while those of less important criteria are overrated. Therefore, to avoid subjectivity pairwise comparison method is used when allocating weights [9, 10]. Weight coefficients are values of vector r_i calculated according to formula $r_i = 1 / \sum_{i=1}^{n} k_i$, where k is a sum total of pairwise comparison matrix column *i*. Vector $r_i = (a_1; a_2; a_3; a_n)$.

Criterion "Efficiency for business"

Indexes of this criterion are used mostly for quality assessment of data and their presentation in numerical scores. Then it is calculated by additive formula 2.

$$Eb = a_{11} \cdot Iv + a_{12} WEu + a_{13} Oru + a_{14} Cb, \tag{2}$$

where Iv – score of index "Increase in velocity";
WEu – score of index "Work efficiency of user";
Oru – score of index "Optimization of resource utilization";
Cb – score of index "Criticality for business";
a_{11}, a_{12}, a_{13}, a_{14} – weight coefficients.

Cloud computing profitability is to be calculated by formula 3 to determine appropriate resources utilization.

The following formula is used to calculate index "Optimization of resource utilization" (*Oru*):

$$Uh \times (I - \mathrm{Ch_c}) \geq Uh_{DPC} \times \left(I - \frac{\mathrm{Ch_{DPC}}}{\mathrm{Al}} \right), \tag{3}$$

where $\mathrm{Uh_c}$ – spent hours of cloud work;

I – profit;

$\mathrm{Ch_c}$ – cost of 1 hour of work in a cloud;

$\mathrm{Uh_{DPC}}$ – spent hours of Data Processinf Centre;

$\mathrm{Ch_{DPC}}$ – cost of 1 hour of Data Processing Centre work;

Al – average load.

Criterion "Financial benefits" was calculated according to multiple formula (4).

$$Fb = \frac{Ecs}{As}, \tag{4}$$

where Ecs – score of index "Expenditures on cloud services";
As – score of index "Assets saving".

Criterion "Technological priority" is also calculated by additive formula (5).

$$Tp = a_{21} \cdot I + a_{22} \cdot Pamc + a_{23} \cdot Ts + a_{24} \cdot Sa, \tag{5}$$

where I – score of index "Integration";
Pamc – score of index "Possibility of application migration into a cloud";
Ts – score of index "Technological stack";
Sa – score of index "Application style";
a_{21}, a_{22}, a_{23}, a_{24} – coefficients of impact degree.

Criterion "Work reliability and information safety" is calculated by multiplicative formula (6).

$$Is = \sqrt[5]{Sd \cdot St \cdot Au \cdot Iu \cdot So}, \tag{6}$$

where *Sd* – relative safety index of data to be stored;
St – index of data safety under data transmission;
Au – index of authentication;
Iu – relative index of users isolation;
So – index of smooth operation;

Criterion "Degree of risk" is calculated by hydride formula (7)

$$Dr = a_{31} \cdot Nli + a_{32} \cdot Ir + a_{33} \cdot Ii + a_{34} \cdot Rd + a_{35} \cdot P, \tag{7}$$

where *Nli* – score of index of addressing to normative and legal issues;
Ir –index score of response to incidents;
Ii – score of incompatibility index;
Rd –index score of privacy and data recovery;
P – score of overpayment index outlined as pay-as-you-go;
a_{31}, a_{32}, a_{33}, a_{34}, a_{35} – weight coefficients.

Criterion "Psychological factor" is calculated by formula (8)

$$Pf = (Ss + Cs + Wi + MIa + \mathrm{Im})/50, \tag{8}$$

where Ss – index of staff satisfaction;
Cs – index of customers satisfaction;
Wi – index of willingness to innovations;
MIa – index of mental and intellectual abilities;
Im – index of motivation.

The general algorithm of calculating the integral index comprises the following phases:

1. Data mining. Qualitative and quantitative indexes are selected with respect to responses of cloud IT-providers, contracts, agreements, and price-lists. It was performed over the first phase of the systematic approach to strategy development.

2. All indexes were assessed in numerical scores by an expert using 0 to 1 scale (Table 1) and in compliance with required indexes and standards. To realize the requirements "Standards and guidelines on application of cloud computing" developed by the Joint Technical Committee in 2014 are to be referred to, such indicators as those of a leader-competitor and counterpart-competitor on the market can also be used.

3. Calculation of weight coefficients for criteria.

4. Calculation of criteria by formulae 2, 4-8.

5. Calculation of an integral effectiveness index of a cloud-based service (K_{ecs}) by formula 1.

Iteration of indexes is eliminated in the method, enabling objective estimation of effectiveness of cloud-based IT-service application.

An application answers the criteria and meets the business-strategy of a company provided that its integral index *Kecs* is more 0.5. Therefore, we can analyze the possibility to transfer them into a cloud.

Table 1. Preference scale of indexes (criteria)

Value of index	Verbal expression of an index (criterion) of cloud-based service effectiveness
1	Effectiveness index is very high (exceeds the standard one twice and more)
1.00…0.75	Effectiveness index is quite high (exceeds the standard one by 75-100 %)
0.75…0.5	Effectiveness index seems to be high (exceeds the standard one by 50-75 %)
0.5	Average effectiveness index (at level of the standard one)
0.5…0.25	Effectiveness index seems to be low (0-25 % lag behind the standard one)
0.25…0	Effectiveness index is quite low (25-50 % lag behind the standard one)
0	Effectiveness index is very low (100% lag behind the standard one)

3. Results and Discussion

The integrated model to assess effectiveness of cloud computing deployment was applied by Public company "Rootelecom".

Let us address the issue of determining the advantages for following services Business administration (Ba), Personnel management (Mpe), Accounting (Acc), Cloud server (CS), Corporate mail (Cm), Microsoft Office (MO), File backup (Fb), Automatic switching center cloud IP-automatic switching center (ASC), Private cloud (PC). We calculate all criteria and integral index *Kecs* for them. We also follow data given by the provider, those ones on internal infrastructure and standards [11].

As a cloud provides we considered "Cloud4Y". It provides the above services as SaaS.

Weight coefficients of cloud IT-service effectiveness criteria

First we calculate weight coefficients and use them in formulae of effectiveness criteria for applications to be assessed. Formulae for effectiveness criteria are given below.

Efficiency for business: $(Eb)=0.51\cdot Iv+0.28\cdot WEu+0.12\cdot Oru+0.08\cdot Cb;$

Technological priority: $(Tp)= 0.68\cdot I+0.15\cdot Pamc+0.09\cdot Ts+0.06\cdot Sa;$

Degree of risk: $(Dr)= 0.22\cdot Nli+0.16\cdot Ir+0.08\cdot I+0.43\cdot Rd+0.03\cdot P.$

The integrated model is as follows:

$K_{ecs}=Is\cdot(0.26\cdot Eb+0.13\cdot Fb+0.06\cdot Tp+0.5\cdot Dr+0.03\cdot Pf).$

Expert assessment of effectiveness indexes

Scores of effectiveness criteria indexes for application «1C: Business administration" are presented in Table 2.

Table 2. Scores of effectiveness criteria and risks for application "Business administration"

Eb		Tp		Is		Dr		Pf	
Index	Value	Index	Value	Index	Value	Index	Value	Index	Value
Iv	0.75	*I*	0.5	*Sd*	0.85	*Nli*	1	Ss	9.5
WEu	0.8	*Pamc*	0.65	*St*	0.85	*Ir*	0.9	Cs	8
Oru	0.5	*Ts*	0.75	*Au*	0.75	*Ii*	0.85	Wi	6
Cb	0.4	*Sa*	0.9	*Iu*	0.9	*Rd*	0.7	MIa	8.5
				So	0.7	*P*	0.7	Ia	10

Calculation of criteria and integral effectiveness index of cloud-based technologies

Effectiveness criteria were calculated for each service. Let us address to application "1C: Business administration" as an example.

We calculate a score for each criterion by using values of indexes in a corresponding model.

$$Eb = 0.51\cdot 0.75 + 0.28\cdot 0.8 + 0.12\cdot 0.5 + 0.08\cdot 0.4 = 0.3825 + 0.224 + 0.06 + 0.036 = 0.7$$

$Fb = \frac{9543}{34487} = 0.28$ Criterion Fb is given score – 0.7.

$$Tp = 0.68\cdot 0.5 + 0.15\cdot 0.65 + 0.09\cdot 0.75 + 0.06\cdot 0.9 = 0.34 + 0.104 + 0.0675 + 0.063 = 0.56$$

$$Is = \sqrt[5]{0.85\cdot 0.85\cdot 0.75\cdot 0.9\cdot 0.7} = 0.81$$

$$Dr = 0.22\cdot 1 + 0.16\cdot 0.9 + 0.08\cdot 0.85 + 0.43\cdot 0.7 + 0.03\cdot 0.7 = 0.22 + 0.216 + 0.1275 + 0.238 +$$

$$+ 0.035 = 0.76$$

$$Pf = (9.5 + 8 + 6 + 8.5 + 10)/50 = 0.84$$

Let us calculate integral index *Kecs* for application "1C: Business administration", using values of criteria in the model:

$$Kecs = 0.81\cdot(0.26\cdot 0.7 + 0.13\cdot 0.7 + 0.06\cdot 0.56 + 0.5\cdot 0.76 + 0.03\cdot 0.84) =$$

$$= 0.81\cdot(0.18 + 0.09 + 0.03 + 0.38 + 0.02) = 0.59$$

Criteria and integral index are calculated the same way for other services. Calculations were performed in software. Ranked calculations for 9 services to be assessed are given in Table 3.

Table 3. Calculations for 9 services to be assessed of Public Company "Rootelecom"

Services to be assessed	Effectiveness criteria						K_{ecs}
	Eb	Fb	Tp	Is	Dr	Pf	
Private cloud 2.0	0.82	0.9	0.5	0.85	0.77	0.92	0.66
«1C: Business administration»	0.7	0.7	0.56	0.81	0.76	0.84	0.59
Cloud IP-Automatic Switching Center	0.84	0.45	0.79	0.75	0.83	1	0.58
«1C: Accounting»	0.65	0.4	0.78	0.85	0.66	0.56	0.53
File backup	0.34	0.15	0.77	0.83	0.68	0.5	0.42
«1C: personnel management»	0.44	0.15	0.76	0.81	0.55	0.46	0.38
Corporate mail	0.28	0.05	0.68	0.78	0.7	0.5	0.38
Virtual cloud server	0.29	0.25	0.63	0.69	0.68	0.5	0.35
Microsoft Office	0.11	0.15	0.68	0.65	0.58	0.42	0.25

Integral index *Kecs* for services Private Cloud 2.0, Business Administration, IP-Telephony and 1C: Accounting is more 0.5. Therefore, these services answer the criteria and meet business-strategy of a company. We can take them into account when analyzing the possibility of transfer/deployment in a cloud. Other applications are not subjected to further analysis.

Conclusion

The system of criteria, as well as integrated model to assess effectiveness of cloud services deployment are presented in the paper. The integrated model of effectiveness assessment based on the system of criteria enables making a decision what applications will be the best possible to meet the requirements of a company business-strategy, evaluate reliability and safety of cloud providers and analyze staff satisfaction even at the phase of strategy formation.

The system of criteria and integrated model were used for calculation of deployment effectiveness of 9 cloud services by Public Company "Rootelecom", providing telecommunication services. 4 services answering business aims and meeting safety requirements were selected on the base of integral index: Private Cloud 2.0, "1C: Business Administration", Cloud IP-Automatic Switching Center, "1C: Accounting".

Acknowledgments

This work was supported by Russian foundation for basic research as a part of the project «Development of mathematical models and the software for an assessment of efficiency and risks from transition to cloudy IT services», the project № 15-01-01251.

References

[1] Jones S 2015 Cloud computing procurement and implementation: Lessons learnt from a United Kingdom case study, *International journal of information management*. Volume: 35 Issue: 6 pp 712-716.

[2] Marston S Li, Bandyopadhyay S, Zhang J, Ghalsasi A. 2011 Cloud computing – The business perspective, *Decision Support Systems*. Volume 51, Issue 1, April 2011, pp 176-189.

[3] Paul P K, Ghose M K 2012 Cloud Computing: possibilities, challenges and opportunities with special reference to its emerging need in the academic and working area of Information Science, *International conference on modelling optimization and computing.* Volume 38, pp 2222-2227.

[4] Amir Mohamed Elamir, Norleyza Jailani, Marini Abu Dakar 2013 Framework and architecture for programming education environment as cloud computing service, *Procedia Technology* 11 pp 1299-1308.

[5] Latifa Ben Arfa Rabaia, Mouna Jouinia, Anis Ben Aissab, Ali Mili 2013 A cybersecurity model in cloud computing environments, *Journal of King Saud University - Computer and Information Sciences.* Vol. 25, Issue 1 pp 63-75.

[6] Razumnikov S V, Zakharova A A, Kremneva M S 2014 A model of decision support on migration of enterprise IT-applications in the cloud environment, *Applied Mechanics and Materials* 682 pp 600-605.

[7] Zakharova A A 2013 Fuzzy swot analysis for selection of bankruptcy risk factors, *Applied Mechanics and Materials* 379 pp 207-213.

[8] Kumar Garg S, Buyya R 2012 SLA-based admission control for a Software-as-a-Service provider in Cloud computing environments, *Journal of Computer and System Sciences*, 78 (5), pp 1280-1299.

[9] Chernysheva T Yu 2013 Preliminary risk assessment in it projects, *Applied Mechanics and Materials* 379 pp 220-223.

[10] Razumnikov SV, Kremnyova M S 2015 Decision support system of transition IT-applications in the cloud enviroment, *International Siberian Conference on Control and Communications (SIBCON): proceedings*, Omsk, May 21-23.

[11] Maricela-Georgiana Avram (Olaru) 2014 Advantages and challenges of adopting cloud computing from an enterprise perspective, *Procedia Technology* pp 529-534.

IOP Conf. Series: Materials Science and Engineering **127** (2016) 012019 doi:10.1088/1757-899X/127/1/012019

Improving Processes of Mechanized Pulsed Arc Welding of Low-Frequency Range Variation of Mode Parameters

Yu N Saraev[1a], S A Solodskiy[2b], O V Ulyanova[3c]

[1a]Doctor in Engineering, professor. Institute of Strength Physics and Materials Science of the Siberian Branch of the Russian Academy of Sciences, Tomsk, Russia;
[2b]PhD in Engineering, associate professor. National Research Tomsk Polytechnic University, Yurga Institute of Technology, Tomsk, Russia;
[3c]teaching assistant. National Research Tomsk Polytechnic University, Yurga Institute of Technology, Yurga, Russia;

e-mail: [1a]litsin@ispms.tsc.ru; [2b]serdgio80@inbox.ru; [3c]olgauljanova@mail.ru

Abstract. A new technology of low-frequency modulation of the arc current in MAG and MIG welding is presented. The technology provides control of thermal and crystallization processes, stabilizes the time of formation and crystallization of the weld pool. Conducting theoretical studies allowed formulating the basic criteria for obtaining strong permanent joints for high-duty structures, providing conditions for more equilibrium structure of the deposited metal and the smaller width of the HAZ. The stabilization of time of the formation and crystallization of the weld pool improves the formation of the weld and increases productivity in welding thin sheet metal.

1. Introduction

Today conventional methods of welding and hardfacing (arc hardfacing, electroslag hardfacing, plasma-jet hardfacing) have exhausted their potential. However production of new promising steel types in industry is being constantly improved and developed. It requires research and development in new technologies increasing performance capabilities of conventional heating sources (arc, plasma, laser radiation) and their hybrid combinations used to produce permanent joints of these steels [1-4].
Pulse technologies of welding and hardfacing [5-9] as well as specialized equipment which is being intensively developed by world leading producers of welding equipment [10-16] offer opportunities for this development. The interest to pulse technologies in welding and hardfacing is aroused by their advanced performance as compared with conventional methods. Pulse technologies provide:

- control of welding, metal transfer and crystallization regardless of welding pool space position with lower values of the main process variables;
- two- or threefold increase of welding pool crystallization rate due to unsteady influence of a heating source on the welding pool reducing molten metal temperature;
- lower deformation in welded products, built-up layers;
- improvement in the quality of welded joints and the coatings applied (improved weld regardless from space position, homogenous chemical composition throughout the coating, structure atomization in the weld and in the heat-affected zone). These advantages are obtained by means of directed crystallization of the welding pool and intensified

hydrodynamic processes in the molten metal which leads to degassing of the welding pool and uniform distribution of alloying element throughout the hotmelt;
- improving mechanical properties of welded products due to considerable reduction of the heat-affected zone and atomization of its structure.

A great number of research works deal with these processes in welding but there are some understudied issues concerning development and application of low-frequency pulse control of welding mode energy parameters.

The aim of the research is finding ways to improve structural behavior and performance characteristics of permanent joints of metal structures by means of pulse technologies of welding and hardfacing low-frequency variation range of mode energy parameters.

2. Development history of low frequency pulse technologies in welding.

A method of arc welding of steel sheets with programmable low-frequency modulation of welding current was first developed by M P Zaitsev [19]. It was followed by a great variety of processes and devices for its application.

A P Petrov and G A Slavin suggested low-frequency range of pulse arc (up to 10hz) which allows periodical changes (regulations) of heat flow fed to the metal and thus control of base metal penetration and weld formation. Such a modulation of the weld current found wide application in argon tungsten arc welding [20].

3. Analytical part

A number of scientific schools in the former USSR (in Kiev, Rostov, St-Petersburg, Moscow, Tomsk, Karaganda) grounded and developed this application of this method of low-frequency modulation in consumable-electrode welding. Thus I I Zaruba and co-authors [21, 22] first showed that current modulation can be an effective way of welding control reducing its cost. Further on F A Vagner, I N Vornovitskiy, F A Hromchenko, R I Dedyuh, V S Sidoruk, T G Shigaev and others [23-30] proved that in many cases current modulation can improve performance characteristics of welded joints and widen capacities of conventional welding and hardfacing.

The researchers mentioned above have determined that in manual arc metal welding under average current values changes in heat supply mode do not influence the effective efficiency of arc heating. However thermal efficiency of base metal penetration and weld dimensions in modulated current welding are higher than those in direct current welding. These values increase as the modulation depth grows (ratio of pulse current to pause current). It leads to increased arc pressure and stimulates its penetration into base metal. Conditions of transferring arc heat power to the base metal improve which provides effective melting of the metal [31, 32].

Works [33-35] prove that automatic control of the arc power by means of modulated current provides considerable stabilization of weld root pass formation which means more stable penetration as compared with continuous arc welding. This advantage of modulated current welding with forced destruction of ligament during the pause is stipulated by the more precise metering of energy supplied into the metal.

Noting a positive influence of low-frequency current on weld metal formation at different space positions it should be pointed out that majority of works in the field of development and application of pulse technologies deal with manual arc welding. However D A Dudko, V S Sidoruk, S A Zatserkovny [36-38] proved that the advantages of pulse technologies described above are valid in mechanized arc welding as well. At the same time low-frequency alternation of welding energy parameters in mechanized welding do not always eliminate the most significant drawbacks of arc welding – weld spattering and process instability especially in CO_2 [39].

To insure regular welding process under winter conditions it is important to take into account specifics of weld development at low temperatures. Low atmosphere temperature can have a negative influence not only on welding process stability but also on the structure and mechanical properties of welding joints. It is caused by quick cooling of the welding pool metal under winter conditions [40-41].

For some high-strength steels, changes in cooling rate caused by dropping temperature at low critical cooling rates makes it possible to get quenched structures in the heat-affected zone and quench cracks [42]. Due to that weld formation under extremely low temperatures requires special technologies. It becomes necessary to control the process of crystallization in weld development.
This approach becomes possible if thermal cycle control takes place at the stage of crystallization of weld metal from melt. It can be achieved by controlling two parameters: heating temperature and, what is more important, exposure time and proper transition of the weld metal from one structural condition to another according to "Iron – Carbon" diagram. Thus to obtain equilibrium, close to the base metal structure of the weld and heat affected zone it is necessary to maintain the required energy and time parameters in the power source –arc – product system so that to provide temperature ranges under which changes in structural and phase conditions go on smoothly [42].
Conventional pulse-arc technologies in the frequency range from 300 to 25hz which are now widely used in welding and assembly production are first of all aimed at controlling electrode metal transfer, increasing penetration depth and improving weld properties at all space positions. But to control structural and phase conditions in the weld metal one should control the crystallization time according to the cycles of phase condition diagram. It can be performed by means of pulse-arc welding in the range from 25 to 5hz. In Russia they are called "low frequency modulation of arc current" [5].

4. Experimental part

Controlling a crystallization temperature mode is possible only by means of arc current alternating in pulse/pause cycles with predetermined frequency and pulse/pause ratio. For example, there is a widely spread type of welding with low-frequency modulation of arc current during the pause performed by means of reducing welding current to the minimal arc current values to allow crystallization of the welding pool [18]. But this method is used in manual metal arc welding. In this case welding is carried out at flat current-voltage characteristic of the arc from power supply with falling external current-voltage characteristic which enables modulating welding current by means of the power supply values [43].
In metal active gas welding with constant electrode feed rate this method cannot be widely used because welding wire tends to dip into the welding pool [6]. It happens mainly when pauses are very long so that melting rate falls but welding wire feed rate remains the same. In this case modulation of voltage and current is possible only in a narrow range of welding energy parameters when at interpulse periods the arc length increases but not until breaking and at interpause periods it reduces but until dipping of the wire into the welding pool.
This approach provides smooth change of the pause current but does not enable reliable pulse/pause ratio control [43, 44].
To obtain a stable welding with low-frequency modulation of the welding current in metal active gas welding it is necessary to provide control of welding energy parameters when working from energy supply with external characteristic which meets the predetermined electrode feed rate at a given interval alongside with control of a corresponding automatic change of the indicated parameters when switching to another wire feed rate or even when it discontinues. Fig. 1 shows a cyclogram of the process described above.
To form a welding pool, welding is carried out with constant welding wire feed rate at rising current-voltage characteristic of the arc and flat external current-voltage characteristic of the power supply until welding pool of a predetermined volume is formed (period T1cycle "Pulse" Fig. 1 and 2). After that for the welding pool to crystallize, a pause time occurs, welding wire movement is stopped, arc burning is switched over to the flat current-voltage characteristic of the arc and external falling characteristic of the power supply (period T2 cycle "Pulse" Fig. 1 and 2).
Such technology involving simultaneous stoppage of welding wire allows controlling the pause time and increases arc current alternation range due to changing current-voltage characteristic of the power supply. The period $T=t_i + t_p$, where t_p is pause time, t_i is pulse time is of primary importance in the suggested type of welding.

The following conditions have to be satisfied:
1. pause time should be enough to enable crystallization of 75 % of the total amount;
2. pause time should provide continuous arc burning (when low-current arc is burning it must not break because of excessive electrode melting. It can be achieved through introduction of an arc current feedback channel and switch over to the falling current-voltage characteristic of the power supply.

5. Research material and methods

To assess the influence of the welding type under the study on the microstructure of welded joints we obtained macro- and microstructures by optical metallography method using optical microscope Olympus GX-71. The following parameters were involved: welding type – MAG, welding wire Sv08GSMT– 1.2 mm in diameter, power supply voltage – 22 V, pulse current – 220 A, pause current – 50 A, cycle frequency – 1.1 hz, pulse/pause ratio – 1.25, air temperature – 15 ˚C.

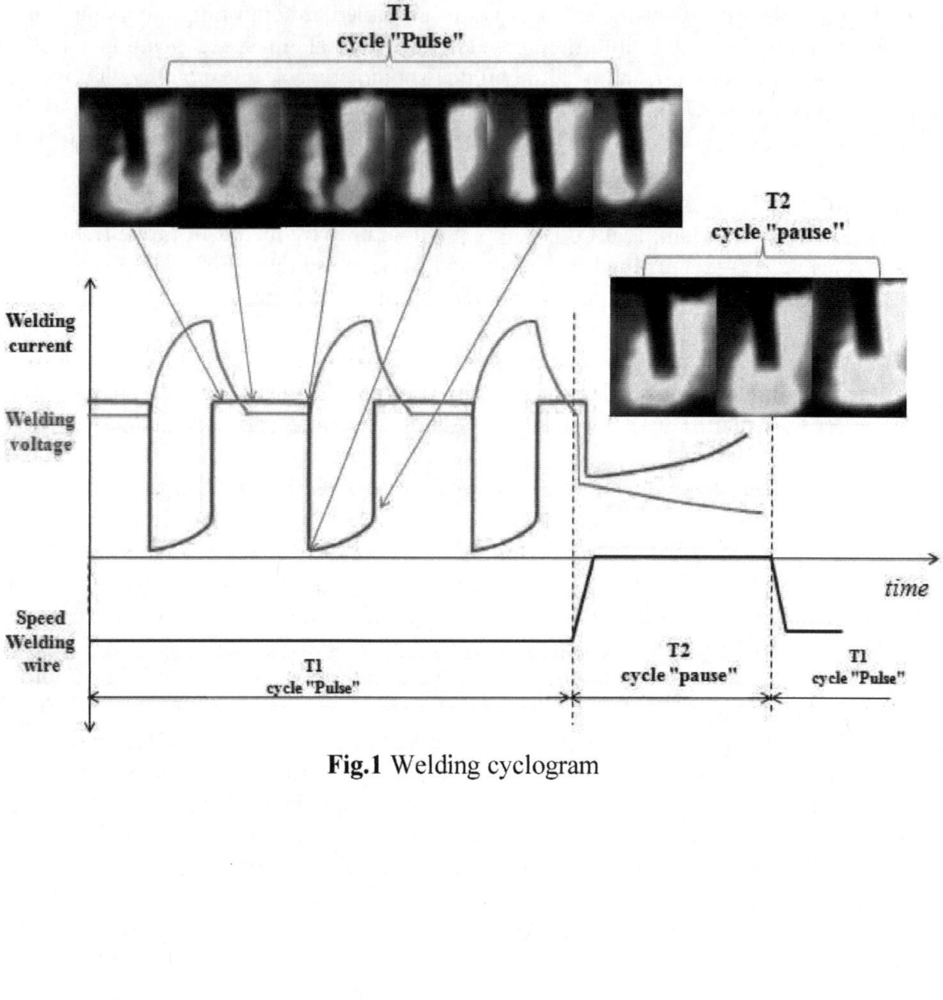

Fig.1 Welding cyclogram

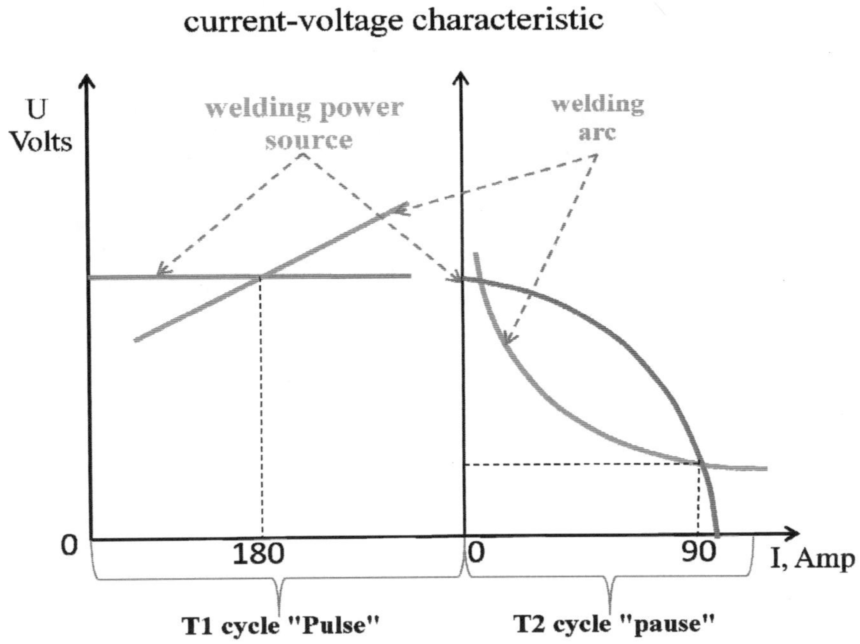

Fig 2. Interdependence of current-voltage characteristic of the arc and of the power supply at different intervals of welding process.

6. Research in microstructure of the weld on C17 samples of 09G2S steel (pipe 159x6).

Initially 09G2S steel has ferritic-pearlitic structure. Pearlite is distributed uniformly (Fig. 3a,b). Pearlite amount and disposition correspond to 1C category according to the National State Standard 5640-68.

Microstructure analysis of sample 1 (low frequency modulation current welding) shows that the heat- affected has ferritic-pearlitic structure. Ferrite grains are in sub-microcrystalline state, ferrite is becoming sorbitic (Fig. 3c). Pearlite amount and disposition correspond to 1B category according to the National State Standard 5640-68. Deposited weld metal structure is ferritic-pearlitic.

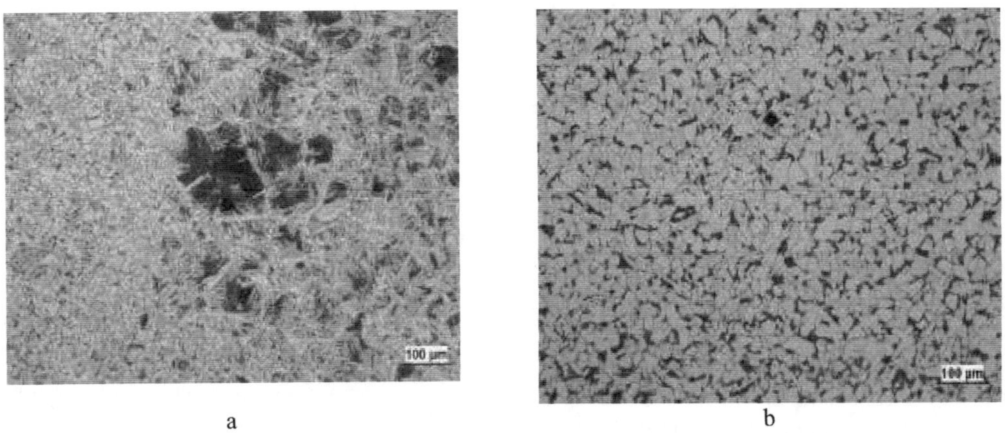

a b

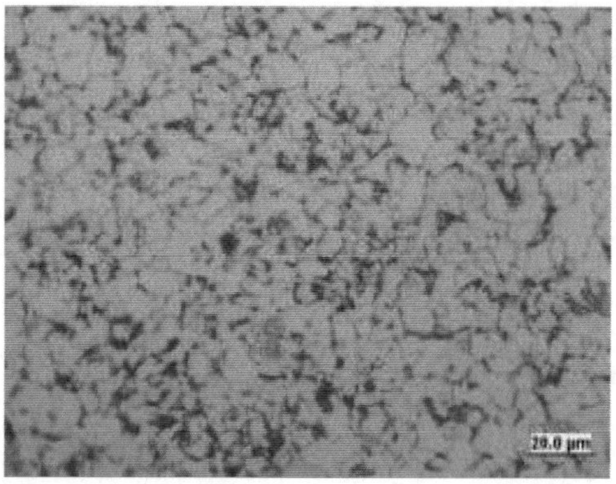

c

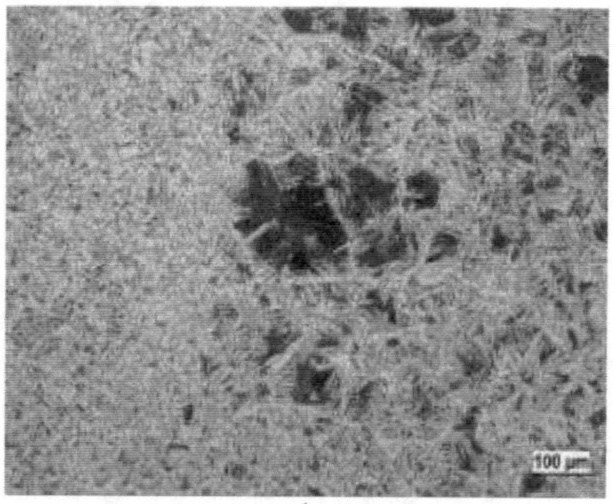

d

Fig. 3 – Weld microstructure: a,b – microstructure of 09G2S steel, c – microstructure of welded joint heat affected zone (low frequency modulation current welding); d – microstructure of welded joint heat affected zone (conventional feed of welding wire)

Sample №2. In the fusion zone of the weld and the base metal there is a small area with Widmanstatten pattern – 3B category according to the National State Standard 5640-68 Fig. 3d). The weld structure is identical to that of Sample 1.

As compared with the upper row, in the bottom row Widmanstatten pattern disappear from the weld and the base metal attached to the weld due to auto thermal treatment.

Differences resulting from the research can be explained by different characteristics of changing cooling temperature in the range of critical temperatures of phase transformations during secondary crystallization (in solid phase) [45].

7. Conclusion

1. Using falling current-voltage characteristic of the power supply during welding pool crystallization increases crystallization time which is advantageous for the welding process. Reduction of current to minimal values during crystallization provides conditions for thermal treatment of the weld.

2. Stabilization of welding pool formation and crystallization time improves the weld and increases labor productivity in welding of thin sheet metals.

3. The suggested technology of MAG welding takes into account time of welding pool formation and crystallization under different temperature conditions of structural-phase transformations which allows a wide-range control of crystallization processes both in liquid and solid phases.

The work was done as a part of Fundamental Scientific Research Program of State Academies of Science of the Russian Federation for 2013-2020.

References

[1] Saraev Yu N, Lebedev V A and Solodsky S A. The development and practical application of adaptive pulse-arc welding in the manufacturing and repair of metal structures responsible function // Applied Mechanics and Materials. Vol. 682 (2014). Pp. 332-338. Submitted: 07.07.2014. Online available since 2014/Oct/27 at www.scientific.net Accepted: 31.07.2014 © (2014) Trans Tech Publications, Switzerland, doi:10.4028/www.scientific.net/AMM.682.332.

[2] Lebedev V A, Maksimov S Yu, Zhernosekov A M and Saraev Yu N Controlling weld geometry by means of welding equipment in mechanized and automatic consumable electrode arc welding. Welding International, 2015. Vol. 29, No. 5, 379–385, http://dx.doi.org/10.1080/09507116.2014.934537.

[3] Paton B E Critical analysis of the current notions of arc self-regulation in consumable electrode welding / B E Paton, V S Sidoruk, S Y Maksimov and Yu Saraev // Engineering: network electronic scientific journal. - 2015 - Vol.3, №4. – P. 38-47.

[4] Chinahov D A Study of thermal cycle and cooling rate of steel 30ХГСА single-pass weld joints // Applied Mechanics and Materials. – Vols. 52-54. – 2011. – p. 442-447. - Mode of access: http://www.scientific.net/AMM.52-54.442.
[5] Saraev Yu N Pulse welding procedures and welding. Novosibirsk, Nauka, 1994, 108 p.
[6] Zhernosekov A M Trends in the development and management of processes of metal transfer in protective gases (Review) // Automatic welding. - 2012. - № 1. - p.33-38.
[7] Saraev Yu N Features of the development of electro-technological processes of welding and hardfacing on the basis of algorithms of adaptive pulse control mode energy parameters // Engineering: network electronic scientific journal. - 2015 - Vol.3, №3. - P. 3-11.
[8] Wagner R A Equipment and pulsed arc welding methods. M: Energy, 1980. 120 p.
[9] Lebedev V A and Saraev Yu N Pulsed impact of mechanized and automatic welding equipment on welding processes and welding systems // Welding production. - 2015. - № 5. - P. 34-44.
[10] Loos A V, Lukutin A V and Saraev Yu N Power supplies for pulsed electro-technological processes. - Publishing and printing company of TPU. - Tomsk, 1998 - 159 p.
[11] Paton B E, Lebedev V A, Pichak B G and Poloskov S I Evolution of pulse welding wire feed for welding and surhardfacing. Welding and diagnostics. 2009. № 3. P. 46-50.
[12] Lebedev V A Tendencies in mechanized welding with a controlled transfer of electrode metal (Review) // Automatic welding. - 2010. - № 10. - P.45- 53.
[13] Lankin Yu N Automatic control of the mechanized welding process with controlled electrode metal transfer (Review) // Automatic welding. - 2007. - № 1. - P. 3-10.
[14] Bondarenko V L Arc welding with pulsed wire feed - SMT process proposed by "Fronius" company // Automatic welding. - 2004. - № 12. - P. 55 - 58.
[15] Saraev Yu N Evaluation of the productivity and quality of repair and reconditioning operations in equipment made of 12CR18N10T1 type steel // Saraev Y, Selivanov Y. – Welding International, 2012, T.26, p.p. 466-470.
[16] Paton B E, Maksimov S Yu, Sidoruk V S and Saraev Yu N On the issue of arc self-regulation in consumable electrode welding // Welding production. - 2014. - № 12. - P.3 -11.
[17] Saraev Yu N Experience from the development and practical application of adaptive pulse-arc welding equipment for construction and repair of main pipelines // Problems of Materials Science. - 2015. - № 1. - P. 210-218.
[18] Saraev Yu N The development and application of adaptive pulse-arc welding methods for construction and repair of pipelines. Group Author of books: IEEE Conference: International Conference on Mechanical Engineering, Automation and Control Systems (MEACS). Locality: Tomsk, RUSSIA, published: OCT. 16-18, 2014 Sponsors: Inst Elect Engineers; Russian Fdn. Basic Res; Tomsk Polytechnic University Tomsk IEEE Chapter; Tomsk Polytechnic university, Institute of Physics high-Tech 2014, Pp. 1-5. Print ISBN: 978-1-4799-6220-4. DOI: 10.1109 / MEACS.2014.6986956. Published: 2015.
[19] Zaitsev M P and Kuleshov O M Means of automation in equipment for arc welding // L: Energy, 1975. – 144 p.
[20] Petrov A V and Slavin G A Research of technological capabilities of pulsed arc // Welding Production. 1966. № 2. P. 1 - 4.
[21] Zaruba I I, Lebedev V K, Sheyko P P et al. Arc welding with modulated current // Automatic welding, 1968. - № 5. P. 70.
[22] Zaruba I I, Lebedev V K, Sheyko P P et al. Welding modulated current // Automatic welding, 1968. - № 11. P. 35 - 40.
[23] Wagner F A Manual arc welding with pulsed arc // Welding production // 1970. - № 3. - P. 21 - 22.
[24] Wagner F A Cyclic changes in the parameters of welding process and their impact on the gas saturation of welds // Power plant construction, 1981. - № 4. - P. 70 - 72
[25] Vornovitskiy I N and Trusov A G Optimization of modes in manual pulsed arc welding by the level of gases in welds // Power plant construction, 1978. - № 4. - P. 51 - 52.

[26] Hromchenko F A, Anokhov A E and Alekhova I A Structure and properties of welded joints made by stationary and pulsed arc // Welding production, 1980. - № 6. - P. 21 - 23.
[27] Mazel A G, Dedyuh R I and Knyazkov A F The arc stability in manual arc welding with modulated current // Impact of welding parameters of modulated current on the process of electrode melting and the depth of penetration in manual arc welding // Welding Production, 1976. - № 4. - P. 9 - 10.
[28] Dudko D A, Sidoruk V S, Zatserkovny S A, Taraborkin L A and Makhlin N M Dependence of the chemical composition of the weld parameters from the manual arc welding the modulated current. - Automatic welding. - 1989. - № 2. - p.27 - 29.
[29] Shigaev T G On terminology in modulated current welding // Welding Production.- 1980.- №7.- P.40.
[30] Mazel A G and Dedyuh R I On the stability of manual arc welding with modulated current // Welding Production, 1978. - № 12. - P. 11 - 13.
[31] Wagner F A Equipment and methods for pulsed arc welding M.: Energia, 1980. - 120 p.
[32] A method for adaptive pulse arc welding. - Invention patent number 2410216 (Russia) / Saraev Yu N - Priority on June 16, 2008
[33] Saraev Yu N, Bezborodov V P, Grigorieva A A, Golikov N I, Dmitriev V V and Sannikov I I Distribution of residual stresses in welded joints in 09G2S steel produced by adaptive pulsed-arc welding. Welding International, 2015. Vol. 29, No. 2, 131–134, http://dx.doi.org/10.1080/09507116.2014.897804.
[34] Saraev Yu N, Gladkovsky S V, Golikov N I and Veselova V E Increase of operational properties of steel structures by methods of adaptive pulse-arc welding for use in low temperature climatic // Welding Production. - 2015. - №11. - P 33-41.
[35] Dudko D A, Lebedev V V, Moshkin V F and Pichak V G Stability of arc welding processes and surfacing with pulse feed of welding wire // Welding Production - 2000. - № 1. - P. 12 - 15.
[36] Sidoruk V C, Dudko D A, Zatserkovny S A and Taraborkin L A Influence of the manual arc welding with modulated current on the primary structure of the weld metal. - Welding Production. - 1988. - № 5. - P.1-3.
[37] Zatserkovny S A Improving the process of manual arc and mechanized welding in carbon dioxide, based on the low-frequency modulation current. - Abstract of the thesis for the degree of candidate of technical sciences. - Kyiv, 1989. - 16 p.
[38] Potapyevsky A G, Saraev Yu N and Chinakhov D A Welding of steel with consumable electrode in shielding gases. Technique and technology of the future: monograph / Potapyevsky A G, Saraev Yu N and Chinakhov D A; Yurga Institute of Technology. - Tomsk: Publishing House of Tomsk Polytechnic University, 2012. - 208 P. ISBN 978-5-98298-620-7.
[39] Larionov V P, Sleptsov O E, Saraev Yu N and Bezborodov V P New approaches to the development of modern technologies of welding and coating to ensure the operational reliability of steel structures and products, operating in conditions of Siberia and the Far North // Bulletin of the Academy of Military Sciences. 2008. № 3. - P. 67-69.
[40] Saraev Yu N Experience in the development and practical application of adaptive pulse-arc welding equipment for construction and repair of main pipelines // Problems of Materials Science. - 2015. - № 1. - P. 210-219.
[41] Chinakhov D A and Agrenich E P Computer simulation of thermo-mechanical processes at fusion welding of alloyed steels // Materials Science Forum. – Vols. 575-578 (2008). – Pp. 833-836.
[42] Patent for invention number 2.293.630 (RF). The method of mechanized welding in CO_2 with low-frequency modulation of the weld pool. Fedko V T, Brunov O G, Solodsky S A, Kryukov A V and Sednev V V The priority of invention July 14, 2005 V23K 9/33. Published on 10.04.2007. Bul.№10
[43] Solodskii S A, Brunov O G, Zelenkovskii A A Automated system for controlling the process of CO_2 welding with pulsed wire feed and welding current modulation. Welding International. 2012. T. 26. № 4. p. 314 – 317.

IOP Conf. Series: Materials Science and Engineering **127** (2016) 012019 doi:10.1088/1757-899X/127/1/012019

[44] Brunov O G, Fedko V T and Solodskii S A Transfer of electrode metal in welding with the pulsed feed of welding wire. Welding International. 2007. T. 21. № 1. C. 50–54.

A Concept of Constructing a Common Information Space for High Tech Programs Using Information Analytical Systems

Alexandra A Zakharova[1,a], **Olga A Kolegova**[1,b] **and Maria E Nekrasova**[1,a]

Yurga Institute of Technology, National Research Tomsk Polytechnic University Affiliate, ul. Leningradskaya 26, Yurga, Russia

e-mail: [a]aaz@tpu.ru, [b]Olga030188@mail.ru

Abstract. The paper deals with the issues in program management used for engineering innovative products. The existing project management tools were analyzed. The aim is to develop a decision support system that takes into account the features of program management used for high-tech products: research intensity, a high level of technical risks, unpredictable results due to the impact of various external factors, availability of several implementing agencies. The need for involving experts and using intelligent techniques for information processing is demonstrated. A conceptual model of common information space to support communication between members of the collaboration on high-tech programs has been developed. The structure and objectives of the information analysis system "Geokhod" were formulated with the purpose to implement the conceptual model of common information space in the program "Development and production of new class mining equipment – "Geokhod".

1. Introduction

Russia's economic transition to an innovation-driven model of economic growth has created all the prerequisites for the development of high technology industries in the economy. According to the experience of more economically developed countries, the development of innovative sectors of the economy is the driving force for social and economic progress and creates competition in economy. Innovative activity of high-tech industrial companies enables a steady growth of the economic system through the widespread introduction of leading science-based technologies and the development of high technology products with high intellectual labor costs.

The innovative activity of high-tech companies is focused on the development of a product innovation that has fundamentally new or improved performance characteristics. As a rule, the development of high-tech products is based on the latest achievements in science and technology. The process of designing and developing high-tech products is accompanied by a series of scientific researches and development projects (R&D).

The process of developing high-tech products comprises a set of projects which are part of high technology program.

A high technology program is made as a set of interrelated projects forming a complex of business processes serving to ensure the effective solution of scientific and technical tasks within the specified period of time and in compliance with limited resources.

High-tech program management is a system of tools, techniques, methodologies, resources and procedures used for the coordinated management of interrelated projects intended to achieve collaborative strategic goals formulated in the program that requires the use of modern software to ensure its proper and effective implementation [2], [3].

Management of high-tech programs is complicated by a number of problems.

First, the innovation-related activity is associated with a risk to a greater extent than any other activities [4]. This can be explained by the fact that high-tech programs are implemented in the field of complex technical systems, and a high-technology manufacture is associated with certain difficulties, such as technical complexity to develop and manufacture products, lots of document turnover, limited resources, constant monitoring from the side of sector ministries, difficulty to set a deadline for implementing innovative projects because of the influence of external factors. Such projects are not governed by any well structured design procedures because they are determined as knowledge-intensive and characterized by high technical risks and unpredictable results in view of the influence of various external factors [5], [6]. While planning and implementing high-tech programs, there is a need to solve a lot of ill-defined problems that require expert and intelligent information processing methods [7], [8], [9].

Secondly, program management is also complicated by participation of several members involved in project activities, including those performing work remotely, and this makes the process of effective management difficult and leads to communication barriers. The lack of effective communication between participants in the program may contribute to a dramatic delay in the time taken to implement the program, declining quality and additional costs. In this regard, creation of a common information space among all project participants is the priority for successful implementation of a high-tech program.

Consequently, there is an urgent need to design an information analysis system capable of solving ill-defined problems relative to management of complex programs, improving the effective communication between program participants and the qualitative implementation of the whole program.

2. Survey of the existing project management systems

Project Portfolio Management (PPM) systems are most commonly used for solving problems. Currently, according to analytical studies, the market of project portfolio management systems is sufficiently stable and mature. On the market there are a huge number of PPM packages, ranging from simple tools that can aid you in project managing within a small company to powerful systems used to effectively manage several related projects, planning, controlling and implementing projects, programs and portfolios.

Analyses made by Gartner, one of the world's leading information technology research and advisory company, providing information technology related insights, allow us to explore the latest trends on the software market in the field of project management. According to the researches of many years delivered by this research company, for 10 years the major players in the market of project management information systems are the following PPM packages:

– CA Clarity PPM is the advanced PPM software solution that includes a set of software applications providing a base for comprehensive management of corporate IT resources through the building of a unified information analysis system for strategic planning and IT services management.

– HP Project and Portfolio Management is the flexible integrated software for project and portfolio management solutions, enabling effective control of resources and finances within a company, to ensure the timely implementation of projects and prioritize the investment s in IT.

– Oracle Primavera Portfolio Management is one of the best and most reliable PPM software solutions, providing the process improvement of strategic program, project and portfolios management with its unique functionality, scalability and unlimited configurability. Powerful security features and a clean user interface make this program become the best product among similar PPM solutions for communication between all project team members using a common information sharing space [10].

– Microsoft Enterprise Project Management (EPM) is an effective toolset, providing the Enterprise Project Management technology that is engineered in accordance with international standards in the field of project management to support all project management functions: project implementation planning and monitoring, resources management and reporting activities.

In the international PPM software market, in addition to the leading project and portfolio management solutions, there are a lot of PPM environments such as Planview, Innotas, Wrike and Basecamp, and others.

Currently, there is no reliable statistics on the percentage of using PPM–based solutions by Russia's companies, and among the leaders providing software products and applications for projects and portfolios management on the Russia's market are the same leading project management system as those on the world market:

It should be noted that the Microsoft Enterprise Project Management solution is the most popular among Russia's customers today as an effective integrated environment for project management, program and portfolio management, management planning, investment portfolio management and innovation management.

According to the analysis compiled by the TAdviser Russian analytical agency on project management systems, Russia's developments such as SpiderProject, Proektnaya Integratsiya (upravlenie proektom), 1C: PM Project Management, including cloud–based services, for instance, Advanta on-line project management system, and others are in operation in addition to leading foreign project portfolio management systems on the PPM market over the past few years.

Any of the PPM – based solutions is a flexible and useful tool, which helps a company manage projects and programs in an easier and more effective way [11], [12].

All above management systems are the state-of-the art solutions for project portfolio management, and are designed for different types of companies depending on their size and objectives. Big companies prefer the powerful PPM software, where a senior level board clearly understands the need of providing support to today's business processes through the use of information systems of management. The Microsoft Enterprise Project Management technology is the most optimal among the existing solutions in terms of their functionality and price level.

The survey of the most popular information systems used for project management has shown that, to ensure comprehensive professional management within the existing PPM – related systems, the basic processes are computerized: workflow management, lifecycle management (LCA), project scheduling and resource planning, budget planning, monitoring of project work implementation with creating Gantt charts, histograms and graphs to show activities required, linear and on-line modeling and planning of projects and programs, risk management, etc.

However, despite such a variety of project management tools, software products are insufficient to automatically support the process of solving ill-defined problems relating to high-tech program management, which appear in an environment of high technical risk levels and uncertainties [13]. In engineering business processes, when the PPM software systems are used, many factors and conditions are not taken into consideration. Such factors and conditions are caused by uncertain external environments, in the long run resulting in an incorrect display of real world environment, where processes of project and program management are implemented. In this connection, it is necessary to use modeling tools for forecasting and analytical control of high-tech program management in addition to the PPM – based systems.

3. Rationale for the use of information and analytical support for high-tech program management

The process of implementing the program on developing a multi-purpose shield-type tunneling machine "Geokhod" has proved that constructing a common information space is critical. The project works are performed within the framework of RF government program "The development of science and technology" [14], [15]. This large-scale program is of a complex hierarchical structure with involving a significant number of program participants and is planned to be implemented in a 5-8 year period.

In order to set up a common information space, it is proposed to develop an information analytical system to manage the program (IAS PKMiSP "Geokhod").

For designing main modules of the information analytical system, the following tasks are identified to perform:

- systematic analysis of such business processes as planning, coordination, monitoring and program implementation support;
- engineering of knowledge models in the field of mining equipment - geokhods and the design and implementation of technology for using geokhods in underground workings;
- development of a distributed database for use in IAS PKMiSP "Geokhod";
- engineering of information processing technology to support the project, which provides collaboration of project members in the common information environment;
- development of a conceptual model of decision support technology for planning, coordination, monitoring and program implementation support;
- development of methodologies and mathematical models for planning the project implementation program;
- development of methodologies and mathematical models for program coordination;
- development of methodologies and mathematical models for use in program monitoring.

A structure of the designed information analytical system PKMiSP "Geokhod" should have several sub-systems to ensure the use of information technology tools and informational support for high tech product program management at industrial companies (see Figure 1). On this figure there is a hierarchical data model showing the multilevel structure in program management. Each level presents a software package that provides program management from the lowest level of product database management to the highest level of strategic management; all systems must be integrated with each other.

IAS PKMiSP "Geokhod" provides the tools for use in decision support management system through all stages of high-tech program: planning, coordination, monitoring and program support.

The basic modules of IAS PKMiSP "Geokhod" are as follows:

1. PLM software is an information system used to manage all aspects of the product development lifecycle. The main component of the PLM system is a PDM system that manages all information about the product. It provides the operational level of program management.
2. Program management information system PMIS is an integrated complete solution with a set of technical, software and methodological tools to support the planning process and improve the effective project and portfolio management. It enables the tactical and partially strategic levels of project management.
3. Information system to support strategic high-tech program management ISSHPM is a set of software / hardware tools and technical solutions to support the decision making process in a complex information space in program management.

The system has several modules:

- The module "Strategic Analysis" is designed for analyzing external and internal environments of high-tech programs. When analyzing the external environment, tools are used to analyze external factors that affect the successful implementation of the program in order to anticipate potential threats, develop plans in the event of unforeseen circumstances. Analysis of the project internal space is carried out so that to identify the strengths and weaknesses in high-tech product management. The resulting analysis in the output module presents environmental factors, distributed in order of their importance in high-tech program strategy management.
- The module "Strategic Choice" is designed to evaluate development strategy options for high-tech programs. Particular projects, resources, suppliers and other are among alternatives. Such ill-defined tasks are solved as the selection of the most optimal resource suppliers and investors as well as resources required to perform the work on projects in

programs, the predictive modeling of timing and cost parameters relating the program on the basis of relevant information available with regard to successful projects, the selection of the most feasible alternative of program models, and others.

- The module "Implementation Control" is designed to collectively evaluate the strategy implementation through the program according to criteria developed during the analyzing process. The inspection results are then used for a new cycle of strategic management.
- The module "Expert Evaluation" is used for performing procedures intended to select experts, evaluate and assess the extent of their consistency in the field of high-tech product management.

ISSHPM provides the strategic level of program management.

The ISSHPM module refers to Decision Support System, and this clearly determines its use as a toolset to support decision-making in planning and program management, decision making techniques in a context of uncertainty, methods of expert assessments [16].

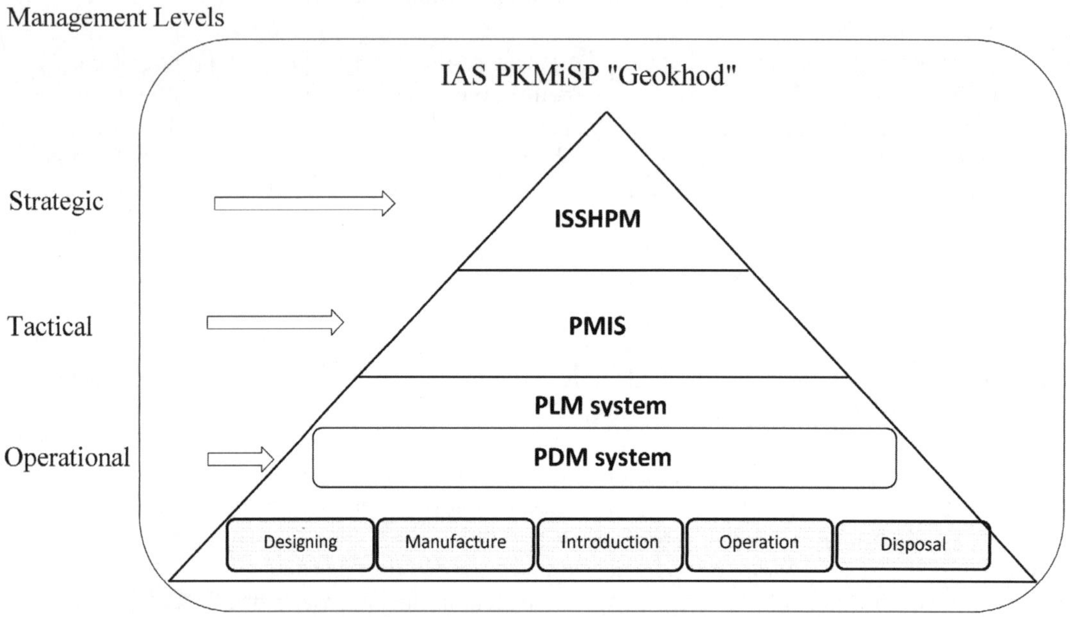

Figure 1. Information Technology for high-tech product program management

IAS PKMiSP "Geokhod" is a toolset designed to support the decision making process in program management, provide all participants in the project with relevant information, construct a common space for sharing information between program members and improve performance of program works through the use of tools for planning, coordination, monitoring and program implementation support in the information analytical system.

4. A conceptual model of a common information space for collaboration between program participants

While implementing large – scale innovative product programs, a great number of stakeholders are usually involved in from both organizations and enterprises, directly engaged in the design and development of high-tech products, and external organizations interested in the project implementation: investors, customers, suppliers, contractors, consultants, and representatives of public organizations.

Across the program organization structure, the following key positions are classified in a hierarchy of authority and responsibilities:

1. A Program General Manager is a person responsible for controlling that project strategies in the program are in line with global business strategies of the program.
2. A Project Portfolio Management Group is a key element in the hierarchy of program management; is responsible for balancing the conflicting opportunities and making the decision on setting priorities, choosing programs and allocating resources to projects within the program, conditioned upon the need of different points of view.
3. A Project Manager is a person responsible for the planning, direction and control of the project during all phases of its project life cycle.

When a large number of participants are involved in the program, the issue of establishing an effective collaboration among them is critical by creating a common integrated information environment.

The common information space must meet the following requirements:

1. Retention of technical and project documentation data in electronic form in order to make an easier access to those data for any participant in the project in accordance with the approved level of access.
2. Formation of a unified database of high-tech products, the filling of which is available at each stage of the program by any participant in project activities.
3. Providing access to the information system via a corporate portal to each project member in accordance with the seniority of staff, experience and competence at any phase of program life cycle. The corporate access, granted to all participants, to the information analytical system is achieved using Portal Technology. Using the portal technology allows the collaboration among participants working remotely or in various organizations, in order to overcome a lack of workplace integrity.
4. Software and hardware integration should be governed by international and relevant industrial standards.

The application of project communication software will greatly simplify the processes of collection, storage and exchange of information among the participants in the program; reduce the time for program managers to obtain the necessary information on the current status of projects and decision-making.

Project Information Space for use in high-tech products manufacture is a specifically structured and interconnected set of computation, information and communication resources and technologies, which operate on the basis of unified principles and techniques to ensure information communication security among all participants, to meet the information demand of project members in accordance with their data assess level.

Figure 2 displays the model of a common information space for communication between the participants in high-tech product programs.

The basic components in the information analytical system are given in the inner circle of the picture to indicate the planning, coordination, monitoring and program implementation support processes (IAS PKMiSP):

1. KB – Knowledge Base in the domain of mining equipment - geokhods.
2. DDB – Distributed Database where all the information about the product, technical and project documentation is stored
3. DDBMS – Distributed Database Management System for IAS PKMiSP
4. DSS – Decision Support System in the program strategy management in a complex IT environment.

The next contour indicates the corporate portal, which serves as a tool for members of the collaboration to communicate with each other. The third contour displays the basic functions of information analytical system:

1. Discussion and approval of documentation in a collaborative manner by the program participants. Setting up a unified system for informing on the works in progress relative to the projects.
2. Automate workflow process at all phases of the program lifecycle.

3. Scheduling, resources and budget planning.
4. Report generation on the results of the project activities at each phase of the program lifecycle.
5. Granting the program participants access to common databases using the secured corporate Extranet
6. Providing a framework for strategic planning of business processes of the program

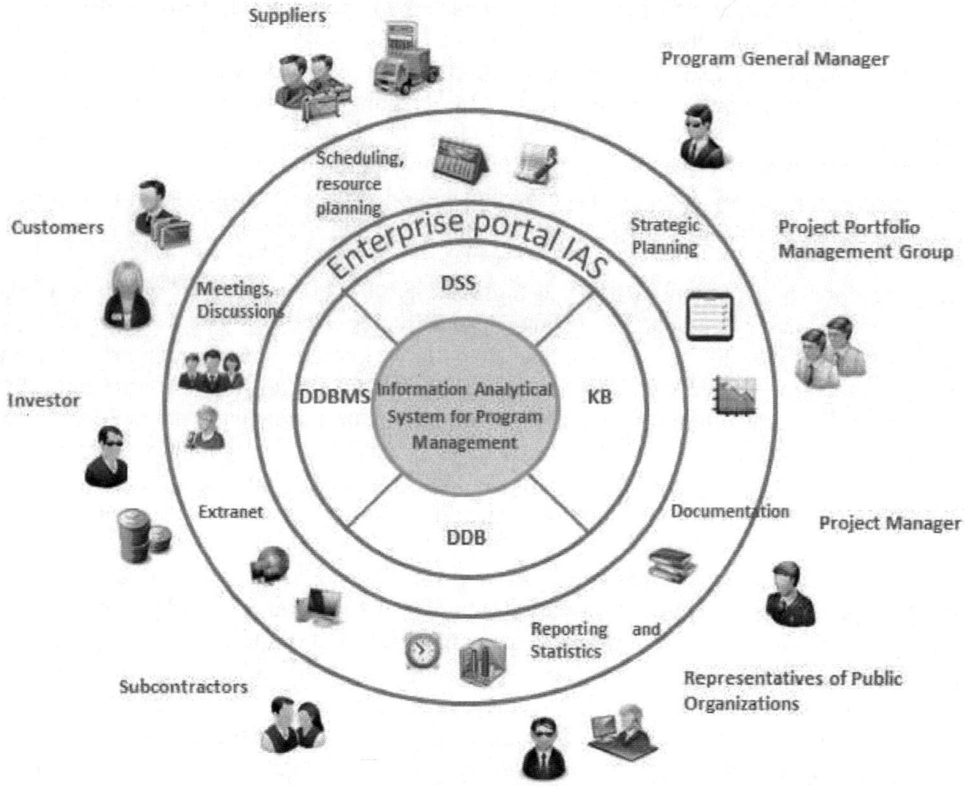

Figure 2. The conceptual model of a common information space for communication between participants in high-tech program

The outer contour displays all information system users.

Conclusion

The proposed concept of the common information space will ensure the improvement of quality of business processes management of high-tech programs and enhance the efficiency of interaction between the participants in the program.

The main advantage of the information analytical system "Geokhod" is using the module to support decision making in program management, which allows you to

– provide senior staff members with effective program management tools for planning, coordinating, monitoring and implementing the program, with giving a visual representation of business processes taking place in the program, as well as provide information required for making decisions on strategic management;

– improve the efficiency of management and decision-making on the program, and enhance program participants' responsibility.

By using information technology, all program participants are granted corporate access to common databases at any phase of the program lifecycle in accordance with the approved levels of access.

Thus, conceptualizing the common information space, based on the engineering of information analytical system used to manage a high-tech program, will ensure data integrity and increase a data access rate for remote participants in the project, enabling decision making to be effective in a context of uncertainty under the external environment.

Acknowledgments
The reported study was funded by RFBR according to the research project № 16-07-00299 a.

References

[1] Sweetman R, O'Dwyer O and Conboy K 2014 Control in software project portfolios: A complex adaptive systems approach, *Springer International Publishing Switzerland* 199 pp 93-104.
[2] O'Dwyer, Sweetman R and Conboy K 2014 Exploring control tensions in is project portfolio management, *35th Int. Conf. on Information Systems "Building a Better World Through Information Systems", ICIS 2014*, Code 110383.
[3] Beringer C, Jonas D and Kock A 2013 Behavior of internal stakeholders in project portfolio management and its impact on success, *International Journal of Project Management* 31 (6) pp 830-846.
[4] Tceplit A, Grigoreva A and Osipov Y 2014 Developing the Model for Assessing the Competitiveness of Innovative Engineering Products, *Applied Mechanics and Materials* Volume 682 pp 623-630.
[5] Korhonen T, Laine T and Martinsuo M 2014 Management control of project portfolio uncertainty: A managerial role perspective, *Project Management Journal* 45(1) pp 21-37.
[6] Kopmann J, Kock A, Killen C P and Gemunden H G 2015 Business Case Control in Project Portfolios An Empirical Investigation of Performance Performance Consequencesand Moderating Effects, *IEEE Transactions on Engineering Management* 62(4) pp 529-543.
[7] Mohagheghi V, Mousavi S M and Vahdani B 2015 New Optimization Model for Project Portfolio Selection Under Interval-Valued Fuzzy Environment, *ARABIAN JOURNAL FOR SCIENCE AND ENGINEERING* 40(11) pp 3351-3361.
[8] Lenfle S 2008 Exploration and project management, *International Journal of Project Management* 26(5) pp 469-478.
[9] Telipenko E V and Zakharova A A 2014 Bankruptcy Risk Management of a Machine Builder, *Applied Mechanics and Materials* 682 pp 617-622.
[10] Oracle Primavera Enterprise Project Portfolio Management // [Digital resource]. – 2015. – URL: http://www.tadviser.ru/index.php/%D0%9F%D1%80%D0%BE%D0%B4%D1%83%D0%BA%D1%82:Oracle_Primavera_P6_Enterprise_Project_Portfolio_Management_EPPM (accessed date– 20.10.2015).
[11] Blichfeldt B S and Eskerod P 2008 Project portfolio management -There's more to it than what management enacts, *International Journal of Project Management* 26 (40) pp 357-365/
[12] Hornstein H A 2015 Relationships between a project management methodology and project success in different project governance contexts, *International Journal of Project Management* 33 (2) pp 291-298.
[13] Rodríguez A, Ortega F and Concepción R 2016 A method for the evaluation of risk in IT projects, *Expert Systems with Applications* 45 pp 273-285.
[14] Efremenkov A B, Aksenov V V and Blaschuk M Y 2012 Force parameters of geohod transmission with hydraulic drive in various movement phases, *Proceedings – 2012 7th International Forum on Strategic Technology, IFOST 2012* doi:10.1109/IFOST.2012.6357716.

IOP Conf. Series: Materials Science and Engineering **127** (2016) 012020 doi:10.1088/1757-899X/127/1/012020

[15] Efremenkov A B 2011 Forming the subterranean space by means of a new tool (geohod), *Proceedings of the 6th International Forum on Strategic Technology, IFOST 2011* 1 pp 348-350.
[16] Zakharova A and Ostanin V V 2015 Formalization model of expert knowledge about a technical index level of engineering products. Paper presented at the IOP Conf. Series: *Materials Science and Engineering* 91(1) doi:10.1088/1757-899X/91/1/012070.

Radon Emission from Coal Mines of Kuzbass Region

V A Portola[1], E S Torosyan[2], V K Antufeyev

Yurga Institute of Tehnology (branch) of Tomsk Polytechnic University
652055 Russia, Yurga, ul. Dostoevskogo, 6

E -mail: [1]portola2@yandex.ru, [2]eltorosjan@mail.ru

Abstract. The article represents the results of a research in radionuclides concentration in coal and rocks of Kuzbass mines as well as radon concentration in operative mines and mined-out spaces. It is proved that radon concentration in mines is considerably higher than in the atmosphere and it rises drastically in the mined-out spaces. It is found out that radon is carried out from mines by ventilation flows and from open pits, generating anomalous concentrations over self-ignition areas.

1. Introduction

Coal mining causes carry-over of hazardous gases released from coal and enclosing rocks into the atmosphere. A great part of the gases released from mines is dangerous due to their toxicity and fire behavior and explosiveness at a certain concentration in the air. Most inflammable gases have been generated in the process of coal rock metamorphism. When coal beds are developed extra gases are released due to mechanical destruction of coal and its low-temperature oxidation. The rate of inflow of different gases into the mine atmosphere can vary considerably through time and space depending on geological and technological conditions.

Radon takes a special place among gases released in mines and then carried over into the atmosphere. Radon is a colorless, odorless inert gas with density 9.9 kg/m^3 and boiling temperature -62 °C. Unlike conventional mine gases radon-222 belongs to a radioactive uranium-radium family and intermediate product of uranium-238 decay. Radon is produced from radium-226 with half-life as long as 1.600 years. The inert gas radon-222 which is produced at its α-decay has half-life as long as 91.68 hours. Annual average concentration of radon in the outside air is from 0.1 to 10 Bq/m^3. Natural radon radiation background in the air outside premises is about 610Bq/m^3 annually in average. Radon decay generates α-particles with total energy 5.482 MeV and γ-radiation with total energy 0.51 MeV [2, 3].

Concentration of radionuclides in soil and rocks depends on their initial activity as well as on mass transfer process in the earth interior caused by underground water movement and tectonic activity. That is why the amount of natural radionuclides in rocks can differ tenfold or even hundredfold. For example, specific activity of uranium -238 in coal vary from 3 to 52010 Bq/kg and its amount in soil is 10-50 Bq/kg [1, 5]. When coal is burnt its organic components turn into gases and volatilize while non-organic components and natural radionuclides remain. Thus concentration of natural radionuclides in coal ash can considerably grow compared with initial coal [2].

Radon generated by radium-226 decay in unbroken coal layers and enclosing rocks moves insignificantly in pores and produces new elements during further decay. Actual mining considerably alters the radium behavior. Due to large surface of coal and rocks in mined-out space some part of radon molecules produced from decayed radium-226 is carried out from inside coal and rocks to the

IOP Conf. Series: Materials Science and Engineering **127** (2016) 012021 doi:10.1088/1757-899X/127/1/012021

surface. Some amount of radon is accumulated on the surface of coal and rocks while the other part is released into the mine atmosphere [6]. Radon is carried out by escaping air into operating mine openings and into the atmosphere. Being inhaled decayed radon is harmful for living things. Thus it is important to determine the amount of radon in the mine air and develop ways to carry it out into the atmosphere.

2. Results and discussion

The content of radionuclides in coal and rock of some Kuzbass mines was studied by means of gamma-spectrometric analysis of collected samples. Gamma activity of samples determined the content of the four radionuclides: caesium-137, potassium-40, radium-226, torium-232. The research showed that the content of natural radionuclides in Kuzbass coal can change 2–4 times. Average volume activity of radium in coal samples is 9.3 Bq/kg. the most radioactive coal is found in the south and north of the region. Radium content in these coals is as great as 20 Bq/kg while in coals of Kemerovo and Leninsk-Kuznetskiy it is not higher than 10 Bq/kg. Experiments showed that the total concentration of radionuclides in enclosing rocks from five to ten times higher than that in coals. Radium activity can reach 50 Bq/kg and potassium-40 is sometimes as active as 1000 Bq/kg. Measurements showed that within one coal field, mine or even coal bed there can be detected considerable changes in concentration of radioactive elements in coal and rocks.

Volume activity of radon in mine atmosphere was determined by means of adsorption method, activated carbon being used as an adsorber. This method is essential when a large number of samples is collected and when low values of radon volume activity are measured. The method has many advantages: efficiency, high sensitivity, portable equipment, availability in any atmosphere. During the research radon concentration in productive mines and mined-out space was determined. Isolated mined-out space was monitored through wells drilled down from the surface or through productive mine faces and also through wall stopings.

Table 1 shows measurements of radon volume activity in mine atmosphere of some productive mine faces of Kuzbass region.

Table 1. Radon volume activity in mine atmosphere

No	Mine, bed	Sampling site	Volume activity, Bq/kg
1	"Pervomayskaya", bed XXYII	Belt road 724	54
2	"Pervomayskaya", bed XXYII	Air roadway	125
3	"Pervomayskaya", bed XXI	Belt road	62
4	"Pervomayskaya", bed XXI	Air roadway	83
5	"Raspadskaya", bed 6-6a	Belt road 5-6-10	61
6	"Raspadskaya", bed 6-6a	Air roadway 3-3-11	82
7	"Raspadskaya", bed 3-3a	Belt road 3-3-11	16
8	"Raspadskaya", bed 3-3a	Air roadway 3-3-11	93
9	"Raspadskaya", bed 5	Belt road 5-5-5	15
10	"Raspadskaya", bed 5	Air roadway 5-5-5	92
11	" Gramoteinskaya ", bed Sychoevskiy 1	Belt road 526	23
12	" Gramoteinskaya ", bed Sychoevskiy 1	Air roadway 526	89
13	"Kolmogorovskaya", bed Baikaimskiy	Belt road 1304	175
14	"Kolmogorovskaya", bed Baikaimskiy	Air roadway 1304	288
15	"Usinskaya",bed III	Belt road 3-3-15	95
16	"Usinskaya",bed III	Air roadway 3-3-15	108
17	"Yagunovskaya", bed Volkovskiy	Belt road 222	82
18	"Yagunovskaya", bed Volkovskiy	Air roadway 222	91
19	"Baidaevskaya" bed 29 a	Belt road 29-31	48

No	Mine, bed	Sampling site	Volume activity, Bq/kg
20	"Baidaevskaya" bed 29 a	Air roadway 29-31	76
21	Named by V.I.Lenin, bed IY-Y	Belt road 0-5-1-7	58
22	Named by V.I.Lenin, bed IY-Y	Air roadway 0-5-1-7	101
23	"Alarda", bed 1	Belt road 4/1	38
24	"Yesaulskaya", bed 29a	Belt road 29-14	15
25	"Yesaulskaya", bed 29a	Belt road 29-16	197

Analysis of the data represented in Table 1 shows that radon volume activity in the air of productive mines does not exceeds the maximum permitted limit but it is considerably higher than in the atmospheric air. But when the air from mined-out spaces is carried over into productive mine faces the maximum permitted concentration of radon in the air can be significantly exceeded.

Table 2 shows data of radon volume activity in the air of mine-out space of productive mines as well as in the air of isolated mined-out spaces.

Table 2. Volume activity of radon in the mined-out space atmosphere

No	Mine, bed	Sampling site	Volume activity, Bq/kg
1	"Pervomayskaya", bed XXYII	Upper face niche 31-71	107
2	"Pervomayskaya", bed XXYII	Upper face niche 724	134
3	"Pervomayskaya", bed XXI	Upper face niche 12-15	74
4	"Raspadskaya", bed 6-6a	Upper face niche 5-6-10	111
5	"Raspadskaya", bed 6-6a	Through wall stopings	178
6	"Raspadskaya", bed 5	Upper face niche 5-5-3	184
7	"Raspadskaya", bed 5	Coming out of the well	309
8	"Raspadskaya", bed 3-3a	Upper face niche 3-3-11	113
9	"Kolmogorovskaya", bed Baikaimskiy	Upper face niche 1304	349
10	"Usinskaya",bed III	Upper face niche 3-3-15	129
11	"Usinskaya",bed III	Through wall stopings 778	273
12	"Usinskaya",bed III	Through wall stopings 726	390
13	"Yagunovskaya", bed Volkovskiy	Upper face niche 222	116
14	"Baidaevskaya" bed 29 a	Well into the face 29-31	2594
15	"Baidaevskaya" bed 29 a	Well into the face 29-31	4127
16	"Baidaevskaya" bed 30	Well into the face 30-35	1077
17	"Baidaevskaya" bed 30	Well into the face 30-35	5636
18	"Baidaevskaya" bed 29 a	Through wall stopings	1144
19	"Baidaevskaya" bed 29 a	Through wall stopings	1066
20	Named by V.I.Lenin, bed IY-Y	Upper face niche 0-5-1-7	138
21	Named by V.I.Lenin, bed IY-Y	Through wall stopings	256
22	"Alarda", bed 1	Through wall stopings	555
23	"Yesaulskaya", bed 29a	Through wall stopings	1580
24	"Yesaulskaya", bed 29a	Through wall stopings	1304
25	"Inskaya", bed Sychevskiy 1	Through wall stopings	4211

Comparing the data represented in Tables 1 and 2 we can see that in mined-out spaces radon volume activity is much higher than in operative mines. In poorly aired areas (upper or lower face niche) radon volume activity is at the maximum permitted level and sometimes higher. In isolated mined-out spaces radon concentration can tenfold exceed the maximum permitted level.

Thus, the experimental research carried out in mines proved that radon volume activity in mine atmosphere can change considerably and often exceed sanitary standards. The most important factors influencing the radon concentration in mine atmosphere are ventilation (an amount of incoming air), natural radioactivity of mined rocks, a dispersion degree of broken-down rock. Examining of the mine atmosphere showed that volume activity of radon daughters in the mine atmosphere can reach 4000–6000 Bq/m^3. The highest concentration of radon is found in mined-out areas which have been isolated for a long time.

Coal mining works promote carrying-over the air from the mined-out spaces onto the earth surface. The mine air containing a great amount of radon is carried out into the atmosphere mostly with ventilation flows. Gas from the mined-out spaces comes out into the atmosphere from wells and cracks in the rock. The analysis of causes stimulating carrying-over of mine gases into the atmosphere represented in [7] proved that the most important factors influencing the intensity of gas transfer are changes in gas pressure between the mined-out spaces and the atmosphere, rock permeability and a stratification depth of gas releasing rocks.

Emission of radon from the earth surface over mined-out spaces was studied at a number of Kuzbass mines. The distance between the mined-out spaces and the earth surface was from 100 to 400 m. Measurements were taken with the help of diffusion collectors with activated carbon installed on the surface. After exposure sorbent activity was determined and radon flows were calculated. Experiments showed that in Kuzbass, a natural level of radon flows from the earth surface over mine fields is about 10–70 Bq/(m^2 s). These values prove that mined-out spaces are properly isolated and radon does not reach the surface. Over steeply pitching coal, rock permeability increases greatly which leads to anomalous zones of radon emission. Radon flows in such zones are as great as 100–500 Bq/(m^2 s).

The research also showed that apart from rock permeability radon emission is effected by coal self-ignition in the mined-out spaces. High temperature in coal self-ignition places causes heat depression which intensifies carrying over of gases from mined-out spaces to the earth surface. Thus on the earth surface over places of coal self-ignition an anomalous emission of radon occurs. It can be used to find and allocate coal self-ignition areas.

Though a rise of temperature does not affect the rate of radon decay, heating can still cause an increase in carrying-over of radon into the atmosphere due to evaporation of inherent moisture from coal and rocks and growth of molecular diffusion coefficient caused be temperature rise. Decrease in radon sulubility factor in water at elevated temperatures can also stimulate radon emission into the air. For example, if water is heated from 10 to 70 °C radon concentration goes down three-fold.

Measurements of radon flows at the earth surface over underground self-ignition areas repeatedly revealed anomalous zones with high radon concentration. Areas with radon concentration of 100–400 mBq/(m^{2}c) were detected and in some places radon concentration was as high as 1000–1500 mBq/(m^{2}c). Radon anomalies at the earth surface coincided with vertical projection of self-ignition places and in few cases they were slightly moved apart from the projection.

3. Conclusions

1. Concentration of radium in the examined coals of Kuzbass region varies from 6.9 to 24.7 Bq/kg and in rocks it can measure from 21 to 48.8 Bq/kg which leads to background volume activity of radon in the mine atmosphere.

2. Background volume activity of radon in the mine atmosphere changes widely and often exceeds the maximum permitted level. The concentration of radon in the mine atmosphere is mostly affected by natural radioactivity of driven rocks, a degree of rock fragmentation, a speed of air flow. The highest radon concentration is found in the atmosphere of continuously isolated mined-out areas, being tenfold higher than the maximum permitted level.

3. Radon is carried out into the atmosphere mainly with ventilation flows. Some part of radon penetrates into the atmosphere through pores and cracks in the rocks. Underground self-ignition considerably intensifies radon emission generating convection gas flows to the earth surface.

IOP Conf. Series: Materials Science and Engineering **127** (2016) 012021 doi:10.1088/1757-899X/127/1/012021

4. Surface monitoring allows detecting locations of underground self-ignition areas judging by the amount of radon flows from the surface.

4. References

[1] *Exposure to natural ionizing radiation* NKDAR The thirty-third session Vena June 25–29 1984
[2] Margulis U Ya *Atomic energy and radiation safety* 2d edition – Moscow: Energoatomizdat 1988 p 224
[3] Kozlov V F Troshkin U S *Reference book on radiation safety* – Moscow: Atomizdat 1967 p 267
[4] V F Torosyan, E S Torosyan, P D Sorokin, A A *Telitsyn Updating of sewage – purification facilities of electroplating enterprises with counterflow ion-exchange filters* IOP Conf Series: Materials Science and Engineering 91 (2015)
[5] *Ionizing radiation: Sources and biological effects* NKDAR Report of 1982 for General Assembly (with appendices) – Vol 2 New-York UN 1982
[6] V F Portola, E S Torosyan, A S Kuznetsova *Indirect Negative Influence of Coal Mine Motor Vehicles on the Environment* Applied Mechanics and Materials Vol 770 (2015) pp 690–694
[7] Portola V A *On gas anomalies over underground gas emission sources* // FTPRPI– 1996 – No6 – pp 110-117
[8] Prutkina M I Shashkin V L *Reference book on radiometric prospecting and radiometric analysis* – Moscow: Energoatomizdat 1984 p 168

Oxidative Activity of Heated Coal Affected by Antypirogens

V F Torosyan [1,a], **E S Torosyan** [1,b], **I F Borovikov** [2], **V A Yakutova**[1]

[1]Yurga Institute of Tehnology (branch) of Tomsk Polytechnic University
652055 Russia, Yurga, ul. Dostoevskogo, 6
[2] Bauman Moscow State Technical University, ul. Baumanskay 2-ya, 5, 105005, Moscow, Russia

E-mail: [a]torosjaneno@mail.ru, [b]eltorosjan@mail.ru

Abstract. The effect of antypirogens on chemical activity of heated coal is studied. It is proved that ammonium sulfate, calcium phosphate, calcium chloride, calcium nitrate and acid fluoride are the most effective antypirogens.

1. Introduction

A great amount of wastes damaging natural environment is one of the most serious problems of coal mining industry hampering its further development. Pollution of air, water, changing soil composition cause degradation of flora and fauna in the area affected by mining, threaten people's health and life. Mine fires lead to a sharp increase of hazardous and harmful effects [1]. Fires reduce coal mining efficiency, deteriorate quality of the produced coal, damage the environment, can cause explosions of inflammable gas and dust, require money for fire suppression.

Spontaneous combustion of coal and coal-bearing rocks occurs when a substance is oxidized with air oxygen and generates heat that raises the rock temperature with radiation heat loss. When the temperature rises the oxidation process accelerates and after some time coal or coal-bearing rocks combust [2]. The research in coal spontaneous combustion discovered that the probability of endogenous underground fire in mines is characterized by such factors as oxidation activity of coal (liability to spontaneous combustion), liability of a coal bed to spontaneous combustion, fire safety in mines.

Oxidation activity of coal is its natural property generated by its origin and genesis conditions. It is determined by the speed of oxidizing reaction of coal.

Research in coal oxidation, experience of spontaneous fire suppression and prevention show that spontaneous fire hazard can be mitigated by means of treatment of coal with different agents.

Antipyrogens are substances which can reduce sorption activity of coal and slow down spontaneous fires. Many organic and non-organic substances possess antipyrogeneous properties. Antipyrogens applied in mining industry should meet the following requirements: they should be non-toxic, cheap, abundant, have great mechanical, chemical or complex effect on coal if added to water or clay pulp in small quantities.

Mechanically active antypirogens are substances which form protective films on the coal surface and prevent air from leaking. They include talc, carbonate of lime, cretaceous and others. Most filming agents are organic compounds: caoutchouc, polyvinylchloride, resins, nondrying oil and others. Gaseous filming agents are carbon dioxide, nitrogen, hydrogen sulfide and others.

Chemically active antypirogens include substances which:

- react with coal in the adsorption layer (hydrogen peroxide, potassium permanganate, potassium chromate);
- educe gases in decomposition (calcium bicarbonate, ammonium carbonate);
- break down generating new chemically active substances (magnesium chloride, calcium chloride, ammonium phosphate, ammonium chloride).

Antypirogens having a combined (mechanical and chemical) effect on coal can be presented by water suspension of calcium hydrate.

For prevention and suppression of fires in coal mines the following substances can be used: calcium chloride, ammonium phosphate, ammonium chloride, potassium permanganate, ammonium nitrate, aluminium sulphate, calcium hydrate, formaldehyde resin, polyacrylimide, rock dust, grind of open-hearth and blast-furnace slag, nonutilizable sand and water from tailing storage facility of aluminum plants, wastes from alkali-works and others.

Research in antypirogens were first initiated and carried out under the direction of academician A.A. Sochinsky.

A great contribution to the research in antypirogen of low temperature coal oxidation was made in the Institute of Anthracides [4, 5]. Water solutions of non-organic compounds were tested. Experiment findings determine a positive effect of ammonium sulphate, ammonium carbonate, ammonium chloride, sodium chloride, calcium chloride, magnesium chloride and prove that these substances can be used as antypirogens.

Antypirogens were tested for effectiveness in prevention of coal self-combustion in many mines of Kuzbass. But due to different methods used for estimating the affect of different substances on coal chemical activity the obtained results were contradictory.

As an example, experiments by V.S. Veselovsky determine that suspensions of calcium hydrate are mostly nonreactive to coal but sometimes can be oxidation catalysts [6]. At the same time A.I. Khrisanfova and V.M. Mayevskaya prove that these suspensions are the most effective of available and cheap antypirogens [7].

2. Results and discussion

Problems of choosing an antypirogen are exacerbated by the fact that the same chemical substances can be either positive or negative catalysts of the oxidation depending on coal composition and properties and a theoretical basis for the choice of antypirogens has not been worked out so far because their effect on coal has not been revealed.

In order to improve prevention of spontaneous fires in coal mines the effect of antypirogens on the chemical activity of heated coal has been researched. Chemical activity of coal was determined with the help of the method developed by Institute of Mining named after A.A. Skochinsky. It is applied for evaluating chemical activity of coals and their liability to spontaneous combustion, it determines the influence of different factors (temperature, humidity, antypirogens) on kinetics of oxidation, evaluates the speed of oxidation when calculating the thermal balance and studies the probability of spontaneous combustion.

Coal test samples weighing 100 gr each were treated by 3 % water solutions of antypirogens:

- sodium oxalate;
- calcium oxide;
- calcium chloride;
- calcium phosphate;
- calcium dihydrogen orthophosphate;
- ammonium sulphate;
- calcium acetate;
- wetting reagent for coal dust "Elfor-M";
- foaming agent PO-6 "Elfor";
- liquid glass;

- calcium nitrate;
- calcium oxalate;
- melting slag;
- clay;
- acid fluoride.

The samples were kept in tightly closed containers, measurements were taken at fixed intervals (each 12 hours) under the temperature of 70 °C. After each measument the container was aired, then sealed again and placed into the same conditions.

As a result it has been found out that sodium oxalate, calcium dihydrogen orthophosphate, foaming agent PO-6 "Elfor", liquid glass, calcium oxalate and melting slag do not have any positive effect on coal. Suspensions of clay and calcium oxide had been reducing chemical activity of coal for 48 hours but then the speed of oxygen sorption increased intensively. Evidently it happens because clay and calcium oxide do not solve in water and after drying down they become more brittle and drop off. While in suspension these substances encapsulate coal pores. But as soon as they run dry coal pores peel off and become penetrable for air.

Solutions of calcium phosphate and ammonium sulfate have given good results. The speed of sorption of oxygen by coal did not exceed 0.15 cm^3/(gr-hour) during the experimental period. Fig. 1 shows correlation of chemical activity of the coal treated by calcium phosphate and ammonium sulfate solutions with untreated coal.

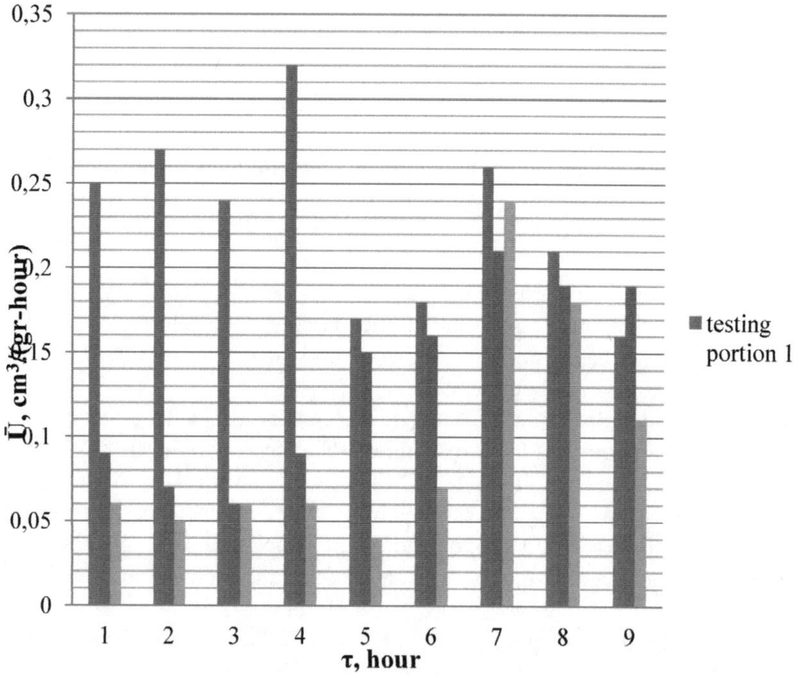

Fig. 1 Chemical activity of coals:
testing portion 1 – untreated coal, testing portion 5 – coal treated by calcium phosphate, testing portion 7 – coal treated by ammonium sulfate

Ammonium sulfate is a colorless transparent salty-tasting crystalline substance. It is classified as ammonium salt of sulfuric acid. Ammonium sulfate is one of the most widely spread mineral fertilizers used in agriculture. It is applied for all agricultural crops (from potatoes to citruses) on black and grey soils. This fertilizer has a very valuable quality – low liability to migration because ammonium kation is absorbed by soil and thus protected from fluviraption. That is why it is recommended to apply ammonium sulfate with irrigation on light soils where the loss of nitrogen from a fertilizer is possible due to migration. Ammonium nitrogen of ammonium sulfate is assimilated by plants. Sulfur is needed to feed all crops as it is found in some important amino acids synthesized by plants.

Ammonium sulfate is also used in viscose fiber production. In biochemistry, reprecipitation by ammonium sulfate is a common method of protein purification. In food industry, it is certified as food additive E517.

Ammonium sulfate is also applied for water chlorination with ammonation. It is introduced into the water a few seconds before chlorine and fixes the chlorine to form chloramines. It reduces chlorinated organics which is bad for human health, cuts down chlorine consumption, lowers water pipe corrosion.

In fire suppression ammonium sulfate is used as an antypurine. In mining it serves a gelling material for protective treatment of the mined-out space being introduced into the explosive to reduce flammability and prevent spontaneous combustion.

The research has also proved that water solutions of calcium acetate, calcium chloride, calcium nitrate, acid fluoride and wetting reagent for coal dust "Elfor-M" can be used as antypurines (Fig. 2).

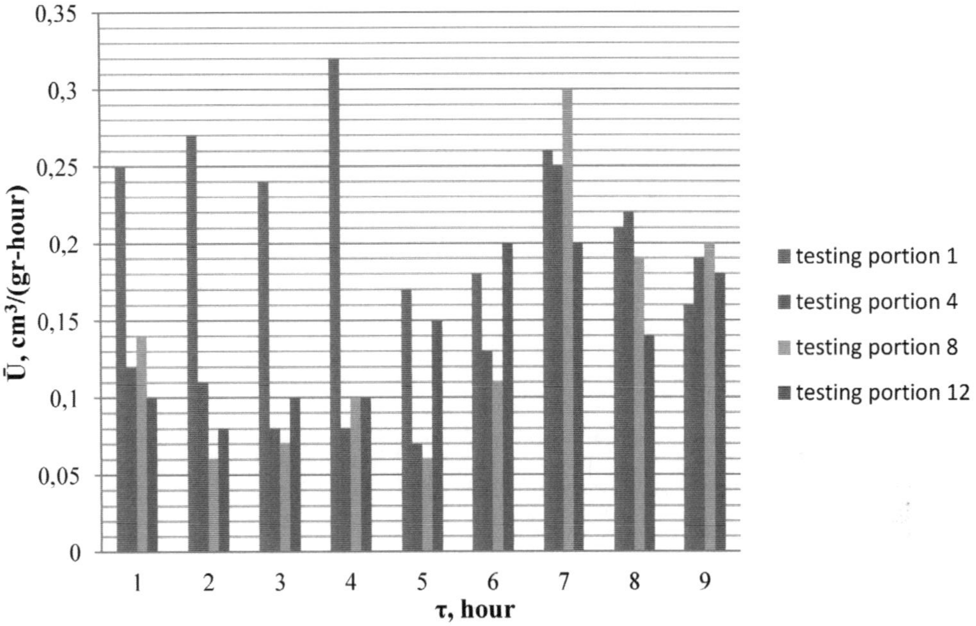

Fig. 2 Chemical activity of coals:
testing portion 1 – untreated coal, testing portion 4 – coal treated by calcium chloride, testing portion 8 – coal treated by calcium acetate, testing portion 12 – coal treated by calcium nitrate

The resulting regularities can determine the periodicity of antypurine supply for fire suppression at rock disposal dumps. Thus ammonium sulfate and calcium phosphate are enough to introduce as often as once a week if there are no fire areas, while water solutions of calcium chloride, calcium nitrate, acid fluoride are introduced every 3–5 days.

3. Conclusions

1. Coal is a complex high-molecular compound. There is no general opinion about coal structure as well as about the origin of self-heating and self-combustion of coal.
2. The research proved that ammonium sulfate, calcium phosphate, calcium chloride, calcium nitrate and acid fluoride are the most effective antypirogens.
3. Water solutions can be applied to treat coal rock immediately prior to loading them into trucks.

IOP Conf. Series: Materials Science and Engineering **127** (2016) 012022 doi:10.1088/1757-899X/127/1/012022

4. References

[1] V F Torosyan, E S Torosyan *Development of mixture "clay – electro-smelting slag" for ceramic brick* Applied Mechanics and Materials Vol 682 (2014) pp 480–484

[2] V F Portola, E S Torosyan, AS Kuznetsova *Indirect Negative Influence of Coal Mine Motor Vehicles on the Environment* Applied Mechanics and Materials Vol 770 (2015) pp 690–694

[3] D I Yakovlev *Coal chemical laboratories* / D.I. Yakovlev – Moscow: Ugletekhizdat 1957 – 516 p

[4] *Inhibitors for prevention of coal oxidation and self combustion* /A I Khrisanova, A K Shubnikov, A N Zakharov, G P Gusev – Moscow: AN SSSR Publishing 1959 – 138 p

[5] A I Khrisanova *Coal storage technology and measures reducing loss of fuel* / A I Khrisanova, V G Litvinov – Moscow: Nedra, 1970 – 164 p

[6] V S Veselovskiy *Self-combustion of industrial products* / V S Veselovskiy, N A Alekseeva, L P Vinigradova. – Moscow: Nedra. 1964. – 254 p.

[7] V M Mayevskaya *Antypirogens for prevention and suppression of endogenous fires* // Izvestiya vuzov Gorny zhurnal – 1961 – № 8 pp 46–61

[8] *Chemist's reference book* / Nikolsky B P and others; 2d edition – Moscow: Chemistry 1966 Vol 1 – 1072 p.

[9] S A Kuznetsova, V V Kozik, A G Malchik *Synthesis of transparent conductive coating In2O3:Sn films from film forming solutions* Applied Mechanics and Materials Vol 682 (2014) pp 401–404

[10] A V Filonov, O A Kireeva *Prospects of Galvanic Sludge Recycling* Applied Mechanics and Materials Vol 770 (2015) pp 709–713

IOP Conf. Series: Materials Science and Engineering **127** (2016) 012023 doi:10.1088/1757-899X/127/1/012023

Environmental Hazard Identification Technique Developing of Territorial Administrations Strategy as Exemplified in Siberian Federal District

S S Timofeeva[1,a], V V Garmishev [1, b], N Yu Lugovtsova [2,c]

[1,a] Doctor of Technical Sciences, Prof.
National Research Technical University. Russia, Irkutsk, Lermontov str., 83
[1, b] PhD, Ass. Prof.
National Research Technical University. Russia, Irkutsk, Lermontov str., 83
Assistant lecturer.
[2,c] Yurga Institute of Technology (affiliate) of National Research Tomsk Polytechnic University. Russia, Yurga, Leningradskaya str., 26

E -mail: [a]timofeeva@istu.edu, [b]diamant1959@mail.ru, [c]lnyu-70583@bk.ru

Abstract. This work, on the example of the Siberian Federal District of the Russian Federation, presents a method for determining mass airbursts of combustion gaseous eco-toxicants, as a technique for assessing the contributory environmental load on the atmosphere. Potential environmental hazards are analyzed by mass of contaminated airbursts. A comparison of specific gross toxicants' emissions in territorial entities of the Russian Federation is made.

1. Introduction

Currently, the scale of human activities' impact on the biosphere is compatible with the scale of geological and geochemical processes. At the same time the environmental fire hazards in industrial plants, urban infrastructure, rural areas and municipalities are among the most frequent and important for their long-term effects; as they are accompanied by airbursts of harmful substances and produce extremely high levels of pollution in some territories.

In Russia, annually, average number of fires is 160.000 with damage over 13 billion rubles. About 11.000 people are killed and more than 12 000 people are traumatized on fires. Fires destroy and damage more than 30 000 pieces of equipment, 127.000 buildings, with a total area of over 6.5 million m^2; 4.8 million m^2 being residential areas, capable of providing accommodation of about 200 thousand people [1,2].

In fire conditions, burning action usually occurs in diffusion mode. Substances and materials do not burn out completely and, along with soot particles, are emitted into the environment in the form of gaseous and liquid products of combustion. Heat fluxes, regulating gas exchange and fire spreading, ensure transfer of pollutants in space. It is found that combustion products contain from 50 to 150 different types of toxic chemical compounds. These agents affect both biocenosis components and human health [3-5]. The most common atmospheric fire pollutants are oxides of carbon, sulfur, nitrogen, chloride and cyanide hydrogen, various classes of hydrocarbons, alcohols, aldehydes, benzene and its homologues, polyaromatic compounds (PAC) and dioxins. Most of these chemicals are harmful to living organisms. For example, dioxins, PAC and others can cause cancer in humans, and sulfur oxides cause the death of vegetation.

In addition, the fire heat factor has a negative effect on humans, flora and fauna (a temperature of 70 °C is critical for humans). In the burning zone temperature can rise to 800–1500 °C, and sometimes higher (firestorm, burning of metals).

Smoke from severe fires causes changes in daylight and air temperature, affects the amount of precipitation. In addition, smoke aerosol and gaseous products, interacting with air moisture, cause acid rains or fogs [6]. Contaminated fumes, dews and rains cause illness and death of plants. Huge smoke air pollutions (forest fires, oil well fires) reduce the amount of solar radiation reaching the Earth's surface, which leads to climate changes lasting several days, weeks, months. These factors affect growth of plants, particularly if they coincide with the growing season [7].

Given the magnitude of environmental effects of fires and urgency of the problem, in this paper a retrospective analysis of fire statistics in regions of the Siberian Federal District is performed; total environmental load on the atmosphere from industrial fires is estimated on priority toxic substances.

2. Objects and methods of research

The Siberian Federal District (SFD) was founded in 13 of May, 2000, and is one of the largest districts of Russia, with the area of 5.114.8 thousand km^2 (30 % of the country), with a population of 20.207.400 people, the population density of 3.9 persons per km^2; the share of urban population is 71.1 %. On the district territory 12 sub-federal entities of the Russian Federation are located: 4 Republics (Altai, Buryatia, Tyva, Khakassia); 3 Territories (Altai, Trans-Baikal, Krasnoyarsk); 5 Regions (Irkutsk, Kemerovo, Novosibirsk, Omsk, Tomsk). Table 1 shows some of social indicators of sub-federal entities of the SFD.

Table 1. Social indicators

Sub-federal entity	Population, thous. people	Area, thous. km^2	Urban population, thous. people	Rural population, thous. people	Population density, people per km^2	Industrial facilities, thous. units
Altai Republic	211.6	92.9	62.3	149.3	2.2	54.3
Buryatia Republic	973.8	351.3	577.4	396.4	2.8	181.8
Tyva Republic	311.7	168.6	163.5	148.2	1.8	75.4
Khakassia Republic	534.0	61.5	374.4	159.6	8.8	127.2
Trans-Baikal Territory	1090.3	431.8	626.3	464.0	2.7	206.4
Krasnoyarsk Territory	2852.8	2366.7	2131.8	721.0	1.3	520.2
Altai Territory	2390.6	167.9	1170.3	1220.3	15.4	325.7
Novosibirsk region	2731.1	177.7	2060.7	670.4	15.1	421.5
Tomsk region	1070.1	314.3	732.1	338.0	3.3	253.6
Irkutsk region	2418.3	774.8	1762.7	655.6	3.4	439.3
Kemerovo region	2734.1	95.7	2348.5	385.6	30.4	488.2
Omsk region	1973.8	141.1	1323.0	650.8	14.9	457.3

In the SFD there are large industrial enterprises of chemical, petrochemical, food, timber, woodworking, pulp and paper industry, energy, machine building, metallurgy, and many others, more than 3.5 million of industrial and technosphere objects.

The SFD takes 30 % of the area of Russia, and a large area of the District is covered with woods; the population is about 20 %. Therefore, the state of fire safety in the District significantly affects fire national safety statistics [8].

For the convenience of research, all objects of technosphere are combined in 4 large groups in view of fire hazard's features.

1. Residential objects – 2196.6 thousand: hostels, residential buildings of state (municipal use), private houses, yurts, garden and dacha communities, mobile residential facilities (trailers), outbuildings.

2. Industrial objects – 473.2 thousand: warehouses of flammable and combustible liquids, flammable gases; timber and lumber, and other combustible materials; factories and warehouses, boiler rooms; critical infrastructure; gas stations; mills; grain elevators; enterprises of cargo and passenger transportation; vehicle maintenance facilities.

3. Public buildings – 795.4 thousand: Commercial and residential (markets, shops, distribution centers); educational institutions (universities, academies, colleges, schools, boarding schools); child care institutions (day care, camps, orphanages; schools of sports, music, arts, additional education); cultural institutions (theaters, museums, philharmonics, libraries) and leisure activities (movie theaters, night clubs, circuses, zoos, gyms); medical institutions (hospitals, clinics, dispensaries, pharmacies, medical and health centers, spas, health resorts); administrative building of the Executives; buildings of tax service, research institutions, banks, post offices, objects of public services, utilities, catering, hotels, objects of social services and other administrative buildings.

4. Other facilities – 35.3 thousand buildings. These are: religious objects, buildings under construction or reconstruction, buildings for agricultural purposes (vegetable stores, hothouses), livestock, farming, poultry, as well as other objects that are not city properties.

Analysis of statistical data submitted in annual reports of Siberian regional center of EMERCOM of Russia, shows that the SFD has 16 fires for every 10 thousand inhabitants; about 12 people are killed in fires per every 100 thousand of inhabitants. Out of every 100 facilities, where fires have occurred, 20 units burn down completely and can not be restored. Average damage caused by a fire is 52.8 thousand rubles.

Table 2. Averaged general indicators of fire hazards in regions of the SFD in 2001–2014

Sub-federal entity	Основные показатели последствий пожаров			
	Fires, thous. per year^{-1}	Direct material damage, mln. rub. per year^{-1}	Killed in fires, persons, per year	Fired objects (buildings) units per year^{-1}
Buryatia Republic	1548.2	148.2	134.5	292
Altai Republic	391.7	19.1	22.7	79
Tyva Republic	531.9	25.7	24.5	112
Khakassia Republic	770.5	37.5	55.6	125
Trans-Baikal Territory	2111.4	102.8	142.5	395
Krasnoyarsk Territory	4784.1	238.9	371.9	339
Altai Territory	4321.3	210.8	314.4	887
Novosibirsk region	4146.3	202.1	301.5	823
Tomsk region	1594.1	77.9	129.1	329
Irkutsk region	4341.5	263.3	343.6	837
Kemerovo region	4574.2	224.2	316.8	902
Omsk region	3171.6	156.1	239.8	760
SFD	32286.8	1706.6	2396.9	6480

Calculation of the mass of burnt material and the amount of toxic substances, emitted to the atmosphere, was made using L.K. Isaeva's method, enhanced by the authors [4,9]. We proposed the use procedural documents for estimating of the mass of burned materials, namely: Technological certificate, fire certificate with the facility layout; fire Description; fire site inspection Protocol; expert's findings; Act of official investigation; Certificate of property damage; reference from insurance companies about material damage; explanatory memorandum of the owner / responsible person; eyewitness testimonies.

When estimating the amount of burnt materials, they were divided into the following groups:

- Wood (construction and decoration materials: logs, timber, lumber, plywood, chipboard and fiberboard, wood sculpture, wood waste: sawdust, wood chips, wood);
- Fossil fuels (coal, lignite, peat);
- Petroleum (crude oil, petrol, diesel, fuel oil, bitumen, oil);
- Rubber products and materials;
- Styrofoam, polystyrene (structural, thermal insulation finishing materials);
- Paintwork materials (paints, varnishes, drying oil, solvents, thinners, hardeners);
- PVC (linoleum, floor tiles, insulation of wires, cables, finishing plastics);
- Cellulose (paper, wallpaper, cardboard);
- Synthetic, wool, cotton, fabrics;
- Products and materials from leather, synthetic leather;
- Polyurethane foam (thermal insulation and composite materials);
- Roofing and waterproofing materials (roofing, synthetic coatings, plastics, mastics);
- Cereals, industrial crops, fodder products for animals;
- Combustible gases.

Table 3 shows the average values of burnt materials' mass in the regions of the SFD.

Table 3. Average mass of materials burnt in fires at technosphere facilities in regions of the SFD (2001–2014)

Object of study	Amount of burnt materials, tonnes per year^{-1}
Buryatia Republic	68704.2
Altai Republic	13509.5
Tyva Republic	19821.3
Khakassia Republic	35594.0
Trans-Baikal Territory	89251.4
Krasnoyarsk Territory	224118.8
Altai Territory	128880.6
Novosibirsk region	109040.7
Tomsk region	63240.5
Irkutsk region	181786.3
Kemerovo region	126631.4
Omsk region	88412.0

Having known the mass of burnt materials and specific emissions of toxic substances during combustion of various materials, we calculated airbursts of toxicants.

3. Results and Discussion

It is established that, during technosphere fires, average of 211.204 tons of pollutants emit annually into atmosphere of towns and cities of the Siberian Federal District. Mass of priority fires' pollutants, emitted into the atmosphere, is shown in Table 4.

Table 4. Mass of priority fires' pollutants, emitted into the atmosphere in the SFD (2001–2014)

Toxicant	Mass of pollution, ton per year^{-1}	Toxicant	Mass of pollution, ton per year^{-1}
Acrolein	32.7	Nitrogen monoxide	1067.2
Ethanol	5.4	Carbon monoxide	57909.3
Ammonia	14.8	Pyridine	1.4
Acetaldehyde	47.5	Propylene	4.3
Acetylene	107.8	Propyl alcohol	10.6
Acetone	132.2	Vanadium pentoxide	$1.8 \cdot 10^{-2}$
Benzpyrene	$2.9 \cdot 10^{-4}$	Soot	35152.6
Benzene	27.2	Hydrogen sulphide	15.5
Ethylene	5.4	Styrene	4.9
Vinyl chloride	6.6	Toluene	5.2
Nitrogen dioxide	380.8	Toluene diisocyanate	4.2
Sulphur dioxide	104.7	Acetic acid	5.9
Carbon dioxide	43131.7	Phenol	21.2
Smoke (ultrafine particles)	72950.1	Formaldehyde	4.3
Cresol	5.5	Phosgene	2.1
Xylene	9.8	Hydrogen chloride	28.2
Naphthalene	6.2	Hydrogen cyanide	23.9

Ecotoxicants are usually divided into four classes of hazard to humans and biota components. Calculations revealed that, as a result of fires in the Siberian Federal District, 9.4 tons of highly toxic substances, 1234.1 tons of highly hazardous substances, 111,033.7 tons of moderately hazardous substances and 98,926.8 tons of little hazardous substances are annually polluted into the atmosphere. Table 5 shows the averages of air pollution from fires by toxic substance hazard categories.

Table 5. Averages of air pollution from fires in cities of SFD classified by toxic hazard (2001–2014)

Object of study	Total pollution, ton per year^{-1}	Air pollution by toxic substance hazard categories, ton per year^{-1}			
		1	2	3	4
Buryatia Republic	12914.0	$4.6 \cdot 10^{-1}$	68.8	6731.2	6113.6
Altai Republic	2303.2	$1.0 \cdot 10^{-2}$	13.1	1190.2	1099.9
Tyva Republic	2859.6	$6.8 \cdot 10^{-2}$	14.5	1488.3	1356.8
Khakassia Republic	5514.6	$1.1 \cdot 10^{-1}$	24.1	2967.2	2523.2
Trans-Baikal Territory	14639.4	$8.1 \cdot 10^{-1}$	93.4	8178.0	6367.2
Krasnoyarsk Territory	38779.2	2.1	221.0	19635.5	18920.6
Altai Territory	24698.4	1.3	144.4	13026.1	11526.6
Novosibirsk region	21583.3	1.2	134.5	10430.2	11017.4
Tomsk region	12726.4	$5.4 \cdot 10^{-1}$	69.8	6770.3	5885.8
Irkutsk region	30211.2	1.7	185.9	16518.7	13504.9
Kemerovo region	26952.3	1.4	160.6	14286.4	12503.9
Omsk region	18022.8	$9.0 \cdot 10^{-1}$	104.0	9811.6	8106.3
The SFD	211204.0	9.4	1234.1	111033.7	98926.8

Let's compare the following data on the mass of air pollutants from stationary emitters in the Siberian Federal District in 2013 [10].

Entity	Mass of air pollutants, thous. ton
Krasnoyarsk Territory	2497.3
Kemerovo region	1356.3
Irkutsk region	685.9
Tomsk region	306.1
Omsk region	213.6
Altai Territory	201.2
Novosibirsk region	195.7
Trans-Baikal Territory	127.1
Buryatia Republic	114.1
Khakassia Republic	90.4
Total	5787.7

Calculations have revealed that uncontrolled fires' airbursts are on average 35–40 % of stationary emitters.

A ranking of SFD entities by specific unaccounted environmental load with toxic fire products has been performed; and it has been found that the highest specific unaccounted load is recorded in Kemerovo region: 0.28 ton/km^2 per year^{-1}

Next, in descending order, Krasnoyarsk and Altai Territories (0.16 and 0.15 ton/km^2 per year^{-1}, respectively) and Novosibirsk region (0.12 tons/km^2 per year^{-1}) are ranged.

In other SFD constituent entities unaccounted specific load ranges from 0.01 to 0.04 ton/km^2 per year^{-1}.

4. Conclusion

Development of new technologies has led to creation and growth of the technosphere, and, with it, a growing number of fires, causing damage to property, wildlife and environment; besides, increasingly threatening lives and health of people.

Russian official reports on State of Environment present data on stationary and mobile air-pollutant emitters; but the contributions of fires in pollution are not taken into account.

Based on the obtained data we can draw the following conclusions:

1. Airburst fire ecotoxicants make an unaccounted contribution to air pollution; they are significant threats for human health and environment.
2. It is established that fires pollute the atmosphere with more than 211 000 tons of pollutants per year.
3. Environmental risks of fires should be considered when developing a territories' administration strategy.

5. References

[1] Social, economic and environmental impact of fires in municipal centers of the Siberian Federal District: analysis, evaluation, forecast: monograph / S Timofeeva [et al.] *FSEI HPE ESI Russian Interior Ministry* – Irkutsk Asprint, 2010 – 169 p

[2] Airburst pollutions from fires and accidents at industrial sites in Baikal region: assessment, forecast, prevention measures: monograph / S Timofeeva [et al.]. – Irkutsk: ISTU Publishing House, 2009 – 192p.

[3] Isayeva L K Environmental effects of air pollution from fires in residential buildings / L K Isayeva, B B Serkov // *Problems of safety in emergency situations* – M.: RISTI, 1992 – No2 pp 39-49

[4] Mal'chik A G, Litovkin S V, Rodionov P V Investigations of physicochemical properties of bottom-ash materials for use them as secondary raw materials // *IOP Conference Series: Materials Science and Engineering.* – 2015 Vol 91 Article number 012081

[5] Isayeva L K Ecological and economic evaluation of damage caused by pollution from fires: tutorial / L K Isayeva – M: USSR Interior Ministry, 1996 – 25 p
[6] Grishagin V M, Safronova A B Solid components of welding fumes as a synthetic composites filler // *Applied Mechanics and Materials* – 2014 – Vol 682 – pp 369-374
[7] Burkanov A K Fires: impact on environment: review Inf. / A K Burkanov, I G Egorov, V V Volohov – M.: Institute of Fire Prevention, 1992. – 18 p
[8] Brushlinskii N N Fire prevention management in Russian Federation entities on basis of fire risks analysis / N N Brushlinskii [et al.] // *Problems of safety and emergencies* – M.: RISTI – 2010 - No3 – pp 104 – 114
[9] Garmyshev V V Air pollution from fires in urban municipalities of Siberian Federal District / V V Garmyshev, S A Hismatulin, S S Timofeeva // *Bulletin of Irkutsk State Technical University* - 2013 No3 (74) – pp 48-53
[10] Government report "On State and Environmental Protection of the Russian Federation in 2013" http://www.mnr.gov.ru/upload/iblock/6c7/gosdokladeco.pdf

IOP Conf. Series: Materials Science and Engineering **127** (2016) 012024 doi:10.1088/1757-899X/127/1/012024

Analyzing the Technology of Using Ash and Slag Waste from Thermal Power Plants in the Production of Building Ceramics

A G Malchik[1,a*], S V Litovkin[1,b], P V Rodionov[1,c], V V Kozik[1,2,d], M A Gaydamak[1,e]

[1] Yurga Institute of Technology, National Research Tomsk Polytechnic University Affiliate,
652050, Yurga, ul. Leningradskaya 26
[2] National Research Tomsk State University, pr. Lenina 36, Tomsk, 634050 Russia

E-mail: [a*]ale-malchik@yandex.ru, [b]protoniy@yandex.ru, [c]rodik-1972@yandex.ru, [d]vkozik@mail.ru, [e] vip.trd777@mail.ru

Abstract. The work describes the problem of impounding and storing ash and slag waste at coal thermal power plants in Russia. Recovery and recycling of ash and slag waste are analyzed. Activity of radionuclides, the chemical composition and particle sizes of ash and slag waste were determined; the acidity index, the basicity and the class of material were defined. The technology for making ceramic products with the addition of ash and slag waste was proposed. The dependencies relative to the percentage of ash and slag waste and the optimal parameters for baking were established. The obtained materials were tested for physical and mechanical properties, namely for water absorption, thermal conductivity and compression strength. Based on the findings, future prospects for use of ash and slag waste were identified.

Introduction

The development of fuel and energy complex and construction industry, building of new towns and population centers, environmental and related social issues directly or indirectly depend on the disposal of coal-burning residuals - fly ash and slag. In the time of the Soviet Union, fly ash and slag were not separated and disposed of, using a hydro transport system, in impoundments and landfills, causing large amounts of slag waste within or close to cities. This problem is particularly pressing in Russia because of harsh climatic conditions, as well as disagreement of production time and potential consumption of coal-burning byproduct solids.

The share of coal burnt for producing electrical energy is 94% in Poland, 93% in South Africa. 78% in India, 77% in Australia, 70% in China, 67% in the Czech Republic, 62% in Greece, 52% in Germany, 50% in the USA, 17% in Denmark, 33% in the UK, over 27% in 15 European countries. The share of coal in energy production was 37% worldwide ten years ago. Such a low share of coal in the energy production was associated with the signing of the Kyoto Protocol, which required reducing emissions of carbon dioxide into atmosphere. However, due to unstable prices for oil and gas and their limiting availability, this share is now about 40%, with an apparent trend to grow [1,2,3].

According to the Energy Strategy of Russia for the period up to 2020 with the fundamental principles thereof approved by the government in 2000, a steady growth of electricity consumption is scheduled as much as 2-3% annually. It is planned to develop environmentally friendly power plants using renewable energy sources, including wind, tidal, geothermal and biofuel power stations, micro hydro power stations, and etc. But totally these sources will give 0.01% of generated energy. Thermal power plants will remain the foundation of the power generation industry, with their relative share in the total power capacity being at 67 - 70%.

The energy production using fossil fuels such as coal, shale oil, peat and other is inevitably accompanied by the formation of ash and slag waste; and fly ash is usually 70 to 90% of total balance in terms of waste solids formed during the coal burning. The share of costs estimated for handling waste is about one third of the cost of energy produced [1].

The problem of processing and recycling ash and slag waste, as well as making researches with the purpose of innovative processing technologies is being tackled in many countries. In Russia, the following experts are involved in dealing with ash and slag waste: I.S. Kozhukhovsky (CJSC "Energy Forecasting Agency"), V.Ya. Putilov (MPEI), V.V. Zyryanov (Nanopowder Technology Ltd.), B.L. Vishnya (Ural Energy Engineering Center), Yu. K. Tselykovsky (All-Russia Thermal Engineering Institute), R.V. Shevtsov (Territorial Generation Company 11).

Currently, no more than 5- 10 % of ash and slag waste is used in various construction and production sector countrywide. The residue is being stored in the ash and slag disposal areas without use. With this, the disposed ash and slag waste is not reduced, and it is likely to increase taking into account the growing demand for electricity and insufficient development of alternative sources for its generation.

Ash and slag materials in terms of their chemical and mineralogical composition are largely identical to natural mineral raw materials. Their use in the industrial sector, construction industry and agriculture is one of the strategic solutions to environmental problems in the area of operation of thermal power plants.

The aim of this work is to study physical and chemical properties of ash and slag waste, to determine its possible use as a secondary energy source to reduce anthropogenic impacts on the environment, to explore the possibility of obtaining ceramic materials with adding ash and slag waste.

Equipment and devices used in studies

Samples for the study were heat treated in a muffle furnace LOIP LF-15/11 – 61; a laboratory mixer BL-10 was used for mixing samples. Particle size analysis was performed using a device Analizette 22 MicroTec Fritsch GmbH (Germany). The samples were tested for ultimate compression strength in accordance with the requirements of GOST 10180-90 using a laboratory press PM-20MG4. The activity of radioactive nuclides was determined using a gamma-radiometer RKG-AT1320. Weighting was performed on a laboratory balance BK-600. The samples were dried in an electric oven SNOL-F-67/350-I2P before baking, and a cubic mould FBS-70 was used for forming the samples.

Results and discussion

A composition of ash and slag waste is determined by the quantitative ratio of contained in the material minerals that are dependent on initial mineralogical composition of fuel used.

The chemical composition of ash gives an idea of the composition of minerals in coal. Coal ash is typically presented by the overwhelming majority (96%) of oxides of silicon, aluminum, iron, calcium and magnesium. It contains sodium and potassium compounds only in small amounts. Some coals and ashes have a little precious metal (gold, silver, and platinum) as well as rare and trace elements.

The main compounds of ash are oxides SiO_2, Al_2O_3, FeO, Fe_2O_3, CaO, MgO. Sulfates $CaSO_4$, $MgSO_4$, $FeSO_4$ are presented in small amounts. Phosphates and alkali metal oxides K_2O, Na_2O are contained in ash to an even less degree. Ash and slag waste may have biogenic (fluorine, manganese, cobalt, lead, copper, etc.). and toxic (boron, vanadium, arsenic, strontium, beryllium, etc.) elements.

Chemical compositions of ash produced in burning different solid fuels vary within a wide range of limits,%: SiO_2 10 – 68; Al_2O_3 10 – 40; Fe_2O_3 2 – 30; CuO_2 2 – 70; MgO 0 – 10; Na_2O and K_2O 0 – 10. Besides, ash contains a small amount of compounds of germanium, vanadium, arsenic, mercury, beryllium fluoride partially passing into water.

Knowing a chemical composition of ash and slag waste is a prerequisite for discussing its properties and making a conclusion on its possible use in various economy sectors.

To determine the chemical composition of ash and slag waste, the ash samples were taken from electric filters and dumped ash according to the procedure RD 34.09.603-88 "Guidelines for the organization of monitoring compositions and properties of ash and slag supplied by thermal power stations to consumers."

The samples were tested to determine ash chemical composition and activity of radioactive nuclides.

The radionuclide activity was determined in accordance with GOST 30108-94 "Building materials and products. Determination of specific activity of natural radioactive nuclei". A gamma-radiometer PKG-AT1320 controlling the radio nuclides of potassium (^{40}K), radium (^{226}Ra) and thorium (^{232}Th) was used to determine the specific activity. The radioactivity measured at the Yurga thermal power plant showed the following results: ^{40}K – 526 Bq/kg, ^{232}Th – 72 Bq/kg, ^{226}Ra – 37 Bq/kg. The specific activity is calculated in accordance with the following formula:

$$A_{eff} = A_{Ra} + 1{,}31A_{Th} + 0{,}085A_K \qquad (1)$$

where A_{Ra} , A_{Th} , A_K are the specific activities of radium, thorium and potassium respectively, Bq/kg.

The calculated value of ash and slag waste activity at the Yurga thermal power plant is 175 Bq/kg that does not exceed the requirements of construction standards, qualifying ash and slag waste as the first class building material; this means it can be used in all types of construction.

Chemical analysis was performed using an X-ray fluorescence spectrometer Quan`X of the Common Use Center at Tomsk Polytechnic University. The results are shown in Table 1.

Table 1 - Chemical composition of ash from Yurga power generating plant

Compound	Ash waste		Electric filter	
SiO2	55,7	56,25	50,4	50,74
CaO	6,8	6,84	13,96	13,57
Al_2O_3	21,83	21,84	20,52	20,6
MgO	1,95	1,65	1,55	1,67
MnO	0,09	0,1	0,1	0,09
Fe_2O_3	7,44	7,4	8,55	8,53
FeO	6,69	6,66	7,69	7,68
K_2O	3,53	3,44	1,35	1,31
TiO_2	1,11	1,28	0,97	0,81
SO_3	0,72	0,68	0,87	0,82
BaO	0,44	-	0,5	0,59
P_2O_5	0,38	0,4	1,08	1,1
SrO	-	0,1	0,16	0,19

The main chemical compounds contained in the ash are oxides of silicon and aluminum; the content of iron oxide present is high as well, the content of calcium oxide in the ash waste samples is lower than that in the electric filter samples. Free calcium oxide is likely to react with the carbon dioxide, which is dissolved in the water used to transport the ash through a pipeline, to form calcium carbonate.

With the purpose to specify the possible use of ash and slag waste, it is required to define such criteria of classification as: acidity index, basicity, silica module and quality factor

The acidity index $M_к$ is calculated on the basis of the following formula:

$$M_к = \frac{SiO_2 + Al_2O_3}{Fe_2O_3 + CaO + MgO} \qquad (2)$$

The acidity index is 2.9 for the electric filter samples and 4.7 for the ash and slag waste samples.
The basicity module M_o is defined according to the base-acid ratio:

$$M_O = \frac{CaO + MgO + K_2O + Na_2O}{SiO_2 + Al_2O_3} \quad (3)$$

The basicity module is 0.2 for the electric filter samples and 0,16 for the ash and slag waste samples.
The silica module M_c is calculated as the ratio of silicon oxide to the sum of aluminum and iron oxides:

$$M_C = \frac{SiO_2}{Al_2O_3 + Fe_2O_3} \quad (4)$$

The silica module is 1.7 for the electric filter samples and 1,9 for the ash and slag waste samples.
The quality factor K shows the ratio of oxides increasing the hydraulic activity to oxides reducing it:

$$K = \frac{CaO + Al_2O_3 + MgO}{SiO_2 + TiO_2} \quad (5)$$

The quality factor is 0.7 for the electric filter samples and 0.54 for the ash and slag waste samples.
The results obtained suggest that the ash and slag waste is of an acidic type of ash. Acidic ashes are characterized by unstable chemical composition, a small amount of free calcium oxide and a high content of silicon oxide. Such types of ash do not have distinct bonding properties which are possible with calcium oxide that forming the hydroxide in water. Drying in air, hydroxide absorbs carbon dioxide from the air and converts into calcium carbonate that makes the product hard.
Particle-size distribution is one of the main indicators of raw materials. The greater the content of micro-dispersed particles, the higher ductility of a material is. Consequently, raw material has a higher bonding property that positively influences strength performance of products made; and a particle size composition is important for determining the adsorption capacity of a material [2,4,5].
The results of the particle-size distribution are represented in Figure 1. The analysis showed that 60% of particle size is 10 to 70 microns.

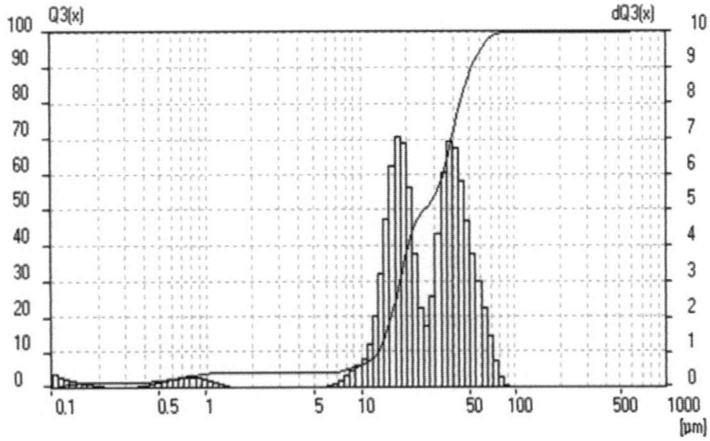

Figure 1 – Particle-size distribution of ash and slag waste

The next stage of the work was to make samples of ceramic bricks with the addition of ash and slag waste. Under laboratory conditions, ceramic bricks with different percentage of ash and slag waste were made by the soft–mud moulding process followed by baking at various temperatures. For illustrative purposes, all the data are summarized in Table 2 and Figures 2, 3 and 4. The figures are the graphs to show the dependency of baking temperatures on proportions.

Table 2 - Physical and mechanical properties of ceramic bricks at various baking temperatures and different percentages of ash and slag waste

Content of ash in clay, %	Heat conductivity, W/m °C				Water absorption, %				Ultimate compression strength, kg/cm^2			
	Temperature, °C				Temperature, °C				Temperature, °C			
	700	800	900	1000	700	800	900	1000	700	800	900	1000
0	0,84	0,85	0,85	0,85	11	10	9	8	60	80	90	110
5	0,83	0,84	0,82	0,85	17	17	15	10	70	96	100	115
10	0,83	0,81	0,82	0,81	18	18	16	12	72	104	110	120
15	0,77	0,79	0,74	0,76	20	19	17	14	75	106	115	125
20	0,77	0,69	0,73	0,71	24	23	18	16	70	90	105	115
25	0,55	0,6	0,57	0,61	27	25	20	18	65	85	95	110

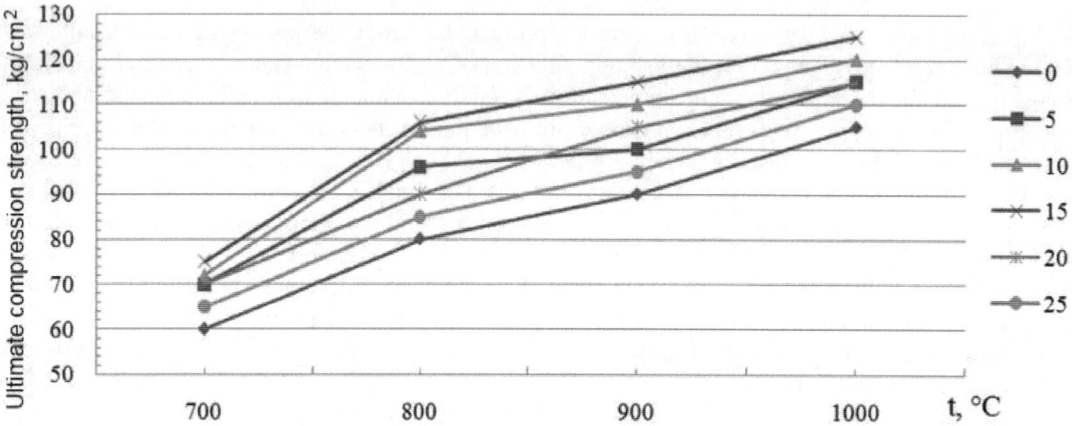

Figure 2 – Backing temperature dependence of the ultimate compression strength.

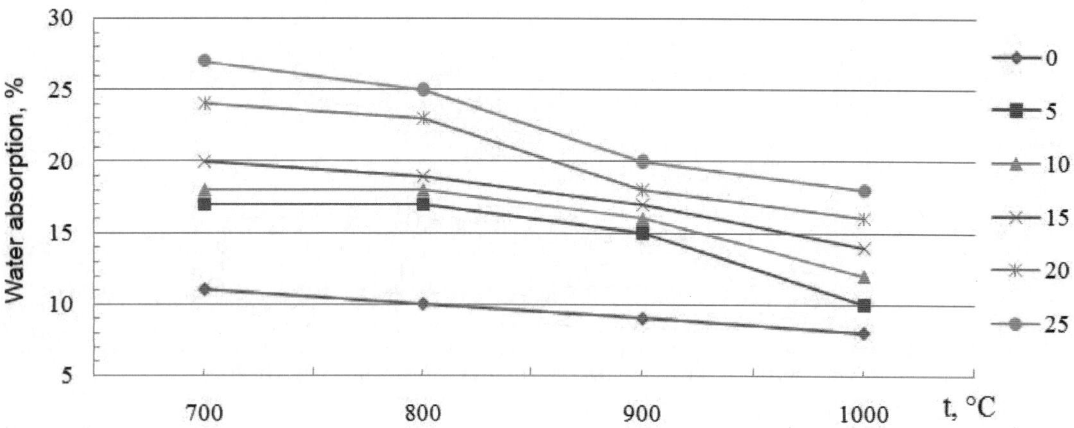

Figure 3 - Backing temperature dependence of the water absorption capacity.

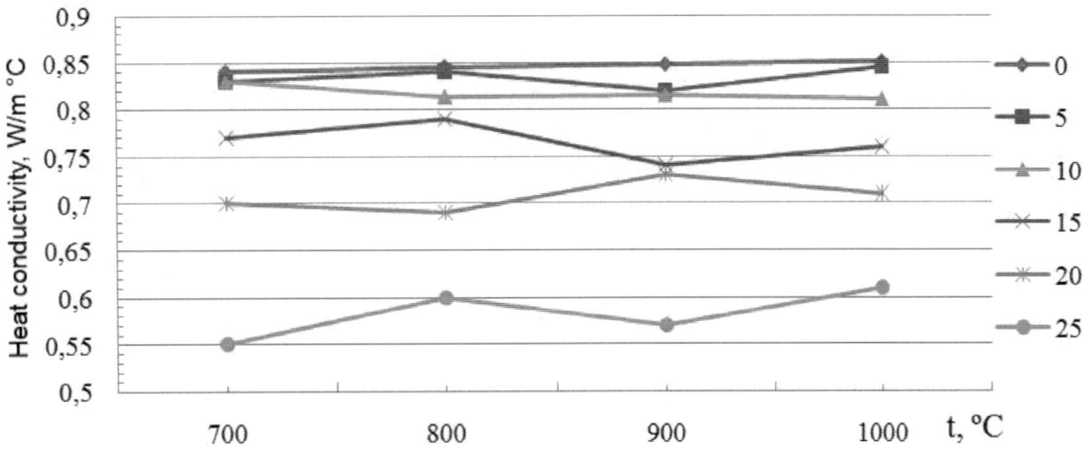

Figure 3 - Backing temperature dependence of the heat conductivity capacity.

Conclusions

The studies found that the heat conductivity, the ultimate compression strength and the water absorption capacity depend on both the amount of ash added and baking temperatures. The greater the ash and slag content in bricks, the lower their heat conductivity is. The water absorption capacity increases with an increase in ash content. The ultimate compressive strength also decreases with increasing the ash content in ceramics. The optimum percentage of adding ash and slag waste is fifteen percent at a baking temperature of 1000 ° C.

It has been established, if ash and slag waste is of an acidic type, it does not have self-sustained bonding properties and require adding curing agents.

The analysis of the particle-size distribution showed that 60% of particle size is 10 to 70 microns; the material is finely dispersed and can be used as a filler in the production of construction mixes and ceramics.

References

[1] Putilin E I Obzornaja informacija otechestvennogo i zarubezhnogo opyta primenenija othodov ot szhiganija tverdogo topliva na TJeS [Survey of domestic and foreign experience in using waste from burning solid fuels by thermal power stations] / E I Putilin, V S Tsvetkov. - Moscow: Soyuzdornii, 2003 - 60 p

[2] Zyryanov V V Zola-unosa – tehnogennoe syr'e [Fly ash – industry-related raw materials] / V V Zyryanov, D V Zyryanov Moscow: OOO IPTz "Maska", 2009 - 320 p.

[3] Dik E P, Mashkovich K I, Vasilchenko Z A, Komponentnyj sostav zoloshlakovyh othodov TJeC // *Novoe v rossijskoj jelektrojenergetike [Chemical composition of ash waste from thermal power plant // New in Russian power sector]* 2003 No.5. – 40 p

[4] Spirin E K, Lugovtsova N Yu, Khomich O I Special features of VP-1Ap anion exchanger behavior under deep silicifi cation // *Russian Journal of Applied Chemistry*. – 2012 – Vol. 85 – p.921-925.

[5] Torosyan V F , Torosyan E S , Sorokin P D, Telitsyn A A Updating of sewage – purification facilities of electroplating enterprises with counterflow ion-exchange filters // *IOP Conference Series: Materials Science and Engineering*. – 2015 – Vol. 91, Article number 012077. – pp 1-8

[6] Bespalov V I, Prirodoohrannye tehnologii na TJeS: uchebnoe posobie [Environmentally freindly technologies at thermal power plants: Workbook] / V I Bespalov, S U Bespalova, M A Wagner. - Tomsk: Publishing House of Tomsk Polytechnic University, 2010 – 240 p.

[7] Putilov V Ya O roli i meste nauki v reshenie problemy obrashhenija s zoloshlakami TJeS [The

IOP Conf. Series: Materials Science and Engineering **127** (2016) 012024 doi:10.1088/1757-899X/127/1/012024

role and place of science in addressing the issue of treating ash and slag from thermal power plants] / V Ya Putilov, I V Putilova. Warsaw: Energoizdat, 2010 – 320 p.

[8] Gavrilov V I Toplivno-transportnoe hozjajstvo i zoloshlakoudalenie na TJeS [Fuel and transport facilities and ash and slag removal at TPP] /V I Gavrilov Moscow: Energoizdat, 1987. 168 p.

[9] Borisenko L F Perspektivy ispol'zovanija zoly ugol'nyh teplovyh jelektrostancii [Prospects for the use of ash from coal-fired power plants] / L F Borisenko, L M Deltsin, A S Vlasov. - Moscow: ZAO "Geoinformmark", - 2001 -68 p.

[10] Tymoshchuk N S Primenenie zoly i shlaka Burshtynskoj GRJeS v zhelezobetonnyh izdelijah dlja dorozhnogo stroitel'stva [Use of ash and slag from Burshtyn TPP in concrete products for road construction / N S Tymoshchuk, I.S. Bobyk. - Moscow: Mir, 1991 – 70 p.

[11] Bazhenov Yu M, Schebenkin P F, Dvorkin L I Primenenie promyshlennyh othodov v proizvodstve stroitel'nyh materialov[Use of industrial wastes in the production of construction materials]. - Moscow: Stroyizdat, 1986 -120 p.

[12] Bozhenov P I Kompleksnoe ispol'zovanie mineral'nogo syr'ja dlja proizvodstva stroitel'nyh materialov [Multiple use of mineral raw materials for the production of construction materials] / P.I. Bozhenov. - Moscow: Stroyizdat, 1963 - 170 p

[13] Tabin Rushad S 2011 Experimental Studies on Lime-Soil-Fly Ash Bricks (International journal of civil and structural engineering, volume 1, № 4)

[14] Obada Kayali 2005 High Performance Bricks from Fly Ash. (USA, Kentucky, Lexington World of Coal Ash, April 11-15)

[15] Henry Liu 2009 Environment Properties of Fly Ash Bricks, (USA in Lexington World of Coal Ash Conference – May 4-7)

[16] Kevan D Sharp. 1993 Fly ash for soil improvement: proceedings of sessions sponsored by Committees on Soil Improvement and Geosynthetics of the Geotechnical Engineering Division of the American Society of Civil Engineers in conjunction with the ASCE Convention in Dallas (Texas, October 24-28)

Technology of Water Purification With Chlorinated Derivatives and Assessment of Risk Associated With Human Exposure to These Substances

S S Timofeeva[1a], DScTech, Prof., Ye A Khamidullina[1b], Cand.Sc., AP, O A Davydkina[1c], magi strand, N Yu Lugovtsova[2d], assistent

[1]Irkutsk National research technical university. Russia, Irkutsk, Lermontova str., 83
[2]Yurga Institute of technology (branch) of National research Tomsk polytechnic university. Russia, Yurga, Leningradskaya str., 26.

E -mail: [1a]timofeeva@istu.edu, [2d]lnyu-70583@bk.ru.

Abstract. In the given paper the authors consider the technology of water purification with consideration to the recommendations of the World Health Organization (WHO), European Union (EU) and standards of developed countries. Carcinogenic Unit Risk (UR) magnitude for people constantly exposed to the analyzed carcinogens in the course of a lifetime is estimated. The authors calculate and evaluate unique carcinogenic risk as a complementary probability of cancer development during the whole life of CR when introducing EU standards into water purification technology.

1. Relevance

One of state policy directions is public health protection and improvement of the quality of life by ensuring uninterrupted water supply and water disposal as well as by reducing negative impact upon water objects by means of improving the quality of waste water treatment [1,2].

At present time continuous deterioration of surface sources of drinking water and practically not improving state of water conditioning objects and centralized water supply systems make these problems extremely important. That is why authorities of the Russian Federation are introducing EU standards and WHO recommendations into the technologies of water purification. Due to this the object of the work is estimation of carcinogenic risks associated with implementation of hygienic standards of EU countries.

In the recent years water resources are exposed to intensive pollution. Water conditioning and water disposal facilities where traditional methods and procedures are applied are not always able to ensure the required degree of water purification. This fully refers to water decontamination – the main barrier against water infections [3].

Water treatment in order to prepare it for drinking is a complex of physical, chemical and biological methods of changing its original composition. The great variety of water treatment methods can be divided into the following basic groups: improvement of organoleptic properties of water (clarification and discoloration, deodorization and other); ensuring epidemiological safety (chlorination, ozone treatment, ultraviolet radiation and other); mineral composition conditioning (fluoridation, defluorination, extraction of heavy metals ions, deferrization, demanganation, demineralization or desalting and other). The method is chosen basing on preliminary study of composition and properties of the considered water resource and their comparison to the customer's requirements [4].

2. Material and methods of research

In the practice of drinking water treatment one of the main methods of treatment ensuring its reliable decontamination and allowing to maintain the sanitary state of waste water treatment systems is chlorination. Alternative technologies of decontamination used for water treatment do not always meet such requirements as prolonging action, low capital costs, formation of toxic agents, safe production process. At the same time chlorination promotes formation of dangerous chemical substances affecting people's health.

Recent research showed that the water contains toxic volatile organohalogen compounds (VHC). These are, mainly, compounds falling into the group of trihalometans (THM): chloroform, dichlorobromometane, dibromochloromethane, bromform and other, having carcinogenic, mutagenic activity and cumulated in living organisms.

For the purposes of harmonization with the recommendations of the World Health Organization (WHO) and European Union (EU) and standards of developed countries in 2002 the Russian Federation corrected hygienic rating for halogen-containing compounds formed in the process of chlorination in the drinking water. As we can see from Table 1 the ratings of ten halogen-containing compounds were harmonized: MPC of four substances were corrected, three of them being carcinogenic, APL of five compounds and MPC of chlorine cyanide were re-established, MPC of chloroform was reduced up to 0.06 mg/l, MPC of carbon tetrachloride was reduced by 3 times and that of chloral hydrate – by 20 times [5].

Pollution with organohalogen compounds is mainly anthropogenic. There are two possible sources of VHC in drinking water:

a) resulting from polluting water sources with industrial waste waters containing VHC. It should be noted that surface water sources, as a rule, contain small quantities of VHC as surface water is characterized by active self-purification processes; besides, VHL are removed from water by surface aeration. The main sources of these substances in the environment are organic synthesis companies, hydrolytic, cellulose and paper mills, woodworking, coke-chemical, varnish-and-paint, pharmaceutical production;

b) formation of VHC in the process of water treatment resulting from reactions between chlorine and substances in the primary water. The main precursors of these toxic compounds are natural humic and fuvic acids. The largest group of organohalogen compounds formed when treating drinking water with disinfecting agents is made up of trihalomethans: chloroform, bromform, dibromochloromethane, dichlorobromometane. Derivatives of bromine are formed when primary water contains bromides or when the chlorinating agent is polluted with bromine [6]. Concentration of the formed VHC is significantly influenced by the amount of plankton in the primary water.

Beside trihalomethans the researchers found chlorinated ketones, chloromethyl cyanides, chloracetic acids, chloral, chloropicrin, chlorophenols. The process of these products formation and their content ratio is influenced by many factors: nature of organic substances in the primary water, bromine content, pH, temperature, nature and amount of chlorinating reagent, time of chlorination, season, also, further transformations of organohalogen compounds in the water-supply system are possible. Influence of the large amount of factors makes it difficult to predict the composition of possible products. It has been shown that presence of humic acids leads to formation of volatile organohalogen compounds, that of fulvic acids leads to formation of chloracetic acids. With Ph increase from 5 to 9c ontent of trihalomethans, particularly chloroform, significantly increases, and content of chloracetic acids, acetone trichloride, acetone dichloride, chloropicrine decreases [6].

Table 1 – Hygienic normatives of halogen-containing compounds in the drinking water which are formed when the water is chlorinated

Compounds	MPC, APL, mg/l before 2002	Corrected normative, mg/l	Potential health impact	Class of hazard
Trihalomethans				
Chloroform	0.2 s.-t.	0.06 s.-t.	Carcinogenic effect	1
Bromodichloromethane	0.03 s.-t.			2→1
Dibromochloromethane	0.03 s.-t.		Hepatotoxic action	2
Bromform	0.1 s.-t.			2
Chlorophenols				
2.4-dichlorophenol	0.002 org.		Change of water taste	4
2.4.6-trichlorophenol	0.004 org.	0.004 s.-t.	Carcinogenic effect	4→1
Chlorophenol	0.001 org.		Water odor appears	4
Chlorine-containing acetic acids				
Chloracetic acid	0.06 s.-t.			
Dichloracetic acids		0.05 s.-t.	Hepatotoxic action	2
Trichloracetic acids		0.01 s.-t.		
Halogen-containing acetonitriles				
Dibromacetonitrile		0.1 s.-t.	Weight loss	2
Dichloroacetonitrile		0.1 s.-t.	Teratogenic effect	2
Trichloroacetonitrile		0.001 s.-t.		1
Other compounds				
Chloral hydrate	0.2 s.-t.	0.01 s.-t.	Hepatotoxic action	2
Tetrachloride carbon	0.006 s.-t.	0.002 s.-t.	Carcinogenic effect	2→1
Chlorine cyanide (for cyanide)		0.035 s.-t.	Neurotoxic action	2→1

The basic concentrations of VHC are formed at the stage of primary chlorination of unpurified water. Most times THM and tetrachlorine carbon are formed. The amount of chloroform usually exceeds the amount of other VHC by 1-3 decades.

In 1980-1990s, when a number of works was published on halogen-containing compounds in the drinking water, the problems associated with effects of these substances on human health were considered least of all. For a long time one of the most important sides of their biological effect – delay effect, namely, carcinogenic one, was underestimated. Although the experimental data on chloroform cancer effect on animals were obtained and it was included into group 2B (factors for which there is limited (or insufficient) evidence of their cancer effect upon humans and almost enough evidence of that for animals) according to the classification of International Agency for Research on Cancer (IARC), in half of the cases the epidemiological survey did not clearly show increase of cancer rate for the people who had been drinking chlorinated water. In the middle of the 1990s studies were completed that made the scientists to see the products of water chlorination previously considered not so dangerous in a new light. Modern methods of chemical analysis allowed identifying several dozens of such compounds. Reliable information on epidemiology of cancer cases of bladder, large and straight intestines, pancreatic gland and brain associated with drinking chlorinated water containing THM. A number of epidemiologic surveys revealed effect of VHM upon the female reproductive system: increasing the frequency of impairments of gestation course, of prenatal development and appearance of congenital malformations caused by chlorinated drinking water with THM concentrations over 80-100 mg/l. So, by 2000, much evidence of unfavorable effect of chloroform on public health had been collected [7].

It has been also established that chlorine used for decontamination of running water effects people at home in several ways: not only through swallowing but also through the lungs, with inhaled air.

The results of research presented in Table 2 demonstrate the fact of chloroform passes from running water into the air of the living space when people take a bath or a shower.

According to the data of the research quite high concentrations of chloroform (167-232 mg/l) were also found in the air layer over the indoor swimming pools [8].

Such criteria as toxic and carcinogenic properties (group B2 according to the classification of the International Agency for Research on Cancer), probability of effecting population as a whole, distribution in the environment and capability of the compound for cross-media distribution determined the choice of chloroform, side product of water decontamination with chlorine-containing compounds, as of one preferred pollutants of drinking water characterized by high medical and social significance [8].

Table 2 – Influence of water temperature when filling the bath and type of room when taking a shower upon chloroform content in bathroom air [7]

Water temperature when filling the bath, °C	Chloroform content		
	water, mg/l	air, mg/m³	
		before	after
		filling the bath with running water	
20	145	5	6
35	160	3	44
Taking a shower in the bathroom	180	4	230
Taking a shower in the shower unit	112	3	296

According to modern international practices the risk assessment methodology is applied for integral assessment of environmental factors effect on personal and public health. Interpreting the quality indices of the drinking water as risk factors will allow forecasting public health and estimate the probability of pathologic processes development determined by water contaminating substances. It should be noted that there is an opinion [8] that the probability of oncological diseases is not the same not only for different water distribution systems but also for different parts of one water distribution system. Due to this fact the risks associated with chlorination side products in the drinking water may differ for the population of the same district.

We have estimated the risks of general effect of drinking water chloroform upon the people's health. The initial data were the monitoring data of Federal Service on Customers' Rights Protection and Human Well-being Surveillance for Irkutskaya region obtained in 2013 when chloroform was included into the list of substances to be controlled in the drinking water of the systems of centralized domestic water supply in the town of Irkutsk.

To estimate the exposition we completed calculations of chloroform intake from the water of domestic water supply into the human organism by three possible routes: oral I_o; inhalative I_i and through skin (absorbed) I_d. Intake of chemical substances was calculated according to the formulas taking into account effecting concentrations, contact, exposure duration and frequency, body mass and the time of exposure averaging.

The intake established on the base of Federal Service on Customers' Rights Protection and Human Well-being Surveillance data is provided in Table 3.

Unit risk (UR) was assessed as the upper, conservative estimate of carcinogenic risk for a person life-long exposed to constant effect of the analyzed carcinogen in the concentration 1 mg/m^3 (atmospheric air) or 1 mg/l (drinking water).

Table 3 – Average daily doses of chloroform intake under multi-route exposure

Sampling point	Actual concentration (average value, mg/dm^3)	Normative, mg/dm^3	I_o, mg/(kg×day)	I_i, mg/(kg×day)	I_d, mg/(kg×day)
Irkutsk, water supply point 1	0.009	0.06	$11\cdot10^{-5}$	0.24	$3\cdot10^{-5}$
Irkutsk, water supply point 2	0.008	0.06	$9\cdot10^{-5}$	0.21	$3\cdot10^{-5}$

Calculation of unique carcinogenic risk as complementary probability of life-long cancer development risk was estimated with consideration to lifetime average daily dose.

Population carcinogenic risks (PCR) reflecting complementary (to background) number of malignant neoplasms cases which may occur in the course of a lifetime due to effect of the studied factor. Unique and population carcinogenic risks characterize the upper estimate of possible carcinogenic risk for the period corresponding to average life expectation of a person (70 years).

Annual population carcinogenic risk (PRCa) was estimated as calculated number of complementary cancer cases during a year.

The results of calculations of the given kinds of carcinogenic risk are given in Table 4.

Table 4 – Results of carcinogenic risk calculations

Sampling point	UR_i, m^3/mg	UR_o, mg/l	CR_o	CR_i	CR_d	PCRo	PCRi	PCRd	PCRao	PCRai	PCRad
Irkutsk, sampling point 1	0.002	0.0002	$6\cdot10^{-7}$	$2\cdot10^{-3}$	$9.8\cdot10^{-7}$	0.3	1212	0.59	$5\cdot10^{-3}$	17	$8\cdot10^{-3}$
Irkutsk, sampling point 2			$6\cdot10^{-7}$	$1\cdot10^{-3}$	$8.3\cdot10^{-7}$	0,3	606	0.50	$5\cdot10^{-3}$	9	$7\cdot10^{-3}$

When calculating total carcinogenic risks we should take into consideration the differences in the intensity of carcinogenic activity of chemical substances under different intake routes. As the values of carcinogenic potential factors are different for different intake routes risk calculation on the base of total doses is appropriate only for the similar routes.

The obtained risk esteems show that for oral and percutaneous exposure routes the risk is at minimal level and it is perceived by the people as small to negligible not differing from usual everyday risks (De minimis level). Such risk does not require any additional measures to be reduced and its level is only to be monitored at regular intervals. The risk of cancer development under inhaled route of exposure makes $1 \cdot 10^{-3} - 2 \cdot 10^{-3}$ and exceeds the values of maximum allowed risk for the population (for example, for the drinking water WHO sets $1 \cdot 10^{-5}$ as allowed risk and for atmospheric air – $1 \cdot 10^{-4}$). The risk magnitude is acceptable for professional groups and inacceptable for population at large.

The values of population carcinogenic risk characterize the long-term trends of alteration of oncological background at the studied territory. The given kind of risk turned insignificant for oral and percutaneous routes of exposure. The population carcinogenic risk is significant under the inhalative route of chloroform exposure. According to the calculations 606 and 1012 of complementary (to background) cases of malignant neoplasms cases could have developed due to effect of inhaled evaporating chloroform. Unique and population carcinogenic risks characterize the risk magnitude in the course of a lifetime of a person.

The given regularity is true for the magnitudes of annual population carcinogenic risk. Its calculation for one year of life also has a large value under inhalative exposure. Annual population risk makes from 9 to 17 complementary cancer cases during the year. The amount of complementary cancer cases during the year for oral and percutaneous routes of exposure allows neglecting annual carcinogenic risk of drinking water effect upon population.

The values of carcinogenic risks reflect mainly the long-term trend for alteration of oncological background. The trend is formed when all reference conditions accepted by the researcher are met (for example, certain duration and intensity of exposure, stationarity of exposure, certain values of the factors of exposure and other).

Analysis of unique carcinogenic risks distribution showed that the main contribution into formation of total unique risk associated with drinking water chloroform is made by the risk formed under the inhalative route of exposure. This risk requires developing and taking health-improving measures. Planning the risk-reducing measures in this case must be based on the results of in-depth assessment of various aspects of integrated effect and priority ranking in relation to other hygienic, ecological, social and economic problems of the region.

Further in the work estimation of non-carcinogenic risks of the integrated effect of drinking water chloroform was completed.

Characteristics of risks of non-carcinogenic effects development for specific substances is completed on the base of Hazard Quotient calculation:

$$HQ = AD/RfD \text{ or } HQ = AC/RfC,$$

where HQ – Hazard Quotient; AD – average dose, mg/kg; AC – average concentration, mg/m^3; RfD – reference (safe) concentration, mg/m^3.

The maximum permissible concentration of chloroform in the drinking water (0.06 mg/l) was taken as equivalent of reference concentration (RfC). The calculation was completed according to the following formula:

$$HQ = C_{aver.}/RfC,$$

where $C_{aver.}$ – real concentration, mg/dm^3.

Calculation results are provided in Table 5.

Table 5 – Results of Hazard Quotient calculation

Sampling point	Actual concentration (average value, mg/dm^3)	Normative, mg/dm^3	HQ
Irkutsk, facility № 5	0.009	0.06	0.15
Irkutsk, 2nd water lift Yershi	0.008	0.06	0.13

3. Conclusion

If Hazard Quotient (HQ) of the substance applied in water purification technologies to eliminate VHC according to the recommendations of World health Organization (WHO), European Union (EU) and standards of developed countries does not exceed one, the probability of harmful effects caused by everyday intake of chloroform in the course of a lifetime is insignificant and such effect is characterized as permissible.

The conducted study of water purification technology according to new standards with account to full route of real drinking water chloroform intake provide evidence of monitoring necessity not only for carcinogenic but also for general toxic risks of chloroform exposure. Quantitative characteristics of multi-medium risk allows developing optimal managing solutions for its reduction on the base of estimation of all contact media and intake routes with consideration to their contribution into levels of exposure.

4. References

[1] Federal Act dated 07.12.2011 № 416-FZ (ed. 28.12.2013) "On water and sanitation" (with amendments, came into effect after 01.01.2014) // ["Official Gazette of RF", 12.12.2011, № 50, ar. 7358].

[2] Torosyan V F, Torosyan E S, Sorokin P D, Telitsyn A A Updating of sewage – purification facilities of electroplating enterprises with counterflow ion-exchange filters // *IOP Conference Series: Materials Science and Engineering.* – 2015 – Vol. 91, Article number 012077. – pp 1-8

[3] Valuev D V, Semenok A A, Kotova D O, Valueva A V Prospects for of processing car tires // *Applied Mechanics and Materials.* – 2014 – Vol. 682 – pp 75-79

[4] Frog B N Water conditioning and purification: manual for higher school / B N Frog, A P Levchenko – M.: MSU press, 1996 – 680 p.

[5] New reference book for chemist and technologist. Radioactive substances. Hazardous substances. hygienic normative. – SPb: ANO NPO "Professional", 2004. – 1142 p.

[6] Kirichenko V Ye Organohalogen compounds in drinking water and methods of their determination / V Ye Kirichenko, M G Pervova, K I Pashkevich // Rossiysky khimichesky zhurnal. – 2002. - № 4. – pp 18-27

[7] Iksanova T I hygienic assessment of integrated effect of chloroform in the drinking water / T I Iksanova, A G Malysheva [and others] // *Gigiyena I sanitariya* – 2006. - № 2. – pp 18-27

[8] Kofman V Ya Chlor-organic compounds in water and in the air: risk for public health // SanEpidem kontrol – 2015 – No 1 pp 121-128

[9] Dmitrenko Ye A Hygienic assessment of carcinogenic and toxic risks of integrated effect of chloroform / Ye A Dmitrenko // *Gigiyena naselennykh mest* – 2010. - No55 – pp 157-161

Biotesting as a Method of Evaluating Waste Hazard in Metallic Mineral Mining

Y R Lugovaya[1,a], K N Orlova[1,b], S V Litovkin[1,c], A G Malchik[1,d], M A Gaydamak[1,e]

[1] Yurga Institute of Technology, National Research Tomsk Polytechnic University Affiliate
652050, Yurga, ul. Leningradskaya 26

E-mail: ayuliapetkova89@gmail.com, [b]lesta@tpu.ru, [c]protoniy@yandex.ru, [d] ale-malchik@yandex.ru, [e] vip.trd777@mail.ru

Abstract. As the result of identifying content of each chemical element it has been revealed that metallic mineral wastes have a considerable amount of valuable useful metals. Thus, large-tonnage inorganic wastes are considered to be an additional raw source of metal production. This paper highlights the necessity of supplementary biotesting metallic mineral wastes in order to bring into correlation with corresponding hazard classes and facilitate efficient recycling of these wastes in future. It has been found out that determined in this way waste class can be dumped or used after recycling. It has been also indicated that mill tailings are to be stored according to contained metals without messing up dissimilar metal-containing wastes. After winning metals these wastes are similar to the group of inorganic non-metallic wastes and can be used in building material production, for filling mined-out spaces, in road construction etc.

The notion "waste" is defined in Waste Frame Directive as any substance or object utilized by an user on his own or in terms of national legislation [1]. This definition is too generalized so in appendix of the Directive there is a list of certain substances that can be regarded as wastes. This list is constantly changed, supplemented and currently contains more 600 waste types.

The amount of industrial wastes rises every year because of ore-dressing and processing enterprises. Anthropogenic load arising in the course of mineral deposit exploitation includes grounds beneath overburdens, mine dumps and waste banks. Extracted raw materials are delivered to processing plants affecting the environment negatively. Processing plants majoring in iron ore dressing destroy Earth surface via mining, discharge air pollutants into atmosphere, dump contaminated wastewater and produce used fine-grained burden, that is, industrial wastes. Used materials are delivered by hydraulic conveying into tailing dumps. Tailing dumps are aggregated wastes of ore-dressing and processing enterprises. The scales of aggregated material are rather considerable.

Any waste operations produced as the result of mining, processing and storing mineral resources and opencast mine activities are legislatively regulated in Mining Waste Directive.

In terms of up-to-date legislation [1, 3, 4] all industrial wastes are to be classified according to their impact on the environment and mankind. Therefore, all wastes are divided into hazard classes.

Investigation into wastes is necessary for identifying optimal technologies of their processing, storing and utilizing, as well as for compiling standards on waste generation and limits of their licenses, waste certificates.

International classification of industrial wastes accepted in all organizations-members of World Bank of industrial wastes highlights solid (S), liquid (L) and paste-like (F) wastes. Each waste is specified to eight parameters (waste state, basic component of the waste, attending substances, acid or alkaline nature of wastes, inflammability of wastes, and possibility of recycling or mixing up with household wastes for disposal). Industrial wastes are coded according to the System based on International Standard of Industry Classification.

If wastes contain chemical substances in a definite concentration they are singled out in a separate class and called hazardous. However, all waste classifications are conventional and can vary with time. In some cases wastes are classified according to the possibility of their utilization [5].

Resolution of waste problem necessitates, first, separation of so-called large-tonnage wastes, because they have the heaviest contaminating impact on the environment.

In EU countries wastes are principally divided in hazardous and nonhazardous. The key difference between them is the extent of harmful effect on the environment and technologies required for their recycling and utilization. The list of hazardous wastes and decision criteria of their hazardous character are provided in Directive 91/689/EEC on hazardous wastes [1]. The Directive determines 40 hazardous wastes and 14 criteria to consider them hazardous.

An integrated laboratory study is to carry out for identifying the class of waste hazard. Methods for investigation of industrial wastes include:

- quantitative and chemical analysis (component content and concentration);
- toxicological research (toxicity degree);
- biotesting (hazard revealing from the standpoint of ecology, by means of protozoa).

Computational method to put wastes into a hazard class is applied provided quantitative and qualitative composition of a waste is known.

Experimental method on the base of biotesting is applied in the following cases:

- confirmation of nonhazardous class determined computationally is required;
- qualitative and quantitative composition of wastes can't be determined.

To sum up the stated above, if a waste is computationally determined as a nonhazardous one; its nonhazardous nature is to be confirmed via biotesting. Unless the hazard class is experimentally confirmed, wastes are considered non-hazardous substances [7, 8].

Computational determination of waste hazard class in metallic mineral mining

Therefore, our work was focused on determination of waste hazard class computationally and via biotesting. As the object of research we chose the enterprise majoring in iron ore dressing by wet magnetic separation and sintered ore production. The reference material is waste from a tailing dump.

Abagur processing factory specializes in recycling primary concentrate located in Tashelgin, Telbes, Tersin and other groups of West Siberian Plain. The deposits are rich in iron ore minerals and fluxing limestone used for secondary concentrate production. There is an extensive network of railway and automobile approach lines on the factory ground. As the result of enterprise activity large-tonnage inorganic metal containing wastes are produced and their further recycling needs separation and classification.

Table 1 – Composition and content of wastes

Sample	Index	Test results		MPC mg/kg
		%	mg/kg	
Wastes of metallic material mining (wastes of wet magnetic separation)	Moisture	4.77		
	Cadmium		< 0.05	1.0
	Lead	0.00116	11.60	32.0
	Vanadium	0.00370	37.02	150.0
	Manganese	0.13770	1377.00	1500.0

Sample	Index	Test results		MPC mg/kg
		%	mg/kg	
	Copper	0.05126	512.60	132.0
	Arsenic	0.00604	60.40	2.0
	Mercury	0.00002	0.16	2.1
	Cobalt	0.01276	127.60	
	Nickel	0.00857	85.72	40.0
	Zink	0.02431	243.10	220.0
	Petrochemicals	0.00149	14.9	
	Nitrates	0.00275	27.5	130.0
	Cyanides		<0.5	
	Benzapyrene		<0.001	0.02
	Phenol		<0.01	
	Carbon	1.42		
	Hydrogen	0.76		
	Silicon dioxide	41.22		
	Aluminum oxide	9.21		
	Ferrum trioxide	17.04		
	Titanium dioxide	0.44		
	Calcium oxide	12.45		
	Magnum oxide	8.80		
	Phosphor oxide (V)	0.41		
	Potassium oxide	0.62		
	Sodium oxide	0.96		
	Sulphur total	1.650	16500.0	160.0
	Fluorine	0.00020	2.0	6.0

The index "Hazard rate" of waste component for the environment Ki is calculated on the base of data on waste composition and content and from the formula:

$$K_i = \frac{C_i}{W_i}, \tag{1}$$

where Ci – concentration of i-component in the waste (mg/kg of waste);
Wi – coefficient of hazard rate of i-component for the environment, (mg/kg) [8,9].

Consolidated index is calculated from the formula:

$$\sum_{i=1}^{m} K_i = \sum_{i=1}^{n} \frac{C_i}{W_i}, \tag{2}$$

The hazard class of wastes depending on the hazard rate is determined on the base of data from Table 2.

Table 2 – Hazard class - hazard rate correlation (K)

Hazard class	Waste hazard rate, K
I	106≥ K >104
II	104≥ K >103
III	103≥ K >102
IV	102≥ K >10
V	K=10

Results of calculation are in Table 3.

Table 3 – Reference data to calculate waste hazard class

№	Component	C_i (mg/kg)	W_i (mg/kg)	K_i
1	Cadmium	0.050	215.443	0.00023
2	Manganese	1377	5878.016	0.23426
3	Copper	512.6	1930.698	0.26550
4	Arsenic	60.4	803.086	0.07521
5	Nickel	85.72	1550.516	0.05528
6	Benzapyrene	0.001	59.97	0.00002
7	Mercury	0.16	17.013	0.00940
8	Lead	11.6	701.704	0.01653
9	Water	47699.839	1000000	0.04770
10	Phenol	0.01	215.44	0.00005
11	Silicon dioxide	412200	1000000	0.41220
12	Aluminum oxide	92100	1000000	0.09210
13	Titanium dioxide	4400	1000000	0.00440
14	Calcium oxide	124500	1000000	0.12450
15	Manganese oxide	88000	1000000	0.08800
16	Potassium oxide	6200	1000000	0.00620
17	Sodium oxide	9600	1000000	0.00960
18	Phosphorus oxide (V)	4100	1000000	0.00410
19	Cobalt	127.6	1000	0.12760
20	Petrochemicals	14.9	1000	0.01490
21	Carbon	14200	1000000	0.01420
22	Hydrogen	7600	1000000	0.00760
23	Cyanides	0.5	215.443	0.00232
24	Ferrum trioxide	170400	1000000	0.17040
25	Zink	243.1	3727.594	0.06522
26	Vanadium	37.02	2404.099	0.01540
27	Sulphur	16500	8576.959	1.92376
28	Fluorine	2	464.159	0.00431
29	Nitrates	27.5	13111.339	0.00210
	Total:	1000000		3.79309

Consolidated index $\Sigma K = 3.79309$, as the consequence, waste is in the 5th hazard class ($\Sigma K < 10$).

Biotesting determination of waste hazard class

As it was stated above an experimental method based on biotesting is used to confirm nonhazardous waste class defined computationally.

Experimental determination of waste hazard class requires at least two test-objects from various groups (Daphnia and Infusorian, Ceriodaphnia, bacteria, algae etc.). Hazard class revealed on a bio-object, which was the most sensitive to the analyzed substance, is accepted as an ultimate result.

The mostly used objects of bio-tests are as follows: Chlorophyta - Scenedesmus (Scenedesmus quadricauda) and Chlorella (Chlorella vulgaris); Crustacea - Daphnia (Daphnia magna Straus) and Ceriodaphnia (Ceriodaphnia affinis Lilljeborg); protozoa - Infusorian (Paramecium caudatum Erenberg).

Methods of biotesting:

1. Methods to test algae bases on registration of falling chlorophyll fluorescence level and decreasing quantity of algae cells exposed to toxic substances in aqueous extract of wastes and comparison with a reference sample in tests without toxic substances.

The criterion of acute toxicity is reducing quantity of algae cells by more 50 % as compared with the reference quantity in 72 hours.

The experiments are to identify:

1) acute toxicity or inhibiting concentration (further - IC50-72) of particular substances or inhibiting order of dissolution (further - IOD50-72), that is, such quantity of toxic substance, which causes reducing chlorophyll level by more 50% as compared with the reference quantity or decreasing number of algae cells in 72 hours;

2) nonhazardous (conditioning no effect of acute toxicity) concentration (further - NH20-72) of some substances and nonhazardous order of dissolution (NOD - BKR 20-72), that is, such quantity of toxic substance, which causes 20% reducing chlorophyll level as compared with a reference quantity or decreasing number of algae cells in 72 hours [15].

2. Methods to calculate death rate and change in fertility of Daphnia (Daphnia magna Straus, Cladocera, Crustacea) affected by toxic substances of wastes base on their lethality over a particular period of exposure.

The criterion of acute toxicity is death of more 50 % Daphnia for 96 hours in water under consideration provided that the death rate doesn't exceed 10% in reference experiment.

In short-termed experiments aimed at determining acute toxic effect one identifies:

1) acute toxicity or mean lethal concentration (here and further LC 50-96) of particular substances (order of waste dissolution), conditioning death of more 50 % test-organisms;

2) nonhazardous (causing no effect of acute toxicity) concentration of particular substances, conditioning death less 10 % test-organisms (NHC10-96, NHC-96) [6].

To determine acute toxicity of the tested solution the percentage of the dead in water under consideration is calculated in comparison with the reference sample. If the death rate of life forms in the solution is $\leq$ 10%, the tested sample doesn't have an acute toxic effect (nonhazardous order of dissolution). If the death rate is $\geq$50% the tested aqueous extract has an acute toxic impact [16].

Hazard class is identified according to the order of dissolving waste aqueous extract, which doesn't affect life forms as on order of dissolution provided in Table 4.

Table 4 – Dependence of waste hazard class on dissolution order of aqueous extract

Waste hazard class	Dissolution order of hazardous waste aqueous extract without harmful effect
I	>10000
II	10000 – 1001
III	1000 – 101
IV	<100
V	1

The obtained results of bio-tested waste aqueous extract were recorded in the form of tested samples presented in Table 5.

Table 5 – Record form of bio-tests

	Test-object	Observation period	Biotesting results	Order of water dissolution (aqueous extract) without any effect on test-object	Tested sample estimation
	Crustacea *Daphnia magna* Straus FR.1.39.2007.03222	96 hours (4 days)	Daphnia death rate		No pronounced toxic effect
			100% solution = 10%	1	
			50% solution = 0%		
			25% solution = 0%		
			10% solution = 0%		
			NHC_{10-96}=1		

	Test-object	Observation period	Biotesting results	Order of water dissolution (aqueous extract) without any effect on test-object	Tested sample estimation
	Algae *Scenedesmus quadricauda* FR.1.39.2007.03223	72 hours (3 days)	Reducing number of algae		No pronounced toxic effect
			100% solution = 4%	1	
			50% solution = 0%		
			25% solution = 0%		
			10% solution = 0%		
			NHC $_{20\text{-}72}$=1		

Therefore, we can sum up the results of carried out bio-test.

1. The death rate of Daphnia in singly dissolved sample is 10 %, that is the waste under consideration has no toxic impact;
2. The decrease in population of algae in 100% dissolution is 4 %, so there is no toxic effect on life forms.
3. The bio-tests of two waste samples produced in metal mineral mining have revealed that the tested sample has no toxic effect on algae and Daphnia $C_{nonhazardous}$ 1. As the result, hazard class to be recommended is a nonhazardous one.

Conclusions

1. International classification says that wastes of metallic mineral mining are large-or-middle-grained solids; their main component is heavy-density metal that can be used after recycling.
2. As the result of identifying content of each chemical element it has been revealed that metallic mineral wastes have a considerable amount of valuable useful metals. Thus, large-tonnage inorganic wastes are considered to be an additional raw source of metal production.
3. The analysis of obtained data of bio-tests and computations has shown that metallic mineral wastes of tailing dump №3 are environmentally nonhazardous substances.
4. The wastes have a slight impact on the environment. This waste class is allowed to be dumped. Collecting wastes of this class contributes to nature resource preservation because they can be used after recycling.
5. Wastes of this class are to be stored in the way facilitating their further recycling. Tailings are to be stored according to the metals they contain, no messing up different metal containing wastes is allowed. Various dressing methods are used for recycling these wastes, among them is pyrometallurgy, chemical methods and biotesting. After winning metals these wastes are similar to the group of inorganic non-metallic wastes and can be used in building material production, for filling mined-out spaces, in road construction etc.

References

[1] Directive 2008/98/EC on waste (Waste Framework Directive) [Electronic resource]. URL: http://ec.europa.eu/environment/waste/framework/ (Accessed date 25.02.2015)

[2] Iron ore mining in Russia. Mining Wiki – free mining encyclopedia [Electronic resource ресурс].URL: http://miningwiki.ru/wiki/Добыча_железной_руды_в_России (Accessed date 25.02.2015)

[3] European practice of waste treatment: problems, solutions, perspectives// Saint-Petersburg 2005. – 17 p.

[4] "On wastes of production and consumption". Federal law of 24.06.1998 N 89-FL. (edit. 29.06.2015) [Electronic resource]. URL: http://base.consultant.ru/cons/cgi/online.cgi?req=doc; base=LAW;n=173740 (Accessed date 25.02.2015)

[5] Mohan Yellishetty, Vanda Karpe, E.H. Reddy. Reuse of iron ore mineral wastes in civil engineering constructions: A case study// Resources, Conservation and Recycling. Sep 05, 2015 [Electronic resource]. URL: http://www.researchgate.net/publication/228626105_Reuse_of_iron_ore_mineral_wastes_ in_civil_engineering_constructions_A_case_study (Accesses date 25.02.2015)
[6] Management of mining, quarrying and ore-processing waste in the european union Study made for DG Environment, European Commission Co-ordination by P. Charbonnier December 2001. [Electronic resource]. URL: http://ec.europa.eu/environment/waste/studies/mining/0204finalreportbrgm.pdf (Accessed date 25.02.2015)
[7] How are waste materials managed at mine sites? MiningFacts.org: A Resource for Canadian Mining Information - See more at: [Electronic resource]. URL: http://www.miningfacts.org/About-Us/#sthash.NTYBc0z4.dpuf (Accessed date)
[8] Letter of Rostekhnadzor / On hazardous waste rating / Federal Service of Environmental, Technological and Nuclear Supervision d.d. 25.12.2007 № 14-08/4908
[9] Criteria to classify wastes according to their hazards for the environment /Approved by the order of Ministry for Nature Management d.d. 15.06.2001 г. № 511.
[10] Nasibulina, B. M. Wastes as a source of environment pollution / B. M. Nasibulina, T. F. Kurochkina – Astrakhan: «University of Astrakhan» Publishing, 2009. – 110 p.
[11] Nyubauer A. Rapprochement with EU policy on wastes. Brief guide book for partner countries and Russia on European policy of good-neighborly relations. ECOLOGIC – Institute of International and European Environment Policy. August 2007. - 19 p.
[12] Umnov V A Management of wastes in mining industry // Mining informational and analytical bulletin (scientific and technical journal). – 1995. - №5 – 99-106 p.
[13] Garbage recycling // Waste classification (MSW). [Electronic resource]. URL: http://ztbo.ru/o-tbo/lit/texnologii-otxodov/klassifikaciya-vidov-otxodov-tbo (Accessed date 25.02.2015)
[14] Technical resource document extraction and beneficiation of ores and minerals. Volume 3. Iron U.S. Environmental Protection Agency Office of Solid Waste Special Waste Branch. [Electronic resource]. URL: http://www.epa.gov/wastes/nonhaz/industrial/special/mining/techdocs/iron.pdf (Accessed date 25.02.2015)
[15] Methods to determine water toxicity and aqueous extracts from soils, sewage sludges, wastes on the base of the death rate and fertility changes of Daphnia / FR.1.39.2007.03222/Moscow AKWAROS, 2007. – 41 p.
[16] Methods to determine water toxicity and aqueous extracts from soils, sewage sludges, wastes on the chlorophyll fluorescence and population of algae cells / FR.1.39.2007.03223/Moscow AKWAROS, 2007. – 47 p.
[17] Kumar A Conveners Report Published in the souvenir of mines environment and mineral conservation week (Goa and parts of Karnataka); 2000. p 8

IOP Conf. Series: Materials Science and Engineering **127** (2016) 012027 doi:10.1088/1757-899X/127/1/012027

Hierarchical Model of Assessing and Selecting Experts

T YChernysheva[1], M A Korchuganova[2], V V Borisov[3], S L Min'kov[4]

[1, 2, 3] 26, Leningradskaya street, Yurga, Kemerovo region, 652055, Russia
Yurga Technological Institute (affiliated) National Research Tomsk Polytechnic University
[4] 36, Lenin Avenue, Tomsk, Russia, 634050 Tomsk State University

e-mail: chernyshevat@tpu.ru

Abstract. Revealing experts' competences is a multi-objective issue. Authors of the paper deal with competence assessing methods of experts seen as objects, and criteria of qualities. An analytic hierarchy process of assessing and ranking experts is offered, which is based on paired comparison matrices and scores, quality parameters are taken into account as well. Calculation and assessment of experts is given as an example.

1. Introduction

Decision making is relevant for all spheres of human activities, and it's of special importance for social and economic development of subjects. Judgments of several experts are worthwhile to support objectivity and improve quality of decision making process. Collective expert assessment is made with this purpose, whereas a lot of experts can be divided into some sub-groups with respect to the subject of assessment [1, 2], depending on criteria applied in a hierarchy.

A typical procedure to form an expert panel comprises such phases as defining the number of experts, developing formal and professional eligibility standards of experts, and assessing awareness of each expert [2].

2. Review of expert awareness assessment procedures

Expert panel is to consist of experts, responsible for each group of urban social and economic development factors. It's quite logical to assume, judgments of experts, representing the field, which an indicator to be assessed belongs to, are the most valuable ones. For instance, knowledge on labor economy and sociology, labor legislation, mechanisms of labor market functioning and its regulating, as well as working experience are necessary for experts, assessing personnel issues. At the same time, judgments of other members in expert panel can't be ignored, even if their competence in the field is far from being high because various spheres of urban development can't be assessed separately, all issues are to be addressed as interconnected ones. Therefore, weights of expert importance are required for assessing diverse indicators.

Minimal quantity of experts depends on the number of urban functional spheres. For instance, six groups of indicators of urban social and economic development are given in [3]. Formula N = 0.5 (3/a + 5), where 0<a£1 is a parameter, specifying minimal level of assessment error, is suggested to be used for determining minimal quantity of experts in [4].

IOP Conf. Series: Materials Science and Engineering **127** (2016) 012027 doi:10.1088/1757-899X/127/1/012027

Minimal quantity of experts is 4 (if a = 1) in compliance with the above condition. As a rule, at least 7 – 9 experts are necessary for group assessment [3, 5].

An appropriate solution can be found provided that experts are grouped according to the fields: reliability, efficiency and workability of a project. Competence level of each expert is to be determined on the base of all indicators.

A table of expert competence is suggested in [5], which supports selecting experts, assessing their competence within the following functional blocks of urban social and economic development: social and economic, personnel, investment, infrastructural, manufacturing and financial. A constant of variation is suggested to define the measure of agreement of expert assessments. Supposing it exceeds 33% (for rather regular distributions), assessments are not agreed, and experts are to revise them [6].

Social and economic conditions and projects are a vital task, provided that expenditures on assessment are reasonable, experts' judgments are averaged out according to their qualification ("weight"). A hierarchical structure of criteria is suggested for determining weighting factors of experts [7].

Aggregated assessment is calculated from the following formula provided that n experts with various levels of significance are involved in the work

$$a_{ij}^{A} = a_{ij}^{p_1} a_{ij}^{p_2} \dots a_{ij}^{p_n} ,$$

where $a_{ij}^{p_k}$ - assessment of an object done by expert k with a weight factor p_k, here $p_1+p_2+\dots+p_n=1$.

For subjective assessment of experts themselves the following criteria are used: professional experience (practical experience in the field of financial planning, intuition), independent views, working experience in the field, creative attitude to the solution of problems and experience in expert assessments. Supposing working experience of experts is 10, 15 and 20 years, weight of this criterion is 0.222; 0.333 and 0.444, respectively. Having calculated weights of all criteria and averaged them out according to the number, we obtain a generalized weight for significance of expert's judgment.

Basic requirements for an expert are outlined in [4] and include broad mental outlook and practical awareness, scientific papers and practical experience, ability to solve practical problems, independence of standpoints etc. Therefore, assessment of competence and selection of experts is a multi-objective issue.

3. Hierarchical procedure of assessing experts

Method of weighted or weight-averaged sums is most popular and frequently applied by professionals for multi-criteria assessment of alternatives.

We suggest using the following criteria and scales for assessing experts [7]:

1. Level of education: secondary (1 point), secondary special (2 points), higher (3 points), PhD (4 points), Doctor of Science (5 points).
2. Working experience in the subject domain: no experience (0 points), 1 to 3 years (1 point), 3 to 5 years (2 points), 5 to 10 years (3 points), 10 to 20 years (4 points), more 20 years (5 points).
3. Administrative and economic independence in the field: absolutely independent (5 points), acquainted with the company (4 points), employed in the same company but has no direct influence on decision making process (3 points), his/her direct responsibilities include organization (2 points), employed in the same decision making body (1 point).
4. Ability to respond to creative challenges and experience of expert assessing: no experience (1 point), low level ability (2 points), average level (3 points), above the average (4 points), high level ability (5 points) etc.

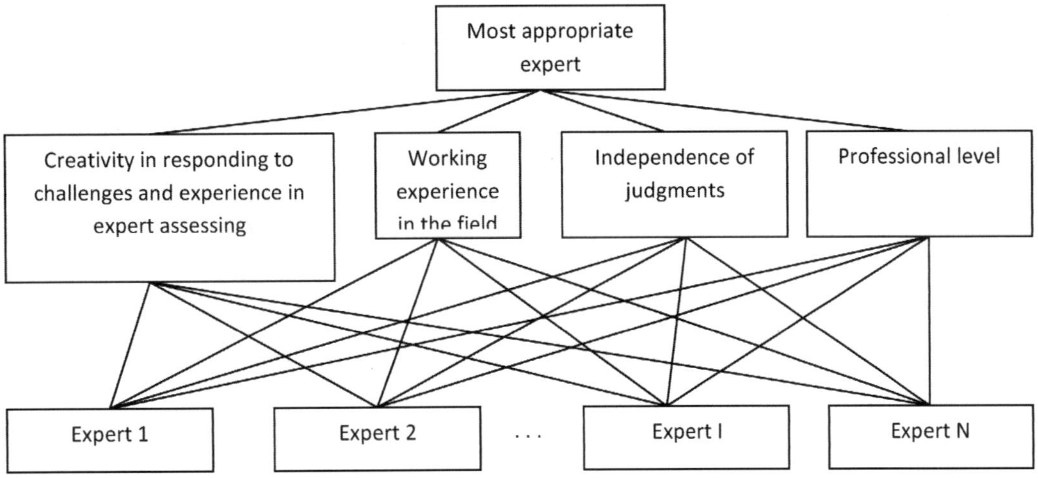

Figure 1. Hierarchy of ranking experts

Criteria of assessing experts can be also averaged out at the level of eigenvectors of paired comparison matrix E (Table 1). Here, the results are equivalent to those obtained at the level of matrix elements provided that homogeneity of formed matrices is sufficient and does not exceed 0.10 [8-13].

Values of priority vector of criterion W are also computed from formula

$$EW = \lambda_{\max} W, \tag{1}$$

where λ_{max} – maximal eigenvalue of matrix E.

Then, comparison matrices for each criterion of experts are to be completed (e.g. for three experts, as it can be seen in Table 3). Priority vectors of professional qualities are designated Wi, i = 1,2,3,4.

Ratio scale is used for paired comparison (Table 2). Tables are completed automatically on the base of paired difference in scores of the same quality when computer processing [7].

Table 1 – Weights of assessment criterion of expert

Assessment criteria of expert	Professional level	Working experience in the field	Independence of judgments	Creativity in responding to the challenges and experience of expert assessing	W
Professional level	1	3	5	3	0.536
Working experience in the field	1/3	1	5	1/3	0.139
Independence of judgments	1/5	1/5	1	3	0.088
Creativity in responding to the challenges and experience of expert assessing	1/3	3	1/3	1	0.136

Table 2 – Ratio scale (level of significance) of a quality

Level of significance	Definition	Explanation
1	Similar significance	Two candidates have a quality of the same level
2	Significance of one professional's quality predominates slightly over that of other professional (weak significance)	The difference in quality level of candidates is one point
3	Considerable or intense significance	The difference in quality level of candidates is two points
4	Obvious or very intense significance	The difference in quality level of candidates is three points
5	Absolute significance	The difference in quality level of candidates is four points
Reciprocal values of above non-vanishing values	Supposing a quality of i candidate has one of non-vanishing values specified above when comparing to a quality of j candidate, a quality of j candidate has a reciprocal value when comparing to a quality of i candidate.	Provided that agreement of qualities was postulated when calculating N numerical values for forming a matrix.

Table 3 – Ranking experts in view of criterion "Professional level"

Professional level	Expert 1	Expert 2	Expert 3	W1
Expert 1	1	3	5	0.72
Expert 2	1/3	1	3	0.18
Expert 3	1/5	1/3	1	0.10

Elements of resultant priority vector of experts are calculated by formula

$$W_{э} = [W_1, W_2, W_3, W_4] \times W, \tag{2}$$

where W_1, W_2, W_3, W_4 – vectors of weights criteria (qualities of experts).
For instance, it was obtained for the set of criteria considered above:
Wэ= {0.672; 0.218; 0.110}.

Values of elements of the calculated vector demonstrate that the first expert is the most appropriate one in view of criterion-based assessment; his/her judgment is the most qualified and final decision is to be made on the base of his/her judgement.

4. Automation of assessment calculations

Information system (IS) of assessing and selecting experts was developed to process labor-intensive and similar calculations. Phases of selecting experts on the base of the IS are outlined in Fig. 2.

The first phase is concentrated on identifying the list of qualities necessary to form a group. Then a questionnaire is completed and testing is carried out. Testing results in assessment of each candidate's qualities; data base of information system is filled in. The following phase is focused on ranking qualities, a precise list of qualities is identified; and assessment of a professional is given. Assessing and selecting a group of experts are main points of the final phase.

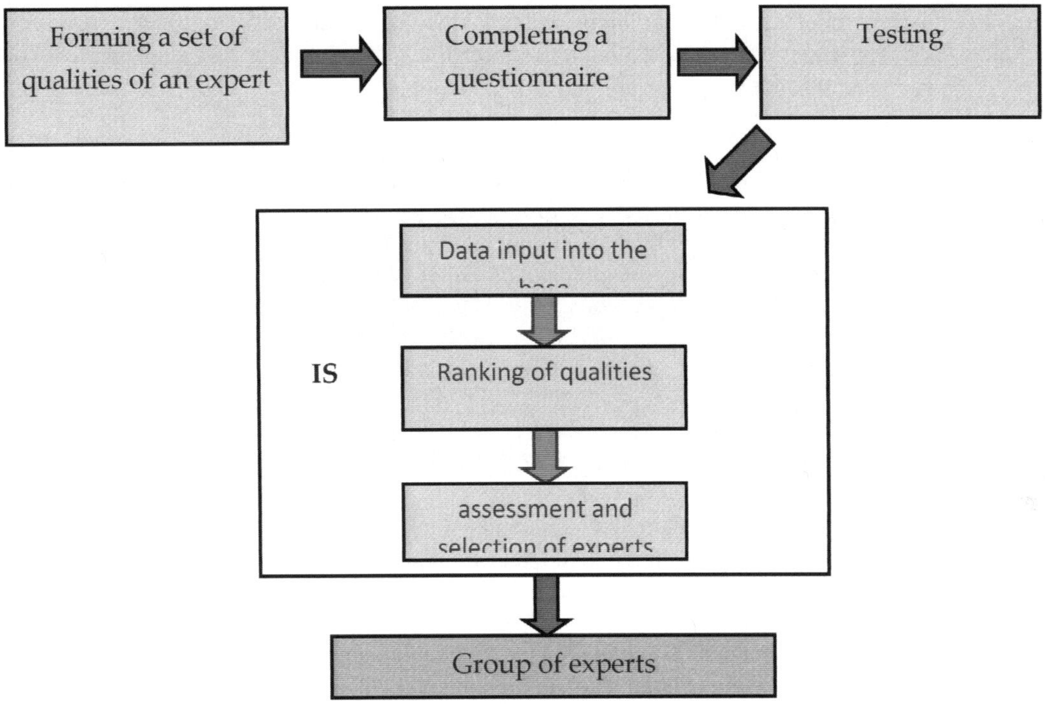

Figure 2. Phases of selecting experts

One of expert assessment procedures is used in the information system – paired comparison of alternatives. It is a procedure of revealing preferred objects when comparing all possible pairs. Qualities of experts are ranked via determining their significance.

Functions of the information system:

– general information, professionally and business relevant qualities are taken into account;
– qualities of professionals are analyzed;
– qualities of professionals are compared;
– business qualities of a professional are assessed via expert procedures;
– experts are ranked on the base of the aggregate of qualities.

Input data:

– personal details of experts;
– information on personal characteristics of experts;
– a list of qualities selected for assessment criteria.

Output data:

– a list of experts meeting the requirements of a user' s enquire;
– a diagram of significant and sufficient qualities of an expert.

A permanent database can't provide well-timed updating of information on experts, as well as data on new experts isn't added in proper time, therefore a questionnaire of a data base can be placed on a company website, where experts have a possibility to edit information on them. This system is developed for a personnel department manager who is responsible for adding data on experts, and it supports decision making by the head of a company, as only most appropriate experts are selected.

5. Conclusion

Determination of experts' competences is a multi-objective issue. Authors of the paper deal with competence assessing methods of experts seen as objects, and criteria of qualities. Four criteria and 5-point scale are suggested for assessing characteristics of experts. An analytic hierarchy process of assessing and ranking experts is offered, which is based on paired comparison matrices and scores, quality parameters are taken into account as well. Algorithm of assessing and selecting experts supports adequate decision making.

References

[1] Mejía-Gutiérrez R, Cálad-Álvarez A, Zuluaga-Holguín D 2013 *Experts' Agreement Support for Distributed Engineering Knowledge Modelling* Communications in Computer and Information Science V 246 CCIS pp 209-223

[2] Leonelli M, Smith J Q 2015 *Bayesian decision support for complex systems with many distributed experts* Annals of Operations Research, 235(1) pp 517-542

[3] Zakharova A A , Ostanin V V 2015*Formalization model of expert knowledge about a technical index level of engineering products* IOP Conference Series: Materials Science and Engineering V 91 pp 1-8

[4] Zakharova A A 2015 *Decision making models on the basis of expert knowledge for an engineering enterprise strategic management* Applied Mechanics and Materials V 770 pp 645-650

[5] Chernysheva T Y, Gnedash E V, Zorina T Y, Lenskaya N V 2014 *Information systems project risk assessment: expert approach* Applied Mechanics and Materials pp 539-543

[6] Silich V, Silich M 2003*Hierarchical Object Model of Variant Choice* Proceedings The 7th Korea-Russia International Symposium on Science and Technology KORUS pp 378-381

[7] Zakharova A A, Chernysheva T Y, Mitsel A A 2010 *Mathematical and software support of strategic decisions in municipal administration* Tomsk

[8] Saaty T L 2008 *Decision making with the analytic hierarchy process* Int. J. Services Sciences V 1 No 1 pp 83-98

[9] Sok H K , Ooi M P-L, Kuang Y C, Demidenko S 2016 *Multivariate alternating decision trees* Pattern Recognition pp195-209

[10] Coman A 1996 *Competence, power and conflict in group decision making Human Systems Management* V 15 pp 245-255

[11] Saaty T L 1994 '*How to make a decision: the analytic hierarchy process*' Interfaces V 24 No 6 pp 19–43

[12] Fox-Young S K 1995 *Issues in the assessment of expert nurses: purposes, standards and methods* Source of the Document Nurse Education Today 15(2) pp 96-100

[13] Srdjevic B, Pipan M, Srdjevic Z, Blagojevic B, Zoranovic T 2015 *Virtually combining the analytical hierarchy process and voting methods in order to make group decisions* Universal Access in the Information Society V 14 Is 2 pp 231-245

IOP Conf. Series: Materials Science and Engineering **127** (2016) 012028 doi:10.1088/1757-899X/127/1/012028

Parametric Modeling as a Technology of Rapid Prototyping in Light Industry

I N Tomilov[1,a], S N Grudinin[2,b] , V D Frolovsky[2,c] and A A Alexandrov [1,d]

[1]Yurga Institute of Technology of Tomsk polytechnic university, TPU affiliate, Leningradskaya str. 26, 652055, Yurga, Russia
[2]Novosibirsk State Technical University, Marx ave. 20, 630073, Novosibirsk, Russia

e-mail: [a]tomilov@corp.nstu.ru, [b]sn_grudinin@mail.ru, [c]frolovsky@asu.cs.nstu.ru, [d]artemka7474@mail.ru

Abstract. The paper deals with the parametric modeling method of virtual mannequins for the purposes of design automation in clothing industry. The described approach includes the steps of generation of the basic model on the ground of the initial one (obtained in 3D-scanning process), its parameterization and deformation. The complex surfaces are presented by the wireframe model. The modeling results are evaluated with the set of similarity factors. Deformed models are compared with their virtual prototypes. The results of modeling are estimated by the standard deviation factor.

1. Introduction

Prototyping is compulsory in the design of a new article. Design of a quality prototype which is maximally alike the future article is a complex scientific and practical problem. Technologies of rapid prototyping have become popular recently [1]. A prototype is generally made on the basis of a solid CAD-model or a model with closed surface contours. Herewith parameterization is of a great importance. Parametric modeling is not as expensive as alternative methods of virtualization and enables to develop new models interactively by changing set parameters. Formally, this approach can be written as follows: development of a new model Ω by deforming an average base model Ψ according to the set of specified parameters ρ. Thus, a new model Ω, after being deformed is to satisfy a set of parameters ρ [2].

For parameterization of objects with a complex geometric shape and presented as wireframe models special technologies are required which take into consideration the information on form and metric characteristics accepted in the object domain. Mannequins used in clothes production are among these objects. The existing CADs of clothes are applied by designers mainly for laying-out and enable to view articles in 2D dimensions only. Systems of three-dimensional scanning and data bases of real object prototypes (man is included [3]) which are being rapidly developed in recent years could be widely applied in clothing industry right up to individual customizing of clothes. However, there are two hindrances obstacle problems. First, the prices of scanning systems are too high for clothing industry. Second, these systems are not integrated into CADs of clothes well enough. On the other hand, models obtained by scanning of real objects have a lot of empirical data and can be applied as the basis for parameterization followed by the design adjusted to the needs of a particular customer.

This approach demands for the development of methods and ways of getting information on the form of the object under modeling. Therefore, the problem of complex object parameterization is an urgent one and needs to be solved by generating new models on the ground of semantic information and evaluation of modeling quality [4, 5].

Thus, two interrelated problems are to be solved: complex objects need to be presented parametrically with further evaluation of parameterization results in order to define the efficiency. Formally, the problem can be stated as follows: there are two original models γ_1 and γ_2, on their basis models α_1 and α_2 are to be developed and specified by parameters π_1 and π_2 respectively; an undefined model $\alpha_1^{'}$, meeting the parameters π_2 is to be developed, and proximity of models $\alpha_1^{'}$ and α_2 is to be evaluated.

2. Description of solution procedure

The basic information for development of a basic model includes 3D model of a female mannequin with a standard form – a right trunk. The model is obtained by three-dimensional scanning with Artec MH 3D Scanner hand scanner [6, 7], providing accuracy of order 10^{-2} mm. The delivery set includes ZScan Lite software, which enables to evaluate the results of scanning and export data into the most accepted format. The most wide-spread format is STL. It is used to keep three-dimensional models of objects in technologies of rapid prototyping, import into the systems of three-dimensional modeling and send to the three-dimensional printer. Figure 1 shows the photos of the object and the result of scanning.

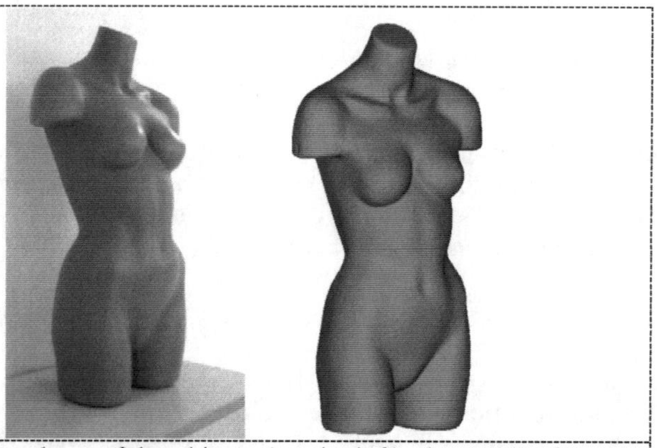

Figure 1. The photo of the objects (on the left), the result of three-dimensional scanning, 10 000 polygons (on the right)

The basic model has a polygon structure: set of points $t_{ij}(x_{ij}, y_{ij}, z_{ij})$, connected into triangular polygons $p_i = (t_{i1}, t_{i2}, t_{i3})$, $i = \overline{1,N}$, $j = \overline{1,3}$, where N – number of polygons. The model doesn't have internal polygons, polygon crosses and holes. The centre-of-mass coordinate of the model $C(x_m, y_m, z_m)$ coincides with the coordinates of origin of the global coordinate system O, when:

$$x_m = \frac{1}{3N}\sum_{i=1}^{N}\sum_{j=1}^{3} x_{ij}\,,\ y_m = \frac{1}{3N}\sum_{i=1}^{N}\sum_{j=1}^{3} y_{ij}\,,\ z_m = \frac{1}{3N}\sum_{i=1}^{N}\sum_{j=1}^{3} z_{ij} \qquad (1)$$

The model is oriented in such a way, that the horizontal plane xOy divides the upper and the low parts of the mannequin, the vertical plane yOz separates the front part from the rear one, and the vertical plane xOz divides the mannequin into two parts along the symmetry axis.

The base model under the deformation is a three-dimensional wireframe model with topology: a set of horizontal sections h_i with the same quantity of equidistant apices $\{v_{ij}\}$ sorted out according to the polar angles: $i=\overline{1,N}$, $j=\overline{1,M}$, where N – number of sections, M – number of points in the sections. The sections are on the levels: $z_i = z_{i-1} + \Delta z$, $z_1 = z_{\min}$, $\Delta z = (z_{\max} - z_{\min})/N$, $i=\overline{2,N}$, where $z_{\max}$ – level of neck base from behind, $z_{\min}$ – level of perineum, N – number of sections. Thereupon, the points of each i section from h_i are arranged in such a way to meet the condition: $\varphi_1 < \varphi_2 < \ldots < \varphi_j < \ldots < \varphi_{M_i}$, where φ_j – is a polar angle of j point of i section.

One of the basic peculiarities of a mannequin form is xOz plane symmetry; therefore, the basic sections are subjected to the symmetrization – calculation of the averages between the left and right halves relative to the axis of symmetry, with the further smoothing – interpolation by third-order splines.

The parameterization of a human body is usually based on anthropometric or medical information: anthropometrical points [2], parameters of clothing industry [4], points of acupuncture [8], and types of spinal curvature [3] and so on. The results of anthropometric research and standards of clothing industry are referred to in paper [2], according to these data the main parameters of human body are height, chest, girth of hips and waist. Measured parameters 1–10 are showed in Fig. 2. To calculate them additional sections $c_k = \{c_{kj}\}$, $j=\overline{1,M}$, $k=\overline{1,5}$ and their girths (4–6) are determined besides the basic ones h_i. The set of anthropometric points is applied for this purpose. Axial displacements Ox (7–9) and Oy (1–3) enable to follow up the position of spinal column. The number of parameters can be increased by making additional sections through the needed anthropometric points.

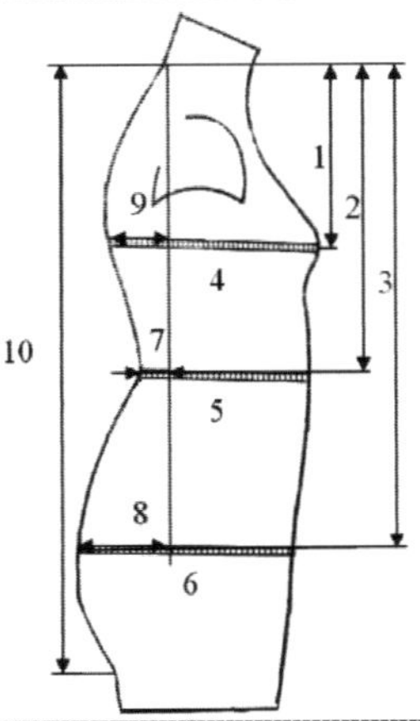

Figure 2. Positioning of the measured parameters

As a girth we see a perimeter of a minimal convex hull of additional section points. K section girth is calculated according to Graham algorithm. The deformation is seen as a change in the position and forms of base sections according to the similar change of additional sections. The deformation process includes several phases.

Axial displacement Oz of base sections proportionally to the changing parameter 10. The distances between base sections remain the same.

Axial displacement Oz of additional sections relative to the neck point according to parameters 1–3.

Scaling of additional sections according to parameters 4–6. Scaling is carried out in the section planes, for this purpose an optimization problem with efficiency function is solved:

$$F(s) = |G'_k - f(q_k \cdot S(s))| \to \min, \tag{2}$$

where G'_k – a new girth, f – function of section girth calculation, q_k – range of k additional section points, $S(s)$ – matrix of scaling on the plane, s – scaling factor. The optimization problem is solved by coordinatewise descent.

Axial displacement Oy of additional sections relative to the neck point according to parameters 7 – 9.

Scaling of base sections according to the new form of additional ones. The base sections which are between sections c_{k-1} and c_{k+1}, are scaled for each k additional section, and herewith, the scale factor s decreases linearly from zero upwards and downwards relative to section c_k.

Developing the surface of quadrangular grid by connecting corresponding apices of adjacent sections, that is by adding edges $\{(v_{ij}, v_{i+1j})\}$, $i = \overline{1,N}$, $j = \overline{1,M}$, where N – number of sections, M – number of points in i-section.

3. Results and Discussion

The results of modeling are evaluated by comparing the base model of an object with the model of another object deformed according to the parameters of the base model. The approaches of three-dimensional object evaluation usually differ in the method of object form description and in the metric, which is to define the degree of similarity. Analyzing the models of a particular class their comparison gets more efficient in terms of peculiarities of this class models. From this point of view the mannequin models are presented quite well by the set of horizontal parallel sections, which are smooth, symmetrical figures. Therefore, to evaluate the model forms 4 factors are applied, which are calculated for each half-section h'_i: perimeter of half-section $S^v{}_i$, perimeter of half-section girth $S^w{}_i$, descriptor of form D_i, area of half-section A_i. The factor, based on the shape function, is also considered [4, 7], that describes the model form on the whole and it is calculated not in view of random points on the surface of mannequin, but according to the points of sections h_i. In this case it is not difficult to choose points on the surfaces of compared models.

Perimeter of i half-section:

$$S^v = \sum_{j=1}^{M'-1} \sqrt{(x_j - x_{j+1})^2 + (y_j - y_{j+1})^2}\,, \tag{3}$$

where M' – number of half-section points, $v_j(x_j, y_j)$ – j point of half-section.

Perimeter of i half-section girth:

$$S^w = \sum_{k=1}^{K-1} \sqrt{(x_k - x_{k+1})^2 + (y_k - y_{k+1})^2}\,, \tag{4}$$

where K – number of girth points, $w_k(x_k, y_k)$ – k point of girth.

Descriptor of i half-section form:

$$D_i = \sum_{j=1}^{M'-1} L_j \cos\alpha_j\,,\ L_j = \left|\overline{O_i v_j}\right|,\ \alpha_j = \left|\frac{180 j}{M'} - \chi_j\right|,\ j = \overline{1,M'}\,, \tag{5}$$

where M' – number of points in a half-section, where χ_j – angle between radius-vector $\overline{O_i v_j}$ and axis $O_i x$ of the coordinate system $yO_i x$ of i section, v_j – j point of half-section, O_i – pole of the coordinate system $yO_i x$ of i section, here $\overline{OO_i} = \frac{1}{M}\sum_{j=1}^{M}\overline{Ov_{ij}}$, where M – number of points in a section.

The axes of the coordinate system of i section $yO_i x$ are parallel to the axes of the global coordinate system yOx.

Area of i half-section:

$$A = \frac{1}{2}\sum_{j=1}^{M'-1} L_j^2 \varphi_j \, , \; L_j = \left|\overline{O_i v_j}\right| , \tag{6}$$

where M' – number of points in a half-section, φ_j – angle between vectors $\overline{O_i v_j}$ and $\overline{O_i v_{j+1}}$, $j = \overline{1, M'-1}$.

Shape function – a set of Euclidian distances $\{\psi_s\}$ between all pairs of points in sections h_i, $s = \overline{1, S}$, $S = NM(NM-1)/2$, where N – number of sections, M – number of points in sections. The compared factor is a density diagram F of the set of Euclidian distances $\{\psi_s\}$.

The factors are compared by calculating the standard deviation of each factor:

$$\sigma = \sqrt{\frac{1}{n-1}\sum_{i=1}^{n}(P^I{}_i - P^{II}{}_i)^2} , \tag{7}$$

where $P^I{}_i$ – i value of factor for the first model, $P^{II}{}_i$ – i value of factor for the second model, n – for shape function – number of intervals in density diagram, in other cases – number of sections.

The algorithms presented in the paper are applied to the models of female mannequins (Fig. 3), containing 50 sections, each section has 200 points. The number of intervals in density diagrams makes 100. Each model was deformed according to the parameters of two others, then the models were compared pair-wise to evaluate the deformation.

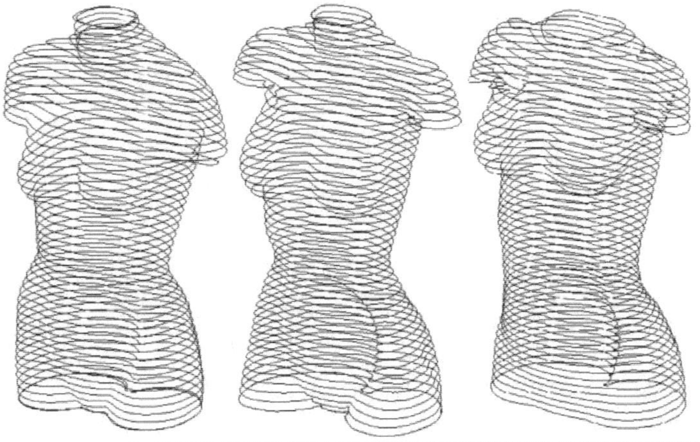

Figure 3. Base models I, II, III (from L to R) of the compared mannequins

The results of factor calculation before and after deformation are shown in Table 1. The obtained results help to conclude that forms of the models after deformation are much more similar to the model-prototypes because of the decreasing standard deviation.

Table 1. Evaluation factors before and after deformation

Pair of compared models	Criterion of comparison (SD) according to the factors				
	S^v	S^w	D	A	F
			Before deformation		
I – II	0.3581	0.3381	9.6928	15.4229	15829.7847
I – III	0.3268	0.2798	8.0335	12.4381	11412.6043
II – III	0.3405	0.3042	9.3217	14.2123	10315.3431

Pair of compared models	Criterion of comparison (SD) according to the factors				
	S^v	S^w	D	A	F
Change after deformation (in factor values)					
I – II	-0.0716	-0.0532	-0.5869	-1.0777	-4565.5172
II– I	-0.0724	-0.0624	-1.5681	-3.0487	-839.8100
I – III	-0.1029	-0.0650	-1.8809	-2.3758	-4304.8324
III – I	-0.0835	-0.0578	-1.7882	-2.6710	-1103.0592
II – III	-0.0159	-0.0652	-3.4951	-4.2805	-3029.4424
III – II	-0.0437	-0.0746	-2.8760	-3.5257	-3854.9402
Change after deformation (% from the original value)					
I – II	-20.0106	-15.7399	-6.0551	-6.9877	-28.8413
II– I	-20.2189	-18.4673	-16.1786	-19.7676	-5.3052
I – III	-31.5014	-23.2600	-23.4129	-19.1016	-37.71998
III – I	-25.5703	-20.6653	-22.2602	-21.4743	-9.6652
II – III	-4.6961	-21.4486	-37.4945	-30.1187	-29.3683
III – II	-12.8452	-24.5481	-30.8533	-24.8077	-37.3709

Conclusion

The paper provides the description of the approach of development of a base model with the set structure in view of polygon model of the original object. The original parameterization keeping the geometrics of a three-dimensional object model with a complex structure is investigated on the ground of clothing industry parameters and anthropometric research. Factors for evaluation of wireframe models similarity are suggested, models of female mannequins are compared in view of these factors. The described approach can be applied to develop virtual mannequins for the purposes of clothing industry.

References

[1] Gantsetseg B and Frolovsky V D 2013 Object-oriented approach to 3D modeling of parameterization non-standard mannequin *IFOST 2013* pp 219-20.

[2] Baljirsuren G, Kuznetsova A and Frolovsky V D 2008 Parametrical modeling of objects with complex structure and compression data of the basis of generative models *SIBIRCON 2008* pp 227-31.

[3] Frolovsky V D and Ganelina N D 2005 Ant colony approach to defining hamilton cycle on segments *GraphiCon 2005*.

[4] Tomilov I N, Dostovalov D N and Klokova E A 2015 Analysis of the two-chamber shock-absorption system of landing gear based on hybrid methodology *SIBCON 2015*.

[5] Shornikov Y V, Nasyrova M S and Dostovalov D N. Simulation of stiff hybrid systems with one-sided events and nonsmooth boundaries *Lecture Notes in Electrical Engineering* pp 169-86.

[6] Tomilov I, Shornikov Y, Myssak M and Zagidulin A 2012 Numerical simulation of complex physical processes using hybrid systems *IFOST 2012*.

[7] Shornikov Y V, Novikov E A, Dostovalov D N and Myssak M S 2013 Document Visual modeling of dynamical systems by instrumental facilities *IFAC Proceedings Volumes* pp 2185-90.

[8] Korenevskiy N, Alshamasin M, Al-Kasasbeh R, Anatolevich K and Ionescu F 2015 Prediction and prenosological diagnosis of stomach diseases based on energy characteristics of acupuncture points and fuzzy logic *Int. J. of Modelling, Ident. and Control* pp 55-67.

Models Used to Select Strategic Planning Experts for High Technology Productions

Alexandra A Zakharova[1,a], Antonina A Grigorjeva[1,b], Anna P Tseplit[2,b] and Evgenij V Ozgogov[1, c]

[1]Yurga Institute of Technology, National Research Tomsk Polytechnic University Affiliate, ul. Leningradskaya 26, Yurga, Russia
[2] Kuzbass State Technical University named after T.F.Gorbachev, ul. Vesennyaya 28, Kemerovo, Russia

e-mail: [a]aaz@tpu.ru, [b]antonina505@mail.ru, [c]blackjack41@mail.ru

Abstract. The article deals with the problems and specific aspects in organizing works of experts involved in assessment of companies that manufacture complex high-technology products. A model is presented that is intended for evaluating competences of experts in individual functional areas of expertise. Experts are selected to build a group on the basis of tables used to determine a competence level. An expert selection model based on fuzzy logic is proposed and additional requirements for the expert group composition can be taken into account, with regard to the needed quality and competence related preferences of decision-makers. A Web-based information system model is developed for the interaction between experts and decision-makers when carrying out online examinations.

1. Introduction

Modern enterprises manufacture a variety of complex high-technology and science - based products for different industries. Product design developing, production facilities setting, manufacture mastering and production supporting require significant work in the field of researches and development, as well as drawing up accompanying documents [1]. Decisions made in these fields are strategic, because they adjust the organization's direction in future and can have potential critical effects on company's activities. Mistakes made at this level can be fatal for the company. The process of making strategic decisions is always characterized by decision-making in the environment of uncertainty in the context of incomplete and unclear information [2]. In this connection, internal and external experts are invited to justify strategic decision-making [3]

The quality of assessment made by experts directly depends on the quality of selecting an expert group; and experts should have sufficient experience and knowledge in the field of the current trends in science and technology in their industry. Examinations by expert groups increase the objectivity and the accuracy of results, but there is a problem of how to determine an adequate number of experts in the group, as the excessive or insufficient number of consultants may have a negative impact on final outcomes [4]. Even the most competent experts in the examination can make a big difference of opinion within the group and this sometimes results in inadequate examination results [5]. In this regard, one of the objectives when organizing the process of making an examination is to assess the consistency of opinions across the expert group in order to eliminate possible deviations with due consideration of opinions by experts, giving recommendations close to reality. For example, the work in [6] represents a model of coordinating experts based on the assessment of quality of several criteria which helps identify experts having opinions with the maximum deviation from the mean value over the group evaluation.

The more an expert has carried out examinations, the more data on his opinion deviating from the group opinion and / or the actual values of parameters evaluated is obtained to have the more accurate assessment of his competence in a given field of knowledge. So, the expert's knowledge and his experience as a universal parameter are very important in assessing the competence of the expert [7]. But different areas of expertise have their specific aspects, and it is needed to assess the expert's knowledge, practical skills and experience received from conducting examinations in particular and relevant disciplines. On the basis of these data, when preparing for the next examination, a person, making a decision, is able to choose the most experienced and competent experts in the required field. This is especially critical for high-tech enterprises, since decision-making is carried out across a number of function areas of its activity: innovation, marketing, production, research and others.

At the same time, for instance, assessments after reviewing the same function area at different enterprises can have its own characteristics associated with peculiarities of companies assessed, their business processes, leadership personalities and others. For this reason it is important to develop tools towards achieving a more balanced composition of expert groups with the consideration of specific interests of decision makers.

2. Determining the composition of a panel of experts on the basis of self-assessment

It is a common practice to create a group for strategic planning - a consultative and coordinating body attached to the company's administration and providing coordination of management actions, as well as all relevant units involved in the strategic planning of the company. The participation of external experts is very important in this group. [8]

A standard plan to form an expert commission includes such steps as determining the number of experts to involve, outlining functionary and professional requirements imposed on experts, determining the expert commission, determining the extend of competence for each expert. The evaluation of competence for each expert, in turn, can be carried out on the basis of self-assessment, mutual evaluation through peer review or on the basis of actual data resulted from previous examinations.

A expert group should be composed of experts in every of the functional areas of the enterprise. With this, it is logical to assume that opinions of experts with the relevant expertise in the field of work needed should be particularly taken into account. For example, when evaluating performance indicators relating to the human resource, experts should have knowledge of economics and labor sociology, employment and labor legislation, and mechanisms for regulating and proper functioning of the labor market, and they are expected to have experience in this field. At the same time, opinions of other members in the expert group, having insufficient competence in this area, can not be neglected, because the isolation of assessment of certain areas is not accepted, because every problem should be viewed in conjunction with the others. Thus, there is a need to determine the weight of relevance relating to experts in evaluating different performance parameters.

We suggest determining the minimum number of experts according to formula (1).

$$N = 0.5\left(\frac{3}{\alpha} + 5\right) \tag{1}$$

Where $0 < \alpha \le 1$ is the parameter that specifies a minimum error level of examination. Based on this condition, the minimum number of experts is equal to 4 (for $\alpha = 1$).

For evaluating every expert, it is necessary to identify his level of competence for each set of performance indicators.

The basic requirements for an expert are usually as follows: the broad vision and knowledge of the subject area, a quantity of scientific papers written and the presence of practical experience, the ability to meet challenges in a creative way, independence of thoughts and others. Thus, the task to identify the competence of experts is multi-criteria. The weighted sum model is the most popular multi-criteria method for evaluating a number of alternatives.

The following evaluation criteria and rating scores for alternatives are proposed to use:

1. Level of education: Secondary (1 point), vocational (2-4 points), higher (5-8 points), a scientific degree (9-10 points).

2. Consistency of education with the knowledge domain (namely a specific area of functioning) inconsistent (1 point), slight extent (2-4 points), some extent (5-8 points), consistent (9-10 points).

3. Experience in the knowledge domain: none (1 point), little (2-4 points), not very much (5-8 points), much (9-10 points).

4. Administrative and economic independence in this area: none (1 point), low (2-4 points), medium (5-8 points), high (9-10 points).

5. Ability to solve creative tasks and experience in assessments: none (1 point), low (2-4 points), medium (5-8 points), high (9-10 points).

The total evaluation score regarding the level of competence of the s-th expert according to the i-th function module is determined by the formula (2).

$$O\kappa_{is} = \sum_j W_j O_{js} \qquad (2)$$

Where $O\kappa_{is}$ is the evaluation of the competence level of the *s*-th expert according to the *i-th* function module, $s = \overline{1,d}$ $i = \overline{1,n}$

O_{js} is the evaluation of the expert according to the *j-th* criterion, $j = \overline{1,m}$;

W_j is the weight of criterion to evaluate the expert, provided $\sum_j W_j = 1$.

For the purpose of convenience, the obtained evaluations $O\kappa_{is}$ are suggested summarizing in the table Competence of Experts (Table 1).

Table 1

Competence of Experts

Function modules	Experts				max $O\kappa_{is}$
	1	2	...	*d*	
Module 1	$O\kappa_{11}$	$O\kappa_{12}$	...	$O\kappa_{1d}$	max $O\kappa_{1s}$
Module 2	$O\kappa_{21}$	$O\kappa_{22}$	...	$O\kappa_{2d}$	max $O\kappa_{2s}$
...	...	...	...	...	...
Module *n*	$O\kappa_{n1}$	$O\kappa_{n2}$	...	$O\kappa_{nd}$	max $O\kappa_{ns}$

On intersections of rows and columns in this table there are the evaluations of the competence level of the *s*-th expert according to the *i*-th function module.

When making analyses, planning and forecasting future actions to develop a company, this table can be useful for:

1) Determining the most competent experts against particular function module parameters (in the case of an individual method of expert estimations). It will be the expert with the highest estimate for the *i*-th module (max $O\kappa_{is}$);

2) Determining the weights of importance of experts in the case of the group expert examination. The weight of importance of the *s*-th expert according to the i-th function module is defined as:

$$W_{is} = \frac{O\kappa_{is}}{\sum_{s} O\kappa_{is}} \tag{3}$$

At the initial stage of system's operation, when statistically accumulated data is not enough, the self-assessment of experts is very important at this stage. This stage is based on the experience of experts in the field needed, supported by certificates and diplomas to confirm their relevant qualifications. Thus, after setting each expert in terms of rates, we get the table to show the competence of experts which can be used for selecting the most competent experts in each function module. An example to calculate the competence of experts considering self-assessment is given in Table 2.

Table 2

Competence of Experts for the Strategic Planning

Function Module	**Experts**							
	1	**2**	**3**	**4**	**5**	**6**	**7**	Most competent expert
Manufacturing Automation	0.135	0.182	0.138	0.096	0.147	0.149	0.153	No. 2
Engineering support	0.198	0.095	0.122	0.115	0.138	0.168	0.164	No.1
Ecology of production	0.129	0.131	0.156	0.198	0.122	0.139	0.125	No. 4
Manufacturing infrastructure	0.188	0.148	0.152	0.126	0.125	0.122	0.139	No. 1
Manufacturing	0.143	0.126	0.132	0.158	0.144	0.151	0.146	No. 4
Finance	0.144	0.159	0.168	0.193	0.124	0.125	0.087	No. 4
Average	**0.156**	**0.140**	**0.145**	**0.148**	**0.133**	**0.142**	**0.136**	No. **4**

3. Expert selection on the base of decision maker judgments

The decision on recruiting and levels of professional and psychological qualities, which experts are required to have, is made depending on complexity and specific features of a problem under research, and further potential use of results. In most cases, a medium level of professionalism is required that is sufficient for understanding tasks and struggling against conformism, with the ability to establish constructive communication with other group members and stakeholders. When reviews are made in non-industrial sectors and their results do not have a direct effect on manufacturing processes, a level of professionalism expected from candidates is lower objectively.

A decision maker can often have additional requirements as to the composition of an expert team, expert's quality and competence levels. In this case, the decision maker may express his preferences by making up a set of attribute-based suggestions and comments. In this regard, the task is to develop models that take into account the set of special requirements and select the most compliant candidates under these requirements.

The tables with the resulting qualification of competence based on self-assessment can be used to form a rating list of candidates that is used for the final selection according to the additional requirements. A fuzzy model is proposed for the selection of expert candidates based on decision maker's judgments. Fuzzy models allow decision-makers to take into account the attribute-based suggestions in regard to required qualities and competence levels of experts [9-11].

To facilitate the selection of expert candidates for projects we use the multi-criteria method of evaluating alternatives.

To evaluate expert candidates we developed the following fuzzy rules:

d_1: «If a candidate is an experienced researcher having a certain work record and experience as an expert in an economy and engineering related field, then he is suitable (meets the requirements)";

d_2: "If in addition to the above requirements he has intuition, then he is more than suitable";

d_3: "If in addition to the conditions of d_2 he is able to find a customer of INP, then he is perfect";

d_4: "If he has all specified in d_3, except for intuition, then he is a highly suitable";

d_5: "If the candidate is a highly experienced researcher and a good expert able to find a customer, but does not have a work record, then he is still suitable";

d_6: "If he does not have a qualification as a researcher or a proven expertise to conduct relevant works, then he is not suitable".

Analyzing six modules of information gives five criteria used in the decision making process: X_1 – researching capacities; X_2 – a work record; X_3– work experience as an expert; X_4– the possession of intuition; X_5 – the ability to find a customer. Linguistic variables are created for the all criteria [12]. A variable Y "suitability" is also linguistic and is defined in the set $J = \{0;0.1;0.2;...;1\}$

Suitability is defined as $\mu_s(x) = x, x \in J$,

More than suitable as $\mu_{ms}(x) = \sqrt{x}, x \in J$,

Perfect as $\mu_p(x) = \begin{cases} 1, \text{если } x = 1; \\ 0, \text{если } x \neq 1. \end{cases}$

Highly suitable as $\mu_{vs}(x) = x^2, x \in J$;

Not suitable as $\mu_{us}(x) = 1 - x, x \in J$;

In the selection process, the suitability can be found for each of the above according to the compositional rule [13] of interference, and the corresponding point estimate is calculated.

The selection is made among five candidates who received the highest average estimate on competence based on self-assessment results (according to Table 2). Evaluations of experts on the criteria are given in Table. 3.

Table 3

Inputs for the Expert Candidate Selection

Alterna-tives	Criteria -based evaluation									
	μ	Qualified researcher	μ	Work record	μ	Expert Capacity	μ	Intuition	μ	Ability to find a customer
Candidate No.1	0.8	Qualified	0.5	Medium	0.6	Average	1.0	High	0.0	None
Candidate No.2	0.6	Satisfactory	1.0	Long	0.9	Very high	0.3	Низкая	0.5	Average
Candidate No.3	0.5	Medium	0.0	None	1.0	Excellent	1.0	High	1.0	Very high
Candidate No.4	0.1	Bad	0.5	Average	0.7	High	0.0	None	0.8	High
Candidate No.5	0.3	Lower than average	1.0	Long	1.0	Excellent	0.0	None	0.1	Low

When selecting an expert, the variable "suitability" is determined for each of the rules and the corresponding point estimate is calculated according to the compositional rule of interference. The outputs using the fuzzy inference system are shown in Table 4 and it can be used complimentarily by a company's director to verify the right selection of the expert.

Table 4

Outputs using Fuzzy Interference System

Rule (Output)	Alternatives				
	No.1	No.2	No.3	No.4	No.5
Rule No.1 (Suitable)	0.5	0.6	0	0.1	0.3
Rule No.2 (More than suitable)	0.5	0.3	0	0	0
Rule No.3 (Perfect)	0	0.3	0	0	0
Rule No.4 (Highly suitable)	0	0.5	0	0.1	0.1
Rule No.5 (Less suitable)	0	0	0	0.01	0
Rule No.6 (Not suitable)	0.4	0.4	0.5	0.9	0.7
Point estimate	0.553	0.554	0.425	0.298	0.391

Thus, the suitability point estimate for the alternative μ_1 is 0.553, for μ_2 is 0.554, for μ_3 is 0.425, for μ_4 is 0.298, for μ_5 is 0.391. The alternative μ_2 is selected as the best suitable, i.e. decision makers preferred a candidate with the long work record (the degree of preference is 1) and a high expert capacity (the degree of preference is 0.9); while the capacity for researches and the possession of intuition as criteria shows a lesser degree of preference (0.6 and 0.3 respectively). This can be explained by the fact that a capacity for researches is strengthened during the theoretical work and a level of intuition can be obtained with increasing the work record. This decision-making method using the fuzzy inference system is an adaptation of fuzzy logic to the decision-making processes with inputs in the form of point estimates.

4. Structure of informational support for the evaluation and selection of experts

Most expert assessments and reviews can be carried out by experts working remotely via the Internet that allows the reduction of travel expenses and accommodation for experts. It is therefore important to develop a system that helps organize the work of experts on-line for both remote examinations and direct cooperation of expert groups on the ground of enterprises. Web-based systems will make possible to unify the process of assessment and involve experts in different fields of knowledge, as well as to provide easier access to reports for decision-makers. Crowd sourcing is now actively developing and changing, enabling you to obtain service needed for solving many tasks. [14]

Figure 1 shows the interaction of key roles in the system. A decision maker specifies requirements for an expert relating to the examination and assessment process on the server and views the most competent expert candidates to be selected for a job in the expert group. The experts, in turn, interact with each other, with self-assessing and assessing their colleagues in the process of examination, and carrying out the tasks assigned, and the results obtained after such interactions are stored on the server and can be reported to the decision maker. The most competent experts can be selected with regard to the results in three basic components: self-assessment, peer assessment and assessment of the expert compiled on the basis of examinations and reviews he has taken part in. Already after the first expert assessment, the server will store statistical data and select the most competent experts in the relevant field according to a particular algorithm. The decision maker also can decide independently what kind of data should be taken into consideration for selecting the most competent experts to build a team. This can also be an average reading over all three components: self-assessment, peer assessment and the data on the basis of the jobs completed. In turn, the decision maker can trust the algorithm to select the most competent experts, or to amend a composition of the expert team on account of personal preferences. It is up to the person in charge to have a final word in the selection of experts.

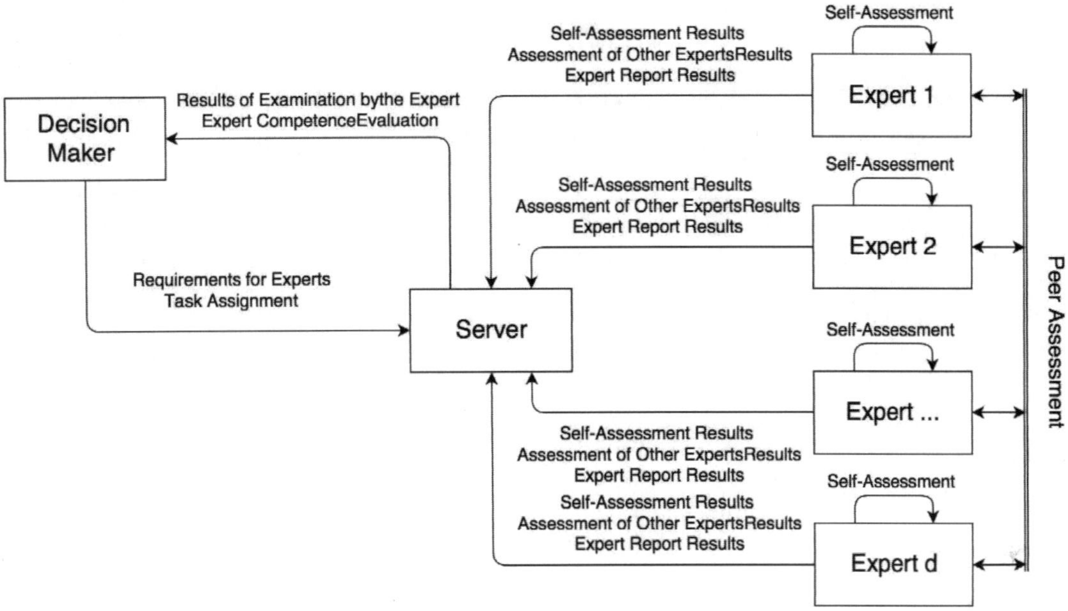

Figure 1. Web-based communications model for interaction between experts and decision makers

Conclusion

The process of reviewing with the purpose to assess high-technology enterprises requires the involvement of internal and external experts, the selection of experts for such a job is a complicated and important task. The proposed in this article model is used for the selection of experts based on determined table – driven competence levels and enables stakeholders to build up a group of experts according to the selected competence-related criteria in specific functional areas of expertise. The results of self-assessment, peer assessment or assessment on earlier examinations made by experts can be summarized in the competence tables. The expert group formed on the basis of the competence tables can be approved a decision maker without any changes. In case of additional requirements of the decision maker for the composition or competence levels of experts, the fuzzy model serves the purpose to select experts on the basis of decision maker judgments. This model allows us to take into account qualitative preferences of decision-makers relative to their needs and interests.

With a view to provide the online examinations by experts, it is relevant to develop a Web-based information system. The information system enables decision-makers to form a group of experts with the highest competence levels within the needed areas of expertise. The stored database of expert candidates with their profiles and examination records are used to determine the competence levels of expert candidates over the three analytical components: self-assessment, peer assessment and expert evaluation on the basis of their examination record.

Acknowledgments

The reported study was funded by RFBR according to the research project № 16-07-00299 a.

References

[1] Malyshev N G, Berstein L S and Bozhenuk A V, Fuzzy model of decision making in CAD system, Fuzzy sets in informatics, International Conference, Moscow, 1998. p.44

[2] Tceplit A, Grigoreva A and Osipov Y, Developing the Model for Assessing the

Competitiveness of Innovative Engineering Products, Applied Mechanics and Materials. 682 (2014), pp. 623-630

[3] Zakharova A A and Ostanin V V, Formalization model of expert knowledge about a technical index level of engineering products, IOP Conference Series: Materials Science and Engineering. 91(2015) 012070

[4] Schaffhausen C R and Kowalewski T M, "Assessing Quality of User-Submitted Need Statements from Large-Scale Needfinding: Effects of Expertise and Group Size", Journal of Mechanical Design, Transactions Of the ASME, vol. 137, no. 12 (2015).

[5] Kuznetsov M P and Strijov V V, Methods of expert estimations concordance for integral quality estimation. Expert Systems with Applications, 41(4 PART 2, 2014), pp. 1988-1996.

[6] Ludolph E, "Cardinal errors in expert opinions for the statutory accident insurance", Unfallchirurg, vol. 118, no. 11 (2015), pp. 963-975.

[7] Cannavacciuolo L, Iandoli L, Ponsiglione C and Zollo G, "Knowledge elicitation and mapping in the design of a decision support system for the evaluation of suppliers' competencies", VINE, vol. 45, no. 4 (2015), pp. 530-550.

[8] Zakharova A A, Fuzzy swot analysis for selection of bankruptcy risk factors, Applied Mechanics and Materials. 379 (2013) pp. 207-213.

[9] Frantti T and Mähönen P, Fuzzy logic-based forecasting model, Engineering Applications of Artificial Intelligence, 14(2) (2001) pp. 189-201.

[10] Ivezić D, Tanasijević M and Ignjatović D, Fuzzy approach to dependability performance evaluation, Quality and Reliability Engineering International. 24(7), (2008) pp. 779-792.

[11] Yadav O P, Singh N, Chinnam R B and Goel P S, A fuzzy logic based approach to reliability improvement estimation during product development, Reliability Engineering and System Safety. 80(1) (2003) pp. 63-74.

[12] Zadeh L A, Fuzzy sets, Information and Control. 8(3) (1965) pp. 338-353.

[13] Yager R R and Rybalov A, "Class of weak uni-norm operators and aggregation of fuzzy rules", Annual Conference of the North American Fuzzy Information Processing Society - NAFIPS, (2000), pp. 424-428.

[14] Christoforaki M and Ipeirotis P G, "A system for scalable and reliable technical-skill testing in online labor markets", Computer Networks, vol. 90 (2015), pp. 110-120.

Importance of Resultant Action of the Mining Machine Actuator for Stresses in Impact Zone of a Separate Cutter

V V Aksenov, V Yu Beglyakov, A A Kazantsev, S I Ganovichev

Yurga Institute of Technology, TPU Affiliate, 26, Leningradskaya Str., 652050, Yurga, Russia

E-mail: 55vva42@mail.ru, begljakov@rambler.ru, ak_uti@rambler.ru,

Abstract: Two stress levels are considered in the general pattern of stress-strain state of the rock destroyed by the mining machine. The authors also ground the necessity of considering the interaction of all cutters of the actuating device when calculating the process of cutting with a separate cutter.

1. Introduction

The works are being completed to develop tunneling machines of a new class – geohods [1, 2], to the systems of which special requirements are imposed [1, 2].

When calculating the actuating devices of mining machines we determine the forces of interaction between the cutter and the rock and then the forces, affecting the cutters, are summed up [6]. But these calculations do not take into account the impact of the summed stress upon the stress-strain state in the local impact area of a separate cutter.

In works [4-8] a different approach to forming the initial data and designing actuating devices of the mining machines is grounded. The approach is based upon considering the background stress influence.

2. Results and Discussion

Let us divide the general pattern of stress-strain states in the effective area of the actuating device into two stress levels.

– *the first level of stresses* – local (element) stresses which are formed in the area of contact between the face rock and the cutter, the cutter tooth or other rock-cutting tools;
– *the second level of stresses* – total (background) stresses which are formed by the actuating device (worm screw, cutter head, drum, etc.). The total stresses are the result of summing up the stresses of the first level, all contact (local) influences;

If we don't take the elastic deformation of tools and rock into consideration the contact between them can be considered to be a point one. Theoretically under the point contact the area of contact tends to zero and the stresses, even under small forces, tend to infinity. But, as the stresses cannot exceed the break-down point of the rock, then with increase of the contact effort it is not the stress that is going to grow but the area where the stresses enough to destroy the rock will emerge, in other words the area of destruction will grow larger.

When considering the interaction between the separate cutter and the rock we can see that its impact area is quite small and varies from several millimeters to several dozens of millimeters. That's why such geometrical parameters as the size of the cutter head or worm screw, diameter of the working, presence or absence of a shelf, etc. do nat have much impact upon the stresses developed by a separate cutter in the local area.

In the given case we speak about the stresses caused by the interaction of a separate element (cutter, tooth, etc.) with the face rock, and the stresses caused by other cutters of the actuating device are not taken into consideration. Dependence of local stresses upon the distance to the contact point along optional axis can be expressed through a generalized polynomial of the following type:

$$\sigma = \frac{A_1}{x} + \frac{A_2}{x^2} + \frac{A_3}{x^3} ..... \frac{A_n}{x^n} \text{ when } x > x_0 \tag{1}$$

where x – distance to the contact point, x_0 – distance from the contact point to the point where the stresses achieve the break-down point of the rock (the destruction area), A_i – empirical constants. Figure 1 a shows the diagram of stresses dependency upon the distance to the contact point.

If the rock massif is influenced by other forces including those produced by other cutters of the actuating device, then the value of stress can be obtained through application of the principle of superposition $\overrightarrow{\sigma_\Sigma} = \sum \overrightarrow{\sigma_i}$, where $\overrightarrow{\sigma_i}$ - stresses caused by the i-th force with consideration to its direction.

If we consider the stress-strain state in the local impact area of a separate cutter with the account of other forces, then expression (1) will look as follows:

$$\vec{\sigma} = \overrightarrow{\sigma_\Sigma} + \overrightarrow{\left(\frac{A_1}{x} + \frac{A_2}{x^2} + \frac{A_3}{x^3} ..... \frac{A_n}{x^n} \right)} \tag{2}$$

where $\overrightarrow{\sigma_\Sigma}$ – «background» stress of the rock caused by other forces, *second level of stresses.*

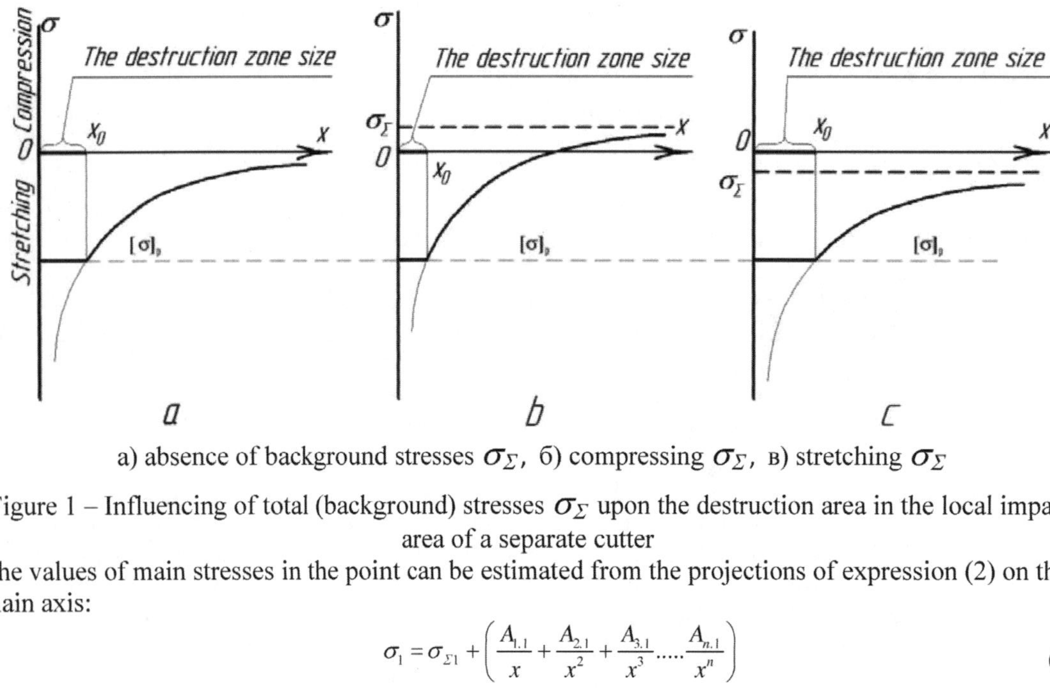

a) absence of background stresses σ_Σ, б) compressing σ_Σ, в) stretching σ_Σ

Figure 1 – Influencing of total (background) stresses σ_Σ upon the destruction area in the local impact area of a separate cutter

The values of main stresses in the point can be estimated from the projections of expression (2) on the main axis:

$$\sigma_1 = \sigma_{\Sigma 1} + \left(\frac{A_{1.1}}{x} + \frac{A_{2.1}}{x^2} + \frac{A_{3.1}}{x^3} ..... \frac{A_{n.1}}{x^n} \right) \tag{3}$$

$$\sigma_3 = \sigma_{\Sigma 3} + \left(\frac{A_{1.3}}{x} + \frac{A_{2.3}}{x^2} + \frac{A_{3.3}}{x^3} ..... \frac{A_{n.3}}{x^n} \right)$$

The modulus of stresses caused by a separate cutter decreases abruptly with moving away from the contact point in all directions. Gradient of these stresses modulus is directed from the rock to the contact point and its modulus decreases even faster.

The gradient of the background stresses modulus (second level stresses) is directed from the rock to the surface and its value is smaller and it does not change so abruptly (Fig. 2). That'swhy background stresses within the local area can be considered a constant value to some extent.

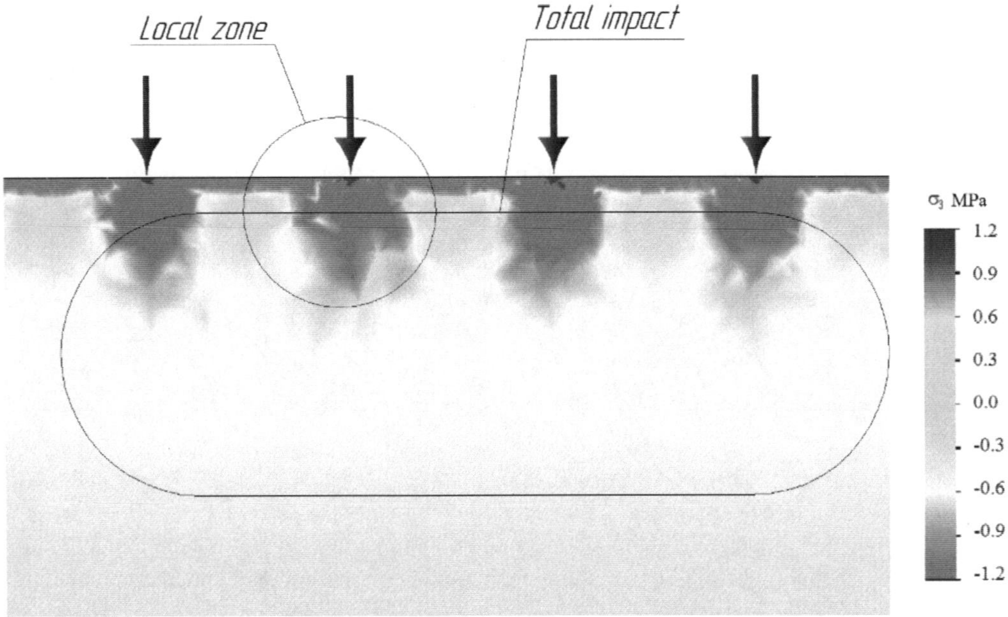

Figure 2 – The field of stress distribution in the area of the actuating device impact

$$grad(\sigma) = -\frac{A_1}{x^2} - \frac{2A_2}{x^3} - \frac{3A_3}{x^4} - ..... - \frac{nA_n}{x^{n+1}} \quad (4)$$

According to Mohr's theory mountain rock is destructed only by shear or stretching [2]. Thus, at least one of the two conditions fulfilled: exceeding the break-down limit for pulling stresses or shear stresses (tangential stresses) which maximal value is determined by main stresses and equal half of the difference between σ_1 and σ_3. (Fig. 3).

Let us suppose that constant force which does not depend upon the destruction process is applied to a point. Let us estimate the impact of background stresses σ_Σ upon the destruction area.

If the rock is destroyed due to exceeding the break-down limit for pulling in the local area, then background stresses will influence the size of destruction area. Figure 1 a shows a curve described by expression (1). The curve illustrates stress distribution in the local area under absence of background stresses. Under background stresses the curve shifts up by σ_Σ (Fig. 1 b) under compressing stresses or down (Fig. 1 c) under pulling stresses σ_Σ. Compressing background stresses σ_Σ lead to reduction of destruction area (Fig. 1 b), and pulling – to its increase (Fig. 1 c).

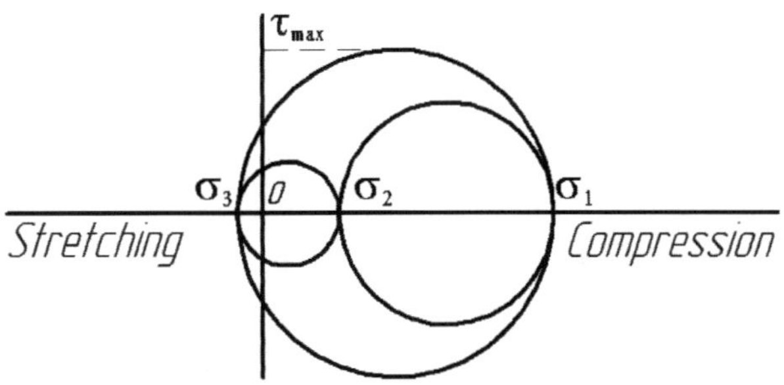

Figure 3 – Mohr's circles

In the cases when main stresses σ_3 do not reach the breaking strength limit, the rock is destroyed by tangential stresses. The value of tangential stresses depends upon the difference between the stresses σ_1 and σ_3 $\tau = 0.5(\sigma_1 - \sigma_3)$. The left side of Figure 4 shows that tangential stresses do not depend directly upon the background stresses. But if we apply the Mohr's circles on the strength certificate (Fig. 4, on the right), then we can judge about the size of destruction area by the limits of segment T cut by the strength certificate curve from Mohr's circles. From the figure we can see that compressing background stresses lead to reduction of destruction area (Fig. 4 d) and pulling – to its increase (Fig. 4 f).

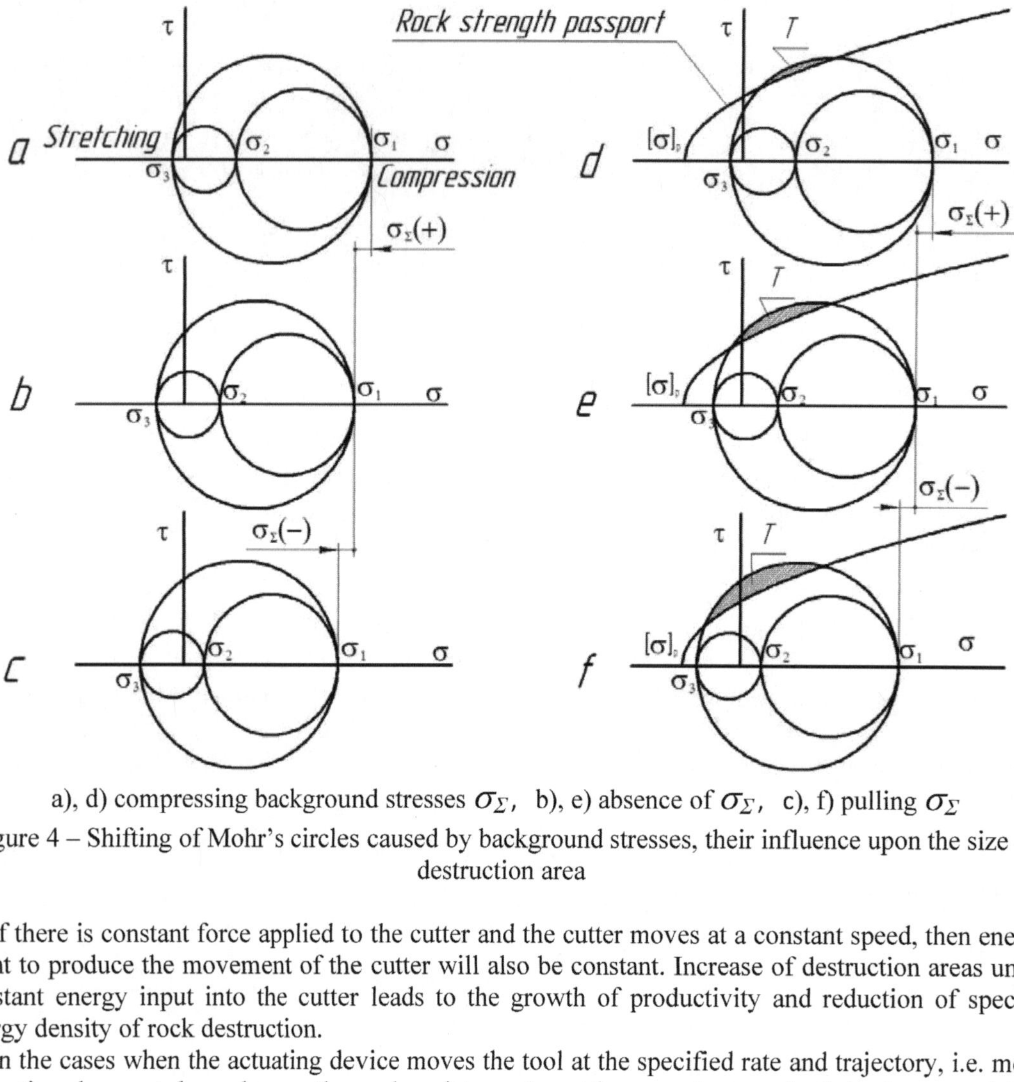

a), d) compressing background stresses σ_Σ, b), e) absence of σ_Σ, c), f) pulling σ_Σ

Figure 4 – Shifting of Mohr's circles caused by background stresses, their influence upon the size of destruction area

If there is constant force applied to the cutter and the cutter moves at a constant speed, then energy spent to produce the movement of the cutter will also be constant. Increase of destruction areas under constant energy input into the cutter leads to the growth of productivity and reduction of specific energy density of rock destruction.

In the cases when the actuating device moves the tool at the specified rate and trajectory, i.e. mode of motion does not depend upon the rock resistance to cutting, constant productivity is ensured. The value of the force applied to the cutter determines the diameter of the large Mohr's circle. Large Mohr's circle touching the line of strength certificate means minimal sufficient cutting force. Figure 5 shows the influence of background stresses upon the value of sufficient cutting force. Shifting of stress towards pulling will reduce the cutting force and, thus, to increase of required energy under the specified productivity.

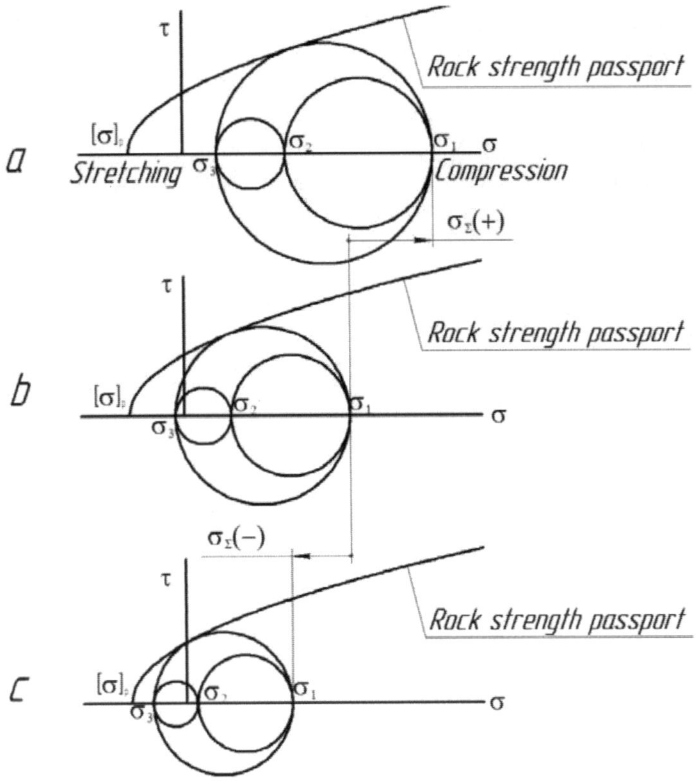

a) compressing background forces σ_Σ, b) absence of σ_Σ, c) pulling σ_Σ

Figure 5 – Influence of background forces upon the cutting force

3. Conclusion

1) the size of the local destruction area depends upon the back ground forces created by other forces in the rock massif;

2) the size of the local destruction area increases under pulling background stresses and reduces under the compressing stresses;

3) energy density of rock destruction decreases under pulling background stresses.

References

[1] Aksenov V.V., Efremenkov A.B., Sadovets V.Yu., Timofeyev V.Yu., Beglyakov V.Yu., Blashchuk M.Yu. Forming the requirements to the main geohod systems // Mining informational and analytical bulletin (Scientific and technical journal). 2009. T. 10. № 12. P. 107-118.

[2] Aksenov V.V., Efremenkov A.B., Beglyakov V.Yu., Blashchuk M.Yu., Timofeyev V.Yu., Sapozhkova A.V. Development of requirements to the main geohod systems // Mining equipment and electromechanics. 2009. № 5. P. 3-7.

[3] Methodology instructions. Boom-type roadheaders. Calculation of operation loading of actuating device transmission. Guideline document 12.25.137-89. – Moscow: Ministry of coal industry of the USSR, 1989. – 51 p.

[4] Beglyakov V.Yu. Justification of parameters of interaction surface between the geohod actuating device and the face rock // doctoral thesis in engineering / Kuzbass State Technical University. Yurga, 2012

[5] Koperchuk A.V., Murin A.V. 2012 *Mining Informational and Analytical Bulletin* Improvement of locked up hydrodynamic coupling for drives of mining machines **OB3** 300-305.

IOP Conf. Series: Materials Science and Engineering **127** (2016) 012030 doi:10.1088/1757-899X/127/1/012030

[6] Murin A.V., Koperchuk A.V. 2011 *Mining Informational and Analytical Bulletin* Reduction of energy losses by locked up hydraulic coupling in drives of mining machines **OB2** 337-343.

[7] Blaschuk M.Y., Dronov A.A., Miheev D.A. 2015 Applied Mechanics and Materials Geokhod Propel Effort Mathematical Model **Vol. 770** 391–396.

[8] Efremenkov A.B., Timofeev V.Y. 2012 7th International Forum on Strategic Technology (IFOST) Determination of necessary forces for geohod movement (IEEE) pp 1–4.

[9] Koperchuk A.V., Murin A.V. 2014 *Applied Mechanics and Materials* Influence of geometrics of synchronization devices of fluid coupling on loading capability **Vol. 682** 499-503.

[10] Koperchuk A.V., Murin A.V., Dortman A. A., Filonov V. V. 2015 *Applied Mechanics and Materials* A change in mechanical behavior of safety fluid couplings when the lockup device is used in its construction **Vol. 770** 279-282.

IOP Conf. Series: Materials Science and Engineering **127** (2016) 012031 doi:10.1088/1757-899X/127/1/012031

Development of Requirements for a Basic Standardized Mathematical Model of Geokhod

V V Aksenov, V Yu Beglyakov, A A Kazantsev, I V Doroshenko

Yurga Institute of Technology, TPU Affiliate, 26, Leningradskaya Str., 652050, Yurga, Russia

E-mail: 55vva42@mail.ru, begljakov@rambler.ru, ak_uti@rambler.ru

Abstract: The article revealed the shortcomings of existing mathematical models geokhods, the necessity of a new approach to modeling the processes of internal and external geokhods interaction, formulated the task of building flexible mathematical models.

1. Introduction

Yurga Institute of Technology, TPU Affiliate has been currently engaged in development of a new generation of geokhods in collaboration with Kemerovo research center of the Siberian Branch of Russian Academy of Sciences [1]. In 2015-16 it is planned to manufacture and test a pilot model of geokhod for tunneling mine workings 3.2m in diameter [1]. Test results will be referred to when taking the decision of a full-scale production.

A number of companies and agencies are interested in developing geokhods with other configurations for tunneling mine workings in various mining geological and technical conditions. Technical orders for research and development have been already formed to design geokhods meeting special technical requirements.

Geokhod is a substantially new tunneling machine. Operation of main geokhod systems and their standards differ essentially from similar systems of all available tunneling machines [1]. Physical configuration and structure of main geokhod systems have been changed, functionality of systems and ways of their interaction have been varied, new layout and design decisions have been constantly made. Substantiating characteristics of available and newly developed design decisions is a burning issue. Mathematical models of force interaction are to be developed for reasoning force, power and design characteristics of geokhods and their systems of various standard sizes and purposes, moreover, interaction of elements, systems and modules, as well as their contact with geological environment are to be studied.

2. Results and Discussion

Various mathematical models describing contacts of geokhod main systems with each other and the geological environment are provided in papers [2…4]. The models are developed for geokhods «ELANG-3» and «ELANG-4» and their particular design characteristics are taken into account (Figures 1 and 2), among them: the structure and geometrical characteristics of a rotation mechanism, types and characteristics of function agents of the main face, technique of building peripheral canals and other particular features. Models presented in papers [4] also involve various characteristics of the geological environment; mining geological and technical conditions are taken into consideration, i. d.

they are intended to solve problems of various force interaction. However, geometrical characteristics of peripheral elements, their function agents and peripheral canals are not considered in those models etc.

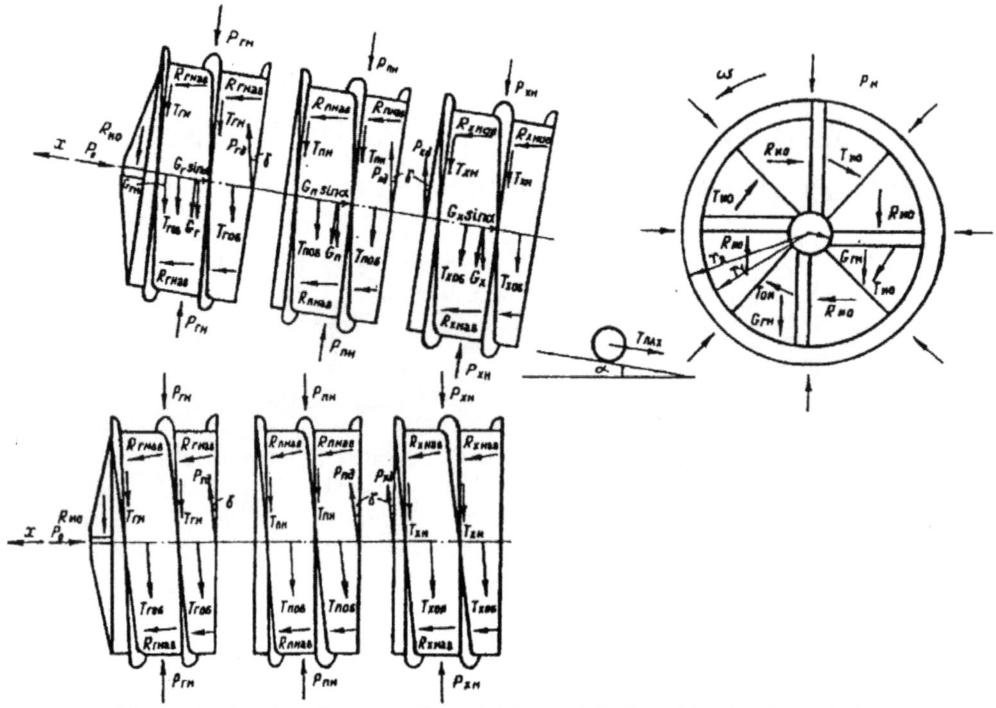

Figure 1 – Design diagram of available model of geokhod «ELANG-3»

The models consist of rather large undividable blocks. Moreover, these models have a significant nomenclature of various characteristics and variables. Characteristics and variables are chosen on the base of element designation and their design features (Figures 1, 2), as the consequence, they can hardly be systemized. When introducing a new variable a new designation being in line with designations of other variables and characteristics is to be created, and one should be convinced a newly created designation has not been used yet. This fact leads to restrictions and does not support team work of researchers, their groups and individuals.

A growing variety of layout and constructive decisions has determined the necessity of developing block-module and unitized machine. Composition and characteristics of geokhod modules make it possible to design diverse modifications on the general base.

In these conditions application of available models gets quite complicated because changing operation principle of one system or the way closely-related systems interact with each other can be reason for adjustment of the whole model. To be more precise, a prevalent approach requires for development of a new model, which will also rely on a certain layout or constructive decision.

A principle of modeling is to be revised when substantiating the characteristics of geokhod and its systems. A new flexible model is necessary, which will be able to take into account specifics of operation and interaction of the systems, make it possible to change their composition and sequence, add new systems or cut down their number, as well as change principles of their work and interaction. It is to be emphasized a particular system is to be changed independently on other elements of the model.

A mathematical model consisting of blocks is to be developed; each of them is a model of a particular system, module or interaction. And the general model is to be adjustable via replacing or changing the blocks. Moreover, adjustment of one block must not necessitate change in other blocks or rearrangement of connections in the model.

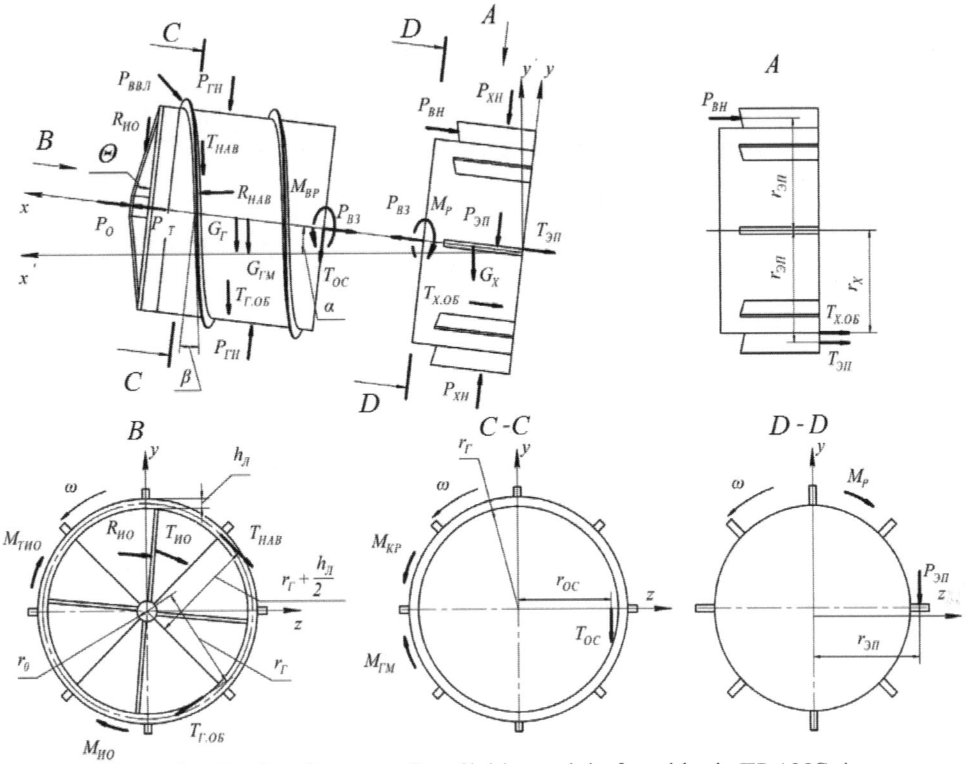

Figure 2 – Design diagram of available model of geokhod «ELANG-4»

To develop a mathematical model of this kind a number of issues are to be solved:

1) Geokhod is to be divided into systems and modules to create a wide range of layout and constructive decisions. The following requirements are to be met when dividing geokhod into modules and systems:

– both modules and systems are to be structurally and functionally independent;

– junctions are to be structurally similar and interchangeability level of systems and modules of the same name within one dimension size or a range of dimensions is to be acceptable;

– level of nesting sub-systems into the systems, units into modules is to be determined etc.

2) A structure, linking modules and systems is to be created when developing a general model of geokhod.

The structure of links is to provide standardized principles of forming input and output characteristics of various systems, modules and units of geokhod.

3) Standardized end connections are to be designed to join modules and systems.

Systems and modules of geokhod differ much in their structure and functions. Variants of geokhod units can be absolutely different. For example, single-thread screw blade and multi-thread propeller propulsive agents are available; principles of their interaction with geological environment and geohkod are significantly different.

Standardized end connections enable substituting various types of outer propulsive agents, function agents and other systems into the general geokhod system without changing the rest of the model. That is, changed or replaced module will be the reason of changes in the corresponding block of the mathematical model and affect the general results of modeling.4) Principles of transforming each system (module) into an independent model, meeting the principles of the general model with input and output data are to be developed and be in line with standardized end connections between systems and modules.

Geokhod consists of more than ten structurally independent modules and systems with separate functions. A team of researchers and engineers working in groups and individually are involved in designing, estimating characteristics of geokhod, and developing its modules and systems.

To provide efficient group interaction agreements are to be done which support uniformity of designations, formulas, methods of design and modeling. Information discussed in groups and data transferred from one group or individual to the others are to be recorded.

5) Development of core of the model.

A core of the model – a constant unchangeable base like a main board in the computer or a tree for forming a 3D model is to be developed to modify the mathematical model of geokhod. In the core of the model there are blocks being models of systems and modules. System and module models will be developed according to the same principles, i.d. they will have their core consisting of models of sub-systems, units etc.

6) Development of "the first generation" flexible models.

Under assumption of extreme forms of model flexibility two opposite levels can be possibly defined:

– "zero flexibility" – model designed for a particular machine with constant characteristics;

– "absolute flexibility" – model, which can be used in any machine, i. d. a model of everything.

Obviously, the zero level is unacceptable. A model with absolute flexibility looks like a utopia, moreover, development of an unlimited flexible model requires a huge amount of time; however, these models are needed urgently. The lack of flexible models of geokhod and their systems is a constraint. If extreme forms of flexibility are not possible something average is required.

Therefore, some restrictions are to be determined:

- a limited number of structural decisions;
- a set of systems, modules and assemblies typical for geokhod etc.;
- a set of standardized connections (branching of the model core);
- a limited number of nesting levels (module/sub-module/unit/…/ element);
- etc.

These restrictions will provide acceptable labor coefficient of model development, but won't support their possibilities. The limits of these restrictions are to meet near-term prognoses of geokhod-construction.

7) Development of plans (prognoses) to develop next-generation models.

First generation models will meet near-term plans, their development and maintenance enable accumulation of useful experience, reveal peculiarities, errors etc.

The team of researchers and engineers involved in designing geokhods already has plans for the nearest and far future. The range of structural and layout solutions in question includes both real ones which have been developed and "fantastic" issues.

Therefore, first generation models will need up-dating soon. So, one should think of developing new generation of models, work out principles and approaches to modeling geokhod and its systems.

8) Development of software.

Formation of new, expandable mathematical models relies on application of excessive mathematical tools, requiring for automated calculations and formation of models. Available CAE-systems are to automate the process of mathematical model formation, but they are multi-purpose tools or software, which can be used to solve a wide range of problems.

To develop mathematical models of geokhods taking into account requirements, restrictions and agreements mentioned above special-purpose software based both on available CAE-systems and on independent platform are necesary.

Conclusion

1) Available mathematical models of geokhods do not solve the issues arising when designing geokhods of new generation.

2) A flexible model capable for expanding and modification is required.

3) Block and module principle of the structure enables development of flexible models, which can be modified and expanded.
4) A system of rules (agreements) to provide independent but coordinated modeling of various systems and modules of geokhod.
5) Models of the first generation, as well as principles of formation models of further generations are required.
6) Special software is necessary for mathematical modeling the interaction process of geokhod with the geological environment and its systems with each other.

References

[1] Aksenov V.V., Khoreshok A.A., Beglyakov V.Y. Justification of creation of an external propulsor for multipurpose shield-type heading machine - geo-walker // Applied Mechanics and Materials. 2013. T. 379. C. 20-23.
[2] Aksenov V.V., Khoreshok A.A., Efremenkov A.B., Timofeev V.Yu. Development of mathematical model of geokhod interaction with geological environment // Mining Informational and Analytical Bulletin (Scientific and Technical Journal). Mining machinery / M.: Publishing «Gornaya Kniga» – 2011.– OB № 2. pp. 79-91.
[3] Blaschuk M.Yu., Dronov A.A., Mikheev D.A. Mathematical model to calculate forces necessary for geokhod movement // Actual problems of present day mechanical engineering: Proceedings of International Scientific and Practical Conference / Yurga Institute of Technology. – Tomsk: TPU Publishing, 2014. pp. 134-139.
[4] Sadovets V.Yu., Beglyakov V.Yu., Efremenkov A.B. Simulation of geokhod movement with blade actuator // Applied Mechanics and Materials. 2015. V. 770. pp. 384-390.
[5] Koperchuk A.V., Murin A.V. 2012 *Mining Informational and Analytical Bulletin* Improvement of locked up hydrodynamic coupling for drives of mining machines **OB3** 300-305.
[6] Murin A.V., Koperchuk A.V. 2011 *Mining Informational and Analytical Bulletin* Reduction of energy losses by locked up hydraulic coupling in drives of mining machines **OB2** 337-343.
[7] Efremenkov A.B., Timofeev V.Y. 2012 7th International Forum on Strategic Technology (IFOST) Determination of necessary forces for geohod movement (IEEE) pp 1–4.
[8] Koperchuk A.V., Murin A.V. 2014 *Applied Mechanics and Materials* Influence of geometrics of synchronization devices of fluid coupling on loading capability **Vol. 682** 499-503.
[9] Koperchuk A.V., Murin A.V., Dortman A. A., Filonov V. V. 2015 *Applied Mechanics and Materials* A change in mechanical behavior of safety fluid couplings when the lockup device is used in its construction **Vol. 770** 279-282.
[10] Methodology instructions. Boom-type roadheaders. Calculation of operation loading of actuating device transmission. Guideline document 12.25.137-89. – Moscow: Ministry of coal industry of the USSR, 1989. – 51 p.

IOP Conf. Series: Materials Science and Engineering **127** (2016) 012032 doi:10.1088/1757-899X/127/1/012032

Substantiating Ways of Load Application When Modeling Interaction of a Multiincisal Mining Machine Actuator With Rocks

V V Aksenov, V Yu Beglyakov, A A Kazantsev, A S Saprykin

Yurga Institute of Technology, TPU Affiliate, 26, Leningradskaya Str., 652050, Yurga, Russia

E-mail: 55vva42@mail.ru, begljakov@rambler.ru, ak_uti@rambler.ru

Abstract: two methods of modeling of interaction between a mining machine working body and rocks are considered; a multi-cutter working body sum impact on rock stress-strain state in a cutter action zone is described; practicability of distributed forces application in math modeling is substantiated.

1. Introduction

Currently geokhods of a new generation are being designed [1]; a specific principle of geokhod operation requires a special working body (WB) [2, 3] There are many different types, configurations and design concepts of WB of tunnel boring machines (TBM). To make the right choice and to determine rational parameters of mining machine WB it is necessary to evaluate the nature of interaction between the instrument and the rock face. It is clear that fabrication and testing of experimental mining machinery is costly, moreover in Russia there is no testing grounds for mining machines.

Using mathematical modeling of interaction between WB and rock face, and analysis of pictures of rock face stress-strain state (SSS) [4-8] give material for research and greatly accelerate creation of the final version of TBM.

Considering a complex nature of stresses in the rock face, it is practical to apply a method of numerical calculations, namely the Finite Element Method (FEM) for math modeling.

Results and Discussion

When modeling interaction between a multi-cutter tool and a face, a force, simulating interaction between rock and each cutter, is to be applied to a model based on **cutter arrangement** and an angle of WB steer about its axis. For WBs of drum-, cutter head- or helical blade-type the FEM gives information on SSS in the rock face for one working body position (rotation angle) and one cutter arrangement.

If we represent a total WB cutting force as equivalent distributed normal and tangential forces, applied to a contact area between a tool and a face, such a force corresponds both to any position of the working body and to each cutter arrangement.

To test the applicability of such approach it is necessary to evaluate similarity of impact both of distributed forces and of a cutter set total force. For this purpose a SS states of a cylindrical rock sample (D = 1200 mm and a height L = 800 mm) have been modeled with various force applications. A comparative analysis of modeling results has been performed.

Figure 1 shows patterns of force application to the models.

The bottom end is fixed and forces are applied to the upper end according to 4 patterns:

1) concentrated force of one cutter in the end center (Fig. 1 b);

2) concentrated forces of a cutter set, uniformly arranged in a circle (d = 893 mm) (Fig. 1в);

3) distributed forces in a circle with a central cutter (D / d = 893/200 mm) (Fig. 1г).

In areas of interaction between the cutter and the rock, forces (normal Pn = 10 кH and tangent Pt = 2.5 кH) and distributed forces (normal qn = 0.303 MPa and tangent qt = 0.076 MPa) are applied, which are equivalent to the total force of the corresponding set of cutters.

Principal force dependences σ_3 on a distance H in the central cutter contact zone (Fig. 1a) are determined for all force patterns.

Table 1 shows results of numerical modeling. Table 2 shows pictures of SSS in the cutter area.

Figures 2, 3, 4 show graphs of results of distributed and point force applications.

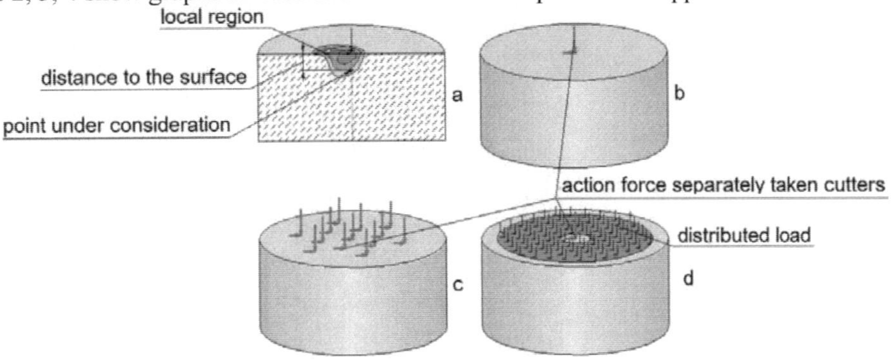

a) location of the area considered, b) one central cutter, c) uniformly distributed cutter set, d) central cutter force and distributed forces

Figure 1 - Patterns of total force applications

Table 1 – Results of SSS numerical modeling in a local area

H (mm)	Minimum values of principal forces σ_3 at a distance H from the surface in central cutter contact area (MPa)					
	σ_{3i}	$\sigma_{3\Sigma}$	$\delta\sigma_{3\Sigma}$	σ_{3iq}	$\delta\sigma_{3iq}$	σ_{3q}
0	-7.275	-7.335	-0.060	-7.341	-0.066	-0.070
1	-5.748	-5.795	-0.047	-5.802	-0.054	-0.056
2	-4.737	-4.785	-0.048	-4.791	-0.054	-0.053
4	-3.076	-3.117	-0.041	-3.122	-0.046	-0.051
8	-2.121	-2.159	-0.038	-2.164	-0.043	-0.046
16	-1.157	-1.178	-0.021	-1.182	-0.025	-0.029
32	-0.408	-0.418	-0.010	-0.421	-0.013	-0.013
64	-0.134	-0.133	0.001	-0.132	0.002	0.001
100	-0.052	-0.033	0.019	-0.038	0.014	0.014
150	-0.021	0.005	0.026	0.001	0.022	0.022

H – distance depthward from the surface
σ_{3i} – one cutter force
$\sigma_{3\Sigma}$ – total force of 13 cutters
σ_{3iq} – one cutter force and distributed forces
σ_{3q} – stress of distributed forces
$\delta\sigma_{3\Sigma} = \sigma_{3\Sigma} - \sigma_{3i}$ and $\delta\sigma_{3q} = \sigma_{3iq} - \sigma_{3i}$ effect of total and distributed forces

IOP Conf. Series: Materials Science and Engineering **127** (2016) 012032 doi:10.1088/1757-899X/127/1/012032

Table 2 - Pictures of principal forces σ_3 to evaluate impact of total and distributed forces on local SSS

Scale	Stress patterns		
	One blade (fig. 1 б)	Set of blades (fig. 1 в)	Distributed forces and a center blade (fig. 1 г)
0...0.1 MPa	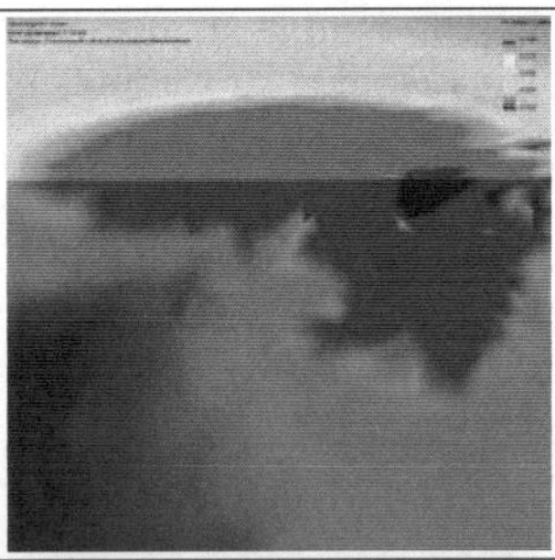	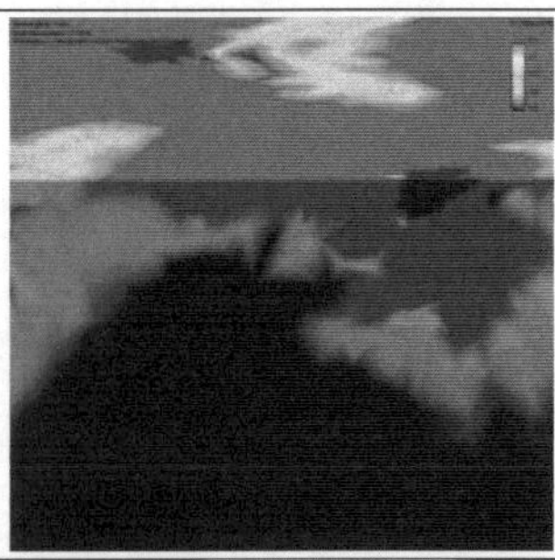	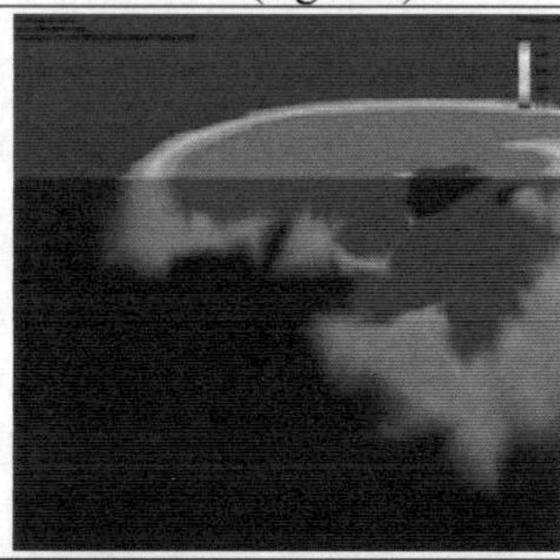
0...0.2 MPa	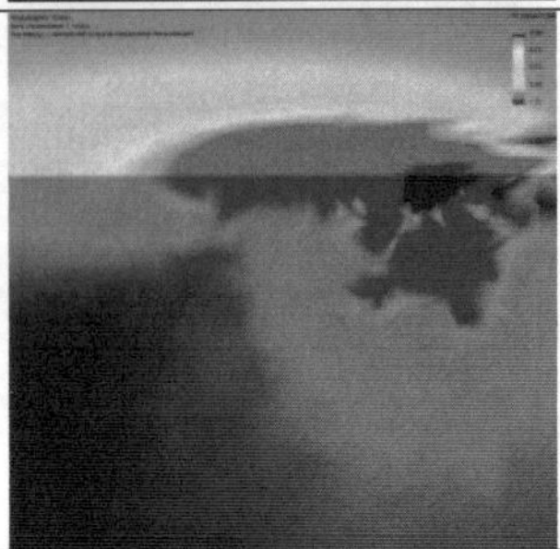	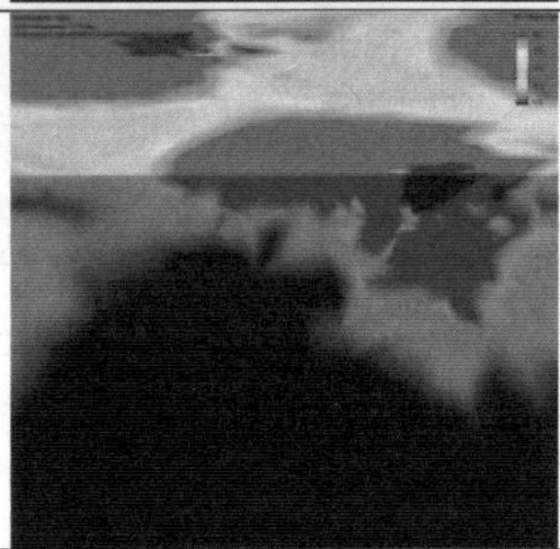	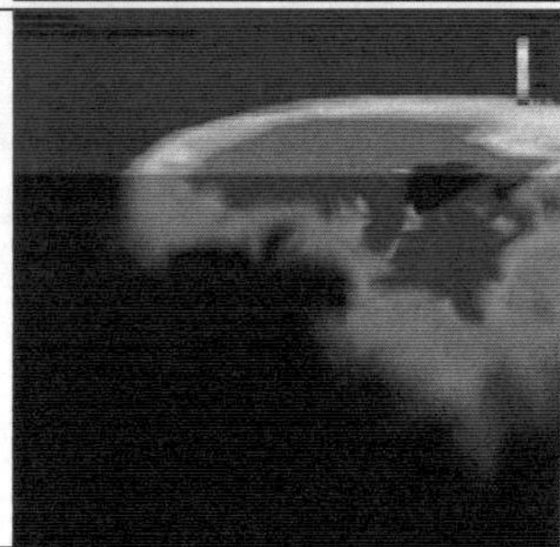
0...0.4 MPa	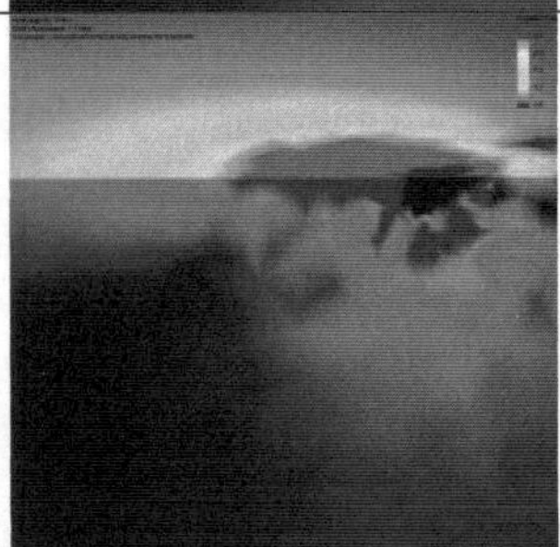	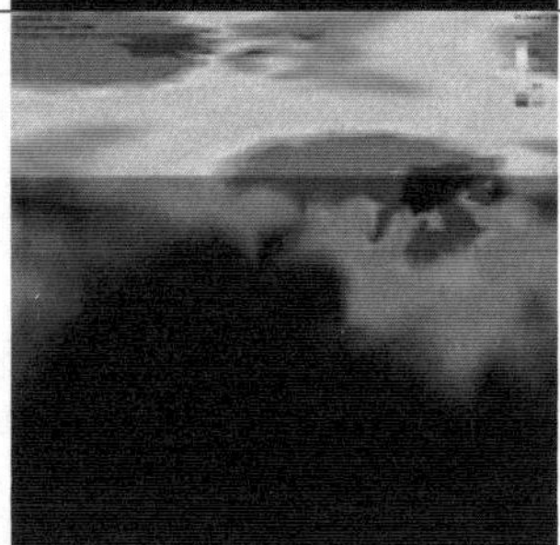	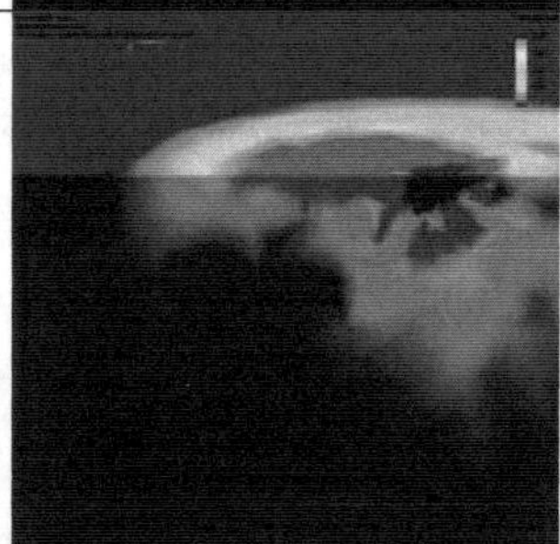
0...0.8MPa	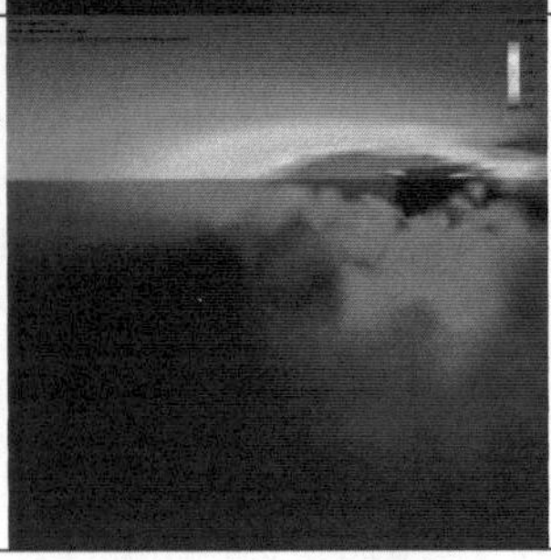	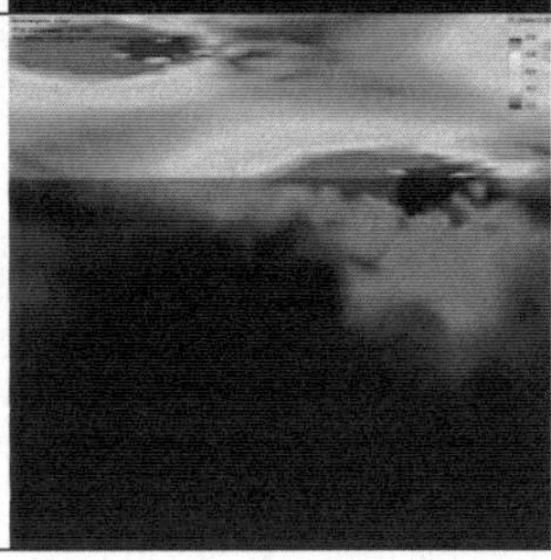	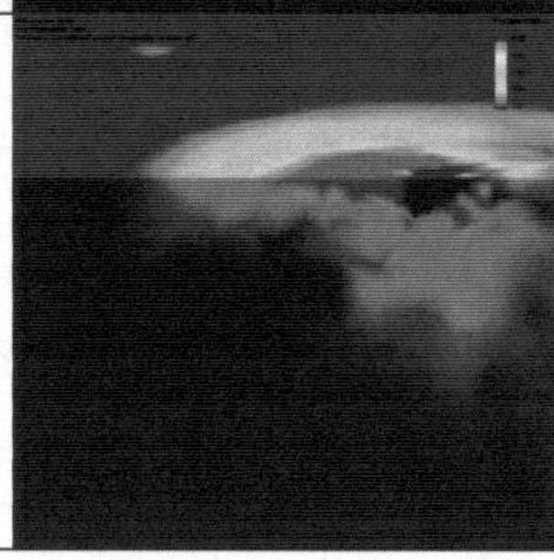

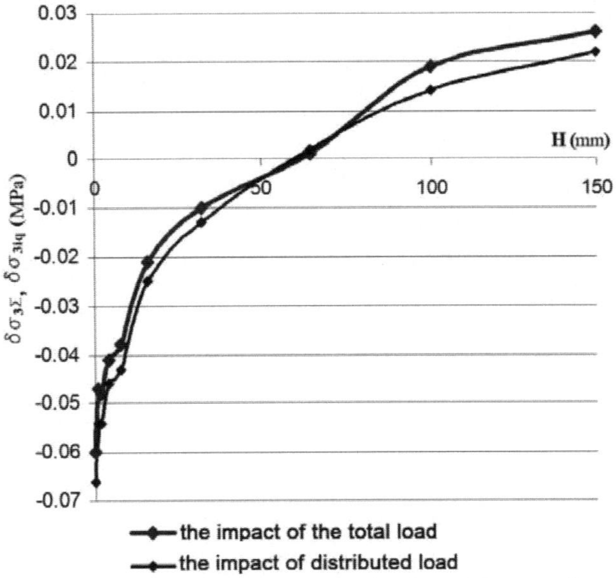

Figure 2 –Total force effect and distributed forces effect

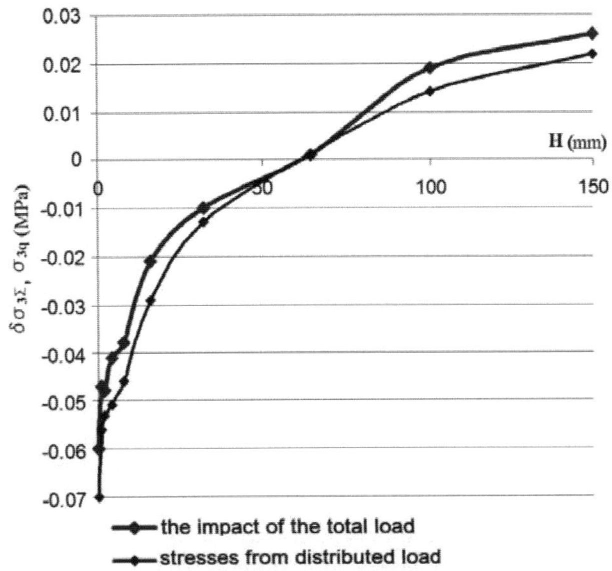

Figure 3 - Total force effect and stress of distributed forces

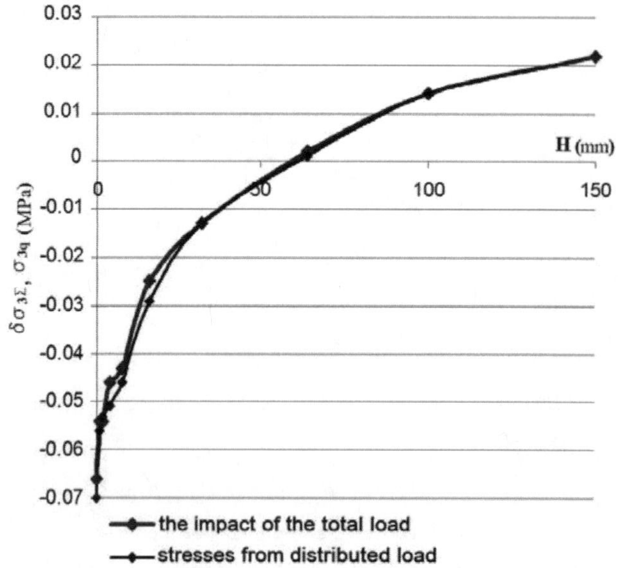

Figure 4 – Effect and stress of distributed forces

Modeling has yielded results for comparison:

1) effect of total and distributed forces on SSS in the central cutter contact area (Fig. 2);

2) effect of total force with applied distributed forces (Fig. 3);

3) effect of distributed forces with its stresses (Fig. 4)

4) Results of comparison of modeling and stress distribution (Table 2) show that a total force of cutters is similar to an equivalent distributed force.

3. Conclusions

1) a distributed force is equivalent to a total force; its impact on local stresses is similar to that of the total force in value and distribution;

2) distributed forces can be a tool for mathematical modeling of interaction between the working body and the rock.

References

[1] Aksenov V.V., Sadovets V.Y, Beglyakov V.Y. Synthesis of geokhod working bodies design solutions // Mining information-analytical bulletin (scientific and technical journal). 2010. Vol. 3. № 12. p. 49-54.

[2] Beglyakov V.Y. Justification of surface parameters of geokhod working body interaction with rock face // PhD thesis / Kuzbass State Technical University. Kemerovo, 2012

[3] Aksenov V.V., A.B. Efremenkov, V.Yu. Beglyakov. The influence of relative distance between ledges on the stress-strain state of the rock at a face // Applied Mechanics and Materials Vol. 379 (2013) pp 16-19.

[4] Methodology instructions. Boom-type roadheaders. Calculation of operation loading of actuating device transmission. Guideline document 12.25.137-89. – Moscow: Ministry of coal industry of the USSR, 1989. – 51 p.

[5] Buyalich G D Anuchin A V Dronov A A 2015 The Numerical Analysis of Accuracy of Hydraulic Leg Cylinder in Modeling Using Solid Works Simulation *J. Applied Mechanics and Materials* **Vol. 770** pp 456-460.

[6] Efremenkov A.B., Timofeev V.Y. 2012 7th International Forum on Strategic Technology (IFOST) Determination of necessary forces for geohod movement (IEEE) pp 1–4.

[7] Burkov P.V., Burkova S.P., Timofeev V.Yu. Analysis of stress concentrators arising during

MKY.2SH-26/53 support unit testing. Applied Mechanics and Materials Vol. 682 (2014) pp 216-223.

[8] Chinakhov D A 2011 Study of Thermal Cycle and Cooling Rate of Steel 30ХГСА Single-Pass Weld Joints *J. Applied Mechanics and Materials* **Vol. 52-54** pp 442–447

[9] Chinakhov D A 2013 Simulation of Active Shielding Gas Impact on Heat Distribution in the Weld Zone, *J. Applied Mechanics and Materials* **Vol.762** pp 717-721.

[10] Koperchuk A.V., Murin A.V., Dortman A. A., Filonov V. V. 2015 *Applied Mechanics and Materials* A change in mechanical behavior of safety fluid couplings when the lockup device is used in its construction **Vol. 770** 279-282.

Calculation of Free Interior Dimensions in Geokhod Transmission With Hydraulic Cylinders

M Yu Blaschuk, A A Dronov, S S Ganovichev

Yurga Institute of Technology, TPU Affiliate, 26, Leningradskaya Str., 652050, Yurga, Russia

E-mail: mby.tpu@gmail.com

Abstract. Analytical expressions to specify overall dimensions of free interior in the geokhod, as well as its relation to design factors of a transmission and geokhod geometry are considered in the paper.

Introduction

Available technologies of mine working based on conventional tunnellers represented mainly by tunneling machines and shields have a number of drawbacks:

- tunneling equipment can't move through the rock mass in any direction of an underground space, i. d. the sphere of utilization is restricted by inclination angles of workings to be done;
- sufficient propulsive forces on the outer propulsive agent and crowding forces on a function agent can hardly be developed to disintegrate hard rocks;
- working in the bottom hole area can't be absolutely safe.

Tunneling machines of this kind are moved by outer propulsive agents – crawler, wheel, wheel-rail or walking brace ones. To develop propulsive forces on the outer propulsive agent and crowding forces on a function agent outer geological environment is not practically used but it is relied on a solid contact surface of geological environment and air or soil of a working (when performing it by a tunneling machine), or surface of a permanent support when shield tunneling. Therefore, a number of studies resulted in development of a geo-winchester technology of mine working, as well as in designing a machine to implement this technology – a geokhod [1].

Geokhods are a new class of tunneling machines distinguished by unconventional engineering solutions and principle of operation [1,2]. The main feature of a geokhod structure consists in two cylindrical sections. The head of the geokhod is burling along the stabilizing end section. The motor rotating the head is to be placed inside the geokhod on the periphery of section coupling. One of the key requirements these motors are to meet is a free space inside a geokhod to transfer the broken rock, to mount other systems, as well as for workers to perform maintenance and repair operations [1,3,4]. Various motors to rotate the head have been already studied [5], it had been revealed rotation transmission with hydraulic cylinders on geokhod chords is the most appropriate variant.

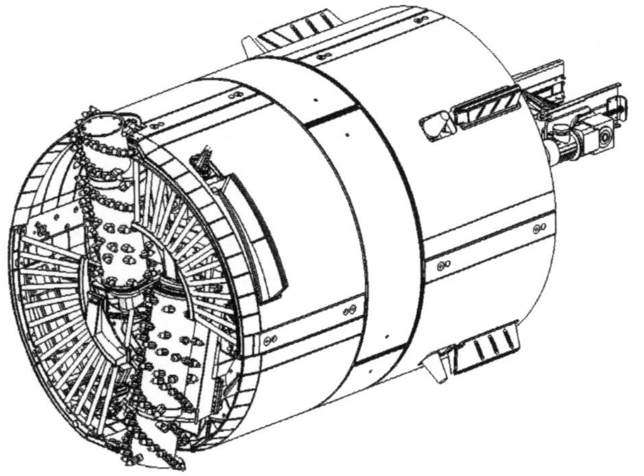

Figure 1. Geokhod

Results and Discussion

If hydraulic cylinders are placed on the circle chords in the space between the head casing and tail shell, design factors of transmission elements (features of hydraulic cylinders) are relevant for free space dimensions inside the geokhod [6]. The layout for calculation of free space inside the geokhod is given in Figure 2.

To eliminate spatial crossing of hydraulic cylinder elements with the head casing and tail shell a minimum positive allowance is necessary between a hydraulic cylinder and corresponding shells. Here, an allowance between a hydraulic cylinder casing and the outer surface of the tail shell is most relevant together with the dimensions $D_{PL.ROD}$, $D_{PL.HC}$, D_{HC} to calculate the maximal probable diameter of tail shell and interior size, respectively.

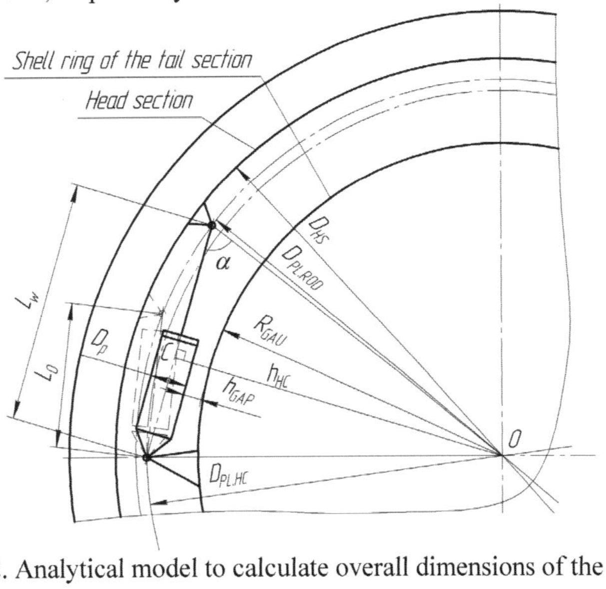

Figure 2. Analytical model to calculate overall dimensions of the interior

As one can see in Figure 2, a case of the hydraulic cylinder is inclined inwards the section, approaching maximally to its longitudinal axis in the end telescoping position under the constructional condition $D_{PL.ROD} \approx D_{PL.HC}$ (tongues of the case and a shaft of the hydraulic cylinder are nearly on the same circle) when the shaft is telescoping.

The shortest (nearest) distance from the longitudinal axis of the head (center O) to the case of the hydraulic cylinder is on the perpendicular OC to the longitudinal axis of the hydraulic cylinder (Figure 2). Then, the outer surface radius of tail shell R_{GAU} is calculated according to equation (1) provided that required allowance h_{GAP} is considered

$$R_{GAU} = h_{HC} - \frac{D_P}{2} - h_{GAP}, \tag{1}$$

where h_{HC} – the shortest distance (perpendicular) from the axis of the section to that of the hydraulic cylinder, m; D_P – diameter of the rod, m.

The distance h_{HC} is calculated according to the equation:

$$h_{HC} = \frac{D_{PL.ROD}}{2} \cdot \sqrt{1 - \left(\frac{\left(\frac{D_{PL.ROD}}{2}\right)^2 + L_W^{\ 2} - \left(\frac{D_{PL.HC}}{2}\right)^2}{D_{PL.ROD} \cdot L_W} \right)^2}, \tag{2}$$

where $D_{PL.ROD}$ – diameter of shaft tongue rotating on the head section, m; $D_{PL.HC}$ – diameter of mounted tongues of hydraulic cylinder on the tail shell, m; L_W – the distance between tongues of the case and the hydraulic cylinder rod under maximal telescoping of a rod.

Substituting h_{HC} calculated in (2) into (1) it is obtained

$$R_{GAU} = \frac{D_{PL.ROD}}{2} \cdot \sqrt{1 - \left(\frac{\left(\frac{D_{PL.ROD}}{2}\right)^2 + L_W^{\ 2} - \left(\frac{D_{PL.HC}}{2}\right)^2}{D_{PL.ROD} \cdot L_W} \right)^2} - \frac{D_P}{2} - h_{GAP}, \tag{3}$$

Taking into account various thicknesses of a case wall in a hydraulic cylinder [7,8] with respect to its structure and dimensions of the rod D_P one has to consider these parameters for the allowance size h_{GAP}. That is why; the distance from the axis of a hydraulic cylinder to the outer shell surface is to be expressed in terms of rod diameter D_P. Then, equation (3) is as follows

$$R_{GAU} = \frac{D_{PL.ROD}}{2} \cdot \sqrt{1 - \left(\frac{\left(\frac{D_{PL.ROD}}{2}\right)^2 + L_W^{\ 2} - \left(\frac{D_{PL.HC}}{2}\right)^2}{D_{PL.ROD} \cdot L_W} \right)^2} - (0.75...1)D_P, \tag{4}$$

Under certain relations of dimensions $D_{PL.ROD}, D_{PL.HC}$ distance h_{HC}, calculated according to equation (2) can be smaller or comparable with dimensions $\frac{D_{PL.ROD}}{2}, \frac{D_{PL.HC}}{2}$. In this case R_{GAU} is dependent on dimensions $\frac{D_{PL.ROD}}{2}$ or $\frac{D_{PL.HC}}{2}$ provided that the necessary allowance h_{GAP} is taken into account (Figure 2). We can calculate equalities $h_{HC} = \frac{D_{PL.HC}}{2}$ and $h_{HC} = \frac{D_{PL.ROD}}{2}$ using Pythagorean theorem

$$h_{HC.ROT} = \frac{D_{PL.HC}}{2} = \sqrt{\frac{D^2_{PL.ROD}}{4} - L_W^2}, \tag{5}$$

$$h_{HC.ROT} = \frac{D_{PL.ROD}}{2} = \sqrt{\frac{D^2_{PL.HC}}{4} - L_W^2}, \tag{6}$$

The dependence of inner overall dimensions (Figure 3) on the number of hydraulic cylinders displays increasing number of hydraulic cylinders 5 to 14 is the reason for extending free interior, which is equal to 66 – 91% of the outer geokhod diameter. The increase in the free space together with the growing number of hydraulic cylinders is possible mainly because the hydraulic cylinders get smaller (D_P – diameter of rod), as well as their arrangement is more compact due to reduced shaft stroke and general extendibility of a hydraulic cylinder.

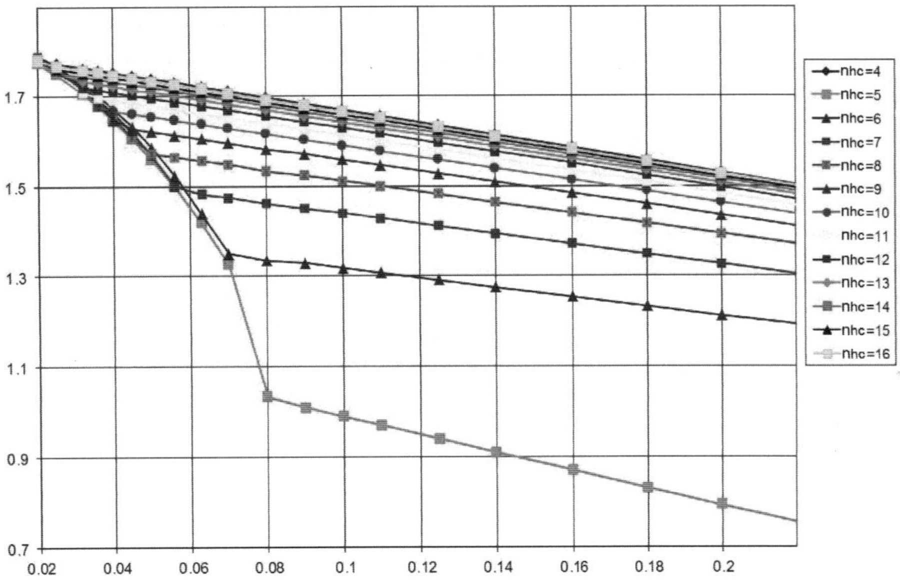

Figure 3. Overall dimensions of the interior in per cents of the head diameter D_{HS} according to the number of hydraulic cylinders n_{HC} .

More precise influence of accepted dimensions of hydraulic cylinders (D_P, L_W) and their number (n_{HC}) on the overall dimensions of free space in the geokhod can be determined with respect to geometrical parameters of hydraulic cylinders in geokhod transmission (Figure 4) [9].

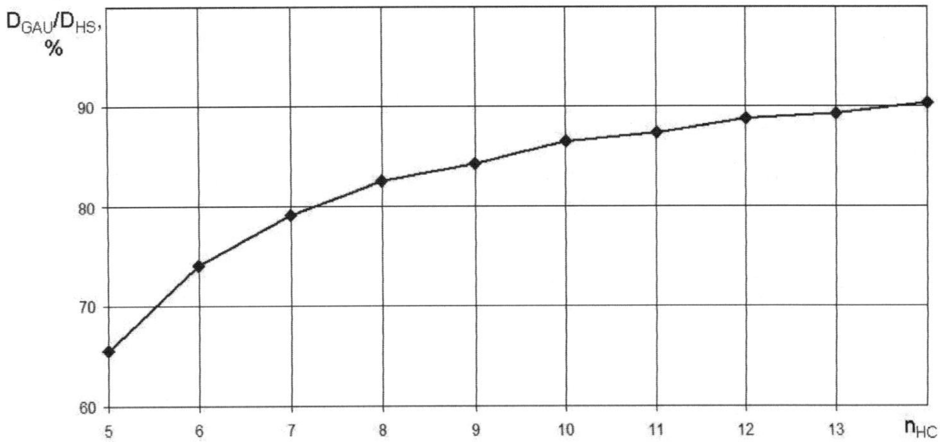

Figure 4 – The relation of distance R_{GAU} to the number of hydraulic cylinders n_{HC} placed on the chords inside the section for some rod diameters D_P .

IOP Conf. Series: Materials Science and Engineering **127** (2016) 012033 doi:10.1088/1757-899X/127/1/012033

These relations are specified for a geokhod with the diameter $D_{HS} = 3.7$ m, and the dimensions of hydraulic cylinders [9] meet the necessary value of the torque moment [10]. The relation is also relevant for other dimensions of a geokhod (Figure 3) as the dimensions of transmission change quite proportionally to changing its dimensions, this fact is obvious in the calculated radius of the inner space (R_{GAU}) of a geokhod with the following size series (Table 1).

Table 1. Geometrical parameters of geokhod transmissions and interior dimensions for the series of tunneling shields

	n_{HC}, items							
D_{HS}=2.1 m; $M_{ROT.REQ}$=0.37 MN·m	**4**	**5**	**6**	**7**	**8**	**10**	**12**	**14**
D_P, m	**0.125**	**0.09**	**0.08**	**0.07**	**0.063**	**0.056**	**0.05**	**0.045**
L_W, m	**1.2**	**1.05**	**0.9**	**0.78**	**0.7**	**0.56**	**0.47**	**0.4**
R_{GAU}, m	**0.49**	**0.66**	**0.73**	**0.79**	**0.83**	**0.87**	**0.90**	**0.92**
D_{GAU}/D_{HS}, %	**60.20**	**65.60**	**73.40**	**79.32**	**82.35**	**87.15**	**88.93**	**90.10**
D_{HS}=2.6 m; $M_{ROT.REQ}$=0.73 MN·m								
D_P, m	**0.16**	**0.125**	**0.1**	**0.09**	**0.08**	**0.07**	**0.063**	**0.056**
L_W, m	**1.5**	**1.27**	**1.1**	**0.96**	**0.85**	**0.69**	**0.58**	**0.5**
R_{GAU}, m	**0.59**	**0.79**	**0.91**	**0.97**	**1.02**	**1.07**	**1.11**	**1.13**
D_{GAU}/D_{HS}, %	**60.11**	**65.48**	**73.18**	**79.22**	**82.05**	**86.97**	**88.76**	**90.50**
D_{HS}=3.2 m; $M_{ROT.REQ}$=1.43 MN·m								
D_P, m	**0.18**	**0.16**	**0.125**	**0.1**	**0.09**	**0.08**	**0.07**	**0.063**
L_W, m	**1.86**	**1.55**	**1.36**	**1.2**	**1.07**	**0.86**	**0.72**	**0.62**
R_{GAU}, m	**0.77**	**0.96**	**1.11**	**1.22**	**1.27**	**1.34**	**1.38**	**1.41**
D_{GAU}/D_{HS}, %	**60.24**	**65.57**	**73.55**	**79.47**	**81.98**	**87.23**	**88.75**	**90.35**
D_{HS}=4.1 m; $M_{ROT.REQ}$=4.43 MN·m								
D_P, m		**0.2**	**0.18**	**0.16**	**0.125**	**0.1**	**0.09**	**0.08**
L_W, m		**2**	**1.7**	**1.5**	**1.35**	**1.1**	**0.92**	**0.8**
R_{GAU}, m		**1.24**	**1.38**	**1.49**	**1.61**	**1.72**	**1.77**	**1.81**
D_{GAU}/D_{HS}, %		**65.42**	**73.42**	**79.24**	**82.22**	**87.02**	**88.53**	**90.27**
D_{HS}=5.6 m; $M_{ROT.REQ}$=10.7 MN·m								
D_P, m					**0.22**	**0.18**	**0.16**	**0.16**
L_W, m					**1.75**	**1.43**	**1.1**	**1**
R_{GAU}, m					**2.09**	**2.25**	**2.34**	**2.35**
D_{GAU}/D_{HS}, %					**82.12**	**87.05**	**88.62**	**90.52**

Conclusion

It has been revealed the maximal dimensions of the free interior in the geokhod is possible if case supports and the rod of a hydraulic cylinder are placed on one circle, i. d. $D_{PL.ROD} \approx D_{PL.HC}$, it can be recommended for designing the transmission.

Provided that the number of hydraulic cylinders is increased 4 to 14 the dimensions of the interior gets extended by 66 to 91% of the outer geokhod diameter.

The radius share of free interior ($D_{GAU} / D_{HS}, \%$) does not practically depend on changing the outer geokhod diameter, as dimensions of a transmission with hydraulic cylinders change proportionally to the change in outer geokhod diameter.

References

[1] Aksenov VV, Khoreshok AA, Beglyakov VY. Justification of Creation of an External Propulsor for Multipurpose Shield-Type Heading Machine – GEO-WALKER. Appl Mech Mater. 2013 Aug;379:20–3.

[2] Chernukhin RV, Dronov AA, Blashchuk MY. The application of the analytic hierarchy process when choosing layout schemes for a geokhod pumping station. IOP Conf Ser Mater Sci Eng. 2015 Sep 14;91:012086.

[3] Aksenov VV, Blaschuk MY, Dubrovskii MV. Estimation of Torque Variation of Geohod Transmission with Hydraulic Drive. Appl Mech Mater. 2013 Aug;379:11–5.

[4] Efremenkov AB, Aksenov VV, Blashchuk MY. Force parameters of geohod transmission with hydraulic drive in various movement phases. In: 7th International Forum on Strategic Technology (IFOST - 2012): Proceedings: in 2 vol [Internet]. IEEE; 2012 [cited 2015 Mar 6]. p. 159–63. Available from: http://ieeexplore.ieee.org/lpdocs/epic03/wrapper.htm?arnumber=6357716

[5] Timofeev VY, Aksenov VV, Galjamova JI. Determination of Parameters of Roller of Wave Transmission with Intermediate Rolling Bodies with Hollow Shaft for Geokhod. Appl Mech Mater. 2014 Oct;682:246–50.

[6] Efremenkov AB. Forming the subterranean space by means of a new tool (geohod). In: 6th International Forum on Strategic Technology (IFOST - 2011): Proceedings: in 2 vol [Internet]. IEEE; 2011 [cited 2015 Mar 6]. p. 348–50. Available from: http://ieeexplore.ieee.org/lpdocs/epic03/wrapper.htm?arnumber=6021037

[7] Buyalich GD, Buyalich KG. Modeling of hydraulic power cylinder seal assembly operation. Taishan Acad Forum - Proj Mine Disaster Prev Control. 2014;167–70.

[8] Buyalich GD, Buyalich KG, Voyevodin VV. Radial deformations of working cylinder of hydraulic Legs depending on their extension. IOP Conf Ser Mater Sci Eng. 2015 Sep 14;91:012087.

[9] Buyalich GD, Anuchin AV, Dronov AA. The Numerical Analysis of Accuracy of Hydraulic Leg Cylinder in Modeling Using Solid Works Simulation. Appl Mech Mater. 2015 Jun;770:456–60.

[10] Efremenkov AB, Timofeev VY. Determination of necessary forces for geohod movement. In: 7th International Forum on Strategic Technology (IFOST - 2012): Proceedings: in 2 vol [Internet]. IEEE; 2012 [cited 2015 Mar 13]. p. 1–4. Available from: http://ieeexplore.ieee.org/lpdocs/epic03/wrapper.htm?arnumber=6357729

IOP Conf. Series: Materials Science and Engineering **127** (2016) 012034 doi:10.1088/1757-899X/127/1/012034

Radial Strains of Double-layer Cylinders in Hydraulic Props of Powered Supports

G D Buyalich[1,2,3,a], **K G Buyalich**[1,b], **V V Voevodin**[1,c]

[1]T.F. Gorbachev Kuzbass State Technical University, Vesennyaya Str., 28, Kemerovo, 650000, Russia
[2]Yugra Technological Institute branch of Tomsk Polytechnic University, Leningradskaya Str., 26, Yurga, 652055, Russia
[3]Institute of Coal of SB RAS, Leningradskiy Ave., 10, Kemerovo, 650060, Russia

e-mail: [a]gdb@kuzstu.ru, [b]konstantin42@mail.ru, [c]vvoevodin@yandex.ru

Abstract. At present a lot of efforts are made to use double-layer power cylinders in hydraulic props of powered supports. To study the response of these cylinders to loads a special finite-element model has been developed and used for investigations into tension effect and double-layer cylinder thickness – radial strain relation under pressure of hydraulic liquid 50 MPa. It has been revealed that double-layer cylinders are distinguished by much lower radial strains in the zone of cup-like sealing elements as if compared with one-layer cylinders, as well as equivalent stresses are lower, and safety factor is higher. The data of the study can be recommended to calculate appropriate geometrical parameters of hydraulic props with respect to lower radial strains of a hydraulic cylinder, which improve its leak-tightness and functioning of cup-like sealing elements. The obtained results can be useful for design and construction of powered supports.

1. Introduction

A hydraulic prop is a hydraulic power cylinder; when it is exposed to loads its radial strains specify the cylinder – piston gap to be sealed affecting efficiency of a seal [1], as the consequence, leak-tightness. A numerical value of this gap is a sum total of a fit up gap, calculated on the base of piston and cylinder fabrication tolerances, and radial strains of cylinder inner surface (*dR*) under hydraulic liquid pressure, dependent on the value of this pressure (*P*), manufacturing technology [2–4], hydraulic prop structure [5, 6], its hydraulic extensibility (*lp*) [7], as well as on the structure of a support [8] and its external environment [9, 10].

2. Methods

A parameter-oriented model on the base of finite-element method was used to calculate radial strains and stresses in elements of the cylinder. This method is an up to date computational approach enabling precise and fast calculating complex designs by means of numerical engineering.

3. Work Description

The results of earlier investigations into radial strains of the main cylinder [7] are as follows: the curve shape is of the form depicted in Fig. 1, which becomes wider at the head end and narrows at the rod end. The following reference points were used to compare assessment criteria of strains lengthwise the cylinder.

P. 1 has maximal radial strains at the piston (dR_1).

P. 2 corresponds with the zone of stable radial strains in the cylinder body (dR_2).

P. 3 agrees with maximal radial strains in the piston zone (dR_3) when compressing the cylinder, which can't exceed the minimal piston – cylinder fit up gap.

P. 4 is in the zone of the first seal at the head end. Radial strains in this point (dR_4) determine the gap to be sealed, which affects sealing efficiency and leak-tightness of the hydraulic prop [1].

P. 5 has maximal radial strains at the bottom of the hydraulic cylinder (dR_5).

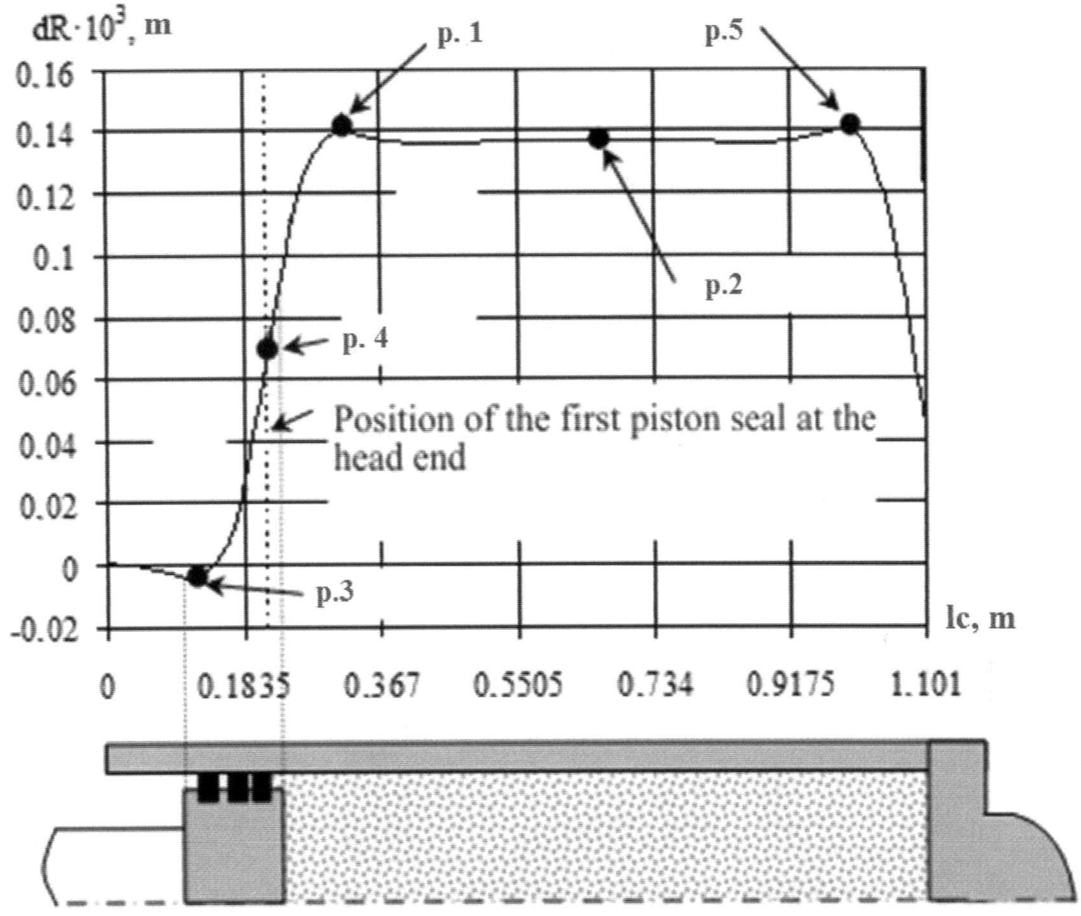

Figure 1. Radial strains lengthwise (lc) the main cylinder (dR) and position of reference points

It was proposed to fabricate the main cylinder of two layers coupled to each other by tension in order to reduce radial strains of the cylinder internal surface, which is necessary to improve operating conditions of cup-like sealing elements and leak-tightness of hydraulic props.

A special finite-element parameter-oriented model was developed to study strain in the main cylinder of the hydraulic prop in the powered support, layout and parameters of this model are outlined in Fig. 2.

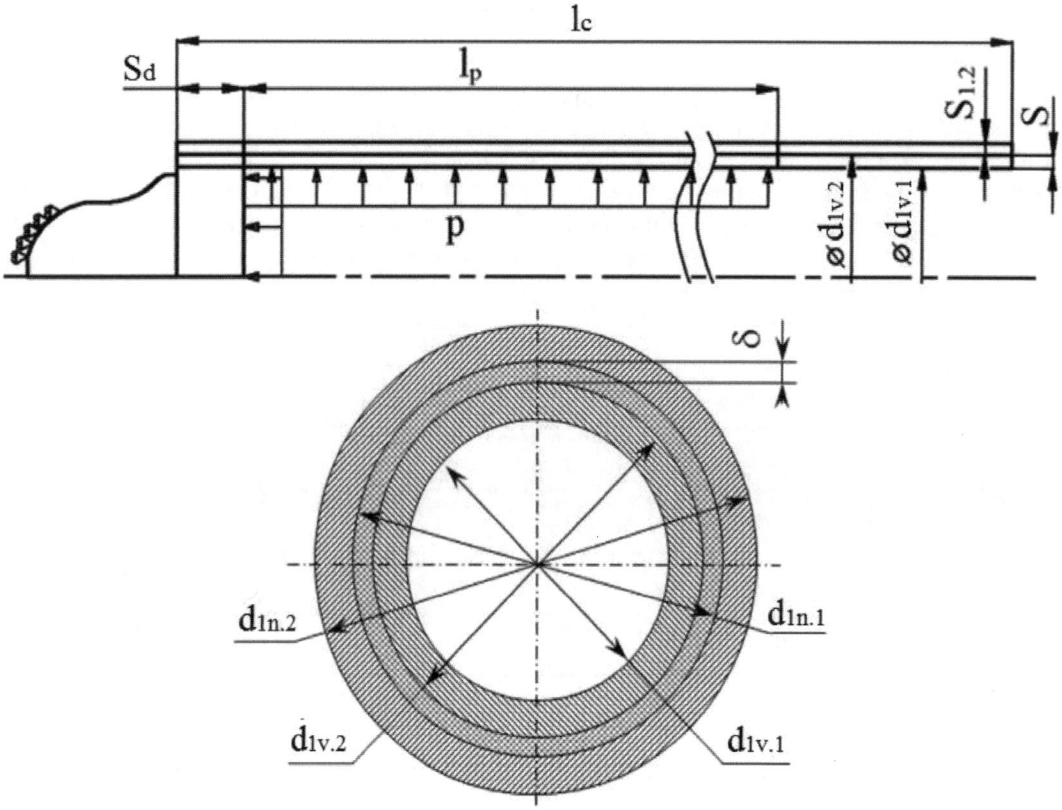

Figure 2. Layout and parameters of finite-element model for computing strains in double-layer cylinder of a hydraulic prop

Steel 9HF was admitted as material for outer cylinder, while steel 30HGSA was thought as material for inner cylinder. Properties of these materials are given in Table 1. Geometrical parameters of the model are presented in Table 2.

Table 1. Properties of materials of double-layer cylinder

Parameter	30HGSA	9HF
Yield point, MPa	490	670
Poisson ratio	0.29	0.345
Young modulus, Pa	$2.15 \cdot 10^{11}$	$2.15 \cdot 10^{11}$

Table 2. Main geometrical parameters of a hydraulic prop

Parameter	Value
Inner diameter of the cylinder ($d_{1v.1}$), m	0.25
Thickness of the cylinder wall (S), m	0.025
Bottom thickness (S_d), m	0.032
Cylinder length (l_c), m	1.115
Extensibility (l_p), m	0.885

Radial strains of double-layer cylinders were calculated according to the developed parameter-oriented finite-element model on the base of planar linear 4-unit forming elements with axis symmetry and contact elements between layers.

The following restrictions were accepted for the research purpose: total thickness of walls – 25 mm; wall thickness relation of inner cylinder to the outer one – 10/15, 12.5/12.5, 15/10, 20/5 mm; fit up tension – 0.104 mm.

As the result, radial strains of inner surface in a single-layer cylinder of a powered support OKP 70 were calculated provided that a hydraulic prop is fully extended (Fig. 3), the same strains were also computed for a double-layer cylinder (Fig. 4). Moreover, strains for various tensions between inner and outer layer of a cylinder are depicted in Fig. 4.

The obtained radial strains in reference point 3 and reference point 4 (dR_3 and dR_4) are given in Table 3.

The following criteria are proposed to measure strains in the main cylinder.

Safety factor with respect to maximal tolerance by bulging cylinder,

$$n_{\Delta_{max}} = \frac{0.15}{dR_4} \tag{1}$$

which is to be more than one to keep operating capacity of a hydraulic prop (a gap to be sealed doesn't exceed the safe one because of leak-tightness).

Safety factor with respect to minimal tolerance by narrowing cylinder at the rod end provided there is no scuffing

$$n_{\Delta_{min}} = \frac{\Delta_{min}}{(-2 \cdot dR_3)} \tag{2}$$

This criterion is to be more than one (in most cases) or negative. When $n_{\Delta_{min}} = 1$ compression strains of a cylinder in reference point 3 are equal to the minimal possible gap dependent on piston and cylinder manufacturing tolerances, therefore, scuffing is possible on the cylinder face if units to be assembled are combined this way. If the values are negative, cylinder – piston gap gets broader relative to the fit up tolerance, as the result, scuffing is less probable than in an unloaded hydraulic prop.

The system of proposed criteria enable assessing operational capacity of a hydraulic prop under critical pressure of hydraulic liquid.

Values of criteria $n_{\Delta max}$ and $n_{\Delta min}$ for powered support OKP 70 with single-layer and double-layer cylinder are given in Table 3.

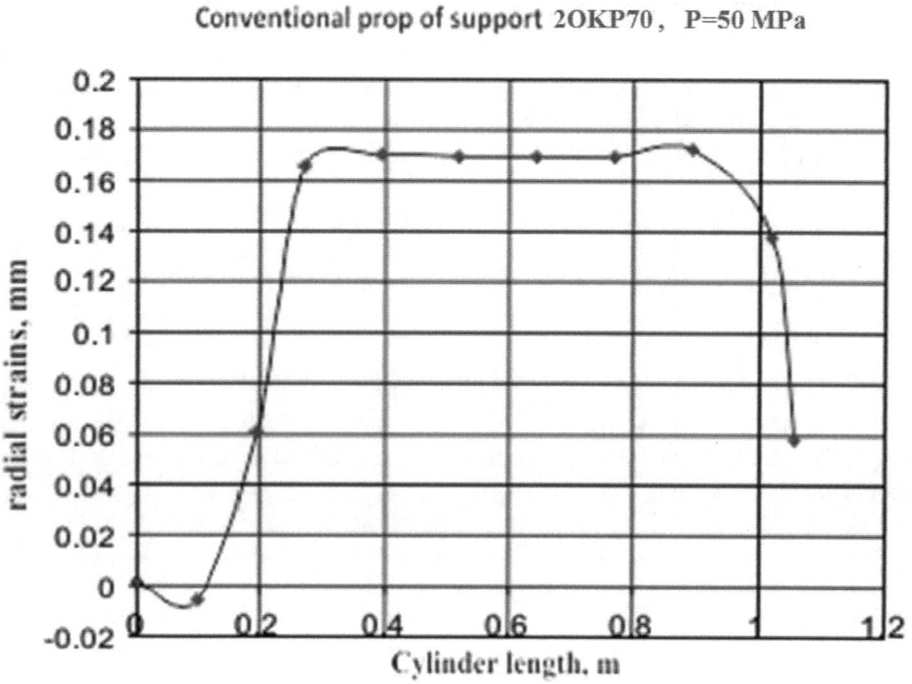

Figure 3. Radial strains of the inner surface in a single-layer cylinder, wall thickness is 25 mm and pressure of hydraulic liquid P=50 MPa

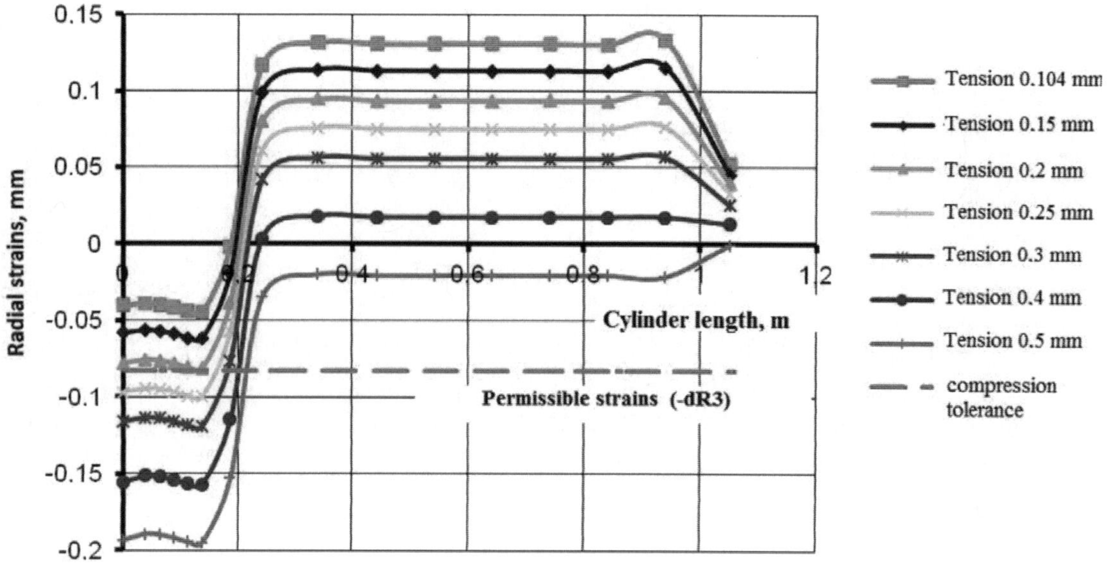

Figure 4. Radial strains of the inner surface in the cylinder, pressure of hydraulic liquid is P=50 MPa, inner cylinder thickness is 15 mm, outer cylinder thickness is 10 mm, tension is various

Double-layer cylinders are advantageous over single-layer ones in view of data presented in Figure 3 and Figure 4, as well as in Table 4, due to the increasing safety factor measured relative to maximal equivalent stresses; dimensions of the hydraulic prop and its steel intensity are equal. In this case either reliability of the hydraulic prop can be improved (e. g. under dynamic load in a longwall, as the consequence, under pressure far beyond the rated one in the head end), or wall thickness can be reduced keeping the same safety factor.

Table 3. Criteria of double-layer cylinders in the hydraulic prop of support OKP70

Thicknesses ratio, mm	10/15	15/10	12.5/12.5	20/5	Conventional
Tension (δ), mm	0.104	0.104	0.104	0.104	-
dR_4, mm	55	65	60	75	85
dR_3, mm	-65	-45	-51	-23	-5
$n_{\Delta max}$	2.73	2.31	2.50	2.00	1.78
$n_{\Delta min}$	0.38	0.56	0.49	1.09	5.00
Maximal equivalent Mises stresses, MPa	220	257	240	284	280
Safety factor	2.23	1.91	2.04	1.73	1.75

In general, double-layer cylinders used in the hydraulic prop of a powered support OKP70 enabled increasing safety factor measured according to the maximal gap as if compared with the single-layer structure, and reducing radial strains of the cylinder near the first seal (Table 3).

4.Conclusions

The following conclusions can be drawn on the base of comparative study on double-and single-layer cylinders in the powered support OKP70. All structures in double-layer cylinders, even those with minimal tension, have lower radial strains if compared with the single-layer cylinder, all other parameters are equal. It supports the seal as it operates in a lower gap to be sealed, therefore, leak-tightness of the prop gets better.

References

[1] Buyalich G D and Buyalich K G 2014 Modeling of Hydraulic Power Cylinder Seal Assembly Operation, *Mining 2014 : Taishan Academic Forum – Project on Mine Disaster Prevention and Control: Chinese Coal in the Century: Mining, Green and Safety, China, Qingdao, 17–20 October 2014*. Amsterdam, Paris, Beijing. Atlantis Press, pp. 167–170

[2] Chinakhov D A, Vorobyov A V, Davydov A A and Tomchik A A 2012 Simulation of active shielding gas impact on heat distribution in the weld zone of consumable electrode welding *In the Proceedings of the 7th International Forum on Strategic Technology IFOST2012,* Tomsk Polytechnic University, **Vol II**, pp 136–138

[3] Chinakhov D A 2011 Study of Thermal Cycle and Cooling Rate of Steel 30ХГСА Single-Pass Weld Joints *J.Applied Mechanics and Materials* **Vol 52–54** pp 442–447
[4] Chinakhov D A 2013 Simulation of Active Shielding Gas Impact on Heat Distribution in the Weld Zone *J. Applied Mechanics and Materials* **Vol 762** pp 717–721
[5] Burkov P V, Vorobiev A V and Anuchin A V 2011 Analysis of Stress Concentrators and Improvement in Designs of Hydraulic Legs *J. Mining Informational and analytical Bulletin (scientific and technical journal)* **Separate issue # 2** pp 172–183
[6] Klishin V I 1995 Inertial means of protecting hydraulic props from dynamic loads *Journal of Mining Science* **Vol 30 # 4** pp: 390–394
[7] Buyalich G D, Buyalich K G and Voyevodin V V 2015 Radial deformations of working cylinder of hydraulic Legs depending on their extension *IOP Conf. Series: Materials Science and Engineering* **Vol 91** 012087
[8] Buyalich G D, Aleksandrov B A, Antonov Yu A and Voyevodin V V 2000 Increasing the Resistance of Powered Support Brackets *J. Journal of Mining Science* **Vol 36** # 5 pp 487–492
[9] Klishin V I and Tarasik T M 2001 Stand tests of hygraulic supports with respect to dynamic loads *Journal of Mining Science* **Vol 37 # 1** pp: 77–84
[10] Klishin S V, Klishin V I and Opruk GYu 2013 Modeling coal discharge in mechanized steep and thick coal mining *Journal of Mining Science* **Vol 49 # 6** pp: 932–940

IOP Conf. Series: Materials Science and Engineering **127** (2016) 012035 doi:10.1088/1757-899X/127/1/012035

Effect of Incorporated Pumping Station Weight on the Rotating Torque of Geokhod Transmission

R V Chernukhin, V Y Timofeev, E G Grigoryeva

Yurga Institute of Technology, TPU, 26, Leningradskaya Str., 652050, Yurga, Russia

e-mail: Chernuhin@tpu.ru

Abstract. The article describes the results of a theoretical research of installation of a pumping station into the tail section of a geokhod. It is considered how an increasing weight of the tail section influences power characteristics of geokhod transmission. The research is based on a mathematical model of geokhod's moving through the underground space.

Introduction

High population density in metropolitan areas, intensive traffic on city roads as well as growth of house building and the necessity of underground utilities all encourage to develop cities' underground spaces. Every year the rate of underground workings increases more and more. In spite of the increase of drivage, rock tunneling is still a relatively labor-intensive and expensive process.

Currently, there are several drivage methods; one of them is shield tunneling. The need to increase the drilling rate and the performance of applied machines as well as the need to reduce the costs conditions such large number of research works on this topic.

The potentials of aspects of improvement of a traditional shield tunneling equipment are exhausted in the extension of its scope and the increase of its effectiveness [1].

An alternative to the shield method is the geowinchester technology [1]. The main technological element of this promising technology is geokhod. A characteristic feature of this mining machine is that it uses the geological environment for moving in the underground space.

The geokhod is notable for an original layout scheme as well as the presence of new functional-in-design assembly parts. Besides, the geokhod construction includes elements which have never been used in mining machines before [2]. The geokhod is a machine built by the modular principle which means that there are several variants for each system or mechanism. Certain circuit designs and structural schemes are chosen depending on the exploitation conditions. It is convenient to chose using multi-criteria methods such as analytic hierarchy process [3].

The construction of a two-section geokhod (figure 1) includes head section 1 and tail section 8 connected with each other by a junction module through which the torque is transmitted. The torque is transmitted via hydraulic cylinders 4. Executive elements 2 that destroy the breast, the sinking system as well as executive elements of the propelling device 3 also possess a hydraulic drive. In general, hydraulic drives are widely used in mining machines because they have certain advantages. In works [4, 5], there are the particularities, the advantages and the disadvantages described concerning the use of hydraulic drives in geokhod transmissions. The main disadvantage of hydraulic transmission is possible irregularity of the torque transfer from the tail section to the head section. Within the frames of experimental development, as an alternative transmission variant, there were geokhod transmission layout schemes suggested with harmonic drive with spacing rollers. The main advantages of such drives are the following: regular rotation of the output element, high torques at the output element, large overload reserve and high rigidity of kinematic elements, small size, high efficiency coefficient, low moment of inertia, high acceleration capacity level, high reliability rate and long life. You can find more information about the geokhod transmission with harmonic drive in work [6].

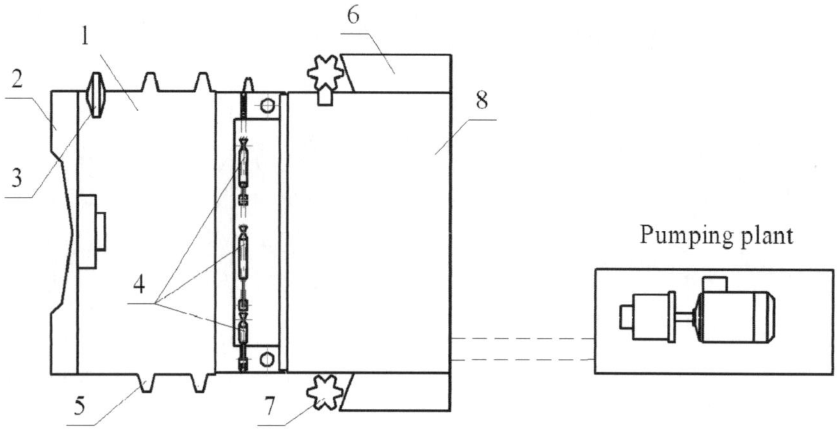

Figure 1. Layout scheme of geokhod.

1 – head section; 2 – actuating element; 3 – actuating element of the propelling device; 4 – hydraulic cylinder of revolution; 5 – helical blade (propelling device); 6 – counterrotation element; 7 – actuating element of the counterrotation element; 8 – tail section.

For the supply of all hydraulic systems of a geokhod, a pumping station is used. The pumping station is also built by the modular principle which means that every module or section supplies its own autonomous hydraulic circuit. This is necessary because geokhod systems work with different output and include hydraulic cylinders or hydraulic motors [7] that work in a simultaneous regime. Pumping station layout arrangements provide embedded schemes as well as mounting outside the tail section.

Up to the present moment, the produced prototype geokhod models had been equipped with a pumping station mounted outside the geokhod body. Of course, it clears some operating space inside the geokhod, however it results in the energy loss in hydraulic lines which can be very long. An alternative is a pumping station, the elements of which are located inside the tail section. Even though this idea has already been suggested, it has not yet been researched how power characteristics of the geokhod drive can change with the increase of the tail section weight by the value of pumping station. Moreover, the geokhods of different diameters have various weights of pumping stations, and the dependence of the pumping station weight on the geokhod diameter is nonlinear.

For now, a constraining factor for the installation of additional equipment, such as a hydraulic tank, on the inner shell of the geokhod consists in technological and economic problems [8, 9, 10].

1. Theoretical basis

The transmission is an important, inseparable part of the geokhod. As distinct from other mining machines, in geokhods, apart from the securing of the torque at the external propelling device and of the moving force, the transmission creates the crowding power at the executive element. Besides, in case the pumping station is mounted "on board", it is the transmission that is loaded the most.

Currently, a new generation of geokhods is in development. In order to define the moving force for geokhods, in work [11] there was a mathematical model suggested of the interaction of geokhods with the geological environment. This mathematical model has been elaborated taking into account current transmissions with hydraulic cylinders. In this mathematical model, there are following main target values defined: P_D – force developed by all hydraulic cylinders of geokhod rotation, and M_{VR} – required torque. The calculated dependences allowed to estimate the values of forces and torques required for the moving of the geokhod (figure 2).

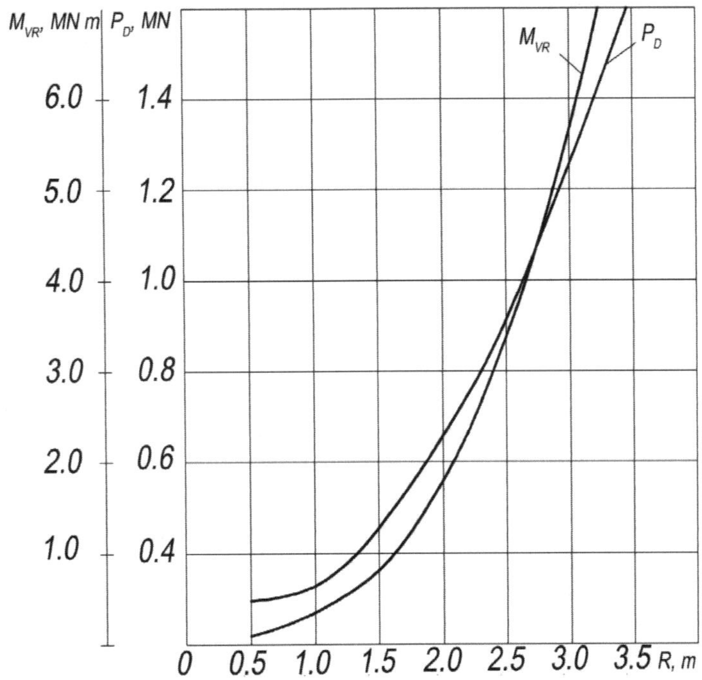

Figure 2. The dependence of moving forces of the geokhod P_D and rotating torque M_{VR} on the geokhod radius

This mathematical model was elaborated for a detached moving mode of the geokhod ELANG-3 (i. e. for the case when geokhod sections move in sequence) and does not count the necessity of continuous rotation of the head section as well as does not count a possible application of a mechanical transmission.

The mathematical interaction model of the hydraulic drive of a geokhod with the geological environment was developed in work [12]. In the improved mathematical model a coupled moving mode is taken into account (i. e. simultaneous moving of geokhod sections). The main target values of this mathematical model are the following: P_{GC} – integral force developed by hydraulic cylinders of rotation for the moving of sections, and $M_{VR.TR}$ – required rotating torque developed by the hydraulic cylinders:

$$P_{GC} = \frac{\Sigma M + \Sigma P \cdot k_1 + T_{G.OB} \cdot r_G \cdot \cos\beta + T_{OC} \cdot r_{OC}}{h_{GC} \cdot \left(1 - \frac{k_1 \cdot f_{TR}}{r_{EP}}\right)} \qquad (1)$$

ΣM – the sum of antitorque moments of geokhod, N·m; ΣP – the sum of motion resistance forces influencing the geokhod, N; k_l – frictional force and geokhod radius coefficient; $T_{G.OB}$ – frictional force at the head section of the geokhod, N; r_G – geokhod radius, m; β – canting angle of the helical blade, degree; T_{OC} – frictional force from the section junction device, N; r_{OC} – radius of the section junction device, m; h_{GC} – distance from the center of the head section to the working line of the hydraulic cylinder rod, m; f_{TR} – factor of the steel friction against the ground; r_{EP} – radius of counterrotation elements, m.

Hereby, as for any other type of mining technique transmission, it necessary to define the following values: required rotating torque M_{VR} gained by the transmission as well as moving force P_T that is resulted from this torque, and R_{NAV}– reaction force at the propelling device:

$$M_{VR} = R_{NAV} \cdot \left(sin\beta \cdot \left(r_G + \frac{h_L}{2} \right) + tg\varphi_{TR}^{P} \cdot cos\beta \cdot \left(r_G + \frac{h_L}{2} \right) \right) - M_Z \tag{2}$$

$$P_T = R_{NAV} \cdot cos\beta \tag{3}$$

$$R_{NAV} = \frac{-M_Z \cdot \frac{f_{TR}}{r_{EP}} - P_Z}{cos\beta - tg\varphi_{TR}^{P} \cdot sin\beta - sin\beta \cdot \left(r_G + \frac{h_G}{2} \right) - tg\varphi_{TR}^{P} \cdot cos\beta \cdot \left(r_G + \frac{h_L}{2} \right)} \tag{4}$$

hL is the height of the helical blade, m; $\varphi_{TR}{}^{P}$ is the friction slope between the ground and the steel, degree; MZ is the sum of antitorque moments of the geokhod, N·m; PZ is the sum of motion resistance forces of the geokhod, N.

This mathematical model is an improved mathematical model of interaction of a two-section geokhod with the surrounding geological environment in the conditions of dry friction and of a continuous navigation mode. This mathematical model counts the continuous translational propeller-driven character of the moving of device sections and is true for all types of transmissions. The mathematical model helps follow the influence of dimensions of geokhods on the value of the required rotating torque with different working angles of gradient α.

The disadvantage of this model consists in the fact that it does not take into account changing force characteristics in the process of exploitation of the geokhod. For instance, the metal body of the geokhod as well as its executive elements are corrodible and exposed to abrasive wear which inevitably affects the force characteristics and the efficiency of the whole machine in the same way as it happens to tunneling shields [13]. Body resistance to corrosion can be improved through the use of special covers described in work [14].

2. Results and discussion

In order to assess the impact of the weight on the rotating torque of geokhod transmission, the weight of the tail section must be increased by the pumping station weight. The weight of the in-built pumping station m_{PS} is calculated with the following equation:

$$m_{PS} = m_{ht} + m_l + \sum_{i=1}^{n} m_{ielm} + \sum_{i=1}^{n} m_{ip} + \sum_{i=1}^{n} m_{iapp}, \tag{5}$$

m_{ht} is the weight of the hydraulic tank; m_l is the weight of power fluid; m_{ielm} is the weight of electric motors; m_{ip} is the weight of pumps; m_{iapp} is the weight of equipment.

The weight of the hydraulic tank with power fluid, taking into account the tank charge level to 90 per cent, can be expressed in terms of the following equation:

$$m_{ht} + m_l = S_{surf} \cdot \delta \cdot \rho_{ht} + 0{,}9 \cdot V_{ht} \cdot \rho_l, \quad (6)$$

S_{surf} is the surface area; δ is the wall thickness of the hydraulic tank, for a steel tank δ = 2..3 mm; ρ_{ht} is the density of hydraulic tank material, for steel ρ_{ht} = 7,800 kg/m^3; ρ_l is the density of the power fluid, kg/m^3.

It should be noted that the weight of pumps and the equipment make only a small part of the cumulated weight of the pumping station, whereas the major part consists of the hydraulic tank with power fluid as well as of electric drive motors.

For geokhods with different diameters, there were dependency diagrams plotted of the pumping station weight and the diameter of the geokhod (figure 3) under the following conditions: the hydraulic tank of the type has a capacity that equals to one *Q*, two *2Q*, and three *3Q* cumulative momentary performances of all pumps. The metal wall thickness of the hydraulic tank equals to 2 mm, the density of the power fluid equals to ρ_l = 0.85 kg/m^3.

Analysis of dependence on figure 4 shows that the weight of the geokhod pumping station takes on large enough values, and in case it is mounted in the tail section, it can significantly affect the transmission parameters.

In equation (1), the weight of the tail section is included into the sum of motion resistance forces of the geokhod ΣP. If we add the weight of the pumping station into this equation, we can plot a dependency diagram of the rotating torque Mt on geokhod diameter. In practice, in equation (1), the dependence of the pumping station weight on geokhod diameter represented in figure 3 is inserted into the sum ΣP.

In figure 4, there is a dependency diagram of the rotating torque of the transmission taking into account the pumping station mounted "on board" (a) and without it (b).

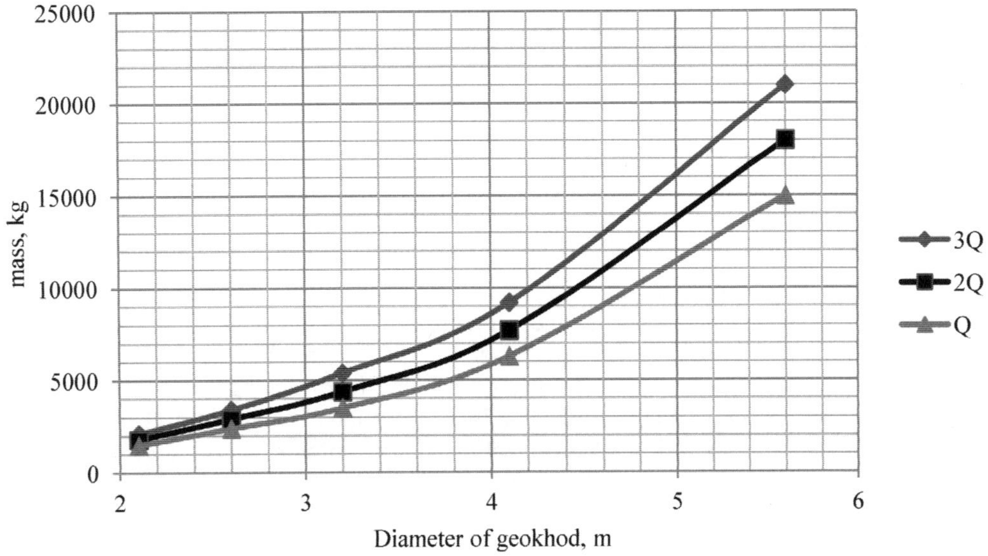

Figure 3. Influence of the geokhod diameter on the weight of the pumping station with hydraulic tanks of the capacity *Q*, *2Q*, and *3Q*

IOP Conf. Series: Materials Science and Engineering **127** (2016) 012035 doi:10.1088/1757-899X/127/1/012035

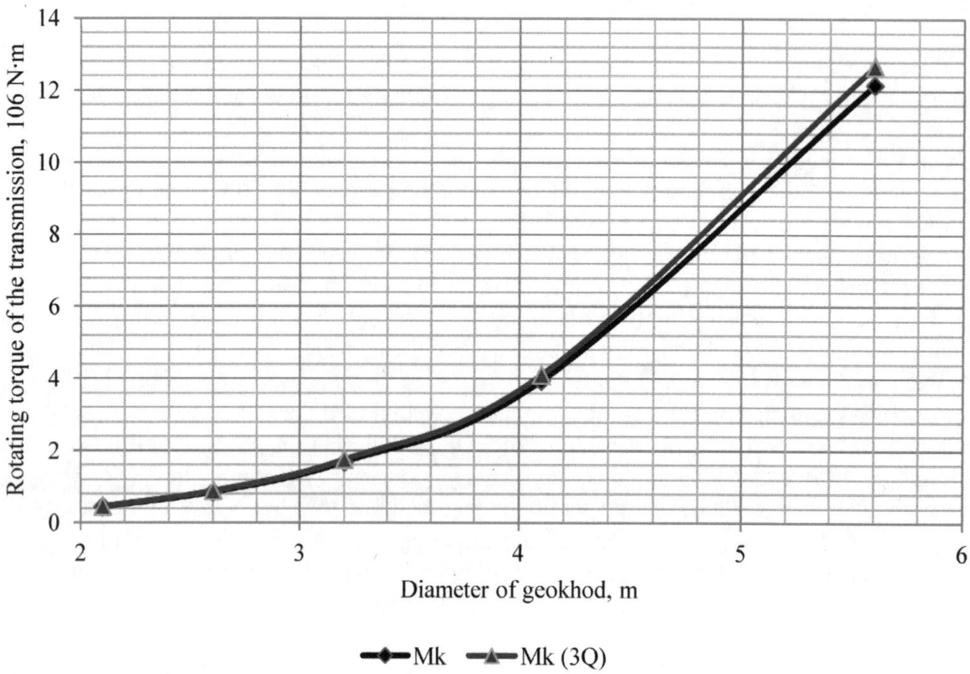

Figure 4. Dependency diagram of the rotating torque of the transmission on geokhod diameter without pumping station Mk as well as with in-built pumping station Mk (3Q) with a hydraulic tank of the capacity equal to three cumulated performances of all pumps

The main factors of weight reducing of the pumping station of the geokhod are the volume reduction of the power fluid and the weight reduction of applied electric motors. The first variant inevitably results in the necessity of use of heat exchange devices in order to maintain the necessary thermal rate.

3. Conclusions

The analysis of the dependence in figure 4 shows that the increase of the rotating torque in the geokhod transmission is insignificant when the pumping station is mounted inside the tail section; the torque growth equals to 4.19-4.72%. It is explained by the fact that the main effort is given to the overcoming of friction forces of the helical blade as well as friction forces by the rotation of the head section against the tail section, while an additional load, i. e. additional weight of the pumping station, does not really affect the process. Thus, the mounting of the pumping station in the tail section does not appear to be any kind of a constraining factor. The impact is insignificant and can absolutely be compensated by means of the assurance factor in the designing process of geokhod transmission.

References

[1] A B Efremenkov Forming the subterranean space by means of a new tool In Strategic Technology International Forum on **1** (2011) 348-350

[2] V V Aksenov, A A Khoreshok, V Y Begljakov Justification of creation of an external propulsor for multipurpose shield-type heading machine–GEO-WALKER. In Applied Mechanics and Materials **379** (2013) 20-23

[3] R V Chernukhin, A A Dronov, M Y Blashchuk The application of the analytic hierarchy process when choosing layout schemes for a geokhod pumping station Materials Science and Engineering (2015)

[4] A B Efremenkov, V V Aksenov, M Y Blashchuk Force parameters of geohod transmission with hydraulic drive in various movement phases International Forum on Strategic Technology Proceedings **2** (2012) 159-163
[5] V V Aksenov, M Y Blaschuk, M V Dubrovskii Estimation of torque variation of geohod transmission with hydraulic drive In Applied Mechanics and Materials **379** (2013) 11-15
[6] V Y Timofeev, V V Aksenov, J I Galjamova Determination of parameters of roller of wave transmission with intermediate rolling bodies with hollow shaft for geokhod In Applied Mechanics and Materials **682** (2014) 246-250
[7] M Y Blashchuk, A A Kazantsev, R V Chernukhin Capacity Calculation of Hydraulic Motors in Geokhod Systems for Justification of Energy-Power Block Parameters In Applied Mechanics and Materials **682** (2014) 418-425
[8] A V Walter, V V Aksenov Determining deviations in geometry of the geokhod shells In Applied Mechanics and Materials **770** (2015) 439-444
[9] V V Aksenov, A V Walter, A A Gordeyev, A V Kosovets Classification of geokhod units and systems based on product cost analysis and estimation for a prototype model production Materials Science and Engineering **91** (2015)
[10] V A Petrova, A A Bakanov, A V Walter Crack Resistance of Weld Seals, Hardening Methods and Composite Tools Quality In Applied Mechanics and Materials **682** (2014) 431-437
[11] M Y Blaschuk, A A Dronov, D A Miheev Geokhod Propel Effort Mathematical Model In Applied Mechanics and Materials **770** (2015) 391-396
[12] V Y Sadovets, V Y Beglyakov, V V Aksenov Development of math model of geokhod bladed working body interaction with geo-environment Materials Science and Engineering **91** (2015)
[13] N Espallargas, P D Jakobsen, L Langmaack, F J Macias Influence of Corrosion on the Abrasion of Cutter Steels Used in TBM Tunnelling Rock Mechanics and Rock Engineering, **48** (2015) 261-275
[14] Z G Kovalevskaya, V A Klimenov, K V Zaitsev Interfacial Adhesion between Thermal Spray Coating and Substrate Achieved by Ultrasonic Finishing In Applied Mechanics and Materials **682** (2014) 459-463

IOP Conf. Series: Materials Science and Engineering **127** (2016) 012036 doi:10.1088/1757-899X/127/1/012036

Improving Performance Of Industrial Enterprises With CGT[1]

I N Dolgih, K A Bannova, N A Kuzmina, A B Zdanova

Tomsk Polytechnic University, 634050, Lenina Avenue, 30, Tomsk, f.(3822)-56-37-89

E-mail: lignain@mail.ru

Abstract. At the present day, a falling in the overall level of efficiency production activities, especially in the machine-building companies makes it necessary to development various actions in the State support, including through the creation consolidated taxation system. Such support will help improve efficiency of activity not only the industrial companies, but also will allow improve economic and social situation in regions where often large engineering factories is city-forming.

1. Introduction

At the present stage of development in the Russian Federation it becomes more urgent ways to the state support of the different sectors and regions through improvement the tax effectiveness. The existing arithmetic formulas of taxable base, speed of tax revenues in the budgets, relationship between taxpayers and tax authorities directly influences to the development of the country. There is the process of redistribution of the income through effective tax administration direction of monetary resources to the social, defensive, nature-oriented sphere. Therefore, a high degree of efficiency the tax base creates the basic conditions for socio-economic development of both enterprises and regions.

2. The conditions of consolidation taxes receivable in Russia

For the purpose of increase in competitiveness of Russian companies on markets of the country and world, development of socio-economic sphere of the country by the Federal Law of the Russian Federation of November 16,2011 N321-FL was introduced the concept of a consolidated group taxpayers. This law is aimed at simplifying paying the tax on profit of the largest taxpayers.

Consolidated group of taxpayers (CGT) is an Association of taxpayers voluntarily on the basis of contract on creation consolidated group of taxpayers for calculation and payment of income tax organizations, taking into account the aggregate financial result of economic activity indicated taxpayers.

Features of formation consolidated tax base in accordance to the Tax code of Russian Federation consist in the following:

---

[1]The study was financially supported by Russian Humanitarian Science Foundation (RHSF) in the frame of the project for scientific studies (Modeling of conditions of the consolidation of tax liabilities to mitigate the conflict of interest of the state and taxpayers), project No. 15-32-01341.

IOP Conf. Series: Materials Science and Engineering **127** (2016) 012036 doi:10.1088/1757-899X/127/1/012036

•The tax base of CGT is defined as the total profits of participants educed by the loss of participants CGT;

•In the provision for doubtful debts receivable is not considered between the participants;

•do not take into account the losses of the CGT participants prior to the consolidation;

•the provisions for warranty repair and service not the sale between the parties.

The advantages of enterprises, combine into a CGT is the possibility of redistribution funds within the CGT. Also there is a possibility of decrease in taxable base due to consolidation, both profits, and losses of the enterprises–members of the consolidated group of taxpayers.

CGN is also exempted from tax administration of transfer transactions and as the result of CGN starts to operate more effectively by means of synergetic effect.

In recent years the engineering industry of Russia took formations of holding companies and financial-industrial groups. This is observed as further development of machine-building companies and holding companies created in the previous years (United Machine Building Plants, Power machine corporation, "New programs and concepts" and others), as well as the formation of new groups established on the basis of capital, earned in other industries. The most is noticeable expansion into mechanical engineering metallurgy companies which result became the formation of powerful financial-industrial groups in the automotive industry "RusPromAvto" and a group of automakers companies "Severstal". United Machinery Factory (UMF Group) is one of the largest Russian industrial holding companies in the field of heavy engineering. It was formed through the merger "Uralmash" with "Factory mechanical engineering of welded constructions" which two years later joined by "Izhora Factories ".

It was formed through the merger "Uralmash" with Factory mechanical engineering of welded constructions" which two years later had joined "Izhora Factories". Nowadays the enterprises are located in Russia and Czech Republic.75% of shares Other Machinery Building Plant are the property of "Gazprombank" and the rest is in free float on stock exchanges in Russia, Germany and the United States.By Russian companies Other Machinery Building Plant is JSCo "Izhora Factories", OOO "Ocher Machinery Building Plant-Special Steels", LLC "Ocher Machinery Building Plant-Foundry Production" and LLC "IZ-KARTEKS"

JSCo "Scientific and Production Corporation" "Uralvagonfactory" named after FE Dzerzhinsky "was created on the basis of the Ural Railway Car factory in Nizhny Tagil, Sverdlovsk region. Since 2009 one hundred percent of shares of the corporation owned by the government. The corporation enters 25 industrial enterprises of mechanical engineering and research organizations. In order to raise technological level of production Corporation acquired a controlling stake in the French company Sambre et Meuse, which supplies details vehicle devices.

Despite all the attractiveness of the consolidated taxation system, which is essentially a preferential treatment, allowing to increase the effectiveness of activity due to synergetic effected while none of the mechanical engineering holding companies are not united in the CGT.

To date, the main conditions that prevent general association of industrial enterprises in the tax consolidated group of taxpayers is a high threshold conditions for entry in to the following groups:

1. The participation in the authorized (share) capital of the holding company in the dependent should range from 90 per cent

2. The total amount of VAT, excise tax, corporate income tax and the tax on extraction of mineral resources-not less than 10 billion rubles

3. The total amount of proceeds from the sale of goods, products, work delivery and service provision, as well as others-not less than 100 billion rubles.

4. The total value of the assets-not less than 300 billion rubles.

Such conditions are primarily intended to consolidate holdings in the field of oil and gas production, financial division. Only a few steel holdings are comply with the regulatory requirements of the legislation in this area (for example the Civil Code "Oskol Metallurgical Combine")

In order to association in the CGT it was available and other industrial holdings is necessary to lower the barrier to entry for participation in the authorized capital of up to 70% and the amount of taxes paid to 156 bln.rub. For example, OMZ Group owns: public company Uralmashzavod (Sverdlovsk region) was 69.2% of the shares, Izhora plants (St.Petersburg) -73,3%,"Red Sormovo"(Nizhny Novgorod region)–60.7% and other large enterprises.

Reduced barrier to entry will allow this company to create the CGT, and thus improve its performance through optimization of financial flows and taxation.

3. The impact of CGT on regional tax revenues.

A special feature is the high labor intensity of mechanical engineering. Production of machines requires a lot of working time and labor resources. Production of machines is very expensive time and man power. Therefore many branch of mechanical engineering gravitate to areas with a high concentration population, and machine-building enterprises are often local economic mainstays. For example, in the Uralmash factory employs more than 15 thousand people, Izhora factory – more than 16 thousand.

All this leads to a close connection between the financial results of the company and the socio-economic situation of the region where the plant is located or where its management company, as the tax income of the region is directly connected with financial results of the enterprise and the mechanism of distribution of income tax between territories is located.

When you create a CGT share of the profits of each member of CGT is calculated from the rate of utilization labor resources (HR) and non-current assets (AC) as follows:

$$d = \left(\frac{HR_i}{\sum_{i=1}^{N} HR_i} + \frac{AC_i}{\sum_{i=1}^{N} AC_i}\right)\cdot\frac{1}{2}\cdot 100\% \quad (1)$$

where d – the profit share of each participant CGT, %

HR_i – average staffing number or payroll budget employees i-that company-the participant of CGN,

N – number of participants of CGN,

AC_i – the residual value of depreciable property the i-th company-participant CGT.

Thus, the income tax for each participant consolidated group does not depend on magnitude given financial result of the company and of the average number employees (or labor compensation fund) company-the participant CGT, as well as from the remaining value property belonging to him.

Therefore owing to the operating mechanism of distribution of profit at creation of CGN there is a redistribution of tax revenues between the regions, where the main production where the main production facilities CGN and a manpower are concentrated. So, owing to consolidation of taxpayers 2012 in 78% subjects of the Russia there was a growth of receipts on income tax in budgets of subjects of 53 billion rubles. But 18 subjects of the Russia diagnosed decrease in receipts on income tax of 61 billion rubles. In aggregate receipts in budgets of subjects of Russian 2012 decreased by 8 billion rubles [1].

This trend is general reduction tax deduction from the activities of the largest taxpayers in conjunction with the economic crisis in Russia was the reason for the suspension of development the system CGT. According to the Federal Law of 24.11.2014 N366,registered by tax authorities during the 2014 contracts on the establishment of a consolidated group of taxpayers, as well as changes in such contracts associated with the accession to a group of new organizations come into effect until with 1 January 2016 year.

Thus, in 2015 the list of consolidated groups and the composition their members remain unchanged. It will be impossible attract new companies into the acting CGT or to unite large holdings for the purpose optimizing business. In 2014, were registered five agreements on the establishment CGT responsible participants are:JSCo "Oskol Electrometallurgical Combine", JSC "Tander" Bank VTB 24 (JSC),JSC "Sibintertelecom" JSCo intercity and International Telecommunications "Rostelecom" [2,3,4].These organizations in 2015 will not be able to take advantage of CGT, and will calculate the income tax on general grounds. All this will negatively affect the investment attractiveness of Russia.

4. Analysis regional distribution of tax revenues while creating CGT

We will carry out the analysis of efficiency of the operating technique of consolidation tax liabilities on the example of JSCo tax obligations on the example of JSC Oskol Electric Steel Works (OESW).He is a responsible participant CGT, which consists of 5 large enterprises that are engaged in metal processing. But as a single taxpayer is the consolidated group will be operational only in 2016 as a consequence of the aforementioned reasons.

Table 1 Initial data for calculating the profit share of each participant CGT [3]

The names of basic companies CGT	Average number of employees	The residual value of depreciable assets, thousand rubles.	Profit share of each participant CGT,%
JSCo Oskol Electrometallurgical Combine	10890	16787407	28.8
JSCo Holding Company "Metalloinvest"	3	46	0.00
JSCo Lebedinsky Mining processing combine	11418	15177983	27.9
JSCo Mikhailovsky Mining processing combine	8602	6558927	16.4
JSCo Uralskaya Stal	11572	13930506	26.9
TOTAL	42485	52454869	100.0

On the basis of the financial statements for 2014 with the registration of the contract on creation of consolidated group of "OESW" in 2014, it would be possible to talk about the following planned data for 2015 on the calculation income tax in the regional budgets.

Table 2 Analysis of changes in regional tax revenues

The names of basic companies CGT	The planned tax into the regional budget to the creation of CGT (in thousand rubles.)	The planned tax into the regional budget after the establishment CGT (in thousand rubles.)	Absolute change (in thousand rubles.)	The region taxpaying
JSCo Oskol Electro-metallurgical Combine	1433588	6470685	+5037097	Belgorod region
JSCo Holding Company "Metalloinvest"	3861309	-	-3861309	Moscow
JSCo Lebedinsky Mining processing combine	12695358	6268476	-6426882	Belgorod region
JSCo Mikhailovsky Mining processing combine	3649247	3684696	+35449	Kurskaya region
JSCo Uralskaya Stal	828155	6043800	+5215645	Orenburgskay a region
TOTAL	22467656	22467656	0	

Thus, there is a clear trend of redistribution income tax in places where basic production assets and manpower. The tax calculated in the regional budget of Moscow from JSCo "holding company "Metalloinvest" will be reduced to zero. The decrease revenue derived from taxes of the Belgorod region is somewhat reduced thanks to their redistribution. The most from this CGT will benefit the Orenburg region, where despite its low profitableness JSCo Ural Steel", the planned of revenues increasing the budget could amount to5.2 billion rubles.

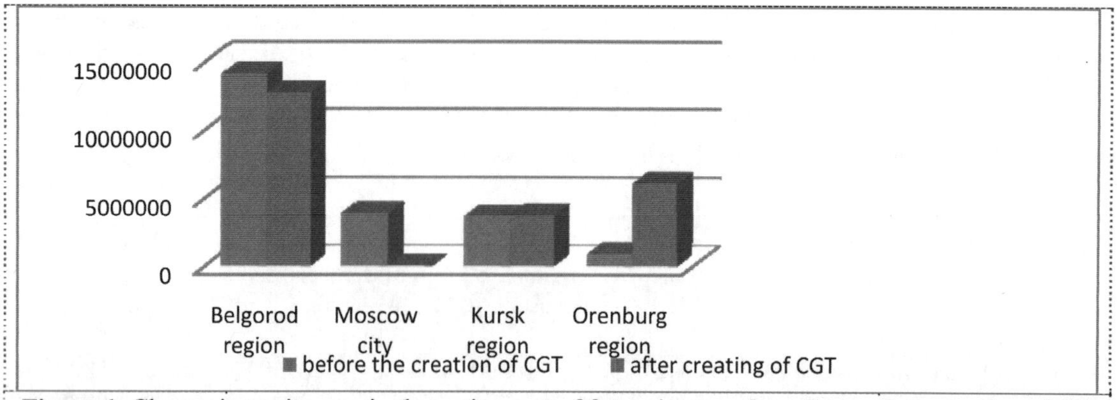

Figure 1. Change in tax income in the region part of future income from the participants given CGT

Decrease in the tax income of the Belgorod region-insignificant. But if JSCo Sokolsky electrometallurgical works and JSCo MGOK were in different regions, the negative effects of the creation of CGT for the territory on which the JSCo MGOK would be more significant: there would be decrease revenues more than doubled. At the same time JSCo MGOK is not a management company and carries out production activities. Thus, the existing approach to distribution of assessable income participants of CGN leads to a situation when after creation of CGN (or entries into CGN of unprofitable or low-profitable enterprise), the regional budget where such enterprise is located, starts gaining additional tax income on income tax from the responsible participant of CGN. At the same time tax revenues to other regional budgets, which operate successful companies are reduced.

5. A new approach to the mechanism distribution of profits participants CGT

To the establishment and operation of the CGT was mutually beneficial to both businesses and regions, you must modify the mechanism of formation and distribution of assessable income.

At the same time, such a mechanism must reflect not only used labor resources in the region and the location of assets there, but also other factors that have contributed to the interest of regional authorities in improving the functioning of the region located in the business, to create conditions for the development of large (anchor) companies.

One of these indicators can act as gross value added, which is the basis of calculation of GVP and the Gross Regional Product. These indicators reflect the socio-economic situation in the country and the region. Also, value added is one indicator of effectiveness of the company. In economic theory, the system of views on value added has a number of directions. Today we can distinguish the following the variety: net value added, economic value added (EVA) [4, 5], corporative value added (CVA),market value added [6,7] and the gross value added. But despite the such a large variety of species, the basis of its definition-is the difference between the total proceeds from sales and the cost of intermediate products [8].Currently being implemented by the Russian Government's strategy of innovative development of the Russian Federation until 2020 also requires increasing the share of products with higher value added. Is therefore consider it appropriate to put in dependence revenue derived from taxes of regions (in terms of income tax on CGT) of the value added value company-participant CGT, situated on the territory of the region. According to the Methodological Recommendations Economic Development Ministry of the Russian Federation added value"...is defined as the difference between the value produced in the period of accounts of goods and services (gross output) and the value of goods and services consumed in the production process (intermediate consumption)» [9].

Thus, in a general form the value added includes:

•Fund for compensation of employees (including contributions and extra-budgetary funds)

•Profits of the enterprise,

•Consumption of fixed capital [10].

The existing model of distribution taxable profit is used only the first and last elements of the indicator of gross added value: labor compensation fund employees and the residual value of depreciable property, which characterizes Consumption of fixed capital. Therefore we suggest to complement the model of one more component-"profit of the enterprise" [11].

When determining its meaning should be guided by the following requirements:

1. It should characterize the economic performance of the enterprise,
2. It must comply with the norms of the Tax Code, in particular Chapter 25, because it will be used in the distribution of income tax,
3. It must not allow doubling of values,
4. Be simple and understandable in a definition or calculation.

Usually in the calculation of added value include the company's net profit that can match the concept of the tax base for income taxes (excluding income taxes).

When the tax base includes the taxable profit from the main activities, as well as non-operating income and expenses. The latter includes many types of income and expenditure, including those that do not form gross value added of the enterprise in the economic sense of this indicator, namely:

- The income from business interest investments in other companies (Article 250 paragraph 1 of the Tax Code) – this profit should be accounted in the added cost of organization paying such income;
- The income in a type of a positive (negative) exchange difference (Art.250 paragraph 2) – this type of income generally characterizes inflationary changes;
- The income in a type of gratuitously received property (works, services) or property rights (Art.250 paragraph 8 of the Tax Code);
- The income in a type of the sums on which in the reporting (tax) period there was a reduction of the authorized (depository) capital (fund) of the organization if such reduction is carried out with simultaneous refusal of return of cost of the corresponding part of contributions (deposits) to shareholders (participants) of the organization (Art.250 paragraph 16 of the Tax Code)
- In the form of the sums accounts payable (liabilities owed to creditors), written off in connection with the expiration of the limitation period (Article 250 paragraph 18 of the Tax Code), etc.

As can be seen, such income and expenses will be distorting real contribution company the formation of the region's Gross Regional Product. Besides, the structure of Gross Regional Product (GRP) is included gross value added by type Russian National Classifier of Economic Activities, i.e. by type of primary activity company, which for tax accounting more in line with the indicators "income from sales" and "expenses from the production and sale".

Therefore, for economically reasonable distribution of income tax between participants of CGN is offered to define an indicator of "business profits" as income from a primary activity of the enterprise according to the 25th chapter of the Tax Code. It will be equal to the difference between sales proceeds (article 249 of the Tax Code) and expenses from production and realization (article 253 of the Tax Code).

Thus, the formula by calculating share in the profit that will have the region in the form of tax revenues on income tax, will have the following form:

$$d = \left(\frac{HR_i}{\sum_{i=1}^{N} HR_i} + \frac{AC_i}{\sum_{i=1}^{N} AC_i} + \frac{P_i}{\sum_{i=1}^{N} P_i}\right) \cdot \frac{1}{3} \cdot 100\% \qquad (2)$$

Where P – the value of the enterprise's profit from operating activities:

$$P = C - S$$

Where C – income from sales (excluding VAT and excise taxes), which are determined accordance with the rules article 249 of the Tax Code and the comply with line 010 of the sheet 02 of the Declaration of income tax;

S – the costs of production and sales, which are defined according to the rules of article 253 of the Tax Code and correspond to the line 030 of sheet 02 of the Declaration of income tax.

The transformations in the calculation formula the share of taxable profits enable to assess the contribution of each participant in the development of CGT all consolidated group.

6. Analysis of the effect new methodology on fairness distribution of profit tax between the subjects of the Russian Federation

Understanding the importance of these changes will give a push to the adoption solutions aimed at increment of added value of companies, not only on the part of management, but also to create conditions for the development of enterprises by the regional authorities, as their tax revenues from CGT become dependent upon effectiveness of the company on region. In turn, this will boost the socio-economic situation regions and the well-being of the population.

Consider application a new method of distribution income tax consolidated group as an example CGT with the responsible party of "OEMK."

Table 3: Analysis of changes in regional tax revenues considering a new method

The names of basic companies CGT	The planned tax into the regional budget before CGT in thousand rubles.	Proportion of tax income in the region after the establishment CGT (new method) thousand rubles.	The planned tax into the regional budget after the creation of CGT(new method) Thousand rubles.	The deviations, thousand rubles.
JSCo Oskol Electro-metallurgical Combine	1433588	24.2	5435550	4001962
JSCo Holding Company "Metalloinvest"	3861309	8.7	1949330	-1911978
JSCo Lebedinsky Mining processing combine	12695358	30.8	6930955	-5764404
JSCo Mikhailovsky Mining processing combine	3649247	18.1	4076491	427244
JSCo Uralskaya Stal	828155	18.1	4075330	3247176
TOTAL	22467656	100.0	22467656	0

The new method makes it possible to neutralize the loss of several regional budgets, which are enterprises with a high share of efficiency; is more economically sound redistribution of tax revenues between the regional budgets.

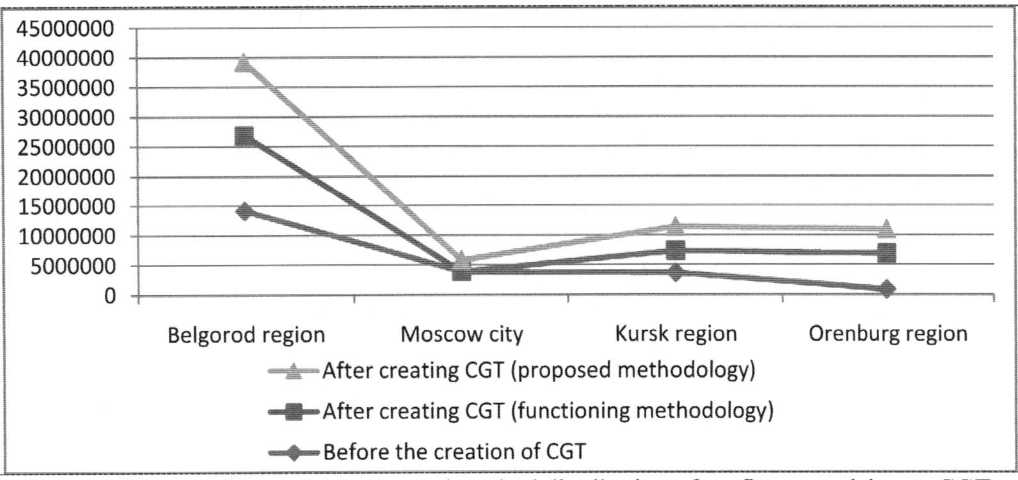

Figure2. Bench-making of the proposed method distribution of profit tax participants CGT.

Conclusion

Enterprise engineering is quite badly need government support, including through the establishment of an effective system of taxation.

The operating system of the consolidated taxation does not allow its application by machine-building holdings owing to the overestimated indicators of conditions entry in the CGT.

Their reduction would allow large industrial organizations to improve performance by reducing the tax base, exemption from tax administration of transfer deals. In addition, the entry of enterprises in the CGT will more effectively redistribute tax revenues between regional budgets. The understanding of a new type distribution of profit tax between the participants CGT must lead association of common efforts between management of the companies and regional authorities on increase of efficiency of functioning of the enterprises which will be expressed not only in creation of additional workplaces and to attraction of investments (the first two indicators of the previous technique),but also receiving return from the mastered capital investments. All this will also contribute to implementation of the overall strategy innovative development of the Russian Federation and to improve the welfare of society.

The study was financially supported by Russian Humanitarian Science Foundation (RHSF) in the frame of the project for scientific studies (Modeling of conditions of the consolidation of tax liabilities to mitigate the conflict of interest of the state and taxpayers), project No.15-32-01341.

References

[1] Carmen Comaniciu, Liliana Bunescu (2013) Taxation in Romania–Some Positive Aspects Procedia Economic sand Finance, Volume 6,2013, pp.615–620

[2] Aziz Jaafar, John Thornton (2015) Tax Haven sand Effective Tax Rates: An Analysis of Private versus Public Europe an Firms The International Journal of Accounting, Volume 50, Issue 4, December 2015 pp. 435–457

[3] Don Herrmann, Tatsuo Inoue, Wayne B Thomas (2003) PREDICTING CONSOLIDATE DEAR NINGSIN JAPAN: THEIN CREMENTAL USEFUL NESSOF SUBSIDIAR YEAR NINGS Advances in International Accounting, Volume 16, 2003, pp. 85–103

[4] Qingyue Meng, Hai Fang, Xiaoyun Liu, Beibei Yuan, Jin Xu (2015) Consolidating the social heal thin suranceschemes in China: to ward sane qui table and efficient health system THELANCET, pp.1484–1492

[5] Anton Orlov An assessment of proposed energy tax reform in Russia: A static general equilibrium analysis EnergyEconomics,Volume50,July2015,pp.251–263

[6] Evi Pappa, Rana Sajedi, Eugenia Vella Fiscal consolidate on with tax evasion and corruption Journal of International Economics, Pages S 56–S 75, 37th Annual NBER International Seminaron Macroeconomics
[7] Peter Berling, Fredrik Eng-Larsson Pricing and timing of console dated deliveries in Financial and environ mental analysis Europe an Journal of Operational Research, Available online 9 October 2015
[8] Marina Purina Tax Accounting in the Russian Federation Procedia Economics and Finance, 16th Annual Conference on Finance and Accounting, ACFA Prague 2015, 29th May 2015, Volume 25, 2015, pp. 127–133
[9] Muhammad Shahid Siddiqui Environ mental taxes and international spillovers: The case of as mall open economy Energy Economics, Volume 48, March 2015, pp. 70–80
[10] Mohammad Al Marzouq, Varun Grover, Jason Bennett Thatcher Taxing the development structure of open source com munities: An information processing view Decision Support Systems, Volume 80, December 2015, pp. 27–41
[11] Noel D. Johnson, Mark Koyama Tax farming and the origins of state capacity in England and France Exploration sin Economic History, Volume 51, January 2014, pp. 1–20
[12] Dolgikh I. N. , Zhdanova A.B. , Bannova K. A. (2015) The Influence of Taxation on Small Enterprise Development in Russia // Procedia - Social and Behavioral Sciences, Volume 166, 2015, pp. 216-221
[13] Bannova K. A. , Dolgikh I. N. , Pokrovskaya N. V. , Zhdanova A. B. (2015) Developing the Competitive Advantage of Companies and Regions by the Creation of Consolidated Groups of Taxpayers // The 26th International Business Information Management Association Conference: Proceedings IBIMA, Madrid, November 11-12, 2015, pp. 834-842.

Research on Operational Characteristics of Imported Rails under Conditions of East Siberian Railway

R A Gizatulin [1,a], N A Kozyrev[2,b], D V Valuev [1,c], A V Valueva [1,d], A Serikbol [1,e]

[1] 652050, Kemerovo region, Yurga, Leningradskaya str.26 Yurga Technological Institute branch of Tomsk Polytechnic University
[2] Siberian State Industrial University, Novokuznetsk, Russia

[b]e-mail: Kozyrev_na@mtsp.sibsiu.ru, [c] email: valuev@tpu.ru

Abstract. In world's manufacturing of railroad rails there are several thermal bonding methods: differentiated waterflow quenching, air-blast or spray quenching, rail head's hardening in polymer solution and bulk oil hardening; rails' service durability is strongly dependent on thermo-hardening methods. Until recently, all of rails, produced at Nizhniy Tagil and Kuznetsk Metallurgical Works, were subjected to bulk oil hardening. Only in 2010-2011, in the Russian Federation, differentiated air quenching of rails was started at the JSC "EVRAZ ZSMK"; and rail heads' hardening in polymer solution - at the JSC "Mechel". At the same time, exploitation of imported rails with differentiated hardening has been performed on Russian railways since 1995. Service durability of these rails in Siberia and the Far North is of considerable interest. In this paper we investigated the quality of P65-type rails, made in Japan, removed after use from curved track section of the East-Siberian railway.

Introduction

A general view of the rail with significant lateral wear up to 15 mm is shown in Figure 1.

Figure 1. Rail macrotemplate

The rails were tracked in the curve section of 297 m radius [1]. Total weight load for usable life was 136 million ton gross. The rail was removed in April, 2013, because of lateral wear.

Results of a control chemical analysis of metal test sample, compared with the specifications of the Russian Federation, are shown in Table 1.

Table 1. Chemical composition

Material	Content of chemical elements, %													ppm
	C	Mn	Si	P	S	Cr	Ni	Cu	Al	V	Ti	Mo	N	O_2
Sample	0.78	0.80	0.64	0.016	0.005	0.51	0.02	0,02	0.002	0.003	0.002	0.006	0.0022	4.6
TU 0921-239-01124323-2007 for rail steel 350LDT	0.72-0.82	0.70-1.20	0.35-1.00	not more		0.30-0.70	not more		not more than 0.005	not more than 0.01	not more than 0.025	not more than 0.02	not more than 0.015	not more than 20

According to the content of chemical elements, the metal sample meets Specification 0921-239-01124323-2007, set for the steel used for manufacturing of 350LDT-category rails.

Results of a fractional gas analysis showed that the highest oxygen mass fraction is in aluminates and aluminosilicates, calcium silicate and magnesium spinel (2.6 and 2.2 ppm respectively) and the lowest oxygen fraction of 1.3 ppm is in silicates [2].

The rail cross-section macrostructure was identified by deep etching in 50% hydrochloric acid solution and evaluated in accordance with the guideline document 14-2R-5-2004 "Qualifier of macrostructure defects of rails, rolled from continuously cast arc-furnace steel blanks". The metal microstructure is satisfactory in terms of centerline segregation (I), points' inhomogeneity (II) and segregation streamers (III). The makrotemplate revealed subtle serpentinous oblique cracks on the head fillet face that form a pattern up to 1 mm deep.

Measuring of tensile properties, impact toughness at +20°C, hardness in the head roll face and in the cross section were carried out on coupons, cut off from the sample, in accordance with State Standard P 51685-2000 and Specification 0921-239-01124323-2007. Two additional coupons were tested for toughness at -60°C. Results of mechanical and hardness tests are shown in Tables 2 and 3.

Table 2. Tensile properties and toughness

Coupon №	Tensile properties				Toughness KCU, at a t°C,	
	σ_{Re}	σ_{Rm}	δ_5	ψ	+20	-60
	H/mm^2		%		J/cm^2	
1- Coupon cut from the non-working head fillet	870	1270	12.0	44.0	20 21	4.8
2 - Coupon cut from the working head fillet (crumple zone)	840	1260	13.0	45.0	12 8.6	4.8
Specifications 0921-239-01124323-2007 for rails 350 LDT	-	Not less 1240	9.0	-	Not less then 15 (1,5)	-

Tensile properties, hardness of the head roll face and of the cross section and impact toughness at +20°C in coupons, cut from the non-working head fillet, meet Specifications 0921-239-01124323-2007 for 350 LDT -category rails.

Table 3. Hardness

Material	Hardness HB							
	Head					rail web	rail foot	
	FCL	10 мм	Fillet		22 мм			
Coupon	404	380	378	-	347	325	339	329
Specifications 0921-239-01124323-2007 for rails 350 LDT	362-400	Not less				Not more		
		341				341	363	

Enhanced hardness at the rail head roll face (HB 404) and diminished impact toughness, determined at + 20°C (8.6-12 J/cm^2) on samples cut from the working head fillet, are caused by strain aging of the rail surface layers during use. Toughness is diminished to 4.8 J/cm^2 at -60°C. This proves that rails, manufactured with differentiated hardening, are unreliable at low temperatures [3-5].

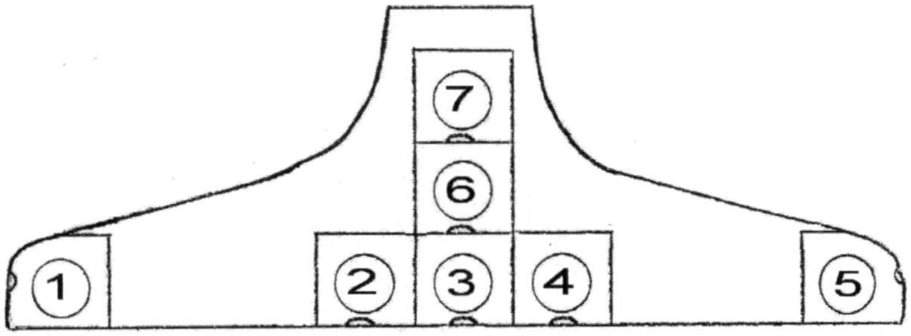

Figure 2. Scheme of coupone's cutoffs

Coupons were cut off according to the scheme shown in Figure 2, to determine tensile properties and impact toughness of the metal at the rail bottom. The coupons were marked with serial numbers (see the scheme). Test results are shown in Table 4.

Table 4. Mechanical properties of the rail cross-section

Coupon №	Tensile properties				Impact toughness KCU, J/cm^2 at t°C	
	σ_{Re}	σ_{Rm}	δ_5	ψ		
	H/mm^2		%		+20	-60
1 (rail foot blade)	630	1020	12.0	58.0	23. 23	9.7
2	780	1200	9.9	28.0	23 19	3.6
3	860	1260	9.5	36.0	15 18	3.6
4	780	1220	11.0	40.0	12 9.8	4.8
5 (rail foot blade)	730	1200	12.0	34.0	23 18	17
6	750	1160	9.5	23.0	9.6 20	7.2
7	670	1090	10.5	25.0	12 15	6.0

The highest values of strength characteristics (σ_{Re} =860, σ_{Rm} =1260 H/MM^2) were obtained on the coupon, cut off at the rail foot, along the vertical axis of the rail sample; they are comparable with the strength properties of the rail head (σ_{Re} = 840-870; σ_{Rm} =1260-1270 H/MM^2). Plastic properties are slightly diminished, in comparison with the head, and are (δ_5= 9.5; ψ=36 %) and (δ_5= 12-13; ψ= 44-45 %) respectively.

Strength properties of the metal are diminished (σ_{Re} = 780; σ_{Rm} =1200-1220 H/mm^2) at some distance from the vertical axis in the direction of the rail foot blades; wherein percent reduction of the area is 28% and 40%, respectively. Percent elongation is diminished to 9.9% and 11% respectively.

In one foot blade the rail foot blade has diminished strength properties (σ_{Re} = 630, σ_{Rm} =1020 H/MM^2), and a percent reduction of area is 58%; while in the opposite foot blade strength properties are somewhat enhanced (σ_{Re} =730; σ_{Rm} =1200 H/MM^2) and a percent reduction of the area is 34%. Percent elongation 12% is equal in both blades.

The coupons, cut along the axis at some distance from the surface, also have diminished strength (σ_{Re} = 670-750; σ_{Rm} =1090-1160 H/mm^2) and plastic properties (δ_5= 9.5-10.5; ψ=23-25 %).

Toughness of the coupons, cut from the foot, at +20°C, is generally in the range of 15-23 J/cm2; two more coupons have toughness 12 J/cm^2, and still other two – 9.6-9.8 J/cm^2, which is comparable to the rail head's toughness.

Toughness at -60°C is diminished from 3.6 to 9.7 J/cm^2, except for one sample (17 J/cm^2); relatively low values are obtained on coupons, cut from the foot blades (9.7 and 17 J/cm^2).

Hardness in the sample's head and foot cross sections was measured by a Rockwell hardness tester «AFFRI 251 VRSD».

Head hardness was measured from the roll face along the vertical axis to 40 mm depth and from the surface of fillets at 45° to the horizontal line, drawn through control points of hardness measuring in fillets' zone, at an interval of 5.0 mm, at 10 mm depth (see Fig.3)

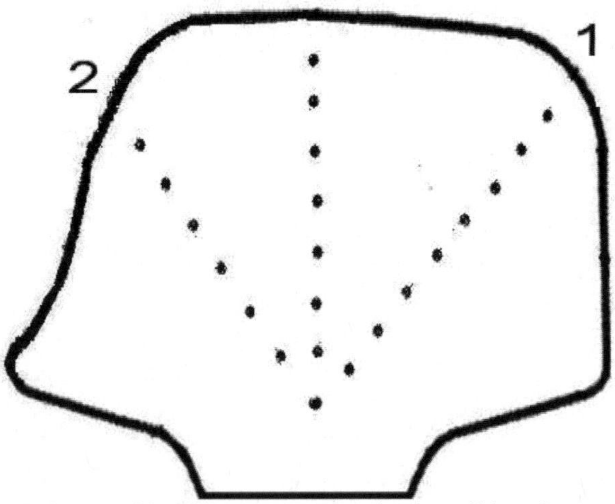

Figure 3. Scheme of hardness measuring in the rail head's cross section

Foot hardness was measured from the foot face along the vertical axis to 35 mm depth and along the horizontal line, drawn through control points of hardness measuring in the rail foot, apart from the vertical axis at an interval of 5.0 mm (see Fig.4)

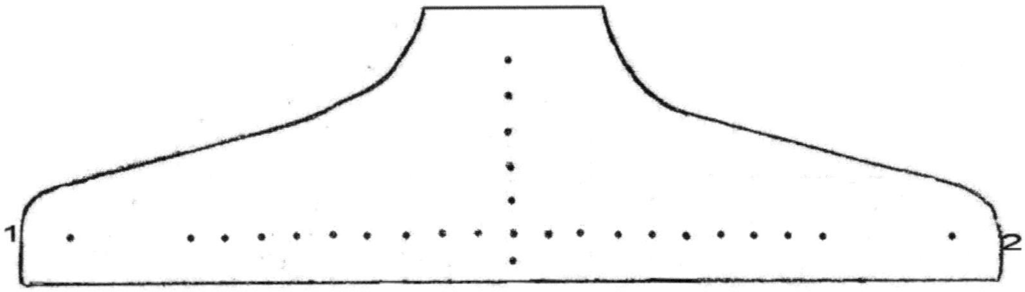

Figure 4. Scheme of hardness measuring in the rail foot's cross section

Table 5. Hardness in the head

Measuring direction	Hardness in the head HRC (HB) at a distance from the surface, mm								
	5	10	15	20	25	30	35	40	45
along the vertical axis	37.3 (361.7)	37.1 (359.9)	36.8 (357.2)	36.3 (352.7)	34.2 (333.8)	33.3 (325.7)	32.7 (321.4)	32.9 (322.1)	-
from fillet 1 (non-working)	38.2 (369.5)	38.1 (368.9)	37.1 (359.9)	36.1 (350.9)	35.2 (342.8)	35.2 (342.8)	34.1 (330.7)	33.3 (325.7)	32.9 (322.1)
from fillet 2 (working)	37.7 (355.3)	37.4 (362.6)	36.4 (353.6)	36.9 (358.1)	35.7 347.3)	35.5 (345.5)	-	-	-

Table 6. Hardness in the foot

M.d.	Hardness in foot HRC (HB) distance from foot surface													
	5	10	15	20	25	30	35	40	45	50	55	60	65	70
1	34,9 (334)	34,5 (337)	33,1 (324)	33,7 (329)	30,6 (303)	33,1 (324)	29,6 (295)	-	-	-	-	-	-	-
7 mm														
2	33,6 (328)	32,9 (322)	31,7 (312)	30,5 (302)	26,9 (275)	25 (263)	24,6 (259)	24,8 (261)	28 (283)	28,1 (284)	28,8 (289)	27 (276)	28,1 (284)	31,4 (309)
3	33,2 (325)	33,3 (327)	32,9 (322)	31,7 (312)	31,2 (308)	25,2 (263)	24,8 (261)	23,5 (253)	28,1 (284)	27,9 (282)	28,4 (286)	28,1 (284)	28,8 (288)	31,7 (312)

Note:
M.d.- measurement direction
1 - along vertical axis
2 - from vertical axis, horizontally towards side 1
3 - from vertical axis, horizontally towards side 2

Hardness in the head cross section diminishes on drawing from the surface. Horizontal hardness towards the blades at a distance of ~ 25 mm from the vertical axis is almost identical: 30.5 (302) - 33.6 (328.4) HRC (HB); at 40 mm distance it diminishes to 23.5 (253) -25 (262.7) HRC (HB); and then enhances to 31.7 (311.6) HRC (HB).

Additionally, hardness was measured in the foot surface spot sample of the template of ~ 330 mm length, after grinding on the depths up to 1 mm. Measurements were performed along rolling direction, by every 10 mm, in five zones: in the axial zone, and at distances of 20 and 40 mm to the left and to the right from the axis. The results are shown in Table 7.

On the sample surface at a distance of ~ 250 mm from the end face, axial zone hardness is almost identical: 35.2 (342.8) - 37.8 (366.2) HRC (HB) to the left and to right up to 40 mm from the axis; farther at a distance of ~ 80 mm hardness diminishes to 23.3 (251.8) – 29.2 (291.6)HRC (HB) in all marked points.

Assessment of nonmetallic contaminations was performed on a microsection, cut off and prepared from the head sample side. Objectionable lines of alumina and titanium nitride, as well as alumina and titanium nitride, cemented by silicates, have not been identified in the sample as well as lines of brittle-fracture complex oxides [6-8].

A single globular inclusion of 10.0 mm diameter is identified. Mainly there are filamentary sulfides scored in 1.5 points on the State Standard 1778-70 scale. Single titanium nitrides occur as light pink crystals of various forms in separate sulfide inclusions. A single plastic silicate of 135 microns length is identified.

Table 7. Hardness on foot base surface

№	Hardness on foot base surface HRC (HB)				
	40mm Left from axis	20mm Left from axis	in axial zone	20mm Right from axis	40mm Right from axis
1	36.6 (355.4)	37.8 (366.2)	36.8 (357.2)	37.3 (361.7)	35.8 (348.2)
2	36.2 (351.8)	37.6 (364.4)	36.7 (356.3)	36.9 (358.1)	35.7 (347.3)
3	36 (350)	37.5 (363.5)	36.2 (351.2)	37.2 (360.8)	35.4 (344.6)
4	36.8 (357.2)	37.3 (361.7)	36 (350)	36.5 (354.5)	35.7 (347.3)
5	36.1 (350.9)	37.3 (361.7)	35.2 (342.8)	37.6 (364.4)	35.9 (349.1)
6	36.4 (353.6)	37 (359)	35.6 (346.4)	36.2 (351.8)	36.1 (350.9)
7	36.2 (351.8)	37.6 (364.4)	35.9 (349.1)	36 (350)	35.7 (347.3)
8	35.8 (348.2)	37.6 (364.4)	36 (350)	36.3 (352.7)	35.6 (346.4)
9	35.7 (347.3)	36.5 (354.5)	35.8 (348.2)	36.3 (352.7)	36.1 (350.9)
10	35.8 (348.2)	36.8 (357.2)	35.8 (357.2)	36.8 (357.2)	35.7 (347.3)
11	36.1 (350.9)	36.9 (358.1)	36.5 (354.5)	36.2 (351.8)	35.7 (347.3)
12	36.2 (351.8)	37 (359)	36.8 (357.2)	37.1 (359.9)	36 (350)
13	36.1 (350.9)	36.8 (357.2)	36.9 (358.1)	37.4 (362.6)	35.7 (347.3)
14	36.9 (358.1)	37.1 (359.9)	36 (350)	37.1 (359.9)	36.1 (350.9)
15	36.3 (352.7)	36.8 (357.2)	37.1 (359.9)	36.9 (358.1)	35.9 (349.1)
16	36.9 (358.1)	36.2 (351.8)	36.4 (353.6)	36.8 (357.2)	35.9 (349.1)
17	36.9 (358.1)	37.3 (361.7)	36.6 (355.4)	37.1 (359.9)	35.6 (346.4)
18	36.5 (354.5)	37.3 (361.7)	37 (359)	36.8 (357.2)	36 (350)
19	36.8 (357.2)	36.8 (357.2)	37 (359)	36.5 (354.5)	36.3 (352.7)
20	36.9 (358.1)	37.1 (359.9)	36.8 (357.2)	36.9 (358.1)	35.3 (343.7)
21	36.3 (352.7)	37 (359)	36.5 (354.5)	37.1 (359.9)	35.3 (343.7)
22	36.3 (352.7)	37 (359)	36.2 (351.8)	37.2 (360.8)	35.2 (342.8)
23	35.2 (342.8)	36.8 (357.2)	36 (350)	36.8 (357.2)	36.3 (352.7)
24	33.2 (324.8)	34.7 (338.3)	36.2 (351.8)	36.2 (351.8)	35.6 (346.4)
25	29.1 (290.8)	33.9 (331.1)	34.7 (338.3)	33.2 (324.8)	33.4 (326.6)
26	26.8 (274.6)	29.1 (290.8)	26.8 (274.6)	23.3 (251.8)	26 (269)
27	27.1 (276.7)	26.4 (271.8)	26.7 (273.9)	25.1 (262.7)	27.3 (278.1)
27	27.8 (281.6)	24.9 (261.4)	23.6 (253.6)	26.1 (269.7)	28.1 (273.7)

№	Hardness on foot base surface HRC (HB)				
	40mm Left from axis	20mm Left from axis	in axial zone	20mm Right from axis	40mm Right from axis
29	28.1 (273.7)	24.4 (258.4)	23.3 (251.8)	26.3 (271.1)	27.3 (278.1)
30	29 (290)	25.6 (266.2)	23.9 (255.4)	26.4 (271.8)	29 (290)
31	29.2 (291.6)	26.5 (272.5)	24.8 (260.8)	27.3 (278.1)	28.9 (289.3)
32	29 (290)	26.6 (273.2)	25.4 (264.8)	26.5 (272.5)	27.3 (278.1)

Metal microstructure was studied it cross-microsections, cut from the head, mid-neck , mid-foot and blades after etching in 4% alcoholic solution of nitric acid. In non-etched microsections of head roll face there are structural integrity failures in the form of serpentinous cracks with depths up to 1.1 mm (Figure 5). Cracks cavities are partially filled with an indiscrete gray mass being a product of corrosion.

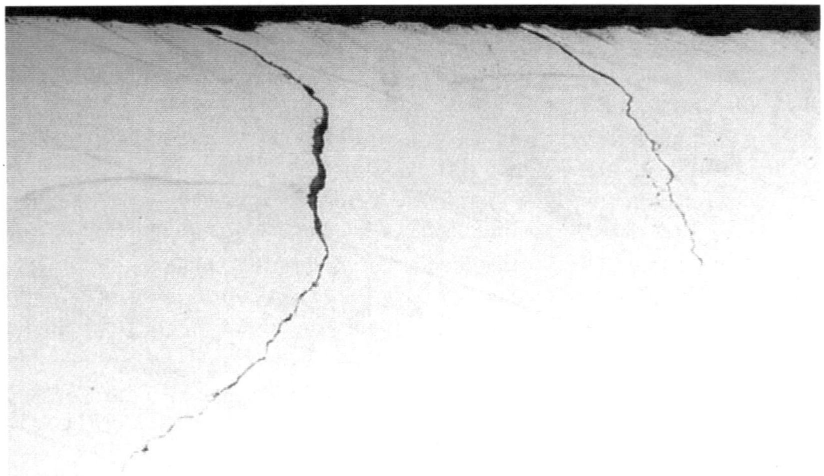

Figure 5. Cracks in the rail head, ×50

After chemical etching of microsections in 4% nitric acid alcoholic solution, structures with deformed grains are observed on cracks' edges; decarbonizing is not observed. On the working surface of the head the structure is deformed; sometimes there are strain aging layers up to 0.065 mm resulting from use.

The rail section metal microstructure is sorbitic and plate-like pearlite, and is typical for differentiated heat-strengthened state, which dispersion diminishes with distance from the head surface. The zone of diminished hardness, identified in the foot, is composed of granular and plate-like pearlite.

Conclusion

Metallographic study of P65-type rail, made in Japan, after use at the East Siberian Railway, has found:

1. According to the content of chemical elements, the metal sample meets Specification 0921-239-01124323-2007, set for the steel used for manufacturing of 350LDT-category rails.
2. The macrostructure of the metal is satisfactory.
3. Tensile properties, hardness of the cross section and impact toughness at +20°C in coupons, cut from the non-working head fillet, meet Specifications 0921-239-01124323-2007 for 350LDT-category rails.

IOP Conf. Series: Materials Science and Engineering **127** (2016) 012037 doi:10.1088/1757-899X/127/1/012037

Enhanced hardness at the rail head roll face (HB 404) and diminished impact toughness, determined at + 20°C (8.6-12 J/cm^2) on samples, cut from the working head fillet, are caused by strain aging of the rail surface layers during use.

Foot surface hardness, measured at spot samples along the length is generally 35.2 (342.8) – 37.8(366.2) HRC (HB); also there are zones with hardness diminished to 23.3 (251.8) – 29.2 (291.6) HRC (HB).

4. Contamination of non-metallic inclusions is insignificant. However, there are objectionable inclusions of exogenous origin (globules of 10.0 micron diameter).

5. The rail section metal microstructure is sorbitic and plate-like pearlite, which dispersion diminishes with distance from the head surface.

6. Subtle serpentinous oblique cracks up to 1 mm deep on the head fillet face as well as significant lateral wear up to 15 mm occurred during use.

References

[1] Kozyrev N A Pavlov V V Godik L A Dementiev V P Rails from elektrosteel Novokuznetsk Polygraph (2006)

[2] Pavlov V V Temlyantsev M V Korneva L V Oskolkova T N Gavrilov V V Defects and quality of rail steel Heat engineering (2006)

[3] Kozyrev N A Main practices of manufacturing of low-temperature resistant rails J Proceedings of Universities Ferrous metallurgy 4 (2011) 31-34

[4] Il'Yaschenko D P Chinakhov D A Danilov V I Schlyakhova G V Gotovshchik Yu M Physical nature of the processes in forming structures, phase and chemical compositions of medium-carbon steel welds J. IOP Conference Series: Materials Science and Engineering. 91 (2015)

[5] Mamadaliev R A Kuskov V N Zemenkov U D Popova A A Influence of high-concentrated heat sources on alloying elements transition into the weld metal J Applied Mechanics and Materials 770 (2015)

[6] Il'Yaschenko D P Chinakhov D A Danilov V I Schlyakhova G V Gotovschik Y M Increasing strength and operational reliability of fixed joints of tubes by MMA welding J. IOP Conference Series: Materials Science and Engineering. 91 (2015)

[7] Zemenkov Yu D Shalay V V Zemenkova M Yu Expert systems of multivariable predictive control of oil and gas facilities reliability J Procedia Engineering 113 (2015), pp. 312

[8] Iljyashchenko D P Chinakhov D A Gotovshchik Y M Influence of power supply energy characteristics upon the stability of MMA process J Applied Mechanics and Materials 756 (2015) 97-100

IOP Conf. Series: Materials Science and Engineering **127** (2016) 012038 doi:10.1088/1757-899X/127/1/012038

A Technology of Wastewater Sludge Treatment

R A Gizatulin[1,a], V V Senkus[2,b], A V Valueva[1,c], A S Baldanova[1,d], I F Borovikov[3]

[1] 652050, Kemerovo region, Yurga, Leningradskaya str.26 Yurga Technological Institute branch of Tomsk Polytechnic University
[2] 654041, Kemerovo region, Novokuznetsk, Kemerovo State University
[3] The Moscow State Technical University of N.E. Bauman, Moscow, 5, 2nd Baumanskaya

[b] email: SenkusVV@suek.ru, [c] email: valueva@tpu.ru, [d] email: ab214286@gmail.ru

Abstract. At many communities, industrial and agricultural enterprises, treatment and recycling of wastewater sludge is an urgent task as the sludge is poured and stored in sludge banks for many years and thus worsens the ecology and living conditions of the region. The article suggests a new technology of wastewater sludge treatment using water-soluble binder and heat treatment in microwave ovens.

Introduction
When urban wastewater is being purified a great amount of sludge is produced (from 0.5 to 1% of the whole amount of wastewater) and stored in sludge banks of wastewater treatment plants.

The wastewater sludge stored in wastewater treatment plants is a complex mixture which consists of biologically harmless substances composed with polluting toxic organic and nonorganic components. Before utilizing wastewater sludge for any purpose it is necessary to treat it.

The purpose of the research

The purpose of the research is determining process parameters to provide effective wastewater treatment and produce a product which is fit for its purpose.

Research methods and material

The research has been carried out in a laboratory of Novokuznetsk Institute (branch) of Kemerovo State University. A press-mould for deliquification has been constructed (Fig.1).

Samples of wastewater sludge with 80% of moisture content were taken at a Novokuznetsk wastewater treatment plant ZAO "Vodokanal" [1].

The following components have been mixed to prepare the bed:

wastewater sludge from Novokuznetsk wastewater treatment plant with 75-80% of moisture content;

sawdust (milled wooden exhaust or straw) which are required to support life and development of sludge strains when the product is moisturized and humus content is increased;

slaked lime for deacidization and elimination of smell;

cellulosic glue used as a bonding.

Table 1 shows percentage content of the bed components.

Table 1 Bed components

Sample №	Content, %			
	Wastewater sludge	Sawdust	Lime	Bonding
1	100	-	-	-
2	95	-	-	5
3	90	5	5	-
4	85	5	5	5
5	80	10	10	-
6	75	10	10	5
7	75	15	10	-
8	70	15	10	5
9	70	20	10	-
10	65	20	10	5
11	65	25	10	-
12	60	25	10	5
13	60	30	10	-
14	55	30	10	5

The stuff was mixed and pressed under 60-100 kg/cm2, the obtained product with moisture content of 10-15% was sterilized by heating it up to 100–200 ° C in microwave ovens during 3-5min which reduced moisture content and incapacitated microbial flora. The end product is packed into waterproof package which provides easy transportation and land application with further dissolution (Fig.1).

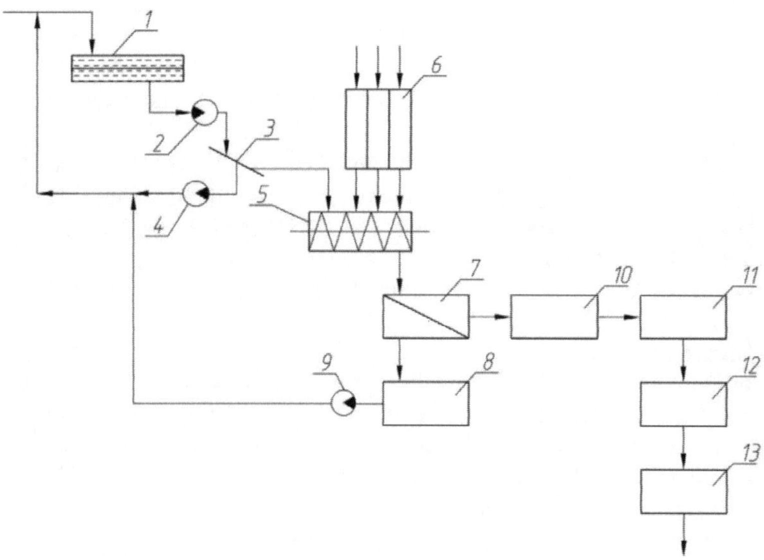

Figure 1. Process diagram of wastewater sludge treatment: 1 – aerotank, 2,4,9, - pumps, 3 – press-filter, 5 – mixing drainage, 7 – roll press, 8 – exhaust water collector, 10 - pressed pellets, 11 – packing machine, 12 – microwave oven, 13 – finished product storage.

Table 2 shows the main characteristics of the end product and the press.

Table 2 The main characteristics of the end product and the press

The main characteristics and dimensions	Values
The product	
Starting material moisture content	76–80 %
Starting material fraction size	up to 1 mm
Pressed pellet shape	round
End product dimensions	22 mm 60 mm
Admitted impurities to the starting material	45 %
Losses, %	до 50 %
Heating time, min	5 min
The press	
Press dimensions -diameter -length	 32 mm 150 mm
Water hole	3 mm
Material	steal

Results and discussion

The end product is shown in Fig.2.

Figure 2. End product

The end product has been tested for strength, moisture resistance, transportation reliability, heavy metals content and coliforms [2].

Samples containing bonding have greater strength than those without bonding. (Fig. 3). Samples 5 and 6 which include 10% sawdust have the greatest strength, sample 6 is stronger due to the bonding.

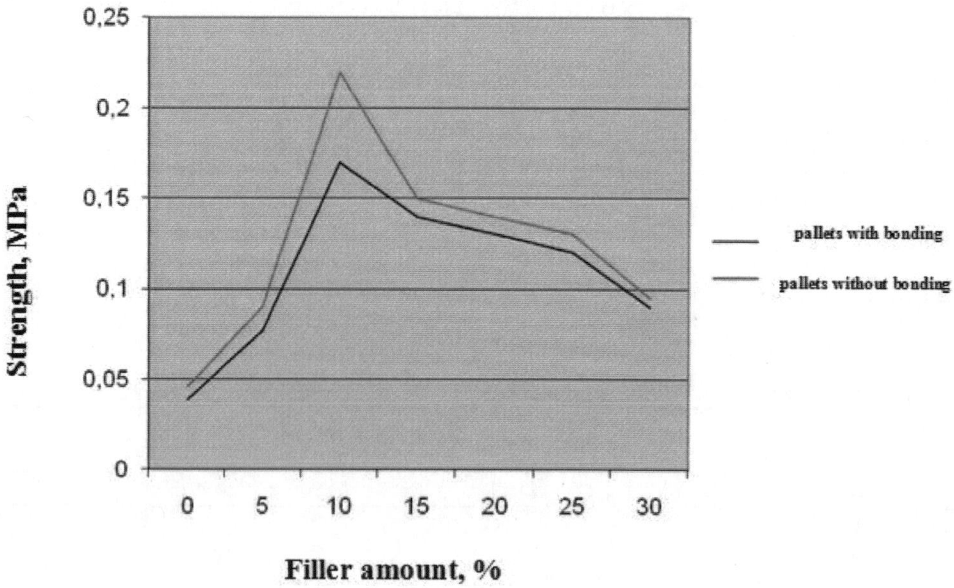

Figure 3. Strength of the end product samples

To test for strength during transportation the samples were loaded into a container and transported down urban and country roads during several hours. Samples № 1, 2, 3, 4, 13, 14 have been damaged and disintegrated. Samples № 5, 6, 7, 8, 9, 10, 11, 12 were not damaged.

To test moisture resistance the samples were dipped into water. After some minutes samples № 1,2 split, the rest of the samples did not change their shape after a month of dipping but they mellowed from a slight mechanical impact [2].

Chemical analysis for heavy metals has been carried out in a Western-Siberian Test Center, OOO "Akvatest" has performed a test for coliforms. The results are shown in Table 3.

Table 3 Chemical composition of wastewater sludge before and after treatment

Substances	Maximum allowable concentration in soil	Wastewater sludge before treatment	Wastewater sludge after treatment
Zink, mg/kg	4000	96.7	93.4
Cadmium, mg/kg	30	0.119	0.09
Lead, mg/kg	1000	11.38	10.2
Copper, mg/kg	1500	4.72	2.9
Mercury, mg/kg	15	0.117	0.08
Arsenic, mg/kg	20	0.41	0.27
Chromium, mg/kg	1200	1.82	0.7
Nickel, mg/kg	400	2.62	2.27
Coliforms in 1 gramm	0	2380000	0

The obtained research results show that adding 10-25% of saw dust or furnace dust into the bed reduces the content of heavy metals to maximum allowable concentration and heat treatment in microwave oven neutralizes bacterial pollution [3-9].

Conclusions

Analysis of the obtained results leads to the following conclusions [3]:

1. The strength of the samples which contain bonding is greater than that of the samples without bonding;

2. Samples with strength less than 0.1MPa are destroyed by transportation and by soaking. Thus the optimal strength of the samples is 0.1Mpa and higher which can be achieved by adding 10-25% of a filler;

3. Tight packing reduces olfaction;

4. Adding saw dust into the bed reduces the content of heavy metals and heat treatment in microwave oven neutralizes bacterial pollution;

5. When using the product as a fertilizer it is recommended to harrow 5-10 days after application or use pallets for winter crops.

References

[1] Potokina M V Wastewater treatment and recycling of sludge and settling J International scientific conference "Sustainable development Rational nature management Health technologies. (2012).
[2] Potokina M V Developing a technology of palletizing of wastewater watered sludge and settling J Modern problems of methodology and innovative activity Conference proceedings GU branch of KuzSU Novokuznetsk (2011).
[3] Potokina M V Senkus V V A method of calculating technology parameters for producing fertilizers from wastewater sludge J AgroEcoInfo 1 (2014).
[4] Galevsky G V, Rudneva V V, Garbuzova A K, Valuev D V, Titanium carbide: nanotechnology, properties, application, J. IOP Conference Series: Materials Science and Engineering 91 (2015)
[5] Il'Yaschenko D P Chinakhov D A Danilov V I Schlyakhova G V Gotovshchik Yu M Physical nature of the processes in forming structures, phase and chemical compositions of medium-carbon steel welds J IOP Conference Series: Materials Science and Engineering 91 (2015)
[6] Valuev D V, Danilov V I,Ivanov Yu F On the causes of formation of non-optimal structures in a pressure-treated low-carbon steel J Russian Physics Journal (2007) 8-12
[7] Kryukov R E, Kozyrev N A, Galevsky G V, Bendre Y V, Goryushkin V F, Valuev D V Some aspects of oxidation-reduction reactions under carbon-bearing flux welding J IOP Conference Series: Materials Science and Engineering. 91 (2015)
[8] Saprykina N A, Saprykin A A, Borovikov I F, Sharkeev Y P Influence of layer-by-layer laser sintering conditions on the quality of sintered surface layer of products IOP Conference Series: Materials Science and Engineering 91 2015
[9] Valuev D V, Danilov V I Reasons for negative formation of structures in carbon steel processing of pressure J Proceedings - 2012 7th International forum on strategic technology, IFOST 2012 2 (2012) 151-154

Influence of the Rigid Connection Between Discs in the Tetrahedral Prisms on Equivalent Stresses When Cutting Work Faces

A A Khoreshok[1,a], L E Mametyev[2,b], A Yu Borisov[3,c], A V Vorobyev[4,d]

[1,2]Prof. dr.-ing., [3]head teacher, [4]candidate of science (engineering), associate professor
[1,2,3]Department of Mining Machines and Complexes/Mining Institute,
T.F. Gorbachev KuzSTU, Russia, 650000, Kemerovo, Vesennyaya street 28,
[1,4]Department of Mining Equipment, Yurga Institute of Technology, TPU affiliate
Russia, 652055, Yurga, Leningradskaya street 26

E-mail: [a] haa.omit@kuzstu.ru, [b,c] bau.asp@rambler.ru, [d] vorob@tpu.ru

Abstract. This paper presents the results of modeling of the stressed state of structural elements of the paired fastening points of the two disc tools to the tetrahedral prisms of the working bodies of the roadheaders of selective action when cutting work faces of heterogeneous structure. The advantages of cooperative mode rotation to separate two disc tools on each of the tetrahedral prisms placed between the axial cutting crowns.

Introduction

Currently, Gorbachev KuzSTU, the Department of Mining Machinery of the Institute of Mining, together with Yurga Institute of Technology, TPU Affiliate, the Department of Mining Equipment, are engaged in researches with the purpose of developing and improving designs of mounting points to hold cutting tools intended for cutting heads in tunneling machinery. These studies are implemented in phases and permits modeling of the stress-strain state of mating elements in points of mounting disk cutters, which are different in designs, supported by brackets and polygonal prisms located on cutting heads of selective roadheaders and geokhods while mining non-uniform in structure working faces [1–15].

2. The device and principle of operation of the working body of the roadheader of selective action with built-in disk tool

The main disadvantage of a number of selective roadheaders with transverse cutter heads is that the performance of the cutting process is low due to the presence of non-crushed rock masses in space between cutting drums, and this requires additional reciprocating rotary movements of the boom in the horizontal plane. To enhance the efficiency of cutting processes performed by a selective roadheader, having cutting heads with cross-axial cutting drums that are separated from each other by a gearbox casing, the technical solution is suggested (Russian patent 136086) that disk cutters mounted in tetrahedral prisms are positioned in the inter-drum space (Figure 1) [4, 5].

The cutter head of a selective roadheader (Figure 1, a, b) comprises a boom 1, a casing of a transfer gear box 2 with two axial cutting drums 3 having bits 4. Tetrahedral prisms 5 with two disc cutters 6 are attached to the cylindrical surface of the gearbox casing 2. By means of six hydraulic cylinders, the cutter boom 1 can move in three directions relative to the stationary roadheader body. Two hydraulic cylinders provide the telescopic axial action in direction n1з, two hydraulic cylinders are used for raising and lowering the boom 1 in the vertical direction along the arrows n2з and n3з, and two hydraulic cylinders ensure the swinging motion of the boom in the horizontal direction to walls of excavation.

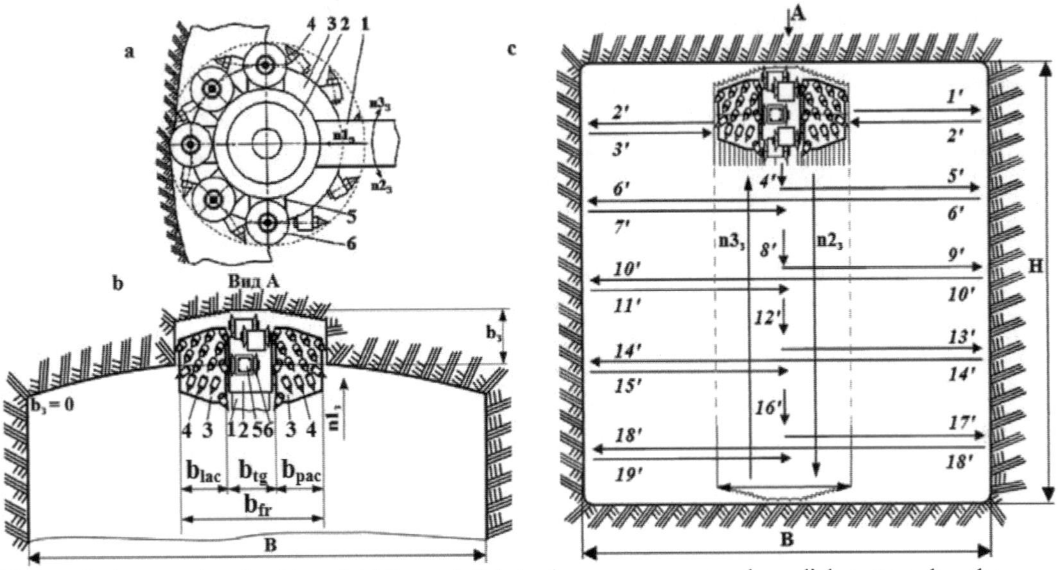

Figure 1. A cutter head with point bits and disk tools: a – a space to place disk cutters; b – the sump in the centre of a working; c – the sequential tunneling cutting process

The inner space of the tetrahedral prism 5 with mounting points to hold the disk tool 6 is closed by a sealing four-sided cover. In this case, the diameter of circle (Figure 1, a) circumscribed about outer edges of the disc tools 6 in a plane perpendicular to the axis of rotation of the cutting drum 3 is not greater than the diameter of circle circumscribed about maximum radius of the outer bits 4 on large basements of the axial cutting drums 3.

Cutting heads of a selective roadheader perform cyclical excavation of a rectangular section with a height H and a width B (Figure 1, b, c). Before starting the excavating operations, the cycle begins with a cut in the central part of a working. When the first cut is made using two axial cutting drums to a desired amount of cut in the direction of the longitudinal axis of development, or the width of web $b_з$ (Figure 1, b), reciprocating cyclic movements of the boom in the vertical plane from the roof to the ground n2з and back, n3з, is combined with a gradual telescoping n1з to the specified widths of web $b_з$. This process is enabled by simultaneous operation of the hydraulic cylinders to raise the boom, the hydraulic cylinders to extend the boom and a continuous rotation of two axial cutting drums. The sump making is carried out over the width of cut front b_{fr}, including the width of cut made by the left b_{lac} and right b_{pac} axial cutting drums, as well as the width b_{tg} of the transfer gearbox casing in the form of a cylinder sector with a coverage angle of $\varphi \leq \pi/2$, to the surface of which the tetrahedral prisms with two disk tools are attached according to a certain arrangement scheme, operating as a pivoting blade installation.

After the sump formed, it is possible to use a trajectory of the boom with the axial cutting drums following along the directions of motion *1'–19'* (Figure 1, c). The processes of crushing are dominated in the direction of motion *1'–11''*, and in the direction of motion *1'–3'*, *17'–19'* there are the processes of crushing, cutting of ragged cuts and asperity on the roof and ground surfaces. After the final cleaning – up the ground from cut masses across the width B of a working, the boom-type roadheader advances towards the face, and using the hydraulic cylinders for telescoping, the boom retracts by an amount of b_3 and the next cut cycle of excavation commences.

The results of research are presented in the works referenced in [4, 5] on modeling the stress state of mating elements in mount points to hold two disk cutters in tetrahedral prisms in decoupled modes of rotation when forming a sump in centre of the working face.

To study the state of stress in the mounting points used for holding the disk tool in the tetrahedral prisms during the operating mode of simultaneous rotation, with a selective roadheader forming a sump in the center of the working, the inventive technical solution (Russian patent 146845) has been proposed that includes a tetrahedral prism with twinned disk tools (Figure 2) [5]. Implementing this technical solution allows the simultaneous rotation of two disk cutters held in each tetrahedral prism, with resulting in reducing their jamming and wearing -out during the sump process made by the cutting head of a boom-like roadheader in rock mass.

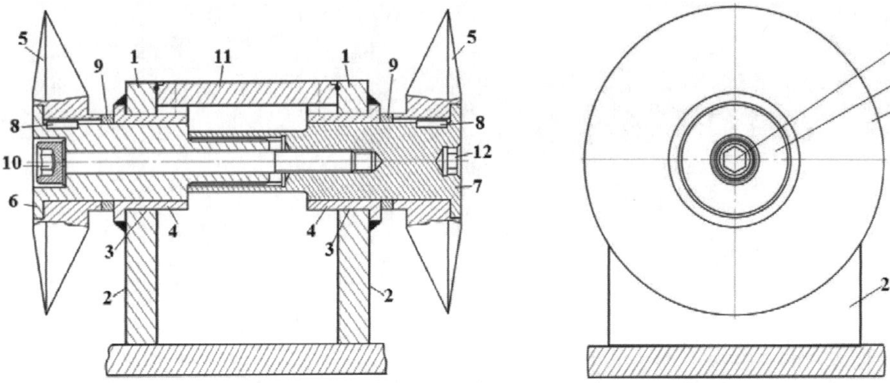

Figure 2. Rigid connection of two disc cutters in a tetrahedral prism

The tetrahedral prism 1 (Figure 2) consists of two working faces 2 having trough cylindrical holes 3 and two side faces. A tubular member 4 is placed inside each cylindrical hole 3. In addition, twin disc cutters 5 are in the form of a single detachable functional modular unit that is able to rotate freely and simultaneously relative to the fixed tubular members 4. The single detachable functional modular unit comprises two axes 6 and 7 with thrust collars, rigidly attached to each other, one axle includes a spline drive and the other contains a spline bush. The axles 6 and 7 have places to provide keyed connection of the disk cutters 5 using the fixing key 8. Distant side washers 9 are placed between the side surfaces of either of two disk cutters 5 and the side surface of the thrust collars of either of two tubular members 4. The axle 6 has a stepped through hole, inside of which a distance screw 10 is located for fixing the detachable functional modular unit. A tetrahedral cap 11 is used to protect the interior space of the tetrahedral prism 1. On the free end of the axle 7 there is a hexagon opening 12 for a wrench used for assembling and dismantling operations.

3. Modeling of the stress state of structural elements of the attachment disk tool for joint rotation mode

Making finite element models (Figure 3) and calculating stress forces P_z, P_y, P_x (Table 1) for the twinned disk tool with mounting points on the tetrahedral prisms are performed using the procedures set out in the research works relating to cutting heads having reversible radial drums and tools with one disc cutter, supported in brackets and triangular prisms, as well as axial drums and tools with two separate disc cutters supported in tetrahedral prisms [1–5].

For calculating parameters of stress state, the following characteristics of mined working faces were taken as initial conditions: coal (σ_{compr} = 12.4; 13.5; 14.8 MPa) and rock (σ_{compr} = 51; 60.6; 78.9 MPa). To simulate the stress-strain state of the mating elements in the mounting points, four possible designs of a disk cutter with a diameter D = 0.16 m (three biconical designs having taper angles: $\varphi = \varphi_1+\varphi_2$ = 25°+5° = 30°; 20°+10° = 30°; 15°+15° = 30° and one conical φ = 0°+30°).

The SolidWorks Simulation package, with considering general problems of solving linear systems using the finite element method (FEM), was used for modeling the stress-strain state of the mating elements in the points of mounting the twinned disk tools to the tetrahedral prisms located on the transfer gearbox casing in the space between the drums of the cutter heads used in selective roadheaders. In discretizating geometric models, the grid with parabolic finite elements (FE) in the form of tetrahedrons was used. The parabolic FE provide a better description of geometry of the model by the grid and increased accuracy of calculations by means of more nodes if compared with linear finite elements. The size of the finite elements was selected so as a further increase in the grid density had no significant effect on the results of calculations. The grid settings: FE size - 10 mm; tolerance - 0.5 mm; automatic grid density was not used. For the calculation, the algorithm FFEPlus was used, as it is more efficient for solving large-scale problems. To describe the interaction of parts in assembly the following contact conditions were used: for rigid connections of elements of the case - the type of contact conditions "related" and for detachable connections - contact condition "no penetration".

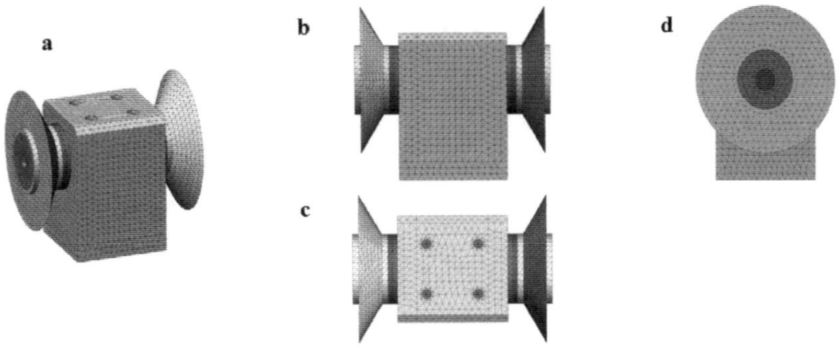

Figure 3. The finite element model of the tetrahedral prism with the twinned disc tools: a - in perspective; b, c, d - three main projections

Table 1. Calculated loads when cutting a work face using the twinned disk tool

Force, kN	Description of a work face in σ_{compr}, MPa					
	Coal			*Rock*		
	12.4	13.5	14.8	51	60.6	78.9
	When h = 0.04 m			*When h = 0.006 m*		
	Biconical disc tool: $\varphi = \varphi_1+\varphi_2$ = 25°+5° = 30°					
P_z	4.946	5.255	5.812	3.548	4.216	5.489
P_y	11.11	11.737	13.054	11.009	13.082	17.033
P_x	0.323	0.314	0.318	4.337	5.154	6.71
	Biconical disc tool: $\varphi = \varphi_1+\varphi_2$ = 20°+10° = 30°					
P_z	5.003	5.286	5.879	3.589	4.26	5.553
P_y	11.239	11.873	13.205	11.137	13.23	17.23
P_x	0.327	0.317	0.322	4.388	5.21	6.788
	Biconical disc tool: $\varphi = \varphi_1+\varphi_2$ = 15°+15° = 30°					
P_z	5.101	5.389	5.994	3.659	4.348	5.662
P_y	11.458	12.105	13.463	11.354	13.492	17.566
P_x	0.333	0.324	0.328	4.473	5.315	6.921
	Conical disc tool: $\varphi = \varphi_1+\varphi_2$ = 0°+30°					
P_z	4.927	5.205	5.79	3.535	4.2	5.469
P_y	11.068	11.692	13.004	10.968	13.032	16.968
P_x	1.739	1.837	2.043	4.321	5.134	6.685

The cutting force P_z, the sump force P_y and side force P_x on the twinned disk cutters (Figure 4) were determined, taking into account the constructive and performance parameters, and characteristics of the worked face σ_{compr}. The calculated forces P_z, P_y, P_x (Table 1) were applied to the finite element model of disk cutters mounted in the tetrahedral prism with the purpose of obtaining the stress state pictures (Figure 5-8) relating to biconical and conical disk tools, with considering the tensile strength under uniaxial compression of rock mass σ_{compr} ranging 12.4 to 78.9 MPa. The following steels were used for making mating elements: 35HGSA, St 45, St 65G, St.15.

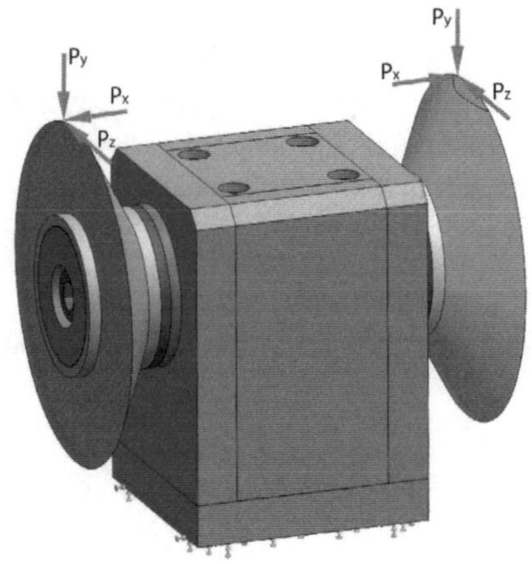

Figure 4. Diagram of applying forces to discs and mounting points of the tetrahedral prism

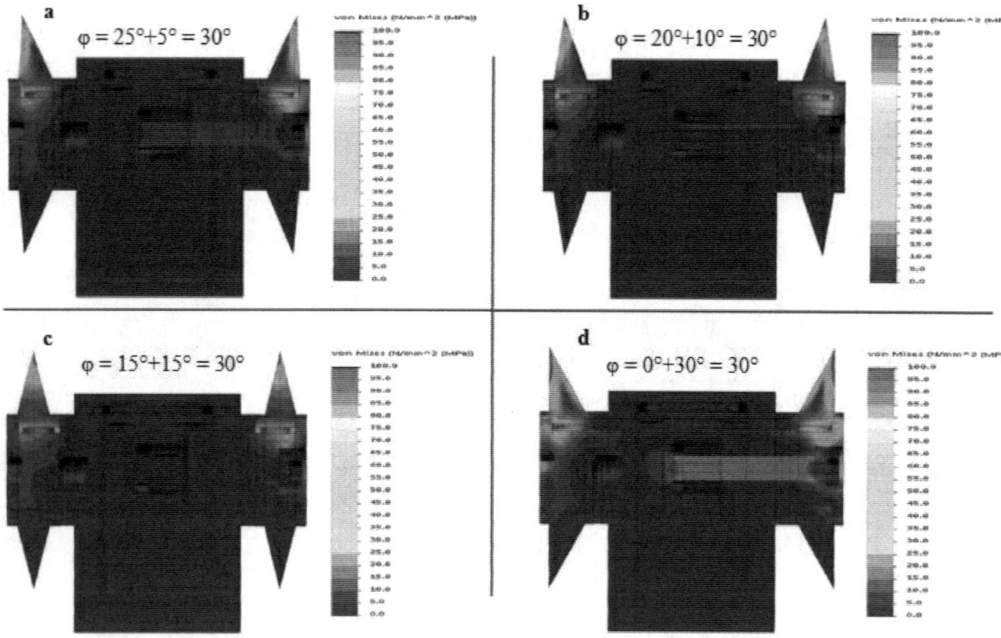

Figure 5. The distribution of equivalent stresses according to von Mises criterion in the mounting points of the twinned disk tools when cutting a coal face σ_{compr} = 12.4 MPa

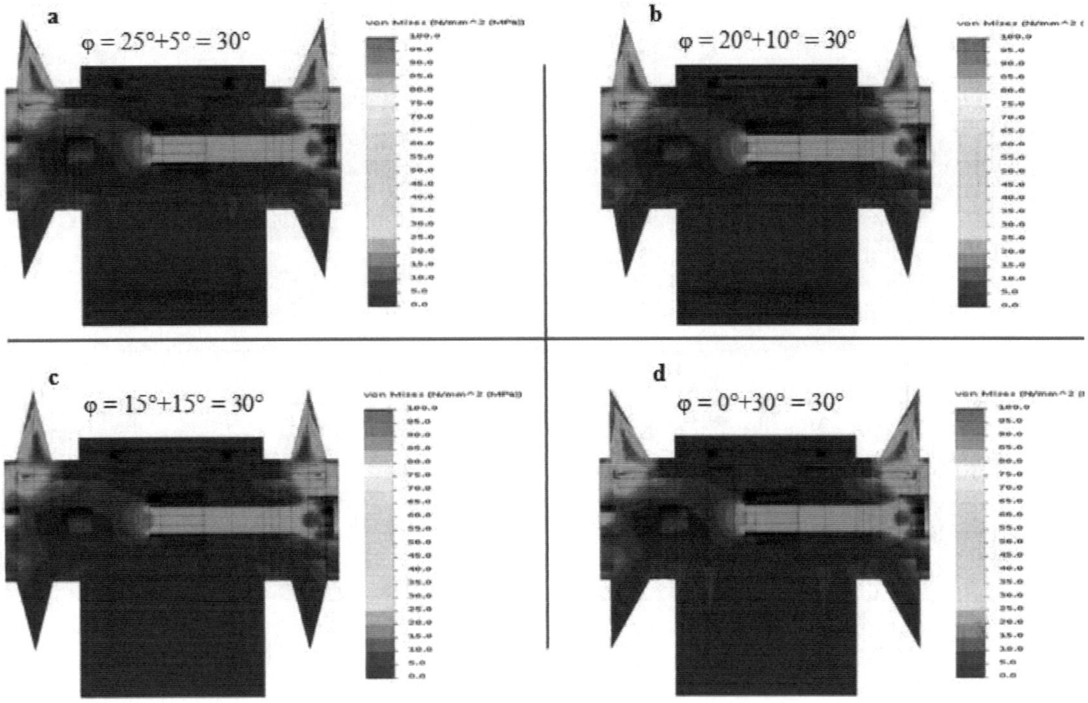

Figure 6. The distribution of equivalent stresses according to von Mises criterion in the mounting points of the twinned disk tools when cutting rock mass σ_{compr} = 51 MPa

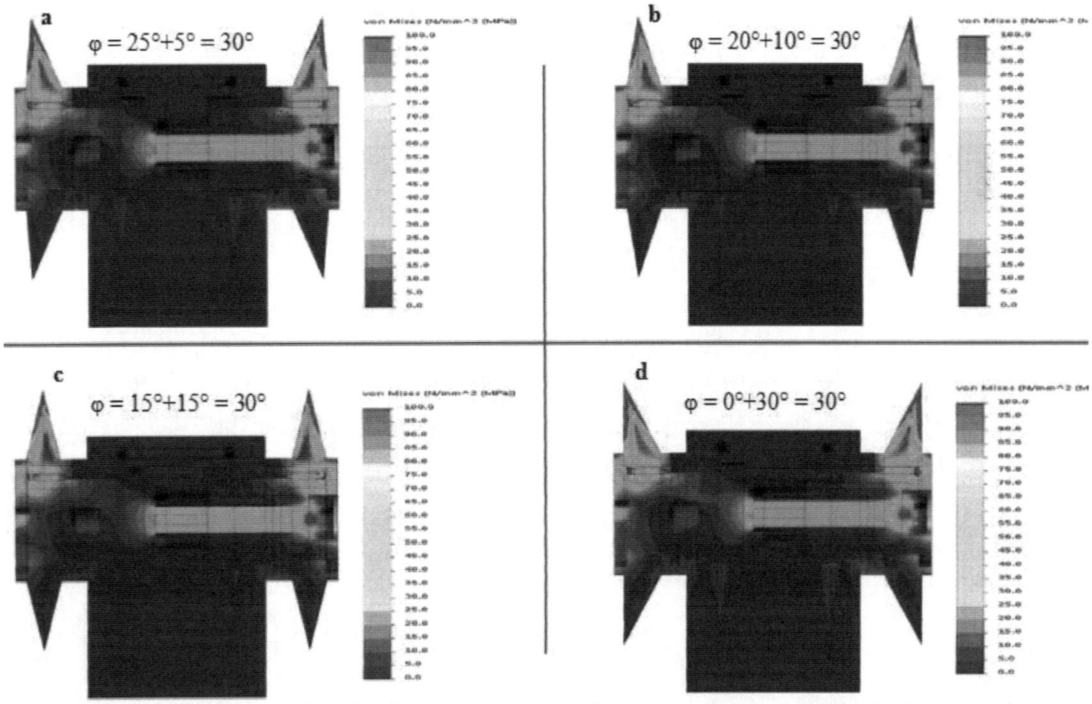

Figure 7. The distribution of equivalent stresses according to von Mises criterion in the mounting points of the twinned disk tools when cutting rock mass σ_{compr} = 60.6 MPa

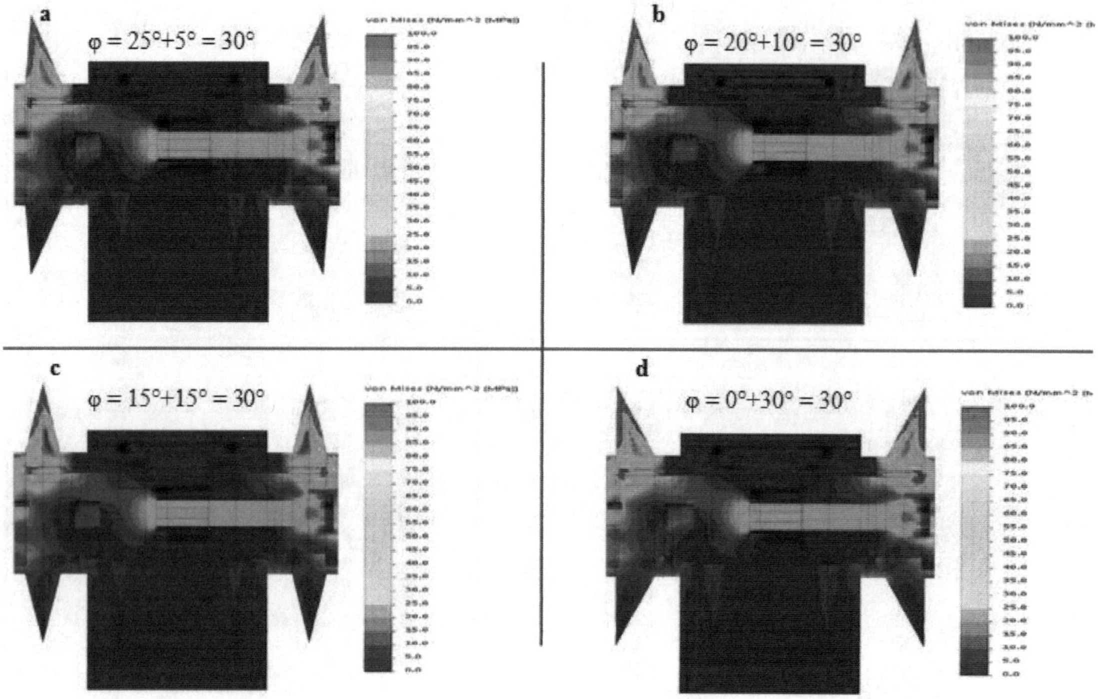

Figure 8. The distribution of equivalent stresses according to von Mises criterion in the mounting points of the twinned disk tools when cutting rock mass σ_{compr} = 78.9 MPa

Figures 9-12 show the dependencies of equivalent stresses according to von Mises criterion σ_{eqv} on the diameter D of the disk tool in the plane of section over the V-flange edge on the part of the fixing screw for the simultaneous rotation mode of disks when cutting: a coal face (1 – σ_{compr} = 12.4 MPa), rock mass (2 – σ_{compr} = 51 MPa; 3 – σ_{compr} = 60.6 MPa; 4 – σ_{compr} = 78.9 MPa).

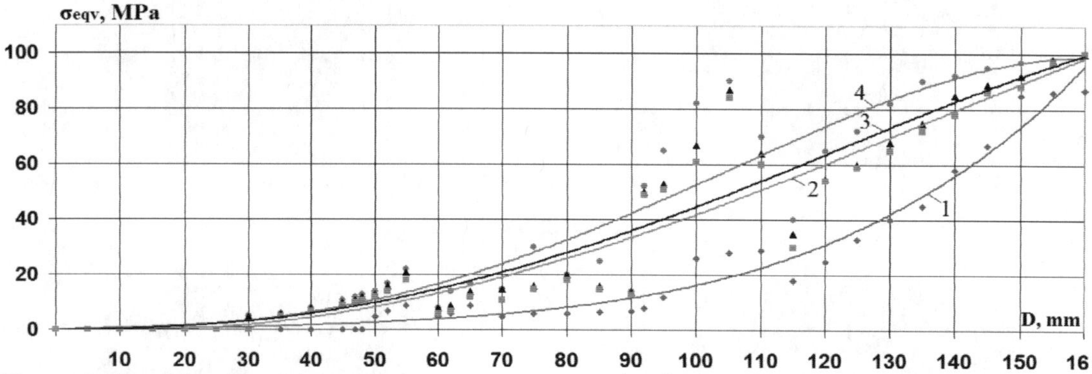

Figure 9. Dependencies of the equivalent stresses σ_{eqv} on the diameter D of the biconical disk tool with the taper angle φ = 25°+5° = 30°

IOP Conf. Series: Materials Science and Engineering **127** (2016) 012039 doi:10.1088/1757-899X/127/1/012039

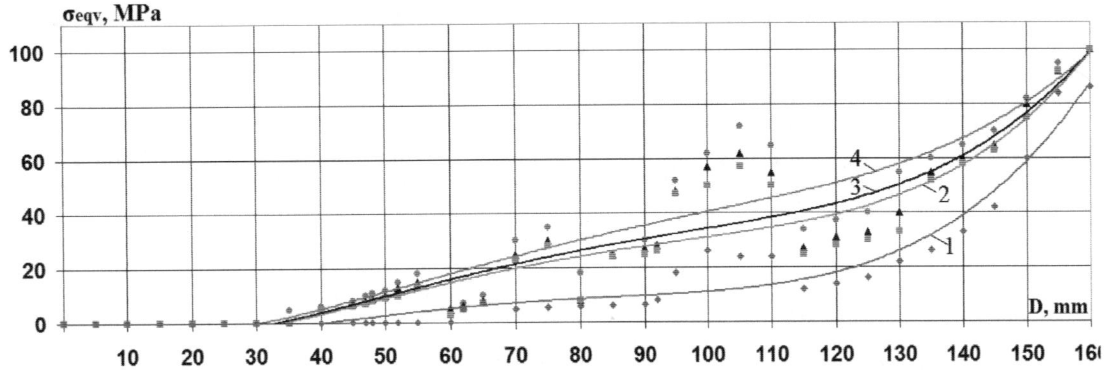

Figure 10. Dependencies of the equivalent stresses σ_{eqv} on the diameter D of the biconical disk tool with the taper angle φ = 20°+10° = 30°

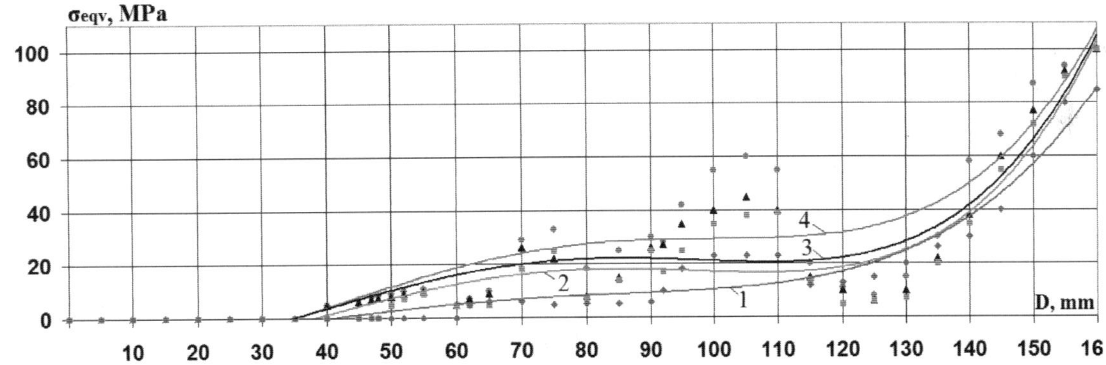

Figure 11. Dependencies of the equivalent stresses σ_{eqv} on the diameter D of the biconical disk tool with the taper angle φ = 15°+15° = 30°

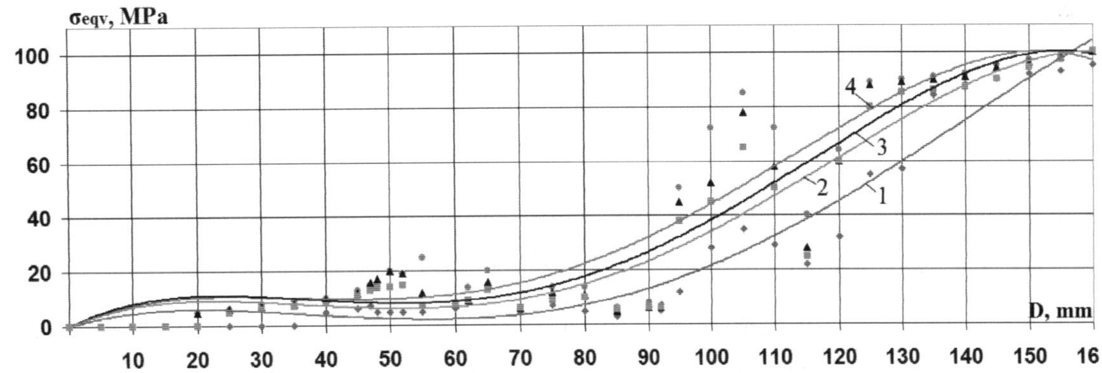

Figure 12. Dependencies of the equivalent stresses σ_{eqv} on the diameter D of the conical disk tool with the taper angle φ = 0°+30° = 30°

Table 2 shows the polynomial dependencies of the equivalent stresses according to von Mises criterion σ_{eqv} on the diameter D of the disk tool in the plane of section over the V-flange edge on the part of the fixing screw of the mount to the prism for the coaxial rotation operation of four disk design options when cutting work faces.

Table 2. Polynomial dependencies of the equivalent stresses σ_{eqv} on the diameter D relating to four design options of twinned disk tools

Taper angles of disks $\varphi = \varphi_1+\varphi_2$, grad	Work faces, σ_{compr}, MPa	Polynomial dependencies	Confidence coefficient of approximation R^2
25°+5°	12.4	σ_{eqv} = 2E-07D^4 - 3E-05D^3 + 0.002D^2 – 0.0106D	0.9711
	51	σ_{eqv} = -8E-08D^4 + 1E-05D^3 + 0.0049D^2 - 0.0918D	0.898
	60.6	σ_{eqv} = -2E-07D^4 + 4E-05D^3 + 0.0025D^2 + 0.0038D	0.8982
	78.9	σ_{eqv} = -4E-07D^4 + 1E-04D^3 – 0.0005D^2 + 0.0579D	0.9037
20°+10°	12.4	σ_{eqv} = 1E-06D^4 – 0.0003D^3 + 0.0244D^2 – 0.5984D	0.9399
	51	σ_{eqv} = 1E-06D^4 – 0.0003D^3 + 0.0339D^2 – 0.7942D	0.8921
	60.6	σ_{eqv} = 1E-06D^4 – 0.0003D^3 + 0.0329D^2 – 0.7628D	0.8877
	78.9	σ_{eqv} = 9E-07D^4 – 0.0003D^3 + 0.0295D^2 – 0.6635D	0.8909
15°+15°	12.4	σ_{eqv} = 1E-06D^4 – 0.0003D^3 + 0.0247D^2 – 0.6035D	0.9423
	51	σ_{eqv} = 2E-06D^4 – 0.0006D^3 + 0.0517D^2 – 1.2264D	0.858
	60.6	σ_{eqv} = 2E-06D^4 – 0.0006D^3 + 0.0523D^2 – 1.1771D	0.8458
	78.9	σ_{eqv} = 2E-06D^4 – 0.0006D^3 + 0.0548D^2 – 1.2626D	0.8041
0°+30°	12.4	σ_{eqv} = -8E-07D^4 + 0.0003D^3 – 0.0246D^2 + 0.6791D	0.9532
	51	σ_{eqv} = -1E-06D^4 + 0.0004D^3 – 0.0355D^2 + 1.0052D	0.9246
	60.6	σ_{eqv} = -2E-06D^4 + 0.0005D^3 – 0.0405D^2 + 1.178D	0.8986
	78.9	σ_{eqv} = -2E-06D^4 + 0.0005D^3 – 0.037D^2 + 1.0852D	0.8808

Figure 13 provides a summary graph to show the dependencies of the equivalent stresses σ_{eqv} on the diameter D of twinned disc tools having various designs held in the tetrahedral prism (three biconical with the following taper angles: 1 – $\varphi = \varphi_1+\varphi_2$ = 25°+5° = 30°; 2 – 20°+10° = 30°; 3 – 15°+15° = 30° and one conical 4 – φ = 0°+30°) when cutting coal faces in mine workings.

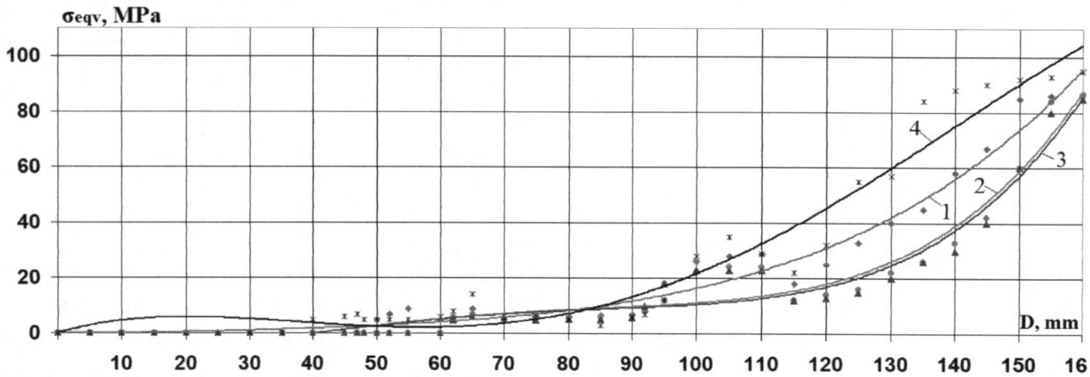

Figure 13. Dependencies of the equivalent stresses σ_{eqv} on the diameter D of disc tools having various designs when cutting a coal face σ_{compr} = 12.4 MPa

The results of research and the technical solutions have been obtained in the framework of the government contract with the Ministry of Education and Science RF, project No. 632 "Investigating parameters of technology and equipment for the choice and development of innovative technical solutions to improve operational efficiency of tunnelling and mining machines in Kuzbass".

Conclusion

Analysis of the dependencies of the equivalent stresses σ_{eqv} on the diameter D in four various designs of the twinned disc tools shows that, when cutting coal faces in mine workings, all of the variants have behaviours in the form of polynomial dependencies with the maximum steepness growing towards the zone of cutting, with the depth of cutting h = 0.04 m and the diameter of the disk V-flange D = 0.16 m. The highest level of equivalent stresses σ_{eqv} is 95 MPa that is indicative for the cutting process in coal faces using the conical rotary tool $\varphi = \varphi_1 + \varphi_2 = 0°+30°$. Relating to the three structural variants of biconical twinned disc tools ($\varphi = \varphi_1+\varphi_2 = 25°+5° = 30°$; $20°+10° = 30°$; $15°+15° = 30°$), the equivalent stresses σ_{eqv} are observed to reduce to 85-87 MPa.

It has been established that, when cutting rock masses, the dependencies of the equivalent stresses σ_{eqv} on the diameter D of the twinned disk tools also have a similar behaviour in the form of a polynomial dependencies with a flatter steepness towards the zone of cutting, compared to the cutting of coal faces, with the depth of cutting h = 0.006 m and the diameter of the disk V-flange D = 0.16 m. In this case, the maximum level of equivalent stresses σeqv for all variants of disk tools is within the range 97-100 MPa.

In general, the trends relating to the stress state of the mounting elements that hold the twinned disk tools, when cutting both coal and rock, are characterized by the following:

- When biconical disk tools are used for cutting, the polynomial dependencies of the equivalent stresses σ_{eqv} on the diameter D of discs are flattening with asymmetry parameters ($\varphi = 25°+5° = 30°$; $20°+10° = 30°$) changing into symmetry parameters ($15°+15° = 30°$) in the area of cutting.
- When conical disk tools are used ($\varphi = 0°+30°$), the maximum steepness growing is observed with respect to the polynomial dependencies compared to the biconical disk tools of diameters close to the area of cutting;

Thus, implementing the proposed technical solution, relating to the mounting elements of disk cutting tools held in tetrahedral prisms to ensure the mode of simultaneous rotation, allows the scope of application to extend and the operational efficiency to enhance when using boom-type selective roadheaders, made in Russia and abroad, with cutting heads having two axial cutting drums in coal mining and rock excavating operations.

References

[1] Khoreshok A A, Mametyev L E, Borisov A Yu, Mukhortikov S G The perfecting of a design of longitudinal axial bits of the road heading machine of electoral operating Mining Equipment and Electromechanicsl **5** (2010) 2-6

[2] Khoreshok A, Mametyev L, Borisov A, Vorobiev A, 2014 Stress-deformed state knots fastening of a disk tool on the drums of roadheaders Mining 2014. Taishan academic forum - project on mine disaster prevention and control. Chinese coal in the XXI century: Mining, green and safety. – Qingdao, China, October 17-20, 2014, Atlantis press, Amsterdam-Paris-Beijing. pp. 177-183

[3] Khoreshok A A, Mametev L E, Borisov A Yu., Vorobev AV 2015 Finite element models of disk tools with attachment points on triangular prisms Applied Mechanics and Materials. vol. 770. pp. 429-433

[4] Khoreshok A A, Mametev L E, Borisov A Yu., Vorobev A V 2015 Stress state of disk tool attachment points on tetrahedral prisms between axial bits Applied Mechanics and Materials. vol. 770. pp. 434-438

[5] Khoreshok A A, Mametyev L E, Borisov A Yu, Vorobyev A V 2015 The distribution of stresses and strains in the mating elements disk tools working bodies of roadheaders IOP Conference Series: Materials Science and Engineering. vol. 91. № 1. p. 012084

[6] Aksenov V V, Walter A V, Gordeyev A A, Kosovets A V 2015 Classification of geokhod units

and systems based on product cost analysis and estimation for a prototype model production IOP Conference Series: Materials Science and Engineering. vol 91. № 1. p. 012088

[7] Aksenov V V, Efremenkov A B, Beglyakov V Yu 2013 The influence of relative distance between ledges on the stress-strain state of the rock at a face Applied Mechanics and Materials. vol. 379. pp. 16-19.

[8] Aksenov V V, Khoreshok A A, Beglyakov V Y 2013 Justification of creation of an external propulsor for multipurpose shield-type heading machine – GEO-WALKER Applied Mechanics and Materials. vol. 379. pp. 20-23.

[9] Chernukhin R V, Dronov A A, Blashchuk M Y 2015 The application of the analytic hierarchy process when choosing layout schemes for a geokhod pumping station IOP Conference Series: Materials Science and Engineering. vol 91. № 1. p. 012086

[10] Efremenkov A B, Timofeev V Y Determination of necessary forces for geohod movement IFOST 2012: Proceedings - 2012 7th International Forum on Strategic Technology, September 18-32, 2012, Tomsk. – pp. 1-4

[11] Li X H, Du W, Huang Z L, Fu W L 2011 Simulation of Disc Cutter Loads Based on ANSYS/LS-DYNA Applied Mechanics and Materials. vol 127. pp. 385-389

[12] Zhao B, Zong X M, He B, Zhang L J 2014 Multi Variable Multi Objective Optimization for the Cutting Head of Roadheader Applied Mechanics and Materials. vols 635-637. pp. 358-364

[13] Zhang Y, Wang X W, Liu H F 2014 Numerical Simulation of Rock-Breaking Process by Disc Cutter in Tunnel Boring Machine Applied Mechanics and Materials. vol 487. pp. 513-516

[14] Li F H, Cai Z X, Kang Y L 2011 A Theoretical Model for Estimating the Wear of the Disc Cutter Applied Mechanics and Materials. vols 90-93. pp. 2232-2236

[15] Zhou Z L, Li X B, Zhao G Y, Liu Z X, Xu G J 2011 Excavation of High-Stressed Hard Rock with Roadheader Applied Mechanics and Materials. vols 52-54. pp. 905-908

IOP Conf. Series: Materials Science and Engineering **127** (2016) 012040 doi:10.1088/1757-899X/127/1/012040

The Load Capability Dependence on Characteristics of Driven Disks in a Lockup Assembly of an Overload Hydrodynamic Coupling

A V Koperchuk, A V Murin, V V Filonov

Yurga Institute of Technology, TPU Affiliate
Tomsk Polytechnic University, 634050, Tomsk, Lenin ave. 30

E-mail: avkop@tpu.ru

Abstract. A new design of a lockup assembly in an overload hydrodynamic coupling is offered, possible design variants of driven disks in a lockup assembly are considered, and results of experimental research into the importance of driven disk parameters for the load capability of a lockup assembly are given.

1. Introduction

Hydrodynamic couplings are used in drives of all currently available machines (belt, chain, scraper and slat conveyors, fans, mixers etc. [1]), as well as in newly developed ones (geokhod is a new class of mining machinery designed for tunneling mine workings [2-7]). A significant drawback of hydrodynamic couplings is energy loss caused by various angular velocities of turbine and pump wheels in steady conditions. This shortcoming can be eliminated via using a lockup assembly. One can find results of research into overload hydraulic coupling with a lockup assembly made as a centrifugal ball clutch in papers [8-11] (Figure 1).

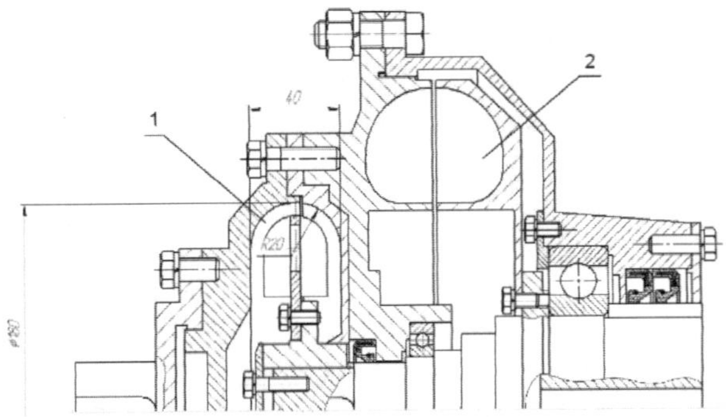

1 – lockup assembly
2 – overload hydrodynamic coupling
Figure 1. The structure of an overload hydrodynamic coupling with a lockup assembly

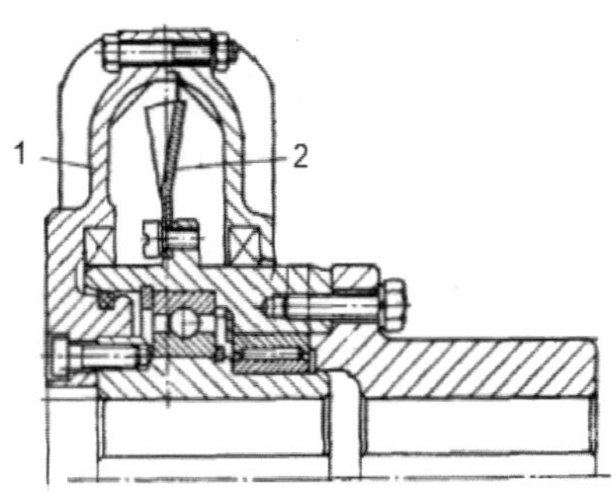

1 – case
2 – driven corrugated disk
Figure 2. Centrifugal ball clutch by Stromag

2. Results and discussions

In available structures of ball clutches with a semi-torus work space (Figure 2) [12] there are some peaks on the inner surface of a work space, which are intended to increase starting and maximal torques of a clutch. There are no peaks in a lockup assembly as it is a hydraulic coupling that provides smooth start of a driven shaft. Since the lower friction torque formed on walls of the work space via contacting with balls a possibility is considered to simplify the form and manufacturing technology of a driven disk, which is usually corrugated. It is proposed to use a flat driven disk with holes (Figure 3a) and a flat disk with peaks (Figure 3b).

The load capability of a solid flat disk in a lockup assembly is theoretically estimated. However, there is no data on using flat disks with holes or peaks in similar assemblies in literature.

We think, moment transferred by disk with holes can hardly be measured theoretically as it is formed not only by friction forces arising between balls and flat segments of the disk but due to tractive resistance of elements in the holes, which are the reason for peaks and hollows in these zones. As the consequence, the load capability of a lockup assembly with disks mentioned above was measured experimentally.

We fabricated disks 3 mm thick with 40 holes 8.5 mm in diameter, ones with 20 holes 17 mm in diameter, and ones with 10 holes 34 mm in diameter. The overall area of the holes was identical on all disks.

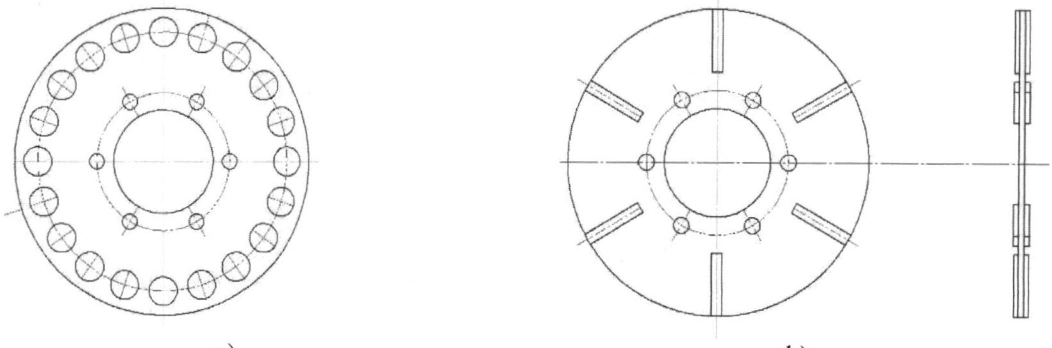

a) b)

Figure 3. The structure of driven disks

Disks with radial peaks were designed the same way as corrugated ones, the data on their fabrication is provided in paper [13]. The height of a peak was accepted almost similar to the ball diameter (3.175 mm) and was equal to 3 mm in the first variant. The height of a peak in the second disk structure was assigned 12 mm under assumption slippage of granular material is impossible on the disk and a case-peak gap is ensured to provide free spreading of balls in the inner space. The length of peaks was similar (35 mm) and assigned with respect to the ring layer thickness of balls in a filled lockup assembly. The number of peaks was the same as that of flutes, i. d. 6 flutes on each side.

Geometrical parameters of the inner space in the lockup assembly (Figure 1):

- Maximal radius of inner space $R_0 = 0.09$ m;
- Radius of torus space $\rho = 0.02$ m.

Geometrical parameters of disks (Figure 3):

- Outer diameter of disks with holes 0.178 m;
- Outer diameter of disks with peaks 0.17 m (to support smooth spreading of filling material in inner space on both disk sides).

Hardened and planished steel bearing balls (steel ShH 15 GOST 801-78) 3.175 mm and 4.763 mm in diameter were used as filling material. It weighted Q = 2.2 kg. The case and disks were made of steel 20 GOST 1050-88 without thermal treatment.

The experiments have demonstrated unloaded driven shaft can't be accelerated without hydraulic liquid in a hydraulic coupling provided that the lockup assembly is filled with balls as specified. The starting torque of the assembly is insufficient to eliminate resistance in bearings, elastic coupling, belt transmission and electrical brake bearings of a test bench. To accelerate the driven section a hydraulic drive was filled up with 1 liter hydraulic fluid (I-20A GOST 20799-88), conforming to 28% of its filling level.

Figure 4 shows tests when accelerating and slowing down the hydraulic coupling filled 28 percent by a semi-torus ball lockup assembly 4.763 mm in diameter and 12 mm high peaks. The result of gauging the dynamometrical assembly is shown in the left section of the figure. The value of torque is given in Nm. Used symbols:

n_1 - rotation frequency of a driveshaft in a hydraulic coupling;
n_2 – rotation frequency of a driven shaft in a hydraulic coupling;
M_2 – torque on the output shaft of a hydraulic coupling;
t - time.

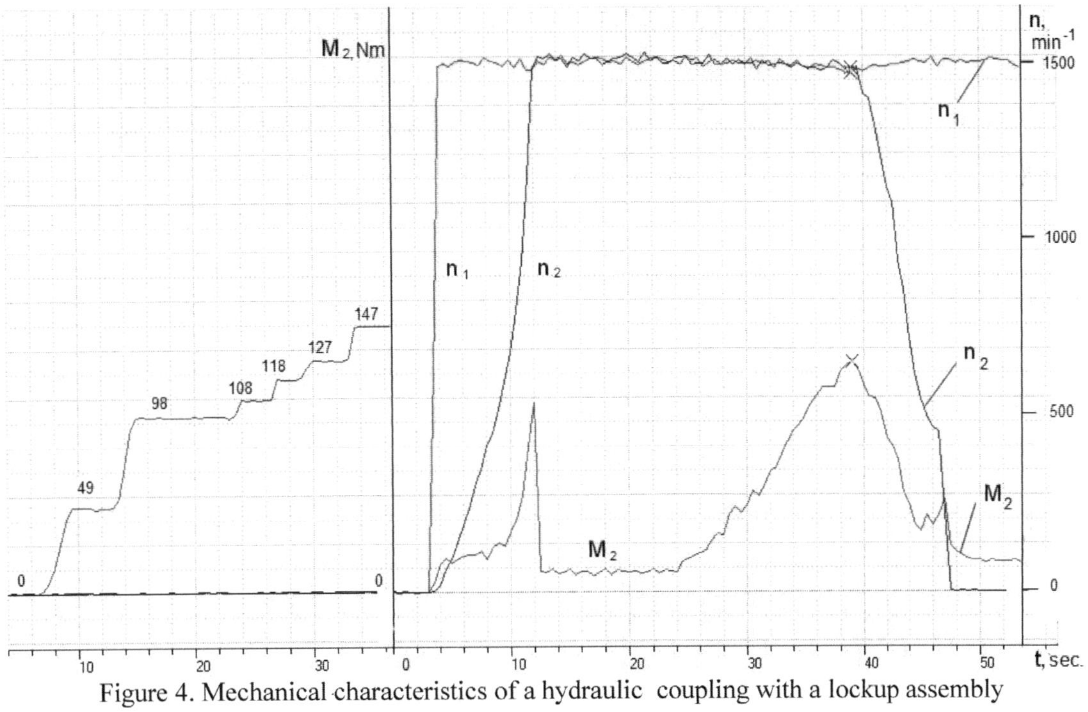

Figure 4. Mechanical characteristics of a hydraulic coupling with a lockup assembly

As one can see from the analyzed experiment recordings:

- hydraulic coupling and lockup assembly running together can accelerate the driven shaft up to the frequency of the driveshaft;
- in these conditions there is no slippage because a hydraulic coupling is not warmed up;
- load is transferred by the lockup assembly only, since a hydraulic coupling can't function without relative movement of pump and turbine wheels.

The analysis of marks resulted from the contact of the lockup assembly with balls has revealed that granular material slips on the surface of 3 mm high disks with holes and peaks (Figure 5). If the peaks are 12 mm high, balls slip, as one has assumed, on the inner part of the lockup assembly case (Figure 6).

a)

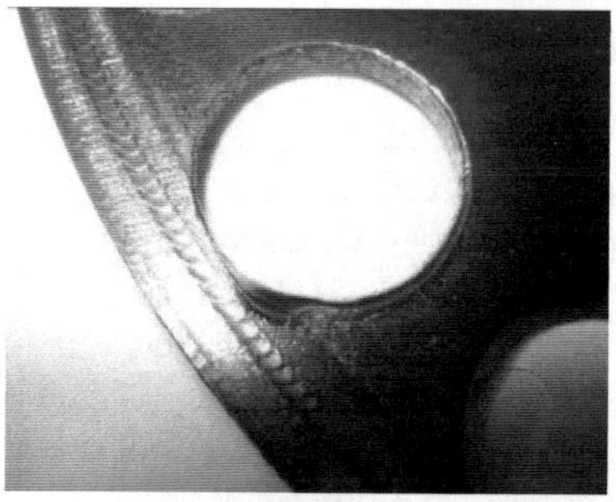
b)

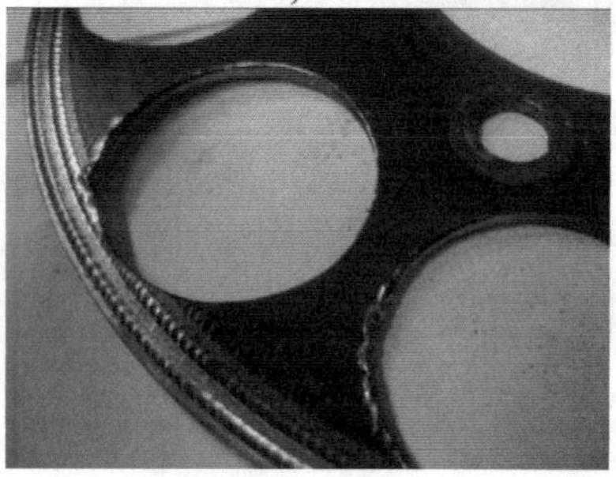
c)

d)

a) a disk with holes 8.5 mm in diameter
b) a disk with holes 20 mm in diameter
c) a disk with holes 34 mm in diameter
d) a disk with 3 mm high peaks

Figure 5. Driven disks with the marks of the contact

Figure 6. Inner surface of lockup assembly case with the marks of contact after using a driven disk with 12 mm high peaks

It should be noted, unhardened disks with holes and 3 mm high peaks are badly damaged under slipping of balls on their surface. If disks with 12 mm high peaks are used there are concentric hollows – result of contact with granular material – on the smooth surface of the case inner space, however, the depth of these hollows is constant for a long time.

Experiments made it possible to draw torque on the output shaft M_2 – slippage S (into lockup assembly) graph, which outlines mechanical statistical characteristics of lockup assembly with various disks (Figure 7). Curves are shown on the same graph for more convenience (Figure 8).

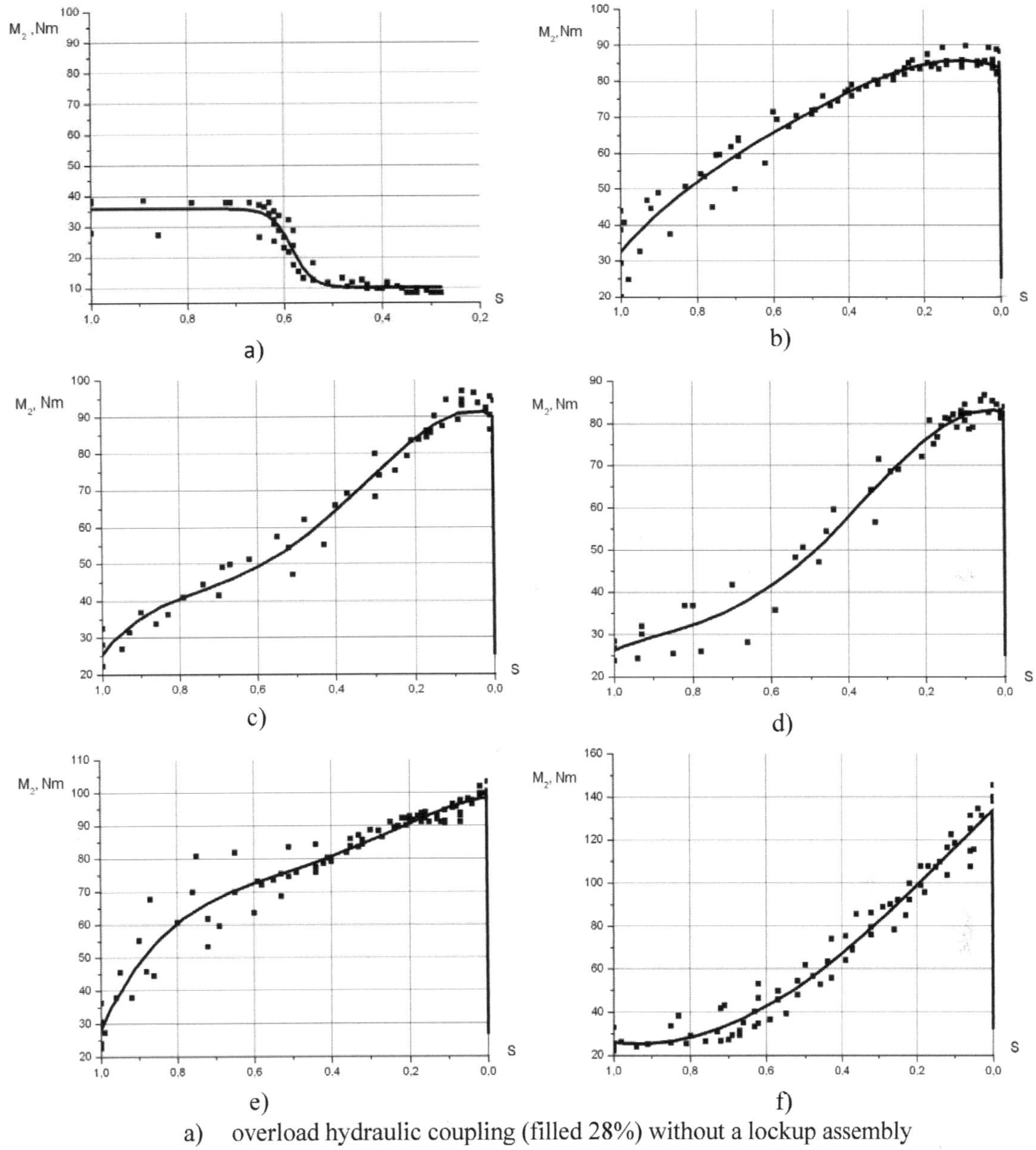

a) overload hydraulic coupling (filled 28%) without a lockup assembly
b) a lockup assembly with disk with 10 holes 34 mm in diameter
c) a lockup assembly with disk with 20 holes 17 mm in diameter
d) a lckup assembly with disk with 40 holes 8,5 mm in diameter
e) a lockup assembly with disk with 12 3 mm high peaks
f) a lockup assembly with disk with 12 12 mm high peaks

Figure 7. Mechanical characteristics of a lockup assembly with a semi-torus work space and various disks

3. Conclusion

The following conclusions can be drawn on the base of analyzed graphs, experimental observations of the research and examination of the lockup assembly components after experiment:

– a lockup assembly with granular filling material has a low starting torque, which doesn't deteriorate starting properties of hydraulic couplings;

– granular filling material slips on the surface of disks with holes and 3 mm high peaks;

– disk with 12 mm high peaks supports the slippage of balls on the inner surface of the case; the loading capacity of a lockup assembly is 1.35 times higher under slippage of filling material on the surface of the case interior;

– disks supporting the slippage of balls on the case surface are recommended to be used in the lockup assembly of an overload hydraulic coupling as they provide higher loading capability and lower additional loads on the drive under slippage S over 0.65.

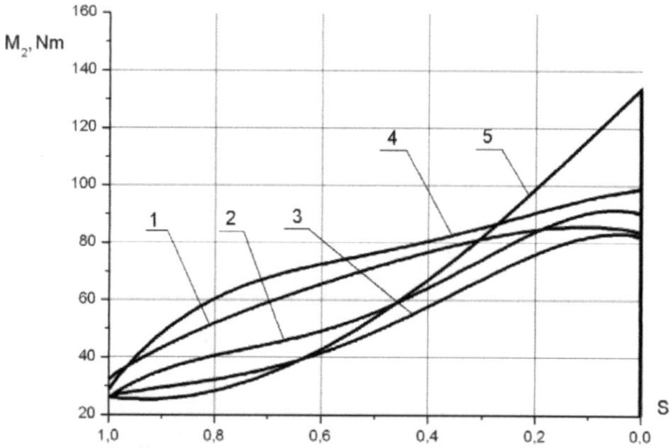

1 – disk with 10 holes 34 mm n diameter
2 – disk with 20 holes 17 mm in diameter
3 – disk with 40 holes 8.5 mm in diameter
4 – disk with 12 3 mm high peaks
5 – disk with 12 12 mm high peaks
Figure 8. Mechanical characteristics of a semi-torus lockup assembly with various disks

References

[1] Aksenov V.V., Efremenkov A.B., Blaschuk M.Yu., Timofeev V.Yu. 2010 *Mining Informational and Analytical Bulletin* Review of transmission in mining machinery **V.3 N12** 55-66.

[2] Aksenov V.V., Khoreshok A.A., Beglyakov V.Y. 2013 *Applied Mechanics and Materials* Justification of Creation of an External Propulsor for Multipurpose Shield-Type Heading Machine – GEO-WALKER **Vol. 379** 20–23.

[3] Efremenkov A.B., Timofeev V.Y. 2012 7th International Forum on Strategic Technology (IFOST) Determination of necessary forces for geohod movement (IEEE) pp 1–4.

[4] Blaschuk M.Y., Dronov A.A., Miheev D.A. 2015 Applied Mechanics and Materials Geokhod Propel Effort Mathematical Model **Vol. 770** 391–396.

[5] Aksenov V.V., Walter A.V., Gordeyev A.A., Kosovets A.V. 2015 *IOP Conference Series: Materials Science and Engineering* Classification of geokhod units and systems based on product cost analysis and estimation for a prototype model production **V. 91** 012088.

[6] Sadovets V.Y., Beglyakov V.Y., Aksenov V.V. 2015 *IOP Conference Series: Materials Science and Engineering* Development of math model of geokhod bladed working body interaction with geo-environment **V. 91** 012085.

[7] Walter A.V., Aksenov V.V. 2015 *Applied Mechanics and Materials* Determining deviations in geometry of the geokhod shells **Vol. 770** 439–444.

[8] Murin A.V., Koperchuk A.V. 2011 *Mining Informational and Analytical Bulletin* Reduction of energy losses by locked up hydraulic coupling in drives of mining machines **OB2** 337-343.

[9] Koperchuk A.V., Murin A.V. 2012 *Mining Informational and Analytical Bulletin* Improvement of locked up hydrodynamic coupling for drives of mining machines **OB3** 300-305.

IOP Conf. Series: Materials Science and Engineering **127** (2016) 012040 doi:10.1088/1757-899X/127/1/012040

[10] Koperchuk A.V., Murin A.V. 2014 *Applied Mechanics and Materials* Influence of geometrics of synchronization devices of fluid coupling on loading capability **Vol. 682** 499-503.
[11] Koperchuk A.V., Murin A.V., Dortman A. A., Filonov V. V. 2015 *Applied Mechanics and Materials* A change in mechanical behavior of safety fluid couplings when the lockup device is used in its construction **Vol. 770** 279-282.
[12] Ryakhovskii O.A., Ivanov S.S. 1991 A reference book on couplings Leningrad, Poly-technique.
[13] Dyachenko S.K., Kirkach N.F. 1962 Safety coupling Kiev, State publishing of technical literature of the USSR.

IOP Conf. Series: Materials Science and Engineering **127** (2016) 012041 doi:10.1088/1757-899X/127/1/012041

Influence of the Conditions of Electrode Position and Heat Treatment on the Structure and Properties of Metallic Coatings

I M Kovenskiy[1] and A A Kulemina[2]

[1,2]Tyumen State Oil and Gas University, 38 Volodarskogo st., Tyumen, 625000, Russia

E-mail: imkoven@tsogu.ru

Abstract. Influence of electrodeposition conditions and heat treatment on the structure and properties of metallic coatings has been considered. It has been that at different values of overvoltage, metals crystallize with a cellular, subgrain or monoblock structure. The final formation of the structure occurs during annealing, in the process of which either polygonization or recrystallization develops in the coatings. Varying the conditions of electrodeposition and heat treatment allows obtaining coatings with functional characteristics for specific operating conditions.

Introduction

Electrolytic deposition of metallic coatings is rather widely used for enhancing the protective and strength properties of mechanical parts. However, reserves related to improvement of these properties by means of heat treatment techniques are hardly ever made use of in practice [1-3]. The paper considers formation of the structure in production, aging and subsequent annealing of most common metallic coatings and the influence of structural peculiarities on the properties of electrodeposited layers.

1. Methods and materials

Electrodeposited chromium, nickel, zinc and cadmium were investigated. The coatings were deposited in industrial and laboratory conditions using standard electrolytes. Polarization curves were constructed for a range of conditions of electrolytic coating deposition. The ratio of E / E_{lim}, where E_{lim} is the overvoltage at the cathode, corresponding to the limiting current density (i_{lim}) was used as the integral parameter taking into account the influence of the conditions of electrodeposition. For convenience, the conditions of metal deposition at $E < 0.33\ E_{lim}$ were considered mild, at $E > 0.66\ E_{lim}$ – hard, and at $E = (0.33\text{-}0.66)\ E_{lim}$ – average.

The obtained samples were annealed in an argon atmosphere, at various time-temperature parameters.

The accelerated corrosion tests were carried out in an environment of 0.1 N sulfuric acid, using the gravimetric method. Before testing, the samples were weighed on an analytical balance with an accuracy of 0.0001g then placed in corrosion cells and kept in this state for 30 days. After that, the corrosion products were removed, weighing was repeated, the thickness of the corroded layer was determined and the corrosion rate was calculated.

The hardness of the coatings was determined using a PMT-3M microhardness tester, at loads up to 2 N, by the method of the restored print.

Electron microscopy studies of the structure were carried out by transmission electron microscopy of thin foils using an EMV-100L electron microscope.

It was previously shown [4] that depending on the conditions of deposition, the studied metals crystallize with a characteristic type of the structure and can be classified as shown in Table 1.

2. Results and discussion

In metals with relatively high melting temperatures (Cr, Ni), deposited under hard conditions close to the limiting current, a typical cellular structure is formed. The cells are volumes, which are relatively free from dislocations separated from each other by dislocation tangles with a high dislocation density. The boundaries between the cells are often so vague and broad that they are commensurate with the size of the cells. In electron micrographs obtained from the cell boundaries, split reflections are observed and the azimuthal misorientation angle is a few degrees. At cathode overvoltages lower than (0.7-0.6) E_{lim} , the boundaries are already flat dislocation walls and, at lowering the cathode polarization, the density of dislocations in the sub-boundaries decreases and the misorientation angle between subgrains decreases to several tens of minutes. Finally, at $E< 0.33\ E_{lim}$, the deposition of coatings is characterized by a monoblock structure.

Table 1.The structure of coatings after deposition.

Group	Metals	Conditions of electrodeposition	Type of the structure	Average grain size
I	Cr,Ni	Hard $E> 0.66\ E_{lim}$	Cellular	10^{-6}
		Average $E= (0.33 - 0.66)\ E_{lim}$	Subgrain	10^{-5}
		Mild $E< 0.33\ E_{lim}$	Monoblock	10^{-4}
II	Zn, Cd	Hard, Average	Subgrain	10^{-3}
		Mild	Monoblock	10^{-2}

In metals with relatively low melting temperatures (Zn, Cd),under mild conditions of deposition, a monoblock structure is formed with individual dislocations or groups of dislocations, not interconnected within the boundaries. At overvoltages higher than (0.3-0.4)E_{lim}, the structure observed is already a subgrain one. It should be noted that to get a cellular structure at electrodeposition of low-melting metals, even at current densities close to the limiting current densities, does not seem possible.The cellular structure is formed only in refractory metals, whose conditions of electrolysis cause significant internal stresses facilitating microplastic deformation at electrocrystallization. It should also be noted, the cellular structure is more easily achieved (i.e., at substantially lower overpotentials) in metals with high values of stacking fault energy.

The structure of coatings of each group of metals is characterized by definite intervals of grain size; the dispersity of the structure is higher in metals with higher melting temperatures, which have been deposited under more hard conditions (Table 1).

Such a significant difference in grain size determines the peculiarities of formation of point defects of the crystal structure in electrodeposited metals belonging to different groups, in particular. It is known, that in electrodeposited metals with a highly dispersed structure, a dimensional vacancy effect appears, which is manifested in a sharp growth of vacancy concentrations, while in metals having a macrocrystalline structure, there dominates the effect of interstitial atoms.

Work [5] presents data on the concentrations of vacancies and interstitials in metals of different groups deposited under different conditions. As it is seen from Table 2, an increase in the deposition potential leads to an increase of point defect concentration in the structure of coatings. Moreover, in all investigated metals, a correlation is observed between the magnitude and sign of internal compression and tension stresses and concentration of interstitials and vacancies, respectively. In addition to that, the experimentally obtained values of internal stresses are in good agreement with values of internal stresses, calculated in accordance with the atomic and vacancy model of their formation . Allowing for this correlation, internal stresses can be used as an effective indicator of the development of relaxation processes associated with migration and annihilation of point defects during aging and annealing of electrodeposited metals.

Table-2. Concentration of point defects and magnitude of internal stresses in coatings.

Group	Metal	i, A/dm^2	C_v	C_{ia}	Magnitude (MPa) and sign of internal stresses	
					calculated	experimental
I	Cr	50	10^{-2}	-	+300	+340
		35	$5{\cdot}10^{-3}$	-	+220	+235
		20	10^{-3}	-	+150	+175
	Fe	50	$8{\cdot}10^{-3}$	-	+280	+310
		35	$5{\cdot}10^{-3}$	-	+220	+225
		20	10^{-3}	-	+150	+190
	Ni	10	10^{-3}	-	+200	+230
		6	10^{-4}	-	+150	+175
		3	10^{-6}	-	+100	+115
II	Zn	4	-	10^{-11}	-12	-15
		3	-	10^{-9}	-20	-18
		2	-	$5{\cdot}10^{-4}$	-26	-20
	Pb	4	-	10^{-10}	-15	-10
		3	-	10^{-7}	-22	-25
		2	-	$5{\cdot}10^{-4}$	-38	-35
	Bi	2	-	10^{-10}	-20	-18
		1	-	10^{-7}	-30	-28
		0,5	-	$5{\cdot}10^{-4}$	-44	-40

Earlier [5], experimental dependences of hardness, electrical resistance, lattice parameter and internal stresses on the time of aging were obtained for electrodeposited metals (schematically shown in Figure 1). On the basis of these data, we'll try to analyze the character of changes in the structure of metals of different groups in the period after electrodeposition.

In metals of the first group, having relatively high melting temperatures, in accordance with the activation energy [6], at room temperature, only those processes are possible which are associated with migration and annihilation of vacancies and hydrogen atoms, whose presence is due to the specifics of the process of electrodeposition of metals belonging to this group. In connection with this, it is possible to distinguish two stages in the aging of coatings.

The first stage, lasting up to 100 hours, is obviously associated with diffusion of the most labile hydrogen. As an interstitial element, hydrogen is located in the interstices of the crystal lattice of coatings and is captured by vacancies - defects having an effective negative charge. In the period after electrolysis, the hydrogen is released from the vacancies and moved from the solid solution to sinks, including pores formed in the process of electrocrystallization, which causes an increase of pressure in them and growth of hardness. Reduction of the crystalline lattice period and electrical resistance at the first stage of aging confirms the dominating role of hydrogen diffusion at the initial stage of aging.

At the second stage, it is the vacancies, less labile than hydrogen atoms that migrate to the sinks. Annihilation of vacancies is indicated by a decrease in internal stress and an increase in the lattice periods, which, however, remain below the equilibrium ones even after prolonged aging, indicating a high concentration of excess vacancies in the coatings. Migrating to the pores, vacancies increase the volume of the latter, reducing the pressure of molecular hydrogen in them. This leads to some reduction in the hardness, but does not cause softening, since structure-ordering processes do not develop to the stage of polygonization, which in metals with a relatively high melting temperature may occur at temperatures above 100 ° C, and is possible only at annealing. Accordingly, the type of coating structure - cellular or subgrain –does not undergo significant changes during aging either.

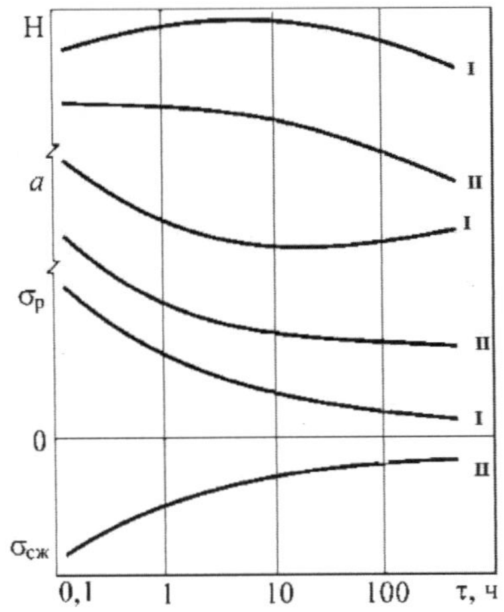

Fig. 1 Nature of the hardness change (H), crystal lattice parameter (a) and internal compression (σ_{com})and tension (σ_{tens}) stresses of electrodeposited metals of the first (Ni, Fe, Cr), and third (Zn, Pb, Bi) groups in the process of aging

Different conditions of coating deposition result in different quantities of dissolved hydrogen and excess vacancies in Co, Ni, Fe, Cr deposits. However, the nature of coating aging processes does not change/remains the same at that. It can only be noted that 1)the harder the conditions of deposition, the higher the concentration of point defects and 2) it takes a longer period of time for aging processes leading to stabilization of properties and relaxation of stresses to occur.

In contrast to metals with high melting temperatures, in which aging processes are limited to stages of relaxation, in low-melting metals of the second group, a complete development of recovery and recrystallization processes takes place. In these metals, in the course of aging time, the magnitudes of lattice parameters decrease due to migration and annihilation of excess interstitials. The decrease is the most intense at the first stage of aging, within 10-100 hours, correlating with a decrease in the electrical resistance of coatings and the magnitude of internal stresses. It can be noted that, in coatings obtained under hard conditions and characterized by a higher concentration of interstitials, the lattice parameter, electrical resistance and internal stresses reach stable magnitudes for a longer period of time than in coatings obtained under mild conditions. Since in low-melting metals, homologous return temperatures lie below the room temperature, during the second stage of aging, processes of redistribution and annihilation of dislocations take place in them, which are characteristic of the second stage of return - polygonization, resulting in enlargement of structural elements and reduction of the hardness of coatings. After polygonisation, in low-melting metals, at prolonged aging time, recrystallization takes place, which is confirmed by metallographic studies [4].

Annealing of coatings leads to transformation of their structure.

If annealing is applied to refractory metals electrodeposited under hard conditions and having, as a consequence, a cellular structure, then at heating the metal, the cells transform into subgrains substantially misoriented relative to each other. In the sequel,the misorientation angles increaseand polygonization thus plays the role of the initial stage of recrystallization. Further annealing leads to structural changes associated with enlargement of recrystallized grains and their abnormal growth. It is obvious that such changes are identical to the stages of collective and secondary recrystallization in deformed metals.

A different situation is observed when annealing refractory metalselectrodeposited under average conditions and having a subgrain structure. In this case, two variants are possible: at relatively low annealing temperatures, polygonization competes with recrystallization but at heating to a higher temperature, it becomes the initial stage of the latter. By analogy with deformed metals [8], it can be assumed that in the first variant,the dislocations are redistributed by climb and cross-slip,and subgrains have low-mobile boundaries with low angles of misorientation, i.e. a structure is formed which is unfavorable for the formation of recrystallization centers.

In the second variant,due to the increasing role of the conservative slip of dislocations, new subboundaries are not strictly normal to the plane of sliding and have higher misorientation angles and a higher mobility than in the first variant.Migration of these boundaries leads to the fact that some of the formed subgrains transform into recrystallization centers.

In refractory coatings deposited under mild conditions and characterized by a monoblock structure, a classical development of polygonization processes is observed during annealing, when redistribution of dislocations leads to formation of subgrains separated by low-angle boundaries, and then, at recrystallization, new grains are formed with a more perfect structure than the initial grains, surrounded by high angle boundaries.

As for electrodeposited low- melting metals whose homologous return and recrystallization temperatures are below room temperatures, then in them, the structural changes characterizing polygonization and recrystallization take place immediately after electrolysis. Stabilization of the structure of low-melting coatings is in practically acceptable time achieved by annealing at temperatures of 100 - 150°C for 1.0-1.5 hours. Upon annealing, the initial structure of low-melting coatings, under the action of recrystallization changes considerably, regardless of whether it was a subgrain or a monoblock structure. New grains with high angle boundaries are formed. Because of that, after annealing, coatings have higher values of strength properties and increased protective propertiescompared to the electrodeposited state (Table 3).

Table 3. Changes in the hardness (HV, hPa) and corrosion rate (Vc mm / year) of electrodeposited low-melting metals at annealing.

Metals	Electrodeposition			Annealing, °C					
	Type of the structure	HV	Vc	50		100		150	
				HV	Vc	HV	Vc	HV	Vc
Zn	Subgrain	0.95	0.27	0.68	0.25	0.50	0.20	0.36	0.15
	Monoblock	0.73	0.20	0.52	0.22	0.40	0.20	0.35	0.15
Cd	Subgrain	0.62	0.25	0.41	0.23	0.32	0.20	0.29	0.18
	Monoblock	0.45	0.20	0.35	0.22	0.29	0.23	0.27	0.17

The question of relationship between polygonisation and recrystallization is of great practical interest primarily for refractory metals deposited with formation of a cellular structure or without it [8]. Indeed, the structure formed in the process of electrodeposition determines either the progress of recrystallization at subsequent annealing or its suppression by polygonization with a corresponding change in properties, i.e. polygonization can be used to prevent recrystallization in the cases, where the latter is undesirable. Consider in this context the character of changes in the mechanical and protective properties upon annealing of coatings having a different type of structure after electrocrystallization (Table 4).

Table 4. Changes in the hardness (HV, hPa) and corrosion rate (Vc mm / year) of electrodeposited refractory metals at annealing.

Metals	Electrodeposition			Annealing					
	Type of the structure	HV	Vc	(0.25-0.30)Tm		(0.30-0.35)Tm		(0.35-0.40)Tm	
				HV	Vc	HV	Vc	HV	Vc
Ni	Cellular	5.1	2.1	6.5	2.9	3.5	2.1	2.0	1.8
	Subgrain	4.5	1.8	5.2	2.5	4.5	2.4	3.1	1.5
	Monoblock	4.0	1.6	3.0	2.0	2.2	1.8	2.0	1.4
Cr	Cellular	7.0	1.6	10.5	2.4	6.2	1.7	4.6	1.3
	Subgrain	6.5	1.4	8.3	2.0	7.2	1.8	4.5	1.2
	Monoblock	5.8	1.1	4.8	1.3	4.6	1.1	4.2	1.0

3. Conclusion

The experimental results of determining the mechanical properties show that annealing of coatings with a cellular structure, at the early stages of polygonization, is accompanied by hardening, which is explained by redistribution of mobile dislocations and pinning them in a more stable configuration by atoms of impurity components that form part of the composition of industrial electrolytes [8]. At later stages of polygonization, the hardness decreases and at recrystallization annealing, softening of the coatings takes place.

If annealing is applied to coatings, which, after electrodeposition, have a subgrain structure, then the hardening effect at polygonization is less pronounced than in coatings with a cellular structure, since coatings with a subgrain structure are characterized by a lower density of dislocations compared with the cellular one. However, as mentioned above, due to the fact that in coatings with a subgrain structure, recrystallization is suppressed by polygonization, softening occurs at higher annealing temperatures.

Finally, in coatings with a monoblock structure, in accordance with the classical development of structural transformations at annealing, a decrease in hardness takes place at all stages of polygonization and recrystallization.

Analysis of the changes in protective properties (Table 4) shows that at the stage of pre-recrystallization annealing of coatings, the corrosion rate takes maximum values. The reason for such a behavior is associated with an increase in the number of corrosion centers in the course of polygonisation, first of all, the defects of the crystal structure: after thermal release of hydrogen from the vacancies, point defects are activated and the configuration of the dislocation structure in coatings with cellular and subgrain structures changes; subgrains separated by low-angle boundaries are formed in coatings with a monoblock structure.

Recrystallization annealing leads to formation of new grains separated by high- angle boundaries with a sharply reduced density of point and line defects,so that the protective properties of coatings increase. It should be noted that improvement of corrosion resistance in coatings with a subgrain structure is achieved at higher annealing temperatures than in coatings with a cellular structure. However, in all cases, the temperature of recrystallization annealing is bound by a limit, above which there occurs development of secondary porosity and deterioration of protective and structural properties.

Thus, by forming a different structure in the process of electrodeposition and subsequent heat treatment metal coatings with improved functional characteristics for particular operating conditions can be obtained.

References

[1] P M Vyacheslavov, G A Volyanyuk Electrolytic Forming (1979) 198
[2] I D Gruev, N I Matveev, N G Sergeeva Electrochemical Coatings for Radioelectronic Equipment (1988) 304
[3] Yu D Hamburg Handbook of Electrodeposited Coatings for Machine Manufacturing (2006) 216

IOP Conf. Series: Materials Science and Engineering **127** (2016) 012041 doi:10.1088/1757-899X/127/1/012041

[4] I M Kovenskiy, V V Povetkin Electrolytic alloys (2003) 288
[5] I M Kovenskiy, A N Venedictov, I A Venedictova and other Aging of metal coating in the process of operation (2012) 101-107
[6] B S Bokstein, A B Yaroslavtsev Diffusion in Metals (20050 362
[7] I I Novicov Theory of Thermal Treatment of Metals (1986) 480
[8] S S Gorelik, S V Dobatkin, L M Kaputkina Recrystallization of metals and alloys (2005) 432

IOP Conf. Series: Materials Science and Engineering **127** (2016) 012042 doi:10.1088/1757-899X/127/1/012042

Modelling and Optimization of Technological Process for Magnetron Synthesis of Altin Nanocomposite Films on Cutting Tools

T D Kozhina

Rybinsk State Aviation Technical University, Rybinsk

E-mail: prorectnir@rgata.ru

Abstract. The paper highlights the results of the research on developing the mechanism to model the technological process for magnetron synthesis of nanocomposite films on cutting tools, which provides their specified physical and mechanical characteristics by controlling pulsed plasma parameters. The paper presents optimal conditions for AlTiN coating deposition on cutting tools according to the ion energy of sputtered atoms in order to provide their specified physical and mechanical characteristics.

Introduction

Modern nanostructured coatings enhance cutting tool wear resistance providing tool lifetime to be increased and metals to be machined at higher speeds. Improvement of technical characteristics (hardness, viscosity) of cutting tools with nanocomposite thin films results in sufficient increase in labor efficiency and reduction of costs of the products manufactured.

Application of modern protective thin films allows employing tools with nanocoatings and refurbishing tools up to 12 times.

The greatest potential for improving the performance characteristics of cutting tools, such as hardness HB, durability σ_v, heat endurance T_Q, abrasion resistance T_{abr}, oxidation resistance T_{ox}, consists in the right choice and calculation of properties of the chemical composition and the structure of the coating for specific operating conditions [1], as well as in improving technology to form nanostructured wear resistant coatings. Cutting tool performance characteristics are mainly determined by behavior of thin films deposited as they interact with the processed surface.

One of the methods to solve the problem to obtain nanocomposite coatings with specified physical-mechanical characteristics, such as indentation hardness H, elasticity modulus E, degree of coating adhesion to substrate HF etc., is improvement of physical and technical processes of their synthesis. Therefore, the leaders in the production of metal-cutting tools such as *Sandvik Coromant, Balzers Aerlicon, Iscar, Mitsubishi, Dormer, Walter*, and others, are actively developing in this direction.

Currently, the most promising methods of coating deposition are the vacuum ion-plasma technologies. Among them one can single out the method of magnetron sputtering, as the drift current of electrons in crossed electric and magnetic fields used in it allows providing uniformity of the coatings applied on large areas of sputtering, as well as forming fine- and superfine-dispersed structures.

One of the main problems of applying thin-film coatings on metal cutting tools by the method of magnetron sputtering is the inability to predict accurately their composition and structure, because there are many factors affecting the properties of the resulting coating [2], [3]. One of the most important parameters of technological process of nanocomposite coating synthesis is the ion energy of sputtering atoms.

Ion energy and ion flow of sputtering atoms have been studied by many Russian and foreign scientists [4] - [8]. However, the majority of works are devoted to the study of influence of the current-voltage characteristics on the magnitude of ion energy and ion flux distribution. Until now there have not been studies considering influence of such pulse parameters as pulse frequency and pause time. Monitoring the amount of ion energy and distribution of ion fluxes of sputtering atoms of the growing thin films by varying the frequency characteristics can be used for simulation and improvement of structure and properties of films with specified physical and mechanical characteristics of the coating and for intensification of magnetron sputtering process. In this connection, it is necessary to determine exactly how the pulse frequency and the pause time affect the physical and mechanical characteristics of high quality coatings obtained.

Analysis of field of application and obtaining of nanocomposite coatings for metal-cutting tools

The most promising method for deposition of thin films is the method of dual unbalanced magnetron ion-plasma sputtering with a closed type field, since it allows synthesis of coatings of almost any composition with high productivity while ensuring a durable adhesion against the substrate, and also allows increasing degree of plasma ionization in the area of coating synthesis on the product. Methods of nanocomposite coating deposition on cutting tools are presented in Fig.1. [3].

As the research showed, currently, the magnetron-sputtering system using unbalanced dual magnetrons with a closed type field operating in a pulsed mode is the perfect system [4].

In this system, cathodes are alternately operated, increasing the density of the ion current near the target. It allows to obtain a high degree of plasma ionization, which has a positive effect on the properties of nanocomposite thin films applied to cutting tools, improves the performance of industrial plants by increasing the size of the "effective" area of the coating deposition, it is possible to synthesize thin films on products of large sizes and complex configurations.

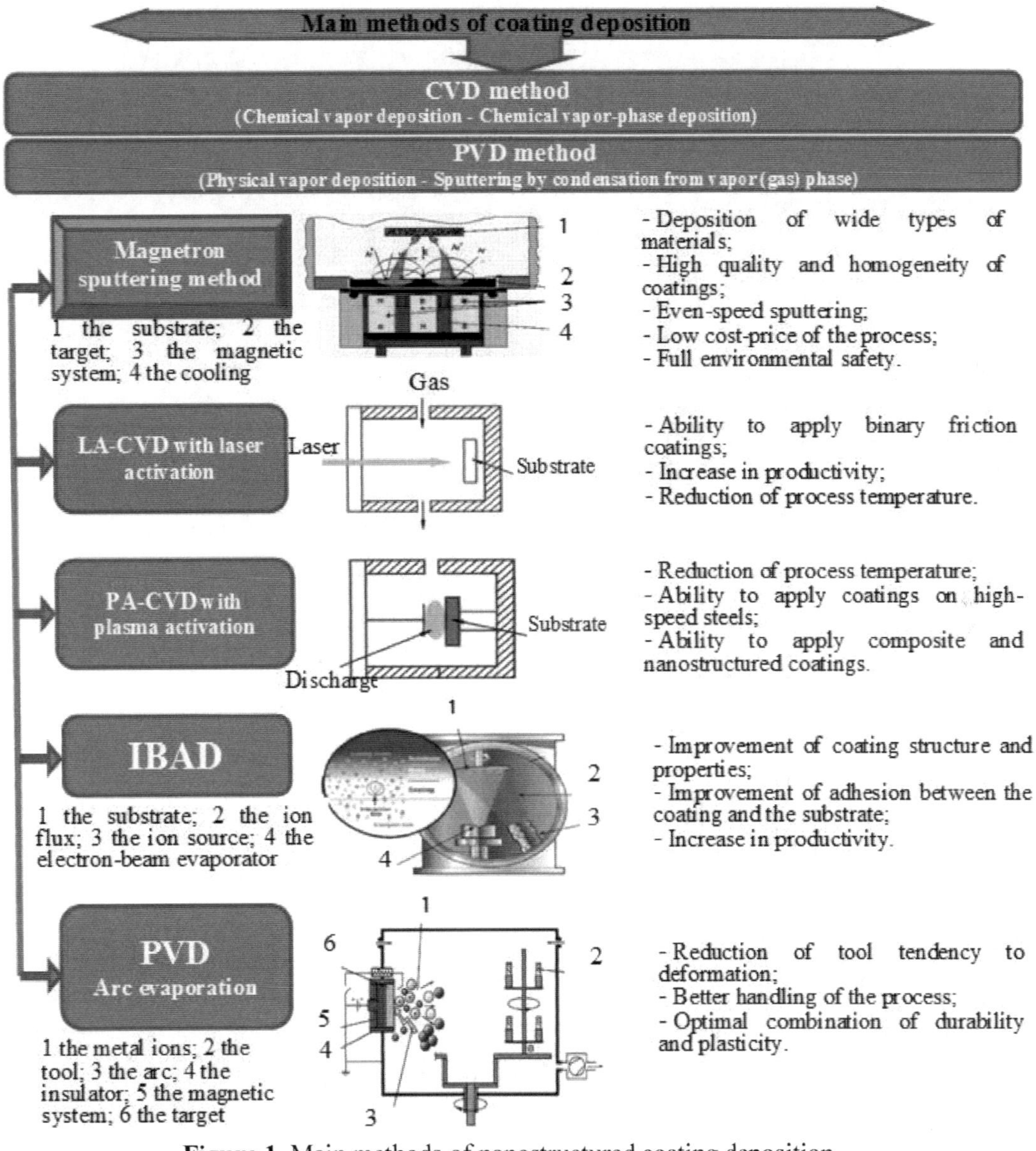

Figure 1. Main methods of nanostructured coating deposition

The results of the analysis of physico-mechanical properties of the coatings show that the mechanical, thermal and tribological properties of hard thin films can be significantly optimized through a well-designed microstructure, driven by the energy of ion bombardment, by varying the pulse plasma parameters [5].

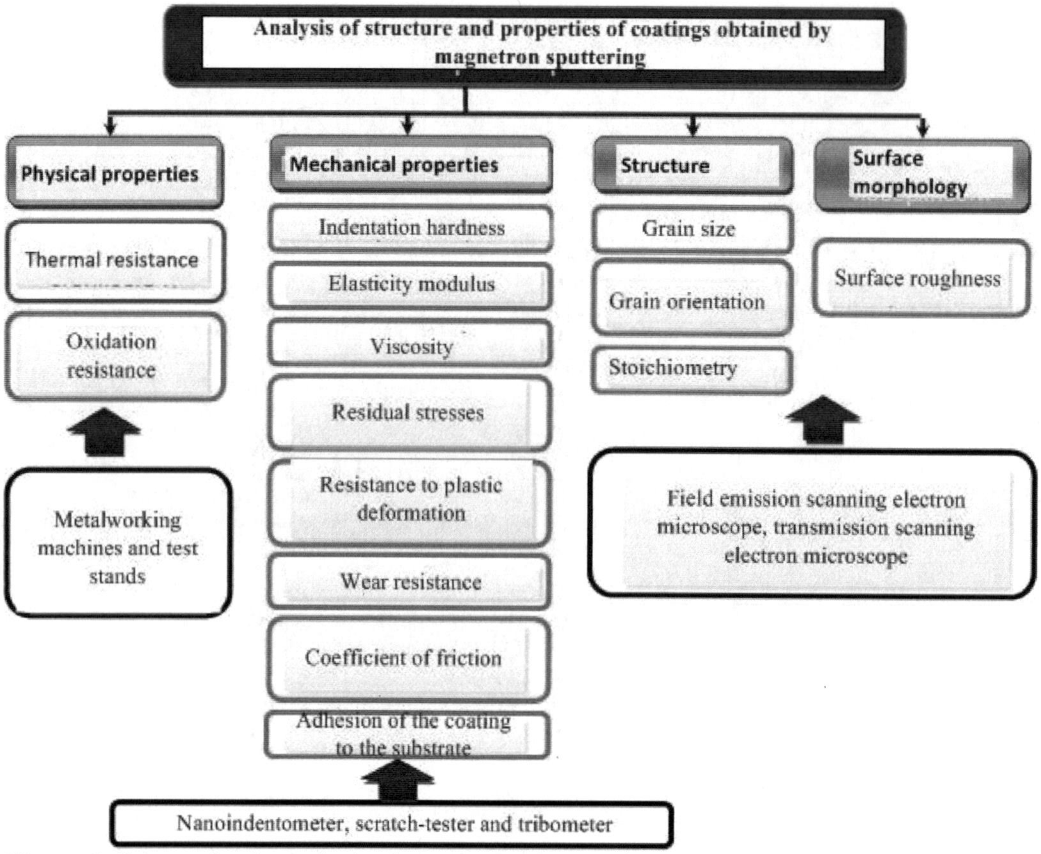

Figure 2. Scheme for the analysis of structure and properties of coatings obtained by magnetron sputtering

Optimization of technological process of AlTiN coating deposition

As a result of research the analysis of the technological process of AlTiN coating deposition was carried out according to the scheme presented in the table format, where P_b is the base pressure, Pa; P_W is the working pressure, Pa; U_d is the discharge voltage on the magnetrons, B; I_d is the discharge current on the magnetron, A; *Bias* is the offset voltage, B/ is the bias current, A.

Table 1 Technological process of *AlTiN* nanocomposite coating deposition

Seq. №	**Operation**	**Time min.**	**Values of assigned parameters**
1	Loading items into the chamber	10	Surface preparation for coating deposition
2	Pumping the vacuum chamber to base pressure	30-60	P_b=0.001Pa
3	Ion cleaning	10	P_b=0.12Pa; U_d =650/650B; I_d =1/0.5A; *Bias*: 800 B/40 A
4	Deposition of the metal sub-layer	30-50	P_W=0.12 Pa; U_d =650/650 B; I_d =6/0.5 A; *Bias*: 40 B/40 A
5	Deposition of reactive layer	50-75	P_W=0.1-0.15Pa; *QN2*=5-15%; U_d =600-800B; I_d =1-6A; *Bias*: 40-50B/40-50A
6	Cooling-down	15	P_w=0.12 Pa
7	Items unloading	10	-

IOP Conf. Series: Materials Science and Engineering **127** (2016) 012042 doi:10.1088/1757-899X/127/1/012042

Based on the model offered by Davis and Vanderslice a mathematical model to measure the ion energy distribution was developed, taking into account the length of free path of an electron in crossed electric and magnetic fields in relation to the process of ionization and frequency of electron-atom collisions:

$$\frac{d_{j_i}}{dU}=\left(\gamma\cdot I_0\cdot\mu_\perp\cdot\varepsilon_0\right)^{\frac{1}{2}}\cdot\left[\frac{\omega}{(\lambda\cdot\upsilon)}\right]^{\frac{3}{2}}\cdot\left[\frac{U\cdot\left(\gamma^{-\frac{1}{2}}-1\right)}{U_k}+1\right], \tag{1}$$

where j_i is the ion flux, A/m^2; I_0 is the discharge current; γ is the Townsend coefficient; $\mu_\perp$ is electron mobility in the direction perpendicular to the magnetic field, m^2/B·sec; ε_0 is the dielectric constant; ω is the cyclotron frequency, Hz; λ is the average track length of the electron, m; υ is the frequency of electron-atom collisions, Hz; U is the electric field strength, B; U_k is the cathode potential drop, B.

In order to account for possible ionization in the magnetron sputtering process made the following assumptions and limitations are taken:

- the mean free path of electrons is considered to be constant;
- edge effects are not considered;
- given that, the degree of ionization of low-temperature plasma of gas discharge is small (of order 10^{-5}), one neglects the collisions of electrons with any particles, besides atoms of the working gas.
- the x-axis is perpendicular to the cathode and directed to the anode, and the y-axis is parallel to the cathode and directed along the magnetic field line;
- within the cathode layer the field is homogeneous;
- acceptable value of ion energy is j (U)≤Uk.

To determine the energy spectrum of the ions, bombarding the cathode unbalanced magnetron, the value of cathode potential drop was determined according to the formula:

$$U_K=1.2\cdot10^{-29}\cdot\left[\frac{\gamma\cdot I_0 T\cdot}{P\cdot B}\right]^{\frac{1}{2}}\cdot\sigma^{-\frac{3}{2}}\cdot\left(\gamma^{-\frac{1}{2}}-1\right), \tag{2}$$

where T is the gas temperature; P is the gas pressure in the chamber; B is the magnetic flux density; σ_i is the cross section of ionization in the magnetron discharge, $\sigma_i=2\cdot10^{-19}$ m^{-2}.

Having solved differential equation (1) for the ion flux density under the initial condition j(U_k)=0 we get the expression describing the energy distribution of the ions bombarding the cathode:

$$j(U)=\frac{\omega\cdot(U-U_k)\cdot\sqrt{\frac{\omega}{\lambda\cdot\upsilon}}\cdot\sqrt{\mu\cdot I_0\cdot\varepsilon_0}\cdot\left(U+U_k-\sqrt{\gamma}\cdot U+\sqrt{\gamma}\cdot U_k\right)}{2\cdot U_k\cdot\lambda\cdot\upsilon}. \tag{3}$$

The results of mathematical simulation proved that the largest contribution to the sputtering process make not high-energy ions, since their density in the flow is small, but the ions with average energy values. Increasing discharge voltage results in increase in the value of the cathode potential drop. As a result, ions are accelerated with greater speed and, consequently, their kinetic energy increases, which is transferred to the target atoms, which has a positive effect on sputtering rates.

Influence of pulse parameters on the properties and structure of AlTiN coating

Performing a research on the influence of pulse parameters on properties and structure of AlTiN coatings the values of the pause time (time of return) and the values of the frequency of the pulses

IOP Conf. Series: Materials Science and Engineering **127** (2016) 012042 doi:10.1088/1757-899X/127/1/012042

varied. Based on mathematical simulation, the following values of process parameters were chosen: pulse frequency 100KHz and 350KHz (the minimum allowed and maximum possible) the pause time of 1.0 μs to 5.0μs.

The results of the analysis of two and three-dimensional surface morphologies of AlTiN films deposited under different asynchronous pulsing conditions, were measured using atomic force microscopy (AFM) and scanning electron microscopy (SEM) (Figure 3). Change of ion energy and ion flux in the plasma by pulsation shows that the increase of ion energy and ion flux leads to a change in AlTiN film texture from (200) to (111). In the study, the films deposited at relatively low ion energy bombardment (e.g., a mode 100/1,0 with total ionic energy 72eV) support orientation (200) to minimize the surface energy. With increase of ion energy and ion flux in the plasma increased strain energy in the film and re-sputtering due to the high ion bombardment cause the growth of the coating in the orientation (111) to reduce the strain energy (for example, 350kHz/1,4μs with a total ion energy of 200eV).

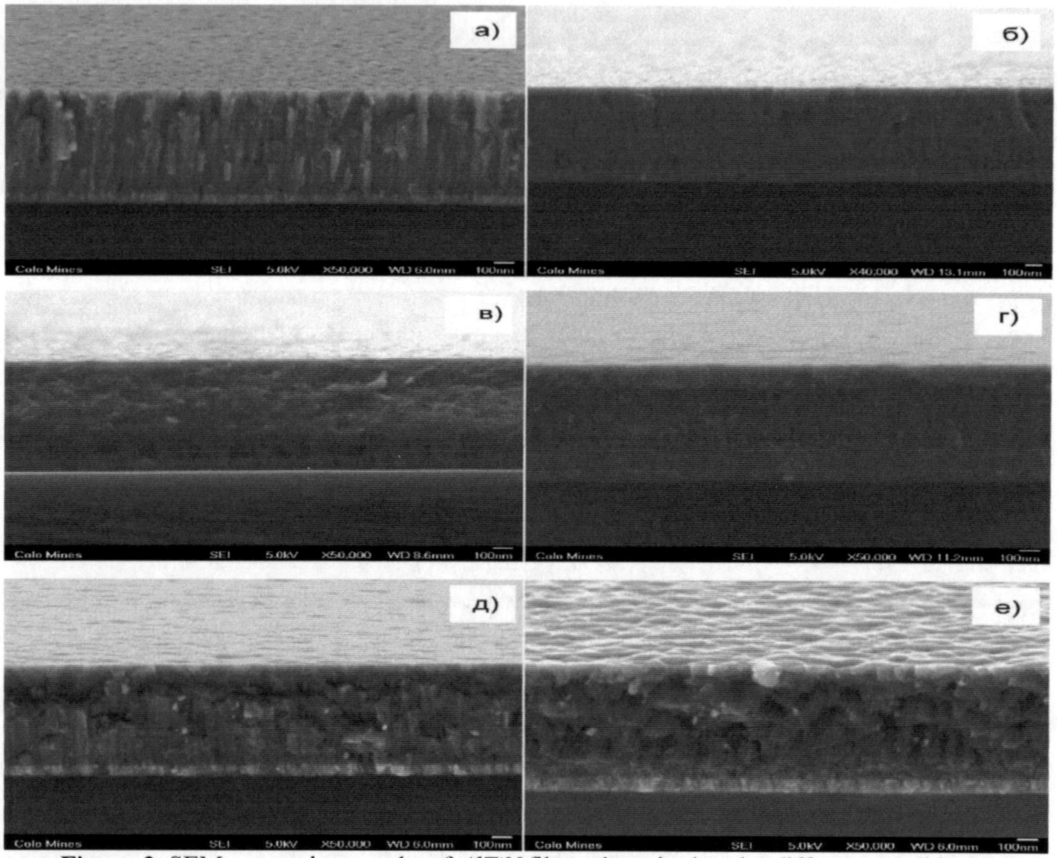

Figure 3. SEM cross micrographs of *AlTiN* films, deposited under different conditions of asynchronous pulsing: а) 100KHz and 1.0μs (72eV); б) 100KHz and 2.5μs (84eV); в) 100KHz and 5.0μs (122eV); г) 350KHz and 0.4μs (177eV); д) 350KHz and 1.0μs (180eV); е) 350KHz and 1.4μs (200eV)

Pulsed ion energy and ion flux have significant impact on the micro-structural development of AlTiN film. The use of controlled ion bombardment to maintain the maximum ion energy at the level less than 120eV and increase the ion flow in the central field of ion energy "B", will increase adatom mobility, reinforce coating density and reduce the size of the film columnar crystals.

IOP Conf. Series: Materials Science and Engineering **127** (2016) 012042 doi:10.1088/1757-899X/127/1/012042

Mechanical and tribological properties of *AlTiN* films obtained under different conditions of deposition are presented in Table 2.

Hardness of the *AlTiN* film has increased from 34 to 41 HPa under growth of the total ionic energy from 72eV (100KHz/1.0μs) to 122eV (100KHz /5.0μs) in the asynchronous pulse mode. This is due to increasing density of the coating and reduction of grain size under conditions of increased ion energy and ion flux bombardment. Strain hardening became significant and excessive, with a total ion energy 177-200eV and the pulse frequency 350KHz.

Table 2 Mechanical and tribological properties of *AlTiN* films

Film properties	Pulse parameters					
	100/1.0	**100/2.5**	**100/5.0**	**350/0.4**	**350/1.0**	**350/1.4**
Al/(Ti+Al)	58.5	57.3	60.4	60.9	67.5	69.7
Film thickness, mm	1.2	0.8	1	14.1	0.9	1.2
Surface roughness, nm	5.45	2.43	1.04	2.16	2.01	2.25
Residual stresses, HPa	-3.4	-4	-5.6	-7.3	-12.2	-12.8
Nanoidentation hardness, HPa	34.3 ±3.8	37 ±3.9	41.1 ±3.3	42.4 ±2.9	43.6 ±2.8	48.0 ±3.6
The Young's Modulus, HPa	370.8 ±26.3	385.6 ±31.9	411.4 ±32.5	432.4 ±30.4	450.5 ±28.4	515.5 ± 43.7
H/E	0.092	0.096	0.01	0.098	0.097	0.093
Friction coefficient	0.38	0.43	0.46	0.46	0.87	0.63
Wear rate	2.9	3.2	3.4	3.9	3.8	4

As a result, the optimal combination of mechanical (hardness) and tribological properties can be achieved only when saving the average ion energy between 70-120eV with high ion fluxes. The work of *Ti* and *Al* targets at 100kHz frequency and pause time of 5.0μs resulted in formation of *AlTiN* films, which showed high hardness (~ 41HPA, viscosity of 0.10) and good wear resistance (coefficient of friction 0.40-0.46; rate of wear $3.4x10^{-6}$ mm^3H^{-1} m^{-1}).

Conclusion

According to the data obtained concerning the influence of the frequency characteristics of the pulse on the properties of the *AlTiN* coating a model was prepared for technological process of magnetron sputtering, providing specified physical and mechanical properties of nanocomposite coatings of metal-cutting tools. Cutting tools coated with AlTiN nanocomposite coating obtained using the developed technology of magnetron sputtering were tested, as well as it was compared with analogues (Fig.4).

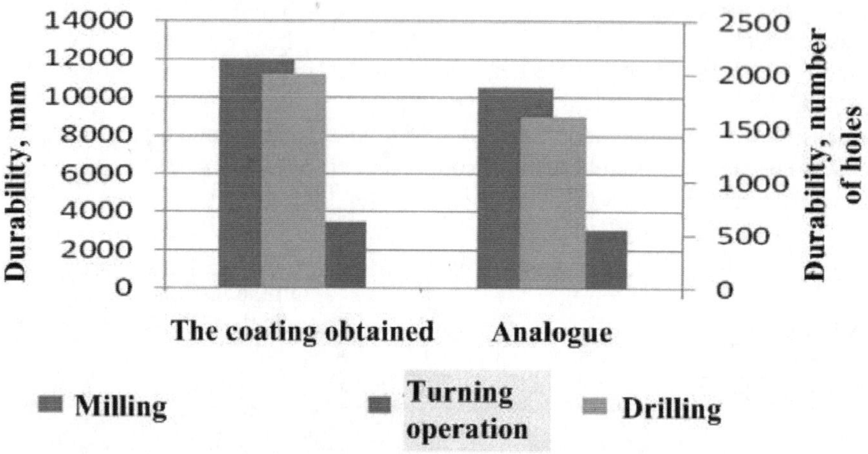

Figure 4. Tests of *AlTiN* coating in comparison with analogues

Based on the research recommendations for improving efficiency of the process of magnetron sputtering were formulated:

- Discharge voltage on the target should be chosen in accordance with the percentage of asymmetric pulse to maintain the reverse positive pulse voltage in the range of 20-80V. It is preferable to use a longer time of return at relatively low pulse frequencies (e.g., 100-200KHz) for reactive pulsed sputtering, for the purpose of obtaining appropriate values of the pulsed ion energy and increasing of ion flux.
- Since pulsed plasma has already provided relatively high values of ion energy in the plasma, it is proposed to use relatively small displacement of the substrate (not more than -50 B) to avoid excessive ion bombardment.

This publication has been produced with the financial support of the Ministry of Education and Science of the Russian Federation, project "Methodological bases of development and creation of functional nanostructures of GTE part, tool and tooling surfaces and coatings" in the framework of the basic part of the state task.

References

[1] T D Kozhina Estimation of physical and mechanical parameters to machine GTE parts based on the analysis of temperature fields in the cutting area Bulletin **3** (2014).
[2] T D Kozhina, V V Nepomiluev, A E Sergeev, Asynchronous dual pulse magnetron-sputtering systems for synthesis of nanostructured coatings **2** (2011) 115-120
[3] A V Kurochkin, M O Mezentsev Study of serviceability of AlSiTiN multi-layer wear resistant coating Reference book Engineering magazine **4** (2009) 62-64
[4] A E Sergeev, V Yu Eroshkov Mechanical and tribological properties of Al-Ti-N films **1** (2011) 87-91
[5] A V Kurochkin, P A Solovyov Technological features of multi-functional nanostructured coatings for cutting tools **3** (2010) 166-172
[6] Alan Richter Aluminum-titanium-nitride tool coatings continue to evolve Cutting tool engineering **1** (2003)
[7] Alan Richter Recipe for enhancement Cutting tool engineering **1** 2005
[8] J John, Brajendra Mishraa, J A Reesc Surface Coatings Technology **202** (2008) 1418 –1436

IOP Conf. Series: Materials Science and Engineering **127** (2016) 012043 doi:10.1088/1757-899X/127/1/012043

Matrix System for Provision of Performance Characteristics of Gte Heavy-Duty parts at the Design Stage of Technological Processes

T D Kozhina[1], V Yu Eroshkov[2]

Solovyov Rybinsk State Aviation Technical University

E-mail: prorectnir@rgata.ru

Abstract. The paper highlights the methods to assign technological conditions for machining providing the required performance characteristics due to control of quality parameters of a surface layer and machining accuracy. Assignment of technological conditions for machining is performed at the design stage of technological conditions in accordance with the methods developed based on the matrix system of analysis and computation of groups of technological and operational factors.

Introduction

Reliability, operational life and efficiency of gas turbine engines are largely determined by the quality of manufacture of their parts. High accuracy along with low part stiffness, strict requirements to the surface quality and physico-mechanical properties of materials, use of heat-resistant hard-to-cut steels and alloys determines application of progressive methods to obtain work parts and their subsequent machining.

Discs are the most typical GTE parts possessing constructive-technological characteristics according to part complexity. Thus, critical compressor and turbine discs of aircraft engines operate with higher loads under high engine speeds. Necessity to reduce engine weight results in complication of the parts form, forcing to make them hollow. Compressor discs are the most complex to produce in terms of machining, as external faces of these parts feature various combinations of cylindrical surfaces, flanges, gearwheels, blade grooves and other surfaces with low stiffness.

Quality of implementation of GTE compressor disk machining technology is the result of interaction of a wide range of factors to a greater or lesser extent influencing performance characteristics of these parts [1], [2]. Inter alia, even non-mating faces of GTE compressor discs are typically thoroughly machined, various methods of hardening technologies are widely used, as well as those of coating deposition and synthesis and chemical treatment. Discs undergo thermal or thermo-chemical treatment, which complicates technological processes of their manufacture sufficiently, enlarges the cycle of their production due to the necessity to introduce additional procedures and to expand the processing route. To provide the required machining accuracy when designing technological processes in aircraft engine building special attention is paid to the choice of technological bases. With the same purpose surfaces of GTE discs are machined in several steps. When choosing machining methods, especially for finishing operations, possibility of work hardening and residual stresses has to be taken into consideration6 as they noticeably influence performance characteristics of these parts.

Besides, manufacture of aircraft engines excels also through special carefulness of control. All critical parts and components go through the hundred-per-cent control with application of special automated means. However, some hard-to-reach places of such parts can only be controlled by indirect methods at the expense of direct control of the tools and the machining process.

Production engineers have to consider all above-mentioned factors when designing technological processes and computing part cutting modes.

1 Assignment of technological conditions for machining based on computation of groups of technological and operational factors

In practice process engineers need to analyze both the external production factors concerning designer's requirements and product prime cost and the internal ones such as technological infrastructure level and equipment capabilities.

Primarily the technical requirements included in the product drawings are analyzed. Thus, the includes requirements to surface roughness R_a, limit endurance σ_v, quality of seating face machining (polishing), methods of work-piece material control etc. These requirements are fundamental when assigning technological conditions of machining.

Furthermore, the technological conditions of machining have to be assigned which not only would provide specified performance characteristics, but also technical-and-economic indexes or the maximal dimensional tool durability, or the required machining accuracy, and when it is determined by the technological process developed then also the given quality parameters of the surface layer or any other variant of the above-mentioned ones.

Taking systematically into consideration all stages of influence of external and internal factors on the surface layer of the part manufactured affecting the machining process allows forecasting technological heredity of machining heredity and manufacturing end products with required performance characteristics without application of destructive testing methods and finishing processes.

As the recommendation based on the conducted computations the process engineer can offer additional technical requirements providing performance characteristics, which can be introduced by the designer into the drawing.

The obtained results allow to introduce corrections into the existing technological processes changing the structure and the content of the machining itself of heavy-duty parts of gas turbine engines.

Let us consider implementation of these propositions in practice.

As a result of series of studies, the method to assign machining modes was developed based on the matrix analysis of the group of technological factors and performance indexes, as well as the multi-level algorithm based on the matrix to optimize technological conditions of machining providing the required performance characteristics. These tools can be applied for various tasks of process engineers and requirements of designers, which they impose during development part construction and technological process of its manufacture depending on the external and internal environmental factors, i.e. on assignment, operating conditions and technological infrastructure [3], [4].

There is not always the necessity to carry out full-scale computations in practice, i.e. to provide performance indexes of parts through surface layer quality parameters control and to fulfill the technical and economical criteria: maximum efficiency, minimal prime cost or both together. In order to simplify computations the matrix was developed for possible variants of computations based on the optimization multilevel algorithm allowing choosing the required machining mode computations in terms of requirements of a developer or a customer.

Examples of the matrixes for possible variants of computations of technological conditions of machining are presented in Tables in Figure 1. In this case, the following conventional signs are used:

C_{min} are the technological conditions of machining, providing minimal prime cost of the process procedure;

P_{max} are the technological conditions of machining, providing maximal efficiency;

C, P are the technological conditions of machining, providing the optimal combination of prime cost and efficiency in the course of the process procedure;

σ_{-1}, y, I, V are the technological conditions of machining, providing the required limit endurance, convergence value of contacting faces, wear intensity, interface gap volume, respectively, i.e. the performance indexes;

T_p are the technological conditions of machining, providing the specified tool life;

Δ_{wear} are the technological conditions of machining, providing the required tool wear (especially important at tool change);

T_p, Δ_{wear} are the technological conditions of machining, providing simultaneously the specified tool life and the specified wear flat.

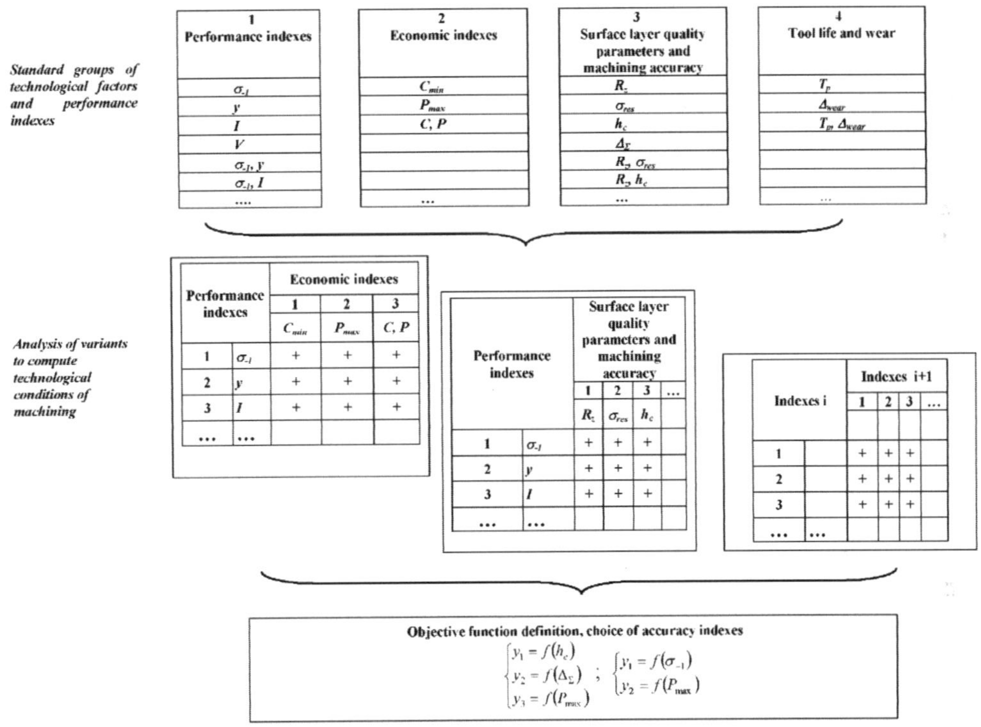

Figure 1. Matrix of groups of technological factors and performance characteristics.

In the rows and columns of the matrix tables (see Figure 2) the possible combinations of various indexes and parameters are presented, which can be needed while machining. At the intersection of rows and columns of the tables by means of conventional signs reasonability of provision of this or that combination of indexes is shown. The sign «+» marks those index combinations which provision is reasonable and can be fulfilled using the optimization multilevel algorithm. Absence of this sign gives evidence that this index combination is not reasonable to be provided because there are direct dependences between some of them.

For instance, development of technological conditions of machining providing the required value of residual stresses and work hardening depth in combination with fulfillment of condition as to the minimal prime cost of machining is reasonable and can be fulfilled by means of the above said algorithm.

On the base of the matrix analysis the conclusion can be drawn, that it is inexpedient to develop the machining modes providing the required values of residual stresses and tool life along with the given values of limit endurance, convergence of contacting faces, wear intensity and interface gap, as the value of residual stresses accordingly to the developed algorithm and the computation method, as well as the tool life depend on the initially specified performance index, and their value is the result of the intermediate in the general computation change, i.e.

$$\sigma_{-1} \Rightarrow f(\sigma_{\text{res}}, \Delta_\Sigma) \Rightarrow f(S, V, t) \quad (1)$$

If we analyze Figure 1, we can detect the following combinations of indexes and parameters for which provision development of technological conditions of machining applying the optimization multilevel algorithm is reasonable and can be fulfilled:

1. Limit endurance (σ_{-1}), minimal prime cost of machining (C_{min});
2. Limit endurance (σ_{-1}), maximal efficiency of machining (P_{max});
3. Limit endurance (σ_{-1}), optimal combination between the prime cost and the efficiency of machining (C, P);
4. Value of convergence of contacting faces (y), wear intensity (I), value of aggregate error of machining (Δ_Σ), minimal prime cost of machining (C_{min});
5. Limit endurance (σ_{-1}), value of convergence of contacting faces (y), interface gap volume (V), value of aggregate error of machining (Δ_Σ), maximal efficiency of machining и (P_{max});
6. Limit endurance (σ_{-1}), value of convergence of contacting faces (y), interface gap volume (V), value of aggregate error of machining (Δ_Σ), optimal combination between the prime cost and the efficiency of machining (C, P);
7. Limit endurance (σ_{-1}), wear intensity (I), interface gap volume (V), value of aggregate error of machining (Δ_Σ), minimal prime cost of machining (C_{min});
8. Surface roughness (R_z), value of residual stresses (σ_{res}), tool life (T_p);
9. Value of convergence of contacting faces (y), interface gap volume (V), value of residual stresses (σ_{res}), value of aggregate error of machining (Δ_Σ), tool wear flat (Δ_{wear}), optimal combination between the prime cost and the efficiency of machining (C, P) etc.

For example, it is necessary to define the technological conditions of machining providing the specified value of aggregate error of machining and depth of work hardening in terms of maximal efficiency or the specified limit endurance and maximal machining efficiency. This is done as follows:

1) before starting the command to the control program is given to choose the objective function, in this case:

$$\begin{cases} y_1 = f(h_c) \\ y_2 = f(\Delta_\Sigma) \\ y_3 = f(P_{\max}) \end{cases} \quad \text{or} \quad \begin{cases} y_1 = f(\sigma_{-1}) \\ y_2 = f(P_{\max}) \end{cases}, \quad (2)$$

2) the computation is conducted in accordance with the method presented above, where the result is assignment of feed, speed and cutting depth (taking into account machining allowance), and, if it is necessary, recommendations are given as to change the initial tool geometry (φ, φ_1, r, α, γ, …),

3) computation of technological conditions can be carried out both in the batch and the dialogue modes in dependence with customer's requirements.

Taking systematically into consideration all stages of influence of external and internal factors on the surface layer of the part manufactured affecting the machining process allows forecasting technological heredity of machining heredity and manufacturing end products with required performance characteristics without application of destructive testing methods and finishing processes. The obtained results allow to introduce corrections into the existing technological processes changing the structure and the content of the machining itself of heavy-duty parts of gas turbine engines.

2 Industrial implementation of the method to define technological conditions of machining and results of experimental research of parts and samples

Let us consider the technological process of GTE compressor disc manufacture as an example of practical application of research results under conditions of industrial enterprises, as well as structure of its change in the context of application of automated system to control technological conditions of machining (Figure 2, Table 1).

Computations show that for various turning operations (to turn faces 1, 4, 5,..., 12, 13, 14, ...) practically one and the same cutting mode is assigned (S=0.2mm/rev, n=48rpm), independently from various cutting length and allowance value Δ. If we use the reference data at that, as can be seen from the table, it does not practically change anything in the process of the turning operation, moreover even in comparison with the factory technology the reference material reveals excessively overrated cutting speeds without taking into consideration the roughness height value of the surface machined specified in the drawing.

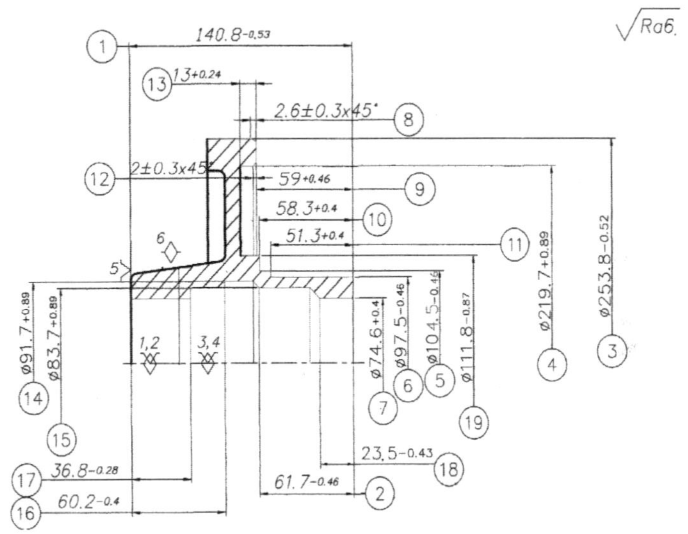

Figure 2. Outline of GTE compressor disc work-piece machining in turning operation.

This is confirmed by factory testing of the compressor disc of one of commercially available GTE products machined in accordance with the existing technology: phosphor glow was detected in the form of longitudinal cracks on the disc body (Figure 3, a).

Table 1 Fragment of technological process of turning operations of 2 stage GTE compressor disc (disc material: VT3-1)

I Existing industrial technology	II Reference data
Tool material – VC8 $\varphi = 45°$, $\varphi_1 = 15°$, $\left.\begin{matrix}\alpha = 5^0 \\ \gamma = 5^0\end{matrix}\right\}$ grinds the worker $r = 0.5$ mm, $l_{mach} = 50$ mm, $t = 1.5$ mm, $S = 0.2$ mm/rev. Quality parameters: $R_a = 2.5$ μm, $\Delta = ^{0.00}_{-0.26}$ mm.	Tool material – VC8 $\varphi = 45°$, $\varphi_1 = 15°$, $\alpha = 5°$, $\gamma = 5°$, $r = 0.5$ mm, $l_{mach} = 50$ mm, $t = 1.5$ mm, $S = 0.2$ mm/rev. Surface machining ←: $V = \frac{125}{T^{0.35} \cdot 0.2^{0.4} \cdot 1.5^{0.2}} = \frac{125}{T^{0.35} \cdot 0.525 \cdot 1.08} = \frac{220.5}{T^{0.35}}$

	n	*S*	*i*	
1) To turn face 1	48	0.2	2	Cutting mode: at $T = 60$ min, $V = 52.6$ m/min = 0.876 m/sec, $n = 200$ rpm, $S = 0.2$ mm/rev.
2) To turn faces 4.5 and R 3.6	48	0.2	2	
3) To turn face 11	48	0.2	2	
4) To turn faces 12, 13, 14 and R 18	48	0.2	2	
5) To turn flat face 10	48	manual		

It must be noted that the crack start its formation from the surface and spreads to the depth, i.e. fatigue comes from the surface. At that compressive residual stresses develop which values varies from –200 to –480 MPa at depth of the work hardening layer more than 40 μm, with degree of work hardening 13-29% and loss of material ductility.

Inspection of mechanical properties carried out on the samples manufactured from the rim and the hub parts gives evidence that after the long-term operation material microstructure on the product (operating time: 4805 cycles) is satisfactory, and the cracks are located on the grain bode, which is typical for fatigue (Figure 3, b). The total length of fatigue cracks is 6-10 mm, their depth is up to 0,7 mm.

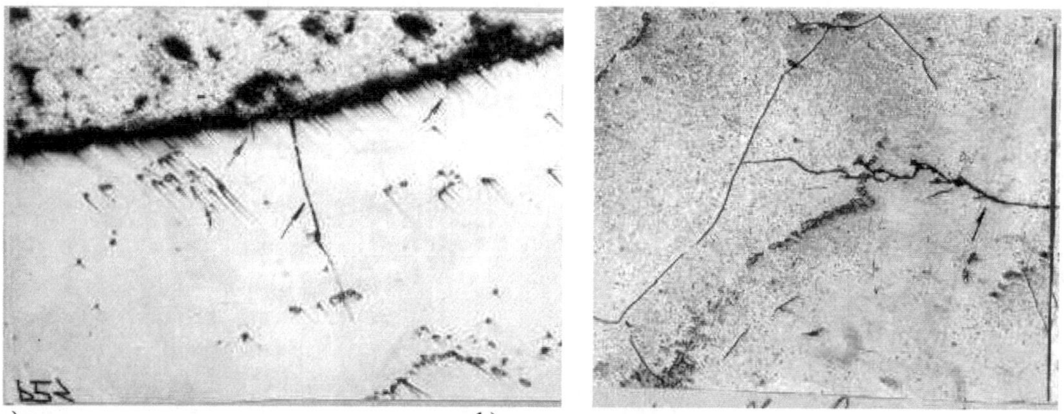

a) b)

Figure 3. View of crack formation in the compressor disc body:
a) photograph of disc microstructure (magnification ×200),
b) view of the crack in the disc micro section (magnification ×200).

Fatigue fractures formed in the discs in the process of testing because the factory mode, presented in Table 6, corresponds to work hardening depth 86 μm, if it is computed according to the method suggested here, and it is proportionately with depth of crack spreading (0.4-0.7 mm). Results of factory testing give evidence that none finishing method will remove the fatigue fracture spreading in depth of 0.7 mm, but will only harden the surface of the part machined, not solving the durability problem at that.

If we make use of the developed computation program for technological conditions of machining (Table 2), not changing the structure of technological process, we can obtain the cutting mode corresponding to the standard value as to the value of cutting speed, and as to the value of roughness height of the surface machined with the considerably lesser one ($R_Z = 9.8$ μm) in comparison with the one specified in the drawing, with value of work hardening layer $h_c = 86$ μm and cutting inaccuracy $\delta_\Sigma = -67.75$ μm [5], [6].

Table 2 Computation according to the developed method and program (tool material VC8 ($\varphi = 45°$, $\varphi_1 = 15°$, $\alpha = 5°$, $\gamma = 5°$, $r = 1$mm, ρ=30 µm, $\delta_{wear} = 100$)

№ of face	R_{out}	t_{cut}	l	Required character-ristics		Design characteristics				Quality parameters			
				R_z	δ_{diam}	S	V	n	t	R_z	δ	h_c	σ_{res}
6	157	2	49,6	10	-20…0	0.247	0.326	39.7	2.77	10	-19.61	180	-270
				10	-10…0	0.247	0.323	39.2	1.78	10	-10	160	-360

Therefore, when machining in accordance with the factory technology the reserves of the process itself are not taken into consideration; and when assigning the cutting mode according to the reference data cutting accuracy is completely excluded.

Testing results of the compressor disc machined in the new modes and inspection of micro-sections showed absence of apparent defects (there are no fatigue fractures spreading in the considerable length) (Figure 4), i.e. they confirmed adequacy of the method developed and accuracy of computations.

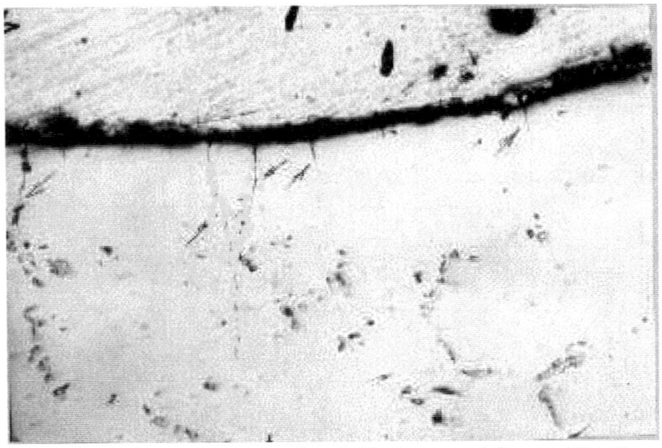

Figure 4. Disc microstructure after LUM-control (magnification×200)

Conclusion

Having regard to the above, it can be concluded that in the existing technology of machining of a standard GTE compressor disc the cutting modes are assigned inconsistently and certain elements of output parameters of machining are considered only indirectly, not to speak of the tool geometry selection procedure. Both the external factors connected with disc operation and the internal ones connected with its machining are not considered. In accordance to the analysis of technological processes and analysis of disc factory tests it should be considered that the modes necessary to machine critical GTE parts, in particular discs have to be assigned strictly controlled subject to the groups of technological and operational factors in order to avoid the defects described above.

In order to solve the above problem the developed mathematical tools can be employed allowing assigning technological conditions of machining. To increase efficiency and reliability of computations a multi-level optimization algorithm was applied which allows carrying out optimization search with usage of additional technological conditions of machining chosen by the process engineer, which increases sufficiently capabilities of the system developed. The distinctive feature of the methods created is the possibility of its application for various output data, various machining methods, various technological procedures and application in various automatic control systems due to the possibility of flexible change of computation algorithm elements.

Analysis of production technologies [5] - [8] implies that the approach suggested to assignment of technological conditions of machining makes possible to create technologies allowing to sufficiently reduce time for introduction of new products and to increase their service reliability and durability due to change in the existing approach to normalization of technological procedures: reduction of pass sequence in the operation; withdrawal of expensive finishing operations; forecasting of part performance properties.

References

[1] V F Bezyazychny, V N Krylov, V A Poletaev Automation of manufacturing technology of gas turbine aircraft engines M Machinostroyeniye (2005) 560 p

[2] T D Kozhina, V Y Eroshkov Engineering fundamentals of provision of required GTE parts' performance characteristics Science journal «Collected scientific papers SWorld» Modern problems and ways of their solution in science, transport, production and education (2013)

[3] T D Kozhina Estimation of physical and mechanical parameters to machine GTE parts based on the analysis of temperature fields in the cutting area Bulletin **3** (2014).

[4] T D Kozhina Logic design system for technological operations with application of computational methods to assign technological conditions of machining based on provision of required performance characteristics Canadian Journal of Science and Education **2** (2014)

[5] Yu S Eliseyev, V V Krymov, A G Boitsov Technology of manufacture of aircraft gas turbine engines M Machinostroyeniye (2003) 512 p

[6] V F Bezyazychny, V A Poletaev, T D Kozhina, V F Bezyazychny, V N Krylov Automation of aircraft gas turbine engine production process M Machinostroyeniye (2012) 560 p

[7] V A Poletaev Gas turbine engine blade automated manufacturing technology M Machinostroyeniye (2006) 256 p

[8] Yu S Eliseyev, V V Krymov, A A Mitrofanov, B P Saushkin Physicochemical machining methods in GTE production under the editorship M Drofa (2002) 655

IOP Conf. Series: Materials Science and Engineering **127** (2016) 012044 doi:10.1088/1757-899X/127/1/012044

Results of Investigative Tests of Gas Turbine Engine Compressor Blades Obtained by Electrochemical Machining

T D Kozhina[1], A V Kurochkin[2]

[1,2]Soloviev Rybinsk State Aviation Technical University

E-mail: prorectnir@rsatu.ru

Abstract. The paper highlights results of the investigative tests of GTE compressor Ti-alloy blades obtained by the method of electrochemical machining with oscillating tool-electrodes, carried out in order to define the optimal parameters of the ECM process providing attainment of specified blade quality parameters given in the design documentation, while providing maximal performance. The new technological methods suggested based on the results of the tests; in particular application of vibrating tool-electrodes and employment of locating elements made of high-strength materials, significantly extend the capabilities of this method.

Introduction

Electrochemical machining is a unique technological method that allows obtaining geometrically complex, shaped surfaces on the products made of materials hard-to-machine by traditional methods. This explains the increasing spreading of ECM as the primary method of profile sizing of compressor blade airfoils of modern gas turbine engines.

However, in addition to wide technological capabilities, the ECM differs from other machining methods through complexity of the main and auxiliary equipment, complexity and high cost of the accessories used (for blade locating and fastening), high energy intensity of processes, hazardous working conditions for personnel and negative impact on the environment.

ECM accuracy is determined by a combination of factors, the main of which are as follows: value of the interelectrode gap and accuracy of its adjustment, localizing capability of electrolyte, its type, concentration and pumping conditions; as well as manufacturing accuracy and correction of the tool-electrode; electrical machining modes; accuracy of the equipment used; electric potential uniformity.

Advanced method to improve ECM accuracy

One of the methods to improve machining accuracy is electrochemical machining with tool-electrode vibration, which is characterized by electrode harmonic oscillations and pulse current pile-up synchronized with them (oscillating electrodes).

For electrochemical machining of gas turbine engine blades, besides the kinematic diagram with electrode vibration, the pulse cyclic and continuous tool-electrode feed diagrams are applied (Table 1).

Table 1 Comparison of kinematic diagrams applied at electrochemical machining of GTE blades

Comparison criterion	**Kinematic diagram**		
	Pulse-cyclic	**Vibrating TE**	**Continuous feed**
Tolerance extremes of blade airfoil dimensions, mm	± 0.08	± 0.02	± 0.1
Performance rate, mm/min	0.4	0.2	0.5 – 0.8
Possibility to take into account irregularities of distribution of allowance and blade deformations in the course of machining	+	+	–

Table 1 demonstrates that the kinematic diagram with electrode vibration provides higher machining accuracy combined with lower performance. Therefore, the kinematic diagram with electrode vibration is the most applicable for blade manufacture with high accuracy and a wide twist angle (Figure 1). However, in order to reduce machining time it is necessary to minimize the value of the allowance removed.

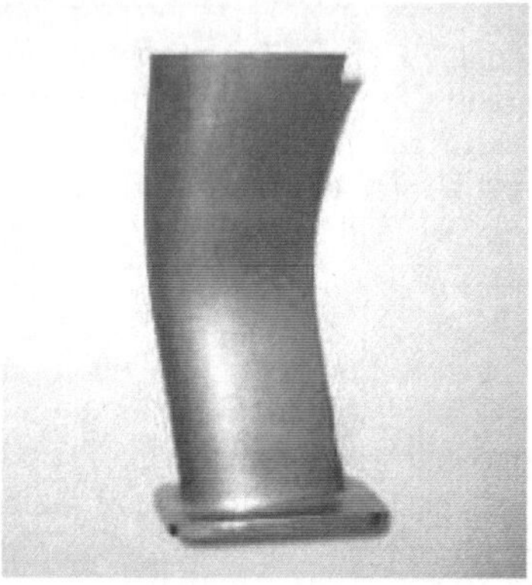

Figure 1. Exterior view of the blade manufactured with application of the kinematic diagram with electrode vibration

Results of compressor blade tests

Tests in electrochemical machining of blades with oscillating tool-electrodes were carried out in two stages. At the first stage, the optimal modes of electrochemical machining with oscillating tool-electrodes were defined. The second stage implied the study of geometrical parameters and roughness of the blade airfoil profile in dependence on the obtained values of machining modes.

Research in order to define ECM modes was conducted on the second stage compressor guide blade (Figure 1) made of TA6V titanium alloy (Table 3) in accordance with the operational definition drawing for machining of airfoil elements at the operation of electrochemical machining (Figure 2) with subsequent inspection of blade airfoil profile geometrical parameters according to Table 2.

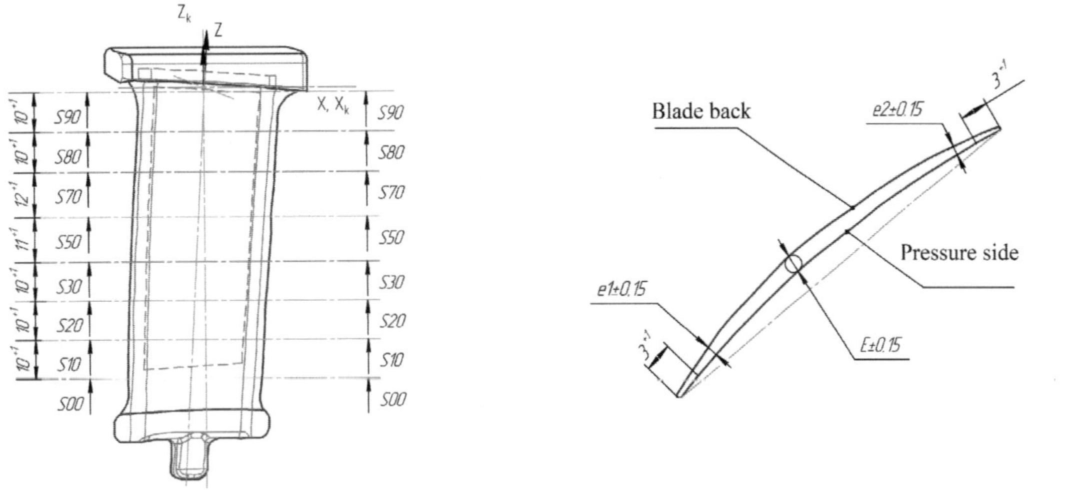

Figure 2. Operational definition drawing of the second stage compressor guide blade airfoil machining

Table 2 Geometrical parameters of airfoil profile of blade semi-products after ECM

Indicator name	**Section**	**00**	**10**	**20**	**30**	**50**	**70**	**80**	**90**
Maximal airfoil profile thickness, mm	Rated value	0.966	1.023	1.084	1.107	1.399	1.930	2.256	2.483
	Tolerance extremes	+0.15 –0.05	+0.15 –0.05	+0.15 –0.05	+0.15 –0.05	+0.15 –0.05	+0.15 –0.05	+0.15 –0.05	+0.15 –0.05
Leading blade edge thickness, mm	Rated value	0.717	0.752	0.786	0.813	0.960	1.268	1.462	1.603
	Tolerance extremes	+0.15 –0.05	+0.15 –0.05	+0.15 –0.05	+0.15 –0.05	+0.15 –0.05	+0.15 –0.05	+0.15 –0.05	+0.15 –0.05
Trailing blade edge thickness, mm	Rated value	0.455	0.446	0.453	0.442	0.539	0.588	0.580	0.637
	Tolerance extremes	+0.15 –0.05	+0.15 –0.05	+0.15 –0.05	+0.15 –0.05	+0.15 –0.05	+0.15 –0.05	+0.15 –0.05	+0.15 –0.05
Airfoil profile roughness, *Ra*, μm	Rated value	3.2							
	Tolerance extremes	± 1							

Table 3 Chemical composition of TA6V alloy

Element	**Sign**	**Percentage of total**
Titanium	Ti	88.37 – 90.62
Aluminum	Al	5.5 – 6.75
Vanadium	V	3.5 – 4.5
Carbon	C	0.08
Iron	Fe	0.3

Specifics of electrochemical etching of titanium alloys considerably complicate implementation of GTE blade airfoil profile machining operation. It is due to the oxide film growth on the part surface, right up to attainment of electric potential of a certain breakdown value, sufficient to break the oxide film and to implement the process of removal of material main allowance. Increase of ECM process efficiency for blades made of titanium alloys can be achieved by means of machine setting.

To define recommended machining modes the research was carried out within the ranges of ECM mode parameters specified in the professional literature [5, Tables 4.3, p. 183] (Table 4).

Table 4 The range of parameters ECM

Process parameters ECM	
Electrical	
Type of current	Direct current
Voltage	5 – 30 V (Constant or a pulse)
Current	50 – 40 000 A
Current density	10 – 500 A/cm^2
Electrolyte	
The type and concentration	
Most commonly used	NaC1 (60 – 240 g/l)
Frequently used	$NaNO_3$ (120 – 480 g/l)
Temperature	20 – 50 °C
Consumption electrolyte	1 l/min/100A
The speed of the electrolyte	1500 – 3000 m/min
The pressure at the inlet of the electrolyte	0.15 – 3 MPa
The outlet pressure of the electrolyte	0.1 – 0.3 MPa
Electrode	
Electrode gap	0.05 – 0.3 mm
Feed rate electrode	0.1 – 20 мм/мин
ECM	
Accuracy of machining	
two-dimensional surfaces three-dimensional surfaces	0.05 – 0.2 mm (0.02 – 0.05 during pulse ECM) 0.1 mm (0.05 during pulse ECM)
Surface roughness	Ra = 0.1 – 2.5 µm
Processing performance	1200 – 2500 mm^3 per 1000 Amps

The first stage of tests implied experimental works in definition of the change range of the most critical parameters of the process and the degree of their impact on the part characteristics, such as electric parameters (voltage, pulse duration, pulse trailing edge phase, form); electrolyte pressure; and vibration cycle.

The aim of this stage was achievement of the maximal performance of the process while providing its stability, reiteration of results, at that the blade quality indexes (maximal airfoil profile thickness, edge thickness, roughness value etc.) had to be as close to the requirements of design documentation as possible. Analysis of design of equipment and tools was carried out also in order to identify possibilities of their further improvement.

The second stage implied ECM modes on the plant with oscillating electrodes based on the group of investigative tests subject to the above-mentioned boundary conditions and capabilities of the equipment used.

As varying parameters of ECM modes, the following ones are used:

1. Vibration cycle (*T*).
2. Pulse voltage (*U*).

3. Pulse duration (t_i).
4. Pulse trailing edge phase (α).

As a result of implementation of the first stage of research, a control program was developed for the experimental plant allowing carrying out Ti-alloy blade electrochemical machining in automatic mode.

The aim of the second stage of the work was research of blade geometrical parameters and specification of the process parameters which values were obtained at the first stage.

Based on the analysis of the data obtained for electrochemical machining of blades made of titanium alloy at the experimental plant with oscillating electrodes the following process parameters were obtained (Table 5).

Table 5 Process parameters of electrochemical machining with electrode oscillation

Name of the parameter	Measuring unit	Value
Parameters of the main allowance removal		
Vibration cycle	microsecond	27 – 33
Electrolyte pressure	atm	1.5
Rinse cycle time	second	10
Number of rinse cycles between sensing	–	1
Rinse time	second	0
Rinse gap in relation to the machining gap	µm	0
Difference of allowances to include aligning	mm	0.1
Machining gap	µm	70
Initial feed rate	mm/min	0.1
Pulse voltage	V	19
Pulse form	–	Bcos
Pulse duration	ms	10
Phase of the pulse trailing edge	degree	240
Parameters of allowance aligning		
Machining gap	µm	70
Initial feed rate	mm/min	0
Pulse voltage	V	15
Pulse form	–	Bcos
Pulse duration	ms	4
Phase of the pulse trailing edge	degree	120

Technological methods to increase ECM quality

GTE compressor blade test results showed that the topical issue hindering improvement of GTE compressor blade ECM process is poor accuracy and, above all, fragility of the equipment used.

As a rule in the process of machining, machine-retaining devices applied during ECM of blade airfoil are located inside the machining cell (plant) and are fully immersed in the electrolyte. In addition to the mechanical impact on the elements of the device, physics of the ECM process causes combined corrosion and electrochemical effects on the device. To the greatest extent the locating elements are damaged immediately contacting the blade work-piece under machining. On the locating elements, centers and prisms, holes and caverns are formed (Figure 3) resulting in rapid loss of machining accuracy, need for replacement and downtime of expensive equipment.

Figure 3. Damage of locating elements in the process of ECM

It is obvious that provision of the lifetime for locating elements of blades in this case is possible only at the expense of selection of the used construction material and, to some extent, by improving design and technology of manufacture of locating elements.

The problem of material selection for manufacture of blade locating elements consists in the need to provide simultaneously chemical resistance, absence of deformations and mechanical damages of locating elements. In addition, much attention is paid to reducing the cost of manufacture and operation of devices, directly affecting the prime cost of products (blades and the engine in general).

Now for the manufacture of locating elements mainly niobium is used having high chemical resistance. The disadvantages of niobium are low mechanical properties (particularly hardness), some difficulties in machining and extremely high cost of the material.

Thus, it is necessary to find the material to replace niobium with comparable chemical stability, high mechanical properties, acceptable machinability indexes and lower cost.

It should be noted that as electrolytes in electrochemical machining the aqueous solutions of neutral salts are used possessing high corrosion activity in relation to vulnerable parts of technological equipment and tooling devices, including locating elements.

The equipment and devices used for locating and fixing the blades can be of different design, however, regardless of the layout of devices there are three main locating elements in their composition being in direct contact with the machined blade: the basic center, the retaining center and the prism.

The locating elements are exposed to the greatest damage during the blade ECM process due to the electrochemical, corrosion and mechanical impact. As practice shows, the most common damages and defects of the device locating elements for blade ECM are the following ones:

- linear wear and crushing of working surfaces,
- etching of working surfaces with formation of holes,
- cracking and material spalling of the locating elements.

As a rule, cracking and spalling set a limit to the possibility to mount the work-piece in the device requiring immediate replacement of locating elements. Crushing and etching of locating reference points results in displacement of the work-piece in relation to the reference target position, in locating error. For the etching is inhomogeneous, as a rule, it results in a twist of the blade relative to the prism (Figure 4), or in a turn relative to the center line (Figure 5). In a number of cases, geometrical summation of these errors can be observed.

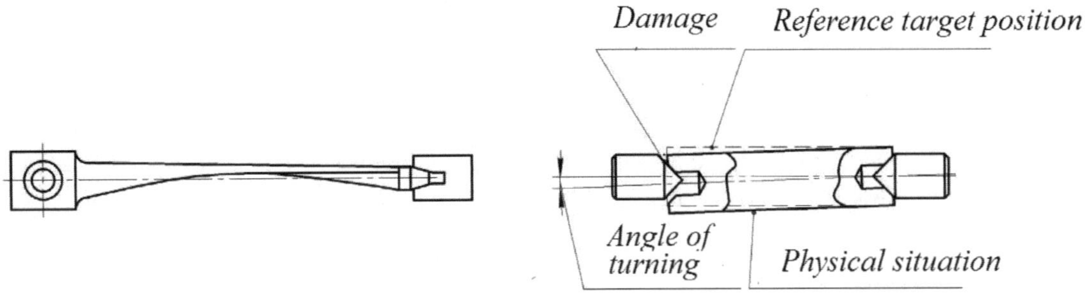

Figure 4. Displacement (turn) of the blade relative to the prism as a result of the center damage

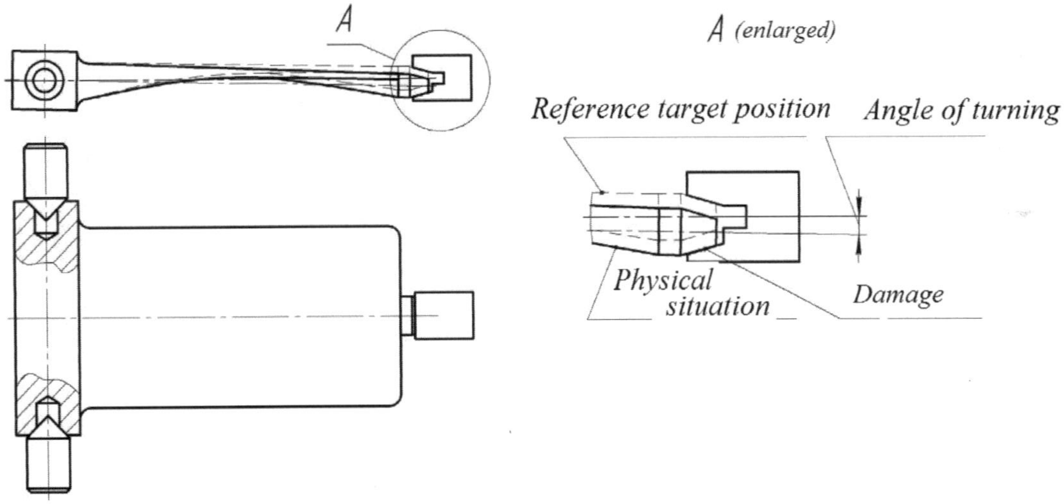

Figure 5. Displacement (turn) of the blade relative to the center line as a result of the prism damage

At present in modern enterprises of aircraft engine building branch to produce locating elements niobium BH-3 GOST 16100-79 [2] is applied, which resistance to anodic attack is provided at the expense of formation of the passive oxide film (Nb_2O_5) under conditions of impact of a corrosive medium (concentrated aqueous solutions of neutral salts). However, this material is rather plastic, which has negative impact both on its serviceability (crushing), and on cutting ability (build-up forming on the tool face, "blunting" of abrasion wheels. Hardness of niobium is 75 HB, which is considerably lower than hardness of the material applied to GTE blade manufacture (270-360 HB). Low hardness of locating elements results in a number of cases in their damage (crushing) by fixing forces, which leads to the need of their periodic diagnostics, replacement and set-up of equipment. Low hardness of this material and its high price lead to rise in cost of ECM process implementation. In addition, it should be noted, that application of special equipment to produce attachment elements of niobium is not needed. Existing multipurpose equipment suffices. The technological process of production of locating elements includes operations of turning and grinding. It is obvious that niobium does not completely meet specifications for materials of locating elements formulated above. However, due to its properties, in particular susceptibility of a protective oxide film, niobium came into widespread acceptance at manufacture of devices for ECM in the absence of alternatives.

The analysis of literature sources [3-6] allowed singling out three main trends in search of optimal materials to produce locating elements in ECM operations:

1. Materials of high hardness (ceramics, leucosapphires).

Hot-pressed nitride ceramics chosen as potential material for manufacture of locating elements is a mixture of aluminum oxide Al_2O_3 and titanium nitride TiN. The main advantages of this material are:

- low electrical conductivity which results in higher resistance to chemical and electrochemical dissolution;
- high hardness (>400 HB), which considerably exceeds hardness of work-pieces of the GTE blades manufactured, which determines absence of deformations and low wear of elements in the process of device operation;
- sufficient workability (ceramics allow obtaining locating elements by hot-pressing original powder blends with subsequent grinding of work-pieces until obtaining drawing dimensions).

Leucosapphire presents a sort of ceramics possessing, among other things, maximal hardness characteristics at shock loading. The main disadvantage of this material is low coefficient of cutting ability. The main method to obtain parts of sapphire is multistep grinding distinguishing by low performance. Reduction of labor intensiveness is possible due to extension of tolerances for locating elements by a factor of 1.2-1.5. However, this will result in need to adjust grommets and housings for locating elements immediately as to the element manufactured. Cost of manufacture of work-pieces and that of subsequent machining of leucosapphires, as well as necessity of the following in-place adjustment of other device elements makes its application unreasonable.

2. Plastic and its derivatives (engineering plastics).

Engineering plastics are polymeric materials used for manufacture of machine parts exposed to impact of aggressive media, in particular, concentrated salt solutions (fixture elements, base members, slides, bearing sleeves etc.). The main features of engineering plastics are chemical resistance and inertness, hydrophobic behavior (determining ease of cleaning from electrolyte), high technological effectiveness, low cost of material. To manufacture locating element prototypes plastic Victrex 450 GL 30 was chosen. Preliminary tests demonstrated intolerable value of crushing of locating elements made of plastics, which can be explained trough its low hardness.

3. Obtaining hard oxide films on locating elements made of metallic materials (titanium alloys).

Locating elements made directly of titanium alloys show certain resistance to etching, however a considerably lower one than that needed for blade batch manufacture.

Indisputable advantage of this material is availability, including possibility to use titanium wastage formed when producing aircraft engines, as well as high mechanical properties.

To protect surfaces of locating elements efforts were undertaken to obtain intentionally oxide films on titanium alloys BT3, BT9, BT20, and BT22. Practical data showed the fact of formation of the most stable and damage-resistant oxide films on titanium alloy BT3.

Based on results of the analysis, comparative tests to determine the selective material for manufacture of blade locating elements in operations of ECM were carried out for three materials:

- niobium (applied at present);
- nitride ceramics (Al_2O_3 + TiN);
- oxidized titanium BT3.

The tests aimed at inspection of serviceability of the materials chosen as locating elements and elimination of materials with low resistance. During tests, the blades were manufactured until occurrence of limiting state (non-serviceability) of locating elements. Tests were conducted for three sets of locating elements, after that the determined indexes of resistance were averaged.

The highest indexes of resistance were achieved for the elements made of oxidized titanium alloy BT3 (aged during 12 hours in temperature 700 – 800°C). Thus, the optimal material for manufacture of blade locating elements in the operation of airfoil ECM is oxidized titanium alloy BT3.

Conclusion

As a result of the investigative tests on the electrochemical machining of blades with oscillating tool-electrodes the features of this process were considered, the optimum process parameters of electrochemical machining of blades made of titanium alloys, ensuring the achievement of quality parameters of the blade specified in the design documentation, while providing maximum performance.

The technological methods under consideration, in particular use of vibrating tool-electrodes and use of locating elements made of high-strength materials can significantly extend the capabilities of this method.

References

[1] Yu S Eliseyev, V V Krymov, A A Mitrofanov, B P Saushkin Physicochemical machining methods in GTE production under the editorship M Drofa (2002) 655
[2] Lifshits, A. L., Kravets A. T., Rogachev I. S. et al. Metal electro pulse machining under the editorship of A. L. Lifshits. – SPB: Let Me Print, 2013 – 156 p
[3] V F Orlov, B I Chugunov Electrochemical shaping M Machinostroyeniye (1990) 240
[4] Yu S Eliseyev, V V Krymov, A G Boitsov Technology of manufacture of aircraft gas turbine engines M Machinostroyeniye (2003) 512
[5] V F Bezyazychny, V A Poletaev, T D Kozhina, V F Bezyazychny, V N Krylov Automation of aircraft gas turbine engine production process M Machinostroyeniye (2012) 560
[6] V A Poletaev Gas turbine engine blade automated manufacturing technology M Machinostroyeniye (2006) 256 p
[7] V G Smirnov Design of technology of final electrochemical machining of gas turbine engine blades in terms of technological heredity (2007)
[8] A R Khamsina Perfecting of technology of GTE part electrochemical machining to improve quality of surface layer and resistance to high-temperature gas corrosion (2010)
[9] E V Smolentsev, Design of electrical and combined machining methods M Machinostroyeniye (2005) 511
[10] Modern electrochemical technologies collected works on the proceedings of All-Russian conference SSTU (2002) 248
[11] T D Kozhina, A V Kurochkin Features of research trials in electrochemical machining with oscillating electrodes **1** (2015) 89-93
[12] T D Kozhina, A V Kurochkin Selection of materials for manufacture of locating elements for GTE compressor blades ECM procedure in order to maintain their manufacture rohrematik cycle **4** (2014) 75-79

IOP Conf. Series: Materials Science and Engineering **127** (2016) 012045 doi:10.1088/1757-899X/127/1/012045

Changes in the Structure and Properties of Welded Joints of Low-Alloy Steels, Subjected to Cyclic Loads

V N Kuskov[1], I M Kovenskiy[2], K V Kuskov[3]

[1,2,3]Tyumen State Oil and Gas University Russia

Abstract. Time-varying loads negatively affect the properties and structure of materials. Structural failures typically occur at loads below the yield point. In this work, fatigue tests of welded joints of low-alloy steels were carried out in an asymmetric cycle at loads of 60and 80% of the yield strength. The stress ratio was 0.8-0.9. On the basis of the results of the tests, equations linking the number of cycles to failure with test parameters were obtained. Such equations can be used for estimating the residual life of elements both under construction and in operation. It has been found that the failure is not instantaneous. Specimens of steels continue to resist variable loads for4000 - 26000 cycles to failure, depending on steel grade and the parameters of the test. Under operating conditions, it gives an opportunity to discover the onset of failure and dispose of the defective part or to replace the entire structure. A standard technique was used to measure the microhardness on the fractured specimens. The distance between the nearest indentations was 0.2 mm. The results of the measurements were plotted in graphs of ahardness change characteristic for all steels under study. A microhardness "step" has been discovered in areas with high dislocation density, as evidenced by x-ray diffraction and transmission electron microscopy. An intermediate stage of the investigation is the development of recommendations for determining the moment of failure of welded constructions with a probability of 95%.

Introduction

In the operation of equipment and engineering products, it has been found that failure of a large number of objects occurs at stresses below the yield point. Repeated exposure of the material to cyclic stresses causes a change in its structure. Microscopic discontinuities appear which subsequently increase in size and become the site of crack initiation and propagation. This is facilitated by repeated alternation of areas with variable stresses. Cyclic loads affect the majority of operating objects: bridge constructions, road and rail transport, metalworking machinery and tools, presses, aircrafts, lifting mechanisms, pipelines, etc. Welded engineering constructions have designed-in structural differences between deposited metal, base metal and heat affected zones. Furthermore, residual stresses form in welded joints and there is a danger of appearance of hardened areas that may lead to cracking. Therefore, investigation of the fatigue of welded structures and their components assumes particular importance [1] .

By now, a great number of works have been published on the fatigue of welded joints. Much has been written about the influence of the welding process, types of the connection, residual stresses, and grades of the deposited and base materials, operating conditions, shapes and sizes of structures. However, studies investigating the effect of structural features on the fatigue characteristics of welded joints are not numerous. One such work was the work of the authors [2], which studied the peculiarities of fatigue failure of pipe steels. This paper is a continuation of this study.

1 Materials and technique of research

Table 1 and 2 shows the chemical composition and mechanical properties of the materials studied.

Table 2: Chemical composition of the steels

Grade of steel	C	Si	Mn	Ni	Cu	Cr	S	P
09G2S	0.087	0.75	1.71	0.07	0.1	0.06	0.025	0.03
17G1S-U	0.18	0.5	1.0	0.15	0.05	0.03	0.022	0.023
16G2AF	0.162	0.41	1.43	0.01	0.07	0.06	0.019	0.018

Table 2: Mechanical properties of the investigated steels.

Strength class of steel	Ultimate strength $\sigma_в$, MPa	Yield strength $\sigma_{0,2}$, MPa	Specific elongation δ, %
09G2S	490	415	29.8
17G1S-U	590	491	20.6
16G2AF	600	450	20.9

Fatigue tests were performed at a ZD-20 test machine. Type IV specimens were fabricated according to GOST 25.502. A variable tensile load from the maximum stress σ_{max} to the minimum stress σ_{min} acted along the axis of the specimen (perpendicular to the weld seam) The frequency of loading cycles was 10 Hz. 7 specimens were used in each series.

Microhardness was measured at a PMT-3 microhardness-testing machine, using standard techniques (load weight 0.981 H). The X-ray diffraction analysis was performed at a DRON-7 diffractometer using cobalt radiation. Electron microscopic studies of the structure were carried out at a TESLA-100 transmission electron microscope using the method of thin foils.

To monitor the quality of the weld, the methods of visual measurement and radiographic inspection were used. To determine the onset of crack initiation as well as its position and depth, the metal magnetic memory method (IKN-8 unit), the ferroprobe («Polus») and electropotential (EPD-8 crack detector) methods were used.

2 Results and discussion

On the basis of the results of fatigue tests [3-5], using the program STATISTICA 6.1 (multiple correlation coefficient 0.95, level of significance 0.01) we calculated regression equations between the number of cycles to failure of specimen N and test parameters : the maximum tensile stress σmax and the stress range Aσ.

For welded specimens of *09G2S* steel

$$N = 1.5768 \cdot 106 + 743.344 \bullet \sigma max - 82834.5135 \bullet A\sigma - 3.0898 \bullet \sigma 2max + 58.6958 + + A\sigma \bullet \sigma max + 1031.6689 \bullet A\sigma 2;$$

For welded specimens *17G1S-U* steel

$$N = 5.1787 \cdot 105 + 251.4986 \bullet \sigma max + 151.026 \bullet A\sigma - 0.6914 \bullet \sigma 2max - - 2.1285 \bullet A\sigma \bullet \sigma max - 164.4628 \bullet A\sigma 2;$$

For welded specimens *16G2AF*steel

$$N = 1.5428 \cdot 106 - 3370.1296 \bullet \sigma max - 21406.875 \bullet A\sigma + 7.1868 \bullet \sigma 2max - - 77.2361 \bullet A\sigma \bullet \sigma max + 526.6146 \bullet A\sigma 2.$$

Figure 1 shows an example of a geometric interpretation of resulting equations, which indicates a complex correlation of the test parameters.

The equations allow us to estimate the residual life of a welded construction on the basis of specified operating parameters, knowing the number of loading cycles at the specified moment of operation [6].

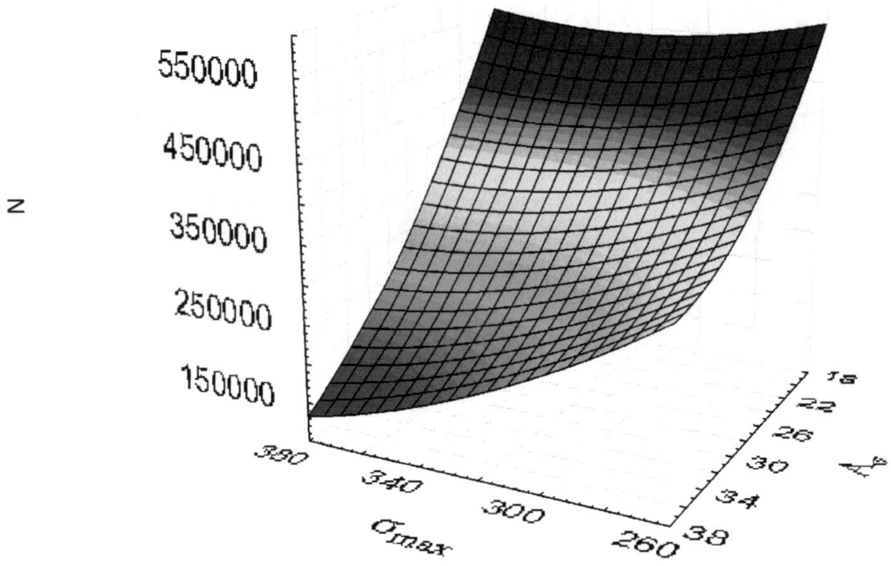

Figure 1. Graphical interpretation of the equation of correlation between the number of cycles to failure of welded 16G2AF steel specimens N, maximum tensile stress σmax and stress range Aσ.

It has been found that structural failure does not occur immediately after the detection of the crack with the help of the above-mentioned devices. Specimens of*09G2S* steel continue to resist variable loads for 6000 - 10000 cycles to failure, whereas 17G1S-U steel can resist for only 4000 - 6000 cycles depending on the parameters of the test. For 16G2AF steel, the number of cycles from the appearance of a crack to the failure of the specimen increases to 18000 - 26000. Under service conditions, it gives an opportunity to detect the onset of failure and dispose of the defective part or to replace the entire structure [7].

After carrying out the fatigue tests, microhardness was measured on the side surface of the fractured specimens (Fig. 2) from the center of the weld to the grips; the distance between the nearest indentations was 0.2 mm.

IOP Conf. Series: Materials Science and Engineering **127** (2016) 012045 doi:10.1088/1757-899X/127/1/012045

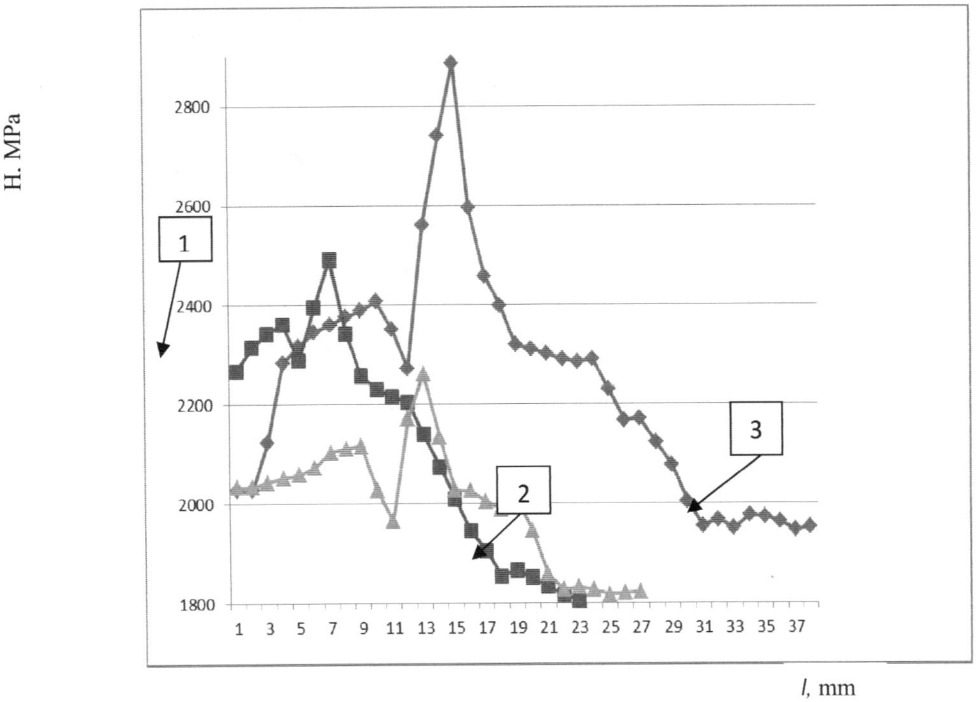

Figure 2. Microhardness of different zones of welded joints
1 -*09G2S* steel; 2 – 17G1S-U steel; 3 – 16G2AF steel

The results of the measurements have shown that the character of variation in the microhardness of all investigated steels is similar. The hardness gradually increases from the center of the deposited metal to the fusion zone and then, in the overheated zone [8]. Farther away, the microhardness increases again, reaching a maximum in the normalization area, and then, goes down to form an intermediate "step" before stabilizing.

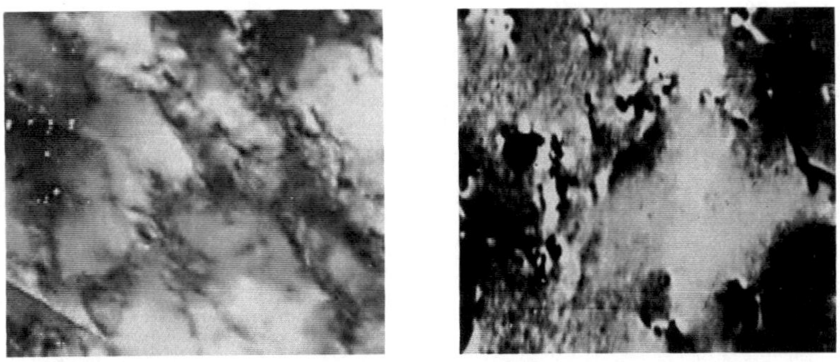

a b

Figure 3. Changes in the density of dislocations (x40000)
a - in the area where the "step" has formed; b - in the base metal near the grips

As evidenced by X-ray analysis, the microhardness "step" is caused by an increase of dislocation density in the zone of its formation. Direct observations of the structure of steels with a transmission electron microscope confirm that the main cause of the increased microhardness is accumulation of dislocations (Figure 3).

Conclusion

1. On the basis of the experimental data on cyclic fatigue life of welded joints of low-alloy 09G2S, 17G1S-U and 16G2AF steels under constant sign loading with cycle asymmetry 0.8 - 0.9, quadratic equations have been obtained, which describe the correlation between the number of cycles to failure of welds in the investigated steels, the maximum tensile stress and the stress range. These equations allow estimating the residual life of a welded construction with regard to designated test parameters (or operating conditions).

2. An increase inmicrohardness to 2070 - 2350 MPa near the heat-affected zone and the presence of a "step", corresponding to the region of fatigue damage accumulation have been found. Transmission electron microscopy has shown that the increase in hardness is caused by an increase in dislocation density.

3. The results of the work, allow creating a technique for determining the residual service life of welded structures under variable loads. This will broaden the base of the fatigue characteristics of materials and allow estimating the feasibility of using this or that steel under specified operating conditions.

References

[1] Zerbst, U., Ainsworth, R.A., Beier, H. Review on fracture and crack propagation in weldments - A fracture mechanics perspective. Engineering Fracture Mechanics. 132 (2014), p. 200-276.

[2] V. N. Kuskov, I. M. Kovenskiy, K. V. Kuskov. The peculiarities of fatigue failure for pipe steels of different strength classes Energy Production and Management 21. (2014). p. 889–894.

[3] K. Arash, H. Reza. Energy dissipation caused by fatigue crack in beam-like cracked structures. Journal of sound and vibration. 363 (2016) p. 247-257.

[4] Kanchanomai, C., Limtrakarn, W. Effect of Residual Stress on Fatigue Failure of Carbonitrided Low-Carbon Steel. Journal of Materials Engineering and Performance. 17 (2008), pp. 879-887.

[5] Zhao, X., Liu, Y., Liu, Y., Gao, Y. Research on fatigue behavior and residual stress of large-scale cruciform welding joint with groove. Materials and Design 63 (2014), p. 593-599.

[6] Kwofie, S., Chandler, H.D. Fatigue life prediction under conditions where cyclic creep-fatigue interaction occurs. International Journal of Fatigue. 29 (2008) p. 2117-2124.

[7] Bhat, S., Patibandla, R. Computational investigation of glare with several cracks and delaminations under monotonic and cyclic loads of constant amplitude. Mechanics of Advanced Materials and Structures 21 (2014), p. 607-630.

[8] Malushin N.N., Valuev D.V., Valueva A.V., Serikbol A., Kinetic study of the effect of plasticity and its role in stress relaxation in the weld speed steel during the martensitic transformation, J. Applied Mechanics and Materials. 682 (2014) p. 58-63.

IOP Conf. Series: Materials Science and Engineering **127** (2016) 012046 doi:10.1088/1757-899X/127/1/012046

Alloying Elements Transition Into the Weld Metal When Using an Inventor Power Source

R A Mamadaliev[1], V N Kuskov[2], A A Popova[3], D V Valuev[4]

[1,2,3,] 625000, Tyumen, 38 Volodarskogo st.,Tyumen State Oil and Gas University
[4] 652050, Kemerovo region, Yurga, Leningradskaya str.26 Yurga Technological Institute branch of Tomsk Polytechnic University

E-mail: mamadaliev_it@mail.ru

Abstract. The temperature distribution over the surface of the welded 12Kh18N10T steel plates using the inventor power source ARC-200 has been calculated. In order to imitate multi-pass welding when conducting the thermal analysis the initial temperature was changed from 298K up to 798K in 100K increments. It has been determined that alloying elements transition into the weld metal depends on temperature. Using an inventor power source facilitates a uniform distribution of alloying elements along the length and height of the weld seam.

1. Introduction

Multi-pass manual arc welding is one of the well-known and common welding techniques used in the industry. Despite its widespread use, problems frequently occur due to a number of drawbacks which result in lack of penetration, possible deformations, slag inclusions, discontinuities, as well as an increased heat affected zone.

Steel 12Kh18N10T is widely used for manufacturing products utilized in aggressive media. However, it is worth mentioning that the main difficulties when welding this steel are its tendency to form crystallization cracks, increased electrode heat and possible loss of corrosion resistance by the weld seams. The base metal for the experiment was selected considering the construction materials applied in the power, oil and gas, instrument-making and other leading sectors of Tyumen region.

The research objective is to calculate thermal cycle curves at varying distances from the heat source and demonstrate temperature influence on alloying elements transition into the weld metal.

2. Materials and methods

Plates 150x50x26 mm in size (length x width x thickness) were butt welded with an OK 61.30 ESAB grade electrode using an inverter power source ARC-200 [1,2]. Welding was performed using the reversed polarity direct current with the amperage specified according to the electrode grade, the number of passes being 5.

In order to imitate multi-pass welding when conducting the thermal analysis on personal computer using «Tpole» software the initial temperatures were taken from 298K up to 798K in 100K increments. The *x*-axis is directed along the weld, the *y*-axis – perpendicular, and the *z*-axis – along the height of the plate.

At the first stage the input data are thermophysical parameters, plate sizes, welding variables. Calculations are performed for each plate separately. Then, according to the given conditions of welding, a computational diagram is determined: a moving (fast-moving) point source on the surface of a semi-infinite body, a moving (fast-moving) point source on the surface of a plane. Values X, Y, Z are given in centimeters, assuming Z = 0, i.e. temperature is calculated on the surface of the welded plates. Based on the obtained results thermal cycle curves were constructed. Figures 1-4 give examples for various initial temperatures.

Thermal cycle curves are constructed in the coordinate system where:

1) Y values are plotted on the axis of ordinates at constant X values, and on the axis of applicates – temperature values (Figures 1 and 4);

2) X values are plotted on the axis of ordinates at constant Y values, and on the axis of applicates – temperature values (Figures 2 and 3).

To determine the chemical composition of the weld metal, electrode rod and coating we used the X-ray fluorescence analysis (XRF) of a ground weld on a spectrometer X-MET 5000 with a software suit X-MET. The excitation source was an X-ray tube with a rhodium anode PW 1404/00 (voltage – 60 kV, amperage – 50 mA). The time it took to identify one element at a depth of 0.1 – 0.5 mm from the surface was approximately 1 minute.

3. Results and Discussion

Figure 1 shows thermal cycle curves at the initial temperature of 298K. The graph lines show heat distribution from the weld to the plate edges (along the *x*-axis). Along the *y*-axis they are symmetrical to the weld axis.

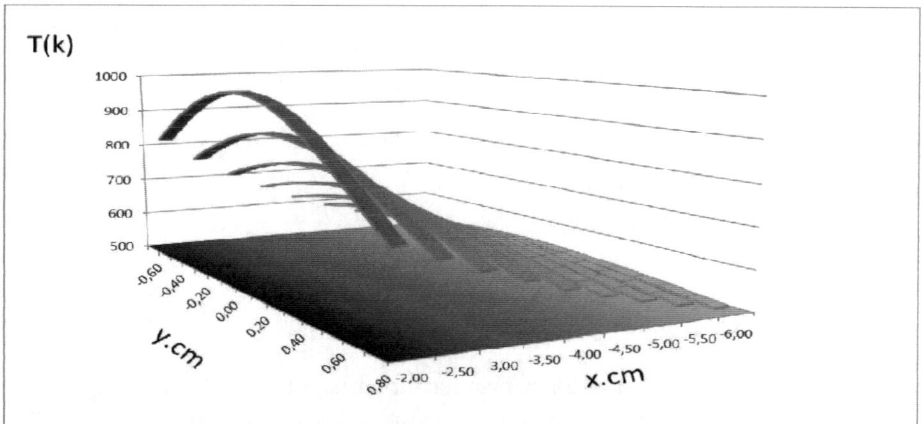

Figure 1. Thermal cycle curves at constant X values and the initial temperature of 298 K.

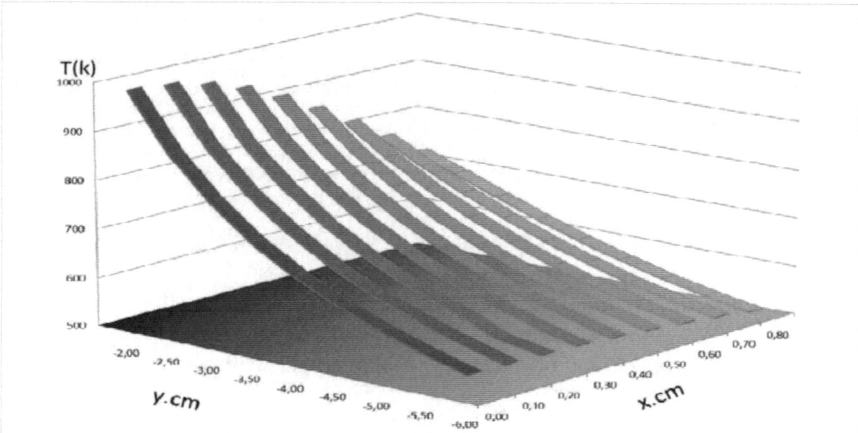

Figure 2. Thermal cycle curves at constant Y values and the initial temperature of 398 K.

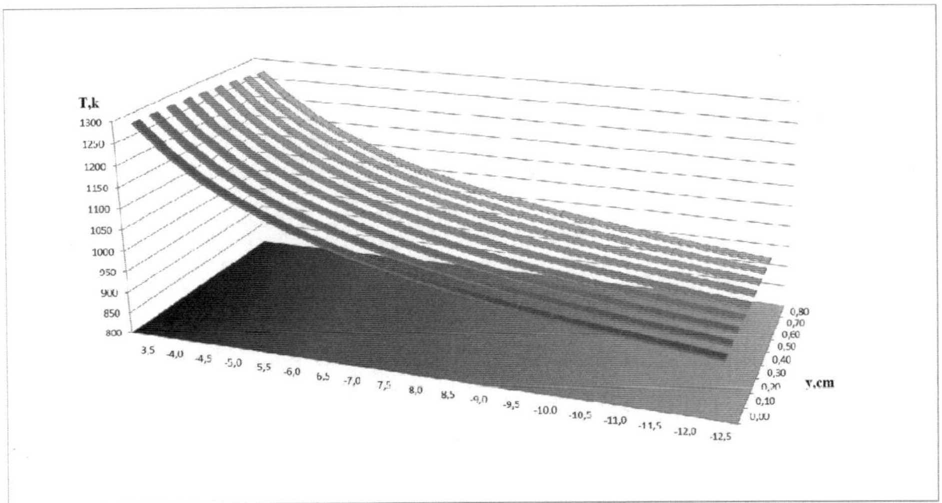

Figure 3. Thermal cycle curves at constant Y values and the initial temperature of 698 K.

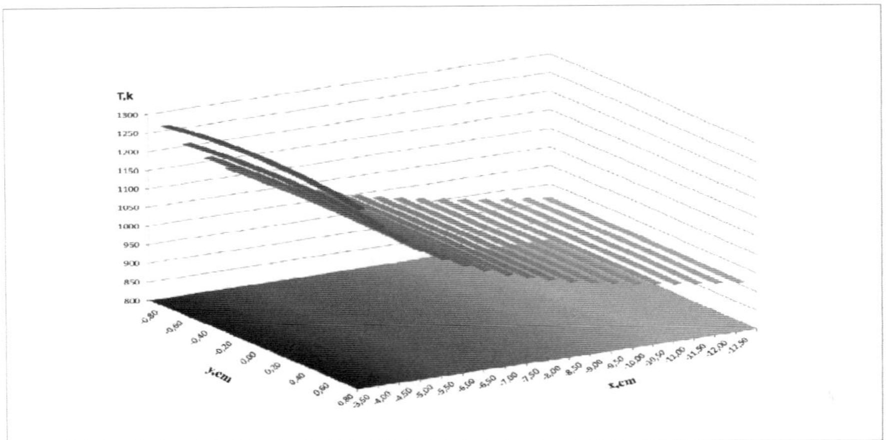

Figure 4. Thermal cycle curves at constant X values and the initial temperature of 798 K.

Naturally, as the initial temperature increases the maximum heat temperatures and heat affected zone width increase as well. From the calculated curves one can evaluate the areas of superheat, hardening, etc. The dependence of alloying elements transition into the weld metal on temperature has been established (Figure 5).

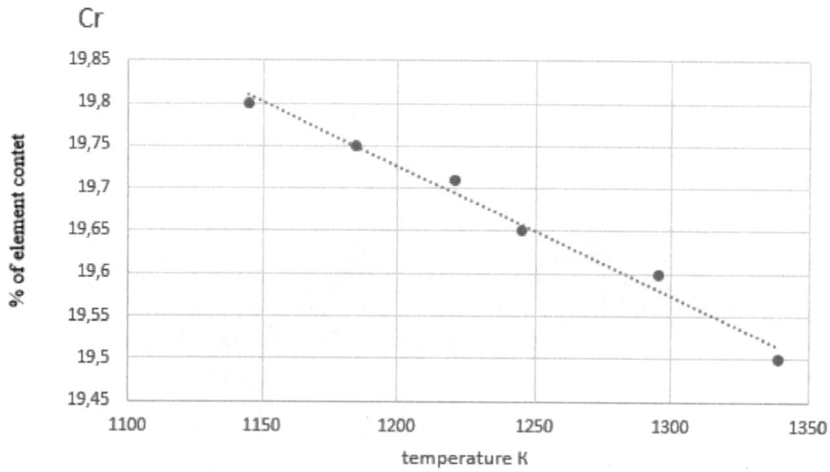

Figure 5. The dependence of chromium content in the weld metal on temperature.

Figures 6 and 7 give the examples of alloying elements distribution along the weld length in its top layers (equal numbers on the *x*-axis correspond to one pass). Chromium and titanium content differs within the measurement accuracy [3]. The content of alloying elements in the top layers is slightly less compared to the middle layers.

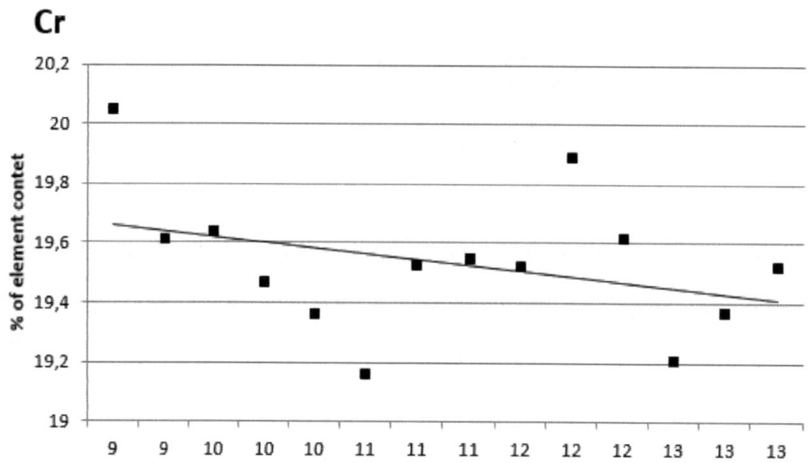

Figure 6. The content of chromium in the weld metal of a multi-layer weld.

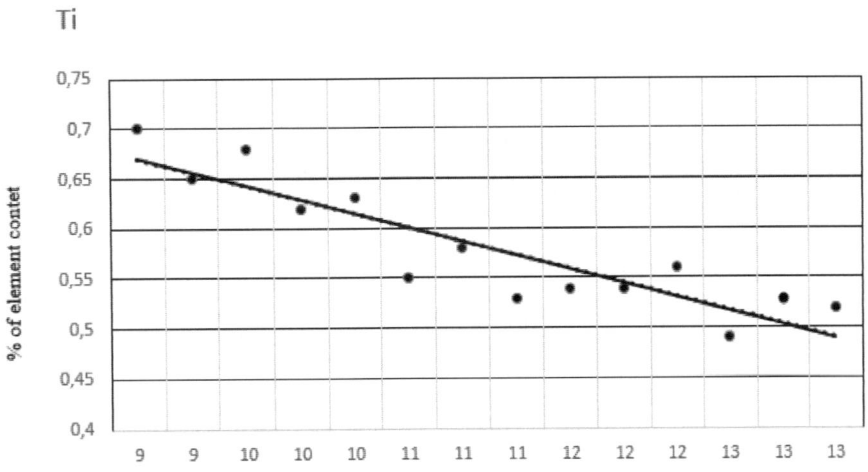

Figure 7. The content of titanium in the weld metal of a multi-layer weld.

Figure 8 shows the example of the structure of weld (cast) and base (austenite) metals. Both the fusion line between them and the layer fusion area can be observed.

Figure 8. The fusion zone microstructure.

4. Conclusion

The results of the investigation allow us to make the following conclusion. Chromium content in the weld metal is from 19.5 to 19.8 %, nickel - from 9.9 to 10.3, which complies with GOST requirements, and manganese and titanium – less than 0.7%. At the same, time the burn-off of manganese and titanium reaches up to 30-50%, although the values of their concentration decrease are 0.3 and 0.2%, correspondingly.

Chromium content in the weld metal decreases insignificantly (1.5%). It is comparable with the measurement accuracy. Changes in cobalt, vanadium and copper concentrations are within the measurement accuracy as well.

Thus, using an inverter power source facilitates a uniform distribution of alloying elements along the length and height of the weld [4, 5].

References

[1] V N Kuskov, R A Mamadaliev, U D Zemenkov Influence of high-concentrated heat sources on alloying elements transition into the weld metal Applied Mechanics and Materials **77** (2015) 19-22

[2] V N Kuskov, R A Mamadaliev, A A Galinskiy The influence of a source of heating for alloying the weld metal upon receipt of the connection of high-alloyed steels Fundamental research **3** (2014) 515-518

[3] V N Kuskov, R A Mamadaliev, A G Obukhov The transition of the alloying elements in the weld metal when welding steel 12X18H10T Fundamental research **11** (2013) 1794-1797

[4] D P Iljyashchenko, D A Chinakhov , Y M Gotovshchik Physical nature of the processes in forming structures, phase and chemical compositions of medium-carbon steel welds J IOP Conf. Series: Materials Science and Engineering **91** (2015)

[5] D P Iljyashchenko, D A Chinakhov, Y M Gotovshchik Calculation of the heat content of the electrode metal droplet wshen applying power supplies for manual arc welding with different volt-ampere characteristic J Applied Mechanics and Materials **756** (2015) 101-104

[6] D P Iljyashchenko, D A Chinakhov, Y M Gotovshchik Investigating the influence of the power supply type upon the properties of the weld joints under manual arc welding J Advanced Materials Research **1040** (2014) 837-844

[7] R E Kryukov, N A Kozyrev, G V Galevsky, Y V Bendre, V F Goryushkin, DV Valuev Some aspects of oxidation-reduction reactions under carbon-bearing flux welding J IOP Conference Series: Materials Science and Engineering. **91** (2015) 1-6

[8] V Kurushina, Y Zemenkov Innovative cyclical development of the Russian pipeline system WIT Transactions on Ecology and the Environment-**2** (2014) 881-888

[9] Yu D Zemenkov, V V Shalay, M Yu Zemenkova Immediate analyses and calculation of saturated steam pressure of gas condensates for transportation conditions Procedia Engineering **113** (2015) 254-258

[10] Yu D Zemenkov, V V Shalay, M Yu Zemenkova Expert systems of multivariable predictive control of oil and gas facilities reliability Procedia Engineering **113 (**2015) 312-315

IOP Conf. Series: Materials Science and Engineering **127** (2016) 012047 doi:10.1088/1757-899X/127/1/012047

Intellectual Control System of Processing on CNC Machines

R Y Nekrasov[1], A A Lasukov[2], A I Starikov[1], I V Soloviev[1] and O V Bekareva[1]

[1] Tyumen State Oil and Gas University, 38, Volodarskogo street, Tyumen, 625000, Russia
[2] Yurga Institute of Technology, TPU affiliate 26, Leningradskaya street, Yurga, 652050, Russia

E-mail: syncler@mail.ru

Abstract. Scientific and technical progress makes great demands for quality of engineering production. The priority is to ensure metalworking equipment with required dimensional accuracy during the entire period of operation at minimum manufacturing costs. In article considered the problem of increasing of accuracy of processing products on CNC. The authors offers a solution to the problem by providing compensating adjustment in the trajectory of the cutting tool and machining mode. The necessity of creation of mathematical models of processes behavior in an automated technological system operations (OATS). Based on the research, authors have proposed a generalized diagram of diagnosis and input operative correction and approximate mathematical models of individual processes of diagnosis.

Introduction

We can talk about the crisis in the metal working industry, having regard to the present state of the global economy, which is in shambles. The enterprises of engineering branch are faced the problem of reduction of production costs. Under such conditions, absence of possibility attracting additional capital becomes impossibility of performing logistical reequipment, which affects the competitiveness of products. Scientific and technological progress can't be stopped, in fact every year there is an acceleration in the rating of its development. All the new discoveries in fundamental and applied areas of knowledge makes it possible to create a fundamentally new equipment. Manufacturing industry is the last limiting link in a long chain from the scientist to the final consumer. It isn't enough to make a discovery also it is necessary to develop technology of processing of the product received as a result of discovery which is possible for processing on the available equipment. At this stage of historical development the leading role in manufacturing industry is played by metal working. Therefore, rates of development of scientific and technical progress, finally, depend on a level of development of opportunities of metal-working industry. The basis of metal-working industry is cutting process. Realization of this process allows to receive everything that variety of vehicles and machinery which so simplify our life. Process of cutting of metals, despite nearly one and a half centenary history of studying, remains misunderstood [1-6]. And it definitely imposes restrictions on possibility of production of the set quality with the minimum material expenses. During an era of competitive struggle between the separate enterprises of engineering branch, advantages are got by the one who is capable to use most effectively resources of the available processing equipment for release of high-quality production. Possibly to achieve the solution of this production task due to increase of effective management of cutting process. There are various approaches to increase of effective management of processing on the metal-cutting equipment. Early studies and the first discoveries in this branch of knowledge were made in the 50th years of the XX century, along with emergence of the first systems of the computer numerical control (CNC) [7-8]. It is necessary to tell that emergence of CTC systems led to emergence of interest in this area. At the beginning increase of effective management of process of cutting was reduced to stabilization of separate output parameters. Was considered, for example, that the technological system will work most effectively if forces of cutting will remain constants in time. Development of CTC systems went on the way of complication of process of management of movement of executive bodies – from position systems to planimetric that, finally, resulted in insufficient efficiency of approach of the management of processing based on stabilization of forces of cutting. In this regard adaptive control systems of processing arose and began to develop. Such systems trace a condition of processing equipment in processing and to a campaign of different detection of errors introduce amendments in the operating trajectory (OT) of movement of executive workers of bodies and the modes of processing. It is necessary to notice that in processing there is a set of various errors. All variety of errors can be reduced to the following reasons:

1. – wear of the rubbing surfaces of processing equipment;
2. – change of spatial situation and a geometrical form of elements of processing equipment, under the influence of the efforts arising in the course of cutting;
3. – change of spatial situation and a geometrical form of elements of processing equipment, under the influence of thermal deformations;
4. – wear of an edge of the cutting tool in processing;
5. – errors of production and assembly of elements, processing equipment;
6. – unforeseen refusals, both separate knots, and processing equipment in general.

All these phenomena arising in technological system doesn't work separately. They works closely intertwining, mutually make an impact at each other thus on the final accuracy of processing. At this stage of development of science and equipment management of processing on CNC machines, it is carried out by means of intellectual systems of adaptive control, i.e. systems controlling technological parameters of system in the course of its work. So, for example, the systems of adaptive control of the RENISHAW company allowing to watch wear of the cutting tool and linear movements of executive bodies of the equipment are well-known. Such systems work in the mode, so-called, real time, i.e. from the moment when the intellectual system of adaptive control determines parameters of a condition of the equipment, till the moment when after processing of these data, changes are made to the operating program (OP), will pass some period, thus the state the equipment will manage to change, thus, we appear in Akhiless's and turtle situation from a known aporiya of Zenon Eleysky. At the moment known control systems of processing on CNC machines allow to consider the changes influencing processing and to bring the corrections compensating errors, caused by such influences as temperature (temperature deformations of a mainspindle and the tool), the cutting force (deformation of working bodies of machines), wear of the cutting tool (shift of an edge in machine coordinates) and tension on the cutting wedge (independent refusal of FRI). But still developed approaches and devices for optimization of process of cutting owing to some difficulties are not introduced. Such complexity, in particular, is that application of the developed approaches assumes intervention in configuration and the software of original expensive processing equipment. In the presence of possibility of failure and removal from guarantee maintenance use of such systems in not certified execution seems unattractive for the enterprise [9-11]. So, for example, transfer of the equipment to management of CNC of system with open architecture is necessary for implementation of the corrections compensating deformations in processing equipment. All aforesaid forces to look for new approach in the solution of this problem which unitary enterprise directly at a creation stage consists in adjustment. In most cases development unitary enterprise for CNC machines is conducted using systems of the automated design (CAM), or carried out by the built-in simulars, directly on the CNC control panel. Therefore, the errors arising at implementation of the operating programs created by CAM system or the CNC control panel on real technological system will be inherited by the processed detail. The wide circulation in modern

IOP Conf. Series: Materials Science and Engineering **127** (2016) 012047 doi:10.1088/1757-899X/127/1/012047

machining production of CNC machines leads to emergence of one more problem demanding the decision – decline in quality of production in use by the choosen technological system . The matter is that any material object is subject to such phenomenon as aging, i.e. gradual deterioration of its initial characteristics. The technological system isn't an exception of this rule, it is also subject to deterioration of the operational characteristics over time, and this process proceeds the quicker, than possibilities of technological system are used more intensively. One of the main operational characteristics of technological system is its ability to provide the demanded dimensional accuracy and quality of the processed surface at the minimum production expenses. The solution of this problem was prompted by technology of construction of high-rise buildings in the conditions of difficult soil when the building is under construction with an inclination so that at the end of construction the building by gravity adopted the vertical provision. This approach can be used in the solution of problems of increasing dimensional accuracy of machining on CNC machines consisting in transformation of the geometrical model created in CAD-system by such change of its form that in processing the errors connected with errors of technological system were compensated. Realization of this approach, requires creation of the neural network mathematical models describing behavior of technological system in processing, distinctive feature which there will be an ability to self-learning ability.

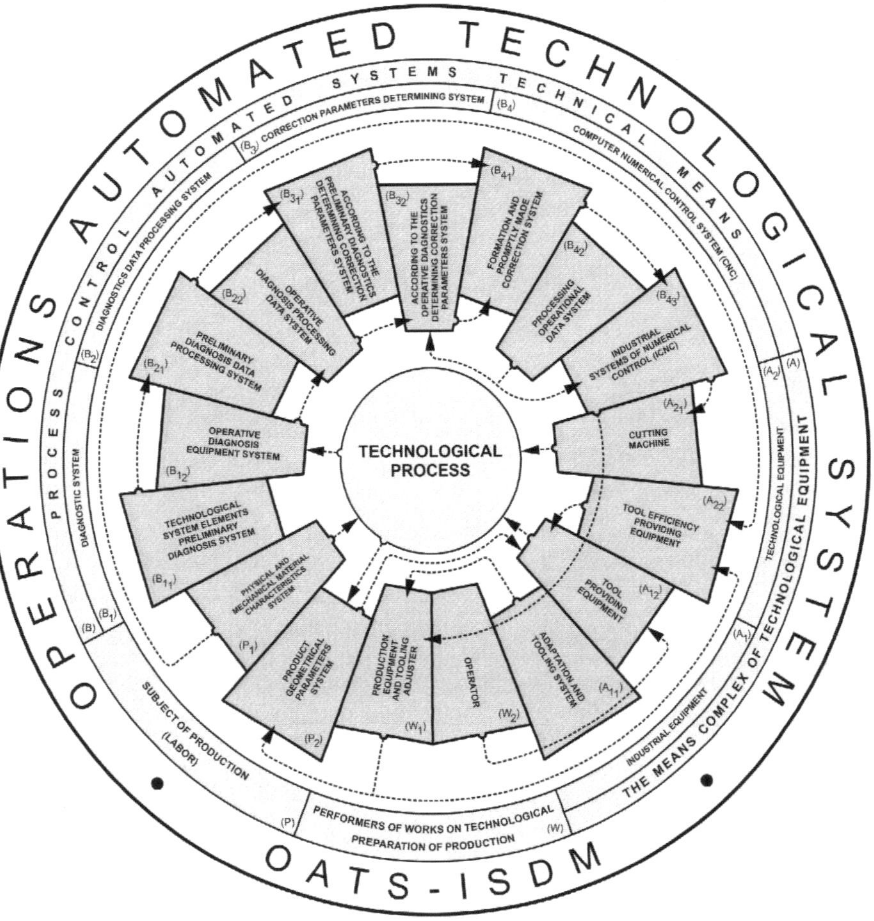

Figure 1. Structure of operations automated technological system (OATS).

2. Research methods

The methodology of transformation of geometrical model of a detail taking into account, the arising errors of processing consists of the following consecutive stages:

1. – creation of geometrical model of a detail by means of systems of solid-state modeling (CAD-system);

2. – definition of a geometrical form of a detail as a result of influence of the thermal and power loadings arising in the course of machining by means of systems of the final and element analysis (CAE system);

3. – modification of initial geometrical model of a detail, based on results of performance of the previous stage.

This methodology creates possibility of input of corrections at a development stage unitary enterprise for the CNC machines. The defined deviations from nominal rates of the sizes are compensated by input of correction coefficients, so that the arising errors were compensated in the course of realization unitary enterprise. The developed methodology is entered in the scheme of the automated technological system of operations (OATS) presented in figure 1 offered by authors.

system, on the basis of self-training neural network model which scheme is submitted in figure 2.

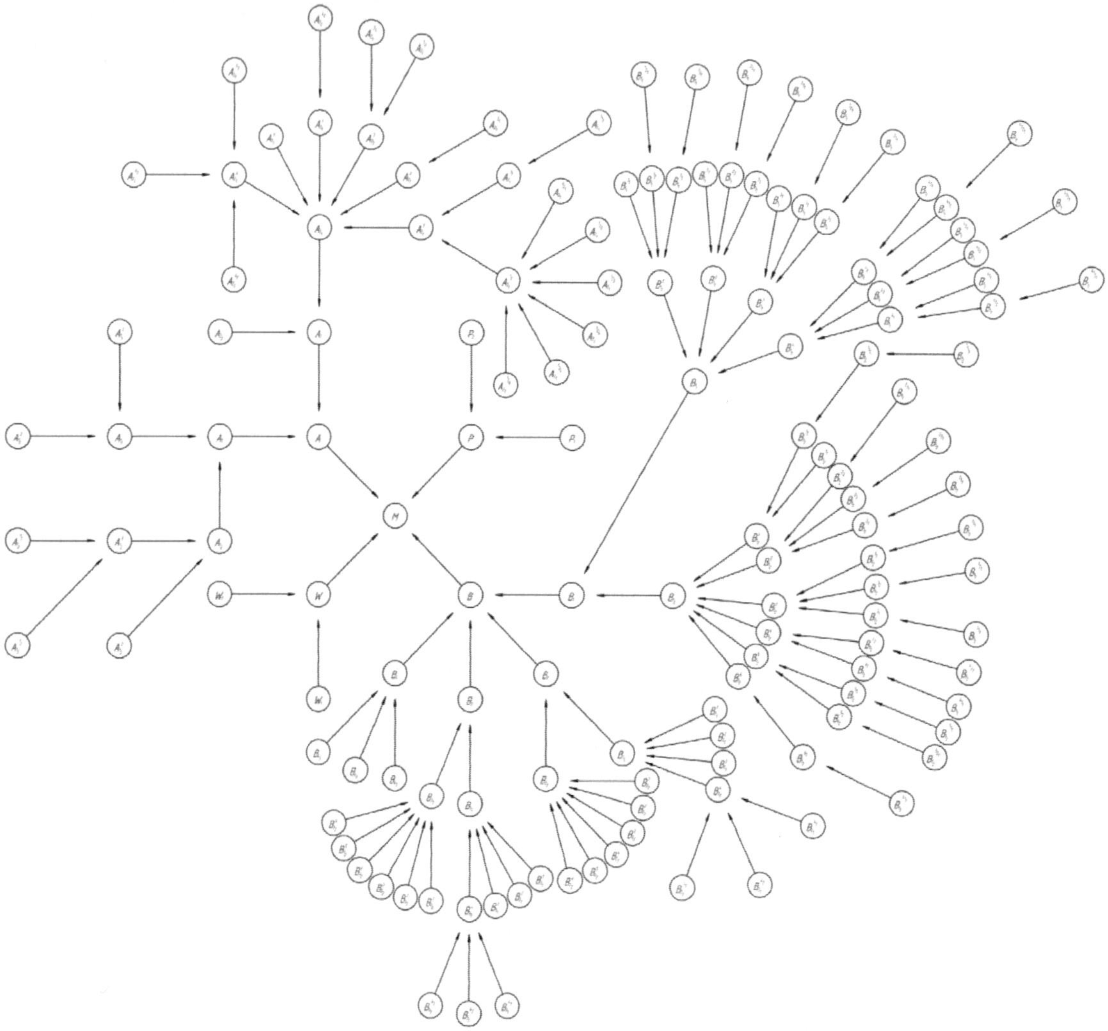

Figure 2. The neural scheme of representation intellectual control system of processing on CNC machines.

1 structure of ATSO represented in drawing isn't full and exhaustive (elements from which ATSO consists are provided in table 1), but sufficient to approach a question of mathematical modeling and development of mathematical models of various processes proceeding in it and necessary for development of an intellectual control system of processing on CNC machines. The problem of realization of this methodology is in the following that for the second stage, i.e. definition of a geometrical form of a detail as a result of influence of the thermal and power loadings arising in the course of machining by means of systems of the final and element analysis (CAM system) data which can't be received without carrying out preliminary and expeditious diagnostics. [12-15] Expeditious diagnostics of researches are necessary, therefore each new unit of processing equipment requires creation of own mathematical models describing behavior in the course of machining. The acceptable solution of this task will be creations of universal mathematical behavior models of technological

The presented scheme of self-training neural network model is reconsideration of structure of OATS represented in figure 2. For creation of this scheme subsystems and elements the making OATS were described, the short list of the most important parts is presented in table 1.

Table 1. Elements the making OATS.

Symbol	Name
M	Operations automated technological system (OATS – ISDM)
P	Subject of production
P_1	Set of mechanical-and-physical material properties of a subject of production (work)
P_2	Set of geometrical parameters of a subject of production (work)
W	Contractors on technological preparation of production
W_1	Serviceman of processing equipment and equipment
W_2	Operator of processing equipment
A	Set of means of technological equipment
A_1	Industrial equipment
A_{1_1}	Set of adaptations and means of industrial equipment
A_{1_2}	Set of means of tool providing
A_2	Processing equipment
A_{2_1}	Metal-cutting machine
A_{2_2}	Set of means of ensuring of operability of the tool
B	Process control of an automated system technical means
B_1	Diagnostic System
B_{1_1}	System of means of preliminary diagnostics of subsystems and elements of technological system
B_{1_2}	System of means of expeditious diagnostics of subsystems and elements of technological system
B_2	System of data processing of diagnostics
B_{2_1}	System of data processing of preliminary diagnostics of technological system
B_{2_2}	System of data processing of expeditious diagnostics of technological system
B_3	System of determination of parameters of corrections

B_{3_1}	System of determination of parameters of corrections according to preliminary diagnostics
B_4	System of numerical program control by the machine (CNC)
B_{4_1}	System of formation and expeditious input of corrections
B_{4_2}	System of operational data on processing
B_{4_3}	System of numerical program control by the machine (PCNC)
$B_{3_1}^{\ 4}$	System of determination of parameters of corrections according to preliminary diagnostics of temperature deformations
$B_{3_1}^{\ 4_1}$	System of determination parameters of corrections according to preliminary diagnostics of temperature deformations from mainshpindel knot
$B_{3_1}^{\ 4_2}$	System of determination of parameters of corrections according to preliminary diagnostics of temperature deformations from a tool head
$B_{3_1}^{\ 4_3}$	System of determination of parameters of corrections according to preliminary diagnostics of temperature deformations of the back center

Summary
The offered approaches to mathematical modeling of the processes proceeding in ATSO allow to solve a problem of creation of an intellectual control system of processing on CNC machines therefore increases of dimensional accuracy of machining. The positive effect from introduction of this system is connected with reduction of the production expenses spent for technological preparation of production of products of engineering production.

References
[1] G S Andreev The stress state of the cutting part of the cutter with periodic cutting Durability of the cutting tool (1969) 28-37
[2] A I Betaneli Communication method of calculating the strength of the cutting part of the tool, Bulletin of engineering. **2** (1965) 16-18
[3] A A Lasukov, P A Chazov, A V Barsuk Investigation on the Elemental Chip Formation Process in Hard-to-Machine Material Cutting Applied Mechanics and Materials **682** (2014) 504-509
[4] V F Bobrov Determination of stresses in the cutting part of the cutting tools High-performance cutting in mechanical engineering (1966) 228-233
[5] A M Geniatulin Holographic method research strained elements of assorted cutters Optical-geometrical methods in the study of deformations and stresses and their standardization, Gorky (1982) 83-84
[6] R Yu Nekrasov, A A Starikov, M A Korchuganov Performance assessment of carbide tooling under thermal and loading conditions Applied Mechanics and Materials **682** (2014) 414-417
[7] A I Afonasov, A A Lasukov Elementary Chip Formation in Metal Cutting, Russian Engineering Research **3** (2014) 152–155
[8] R Yu Nekrasov, I V Soloviev, U S Putilova Laser interferometry method of stress determination in loaded cutting tool parts Applied Mechanics and Materials **698** (2015) 537-541
[9] M B Gordon Methodology and some results of a study of patterns of distribution of the friction forces and tangential stresses on the contact length of the chip from the front surface of the cutter Issues in the application of cutting fluids in machining metals (1965) 41-44
[10] U S Putilova, Yu I Nekrasov, A A Lasukov Loading of the manufacturing systems elements in the process of unsteady mode cutting and the models of their arrangement deviations, Applied Mechanics and Materials **682** (2014) 192-195

[11] R Yu Nekrasov, A I Starikov, I V Soloviev Simulation of technological systems for diagnosis and management machining with CNC Applied Mechanics and Materials 770 (2015) 617-621
[12] R Yu Nekrasov, U S Putilova, A I Starikov, D A Kharitonov Diagnostics of cut-layer deformation and rational tool loading in numerically contolled lathes, Russian engineering research **34** (2014) 826-828
[13] A A Lasukov, A A Mokhovikov Influence of modified layer of tool on stress - Strain state of cutting wedge, Proceedings (2012)
[14] R Yu Nekrasov, A I Starikov, A A Lasukov Entering the operative correcion machining processes CNC Materials Science and Engineering **91** (2015)
[15] A A Lasukov Selection of machining conditions in terms of the temperature dependence of chip formation, Russian Engineering Research **9** (2015) 679–681

Adaptive Control System of Hydraulic Pressure Based on The Mathematical Modeling

A V Pilipenko[1], A P Pilipenko[2], N VKanatnikov[3]

[1] Department of Applied Mathematics and Informatics, State University — Education-Science-Production Complex (State University ESPC), Russian Federation, Orel
[2]Faculty of New technologies and automation of production, State University ESPC, Russian Federation, Orel
[3]Office of Science and Technology, State University ESPC, Russian Federation, Orel

E-mail: kot_nafanya@mail.ru

Abstract. In this paper, the authors highlight the problem of replacing an old heavy industrial equipment, and offer the replacement of obsolete control systems on the modern adaptive control system, which takes into account changes in the hydraulic system of the press and compensates them with a corrective action. The proposed system can reduce a water hammer and thereby increase the durability of the hydraulic system and tools.

Introduction

In the modern economic conditions for the industrial enterprises the matter of great importance is the production of the competitive goods with the minimum costs. Production of products and details by the methods of form changing operations is characterized its metal expenditure economy, the ability to get details of high accuracy. Providing optimal parameters for the equipment operation, in particular due to the use of automation systems achieved the rise of its endurance parameters and production energy efficiency.

A great number of technological processes for pressure metal treatment are performed at the hydraulic equipment. Modern hydraulic presses have automated control systems of electric and hydraulic drives, calibrating, beam deskewing, product quality analysis and tooling conditions. However not all the manufacturing enterprises have an opportunity to replace available hydraulic equipment fleet with a new one. Modern information technologies and technical means of automation allow modernizing any equipment with the minimum costs.

To implement automation systems and optimize equipment operating conditions it is necessary to have a vision of dynamic processes occurring in the systems at performing operations and the connections between the operation of the hydraulic system, electric drive and the main operational unit.

In this article the authors introduce hydraulic automation system different by control algorithms of electric and hydraulic drives allowing to diagnose and keep track of system operation during the technological process online and produce correcting control actions. For the prototype sample used hydraulic press with 630 kN force of LF model 2428A where blanking/perforating dies and roller die stamping were installed.

The peculiarity of hydraulic presses as hydraulic driven machine tools is their equipment with high volumes of hydraulic fluid in cylinders and pipelines hence at their operation may occur dynamic processes of fluid pressure fluctuations in hydraulic system. Matters of hydraulic presses dynamics were under consideration of such scientists as B.V. Rozanov, A.I. Zimin, M.V. Storozhev, A. L. Vorontsov[1] who noted that as a consequence of elastic deformation in the systems of hydraulic presses finds hard to complete the operation with sharp load release as observed a fluid hammer, press vibration, seal failure and beam deskewing. Implementation of adaptive control systems capable to register and decrease dynamic oscillation processes into out-dated hydraulic equipment let us to minimize the costs on production reequipment and also achieve quality increase in production and energy efficiency of technological processes.

The questions of adaptive control of the press were raised by S.S. Oding, I.A. Kretov, A. S. Morse [2] who developed algorithm for records of work material parameters deviations included into the mathematical material model allowing to correct the program of deviations immediately at forming. However in their works they dealt with only technological process of draw-forming and comparison was carried out with the parameters written in the simplified mathematical model.

Adaptive control systems for pneumatic drives on the basis of the mathematical modeling were investigated by Michael Brian Thomas[3]. In his writings he justifies the efficiency for implementation of automated adaptive control systems on the basis of mathematical modeling.

1. Concept

Under the made assumptions the processes of slider movement and press frame in technological processes of form changing are described by a set of the first order differential equations including slider movement equations, fluid pressure changes equations in pressure and drain press lines, speeds of slider, change of pressure pump, movement and speed of frame, change of pressure in the head end at different pump mode (1).

Mathematical modeling of equipment hydraulic operation with regard to hydraulic pump operation parameters significantly decreases the number of assumptions such as fluid compressibility, non-uniformity of fluid supply by the pump and fluid flow through the valve[4]. Modeling of pump gives an opportunity to examine pressure in the head end at different pump modes, to estimate the loads on controls due to pressure in the head end and to optimize control system.

$$F_2(dx_1/dt)-F_2(dx_2/dt)-q_2 \qquad (1)$$

M_1 – slider mass with attached tools; M_2 – frame mass, x_1, v_1, t_1 – movement; speed, and time of slider movement; x_2, v_2 – movement and speed of frame; c_2 – coefficient of foundation bolt firmness; P_{sn} – strain resistance of cutting material; F_1, F_2 – piston area of working and return cavity; Q_h – head end geometric delivery; p_h – pressure in head end; R_h – head end radius; k_l – leakage coefficient; p_1, p_2 – fluid pressure in the working and return cavity of the cylinder; Q_p – pump output; α_1, α_2 – stiffness coefficient of pressure and drain hydraulic press lines; q_2 – fluid flow rate through the throttle; γ – tilt angle of hydraulic pump; x_3 – head end movement in hydraulic pump; ω – shaft angular speed; Q_l – leak flow from head end; Θ – angle of distributor turn; β – additional angle in plane, perpendicular to the plane formed by kinematic neutral point and its projection on a reference pane; d_n – piston diameter; $S_{th}(\Theta)$ – area of window throttling.

As a result of apparent movement at opposing motion of sliding bearing and frame the function $P_{sn}=f(x_1-x_2)$ exists within $(x_1-x_2)=h_M$, where h_M – length of stroke.

Functions $P_{sn}=f(x_1-x_2)$ and $q_2=f(p_2)$ are highly nonlinear.

The solution of the equation system (1) and research of the parameters influence on dynamic processes are carried out by numerical integration on PC.

Deformation stress of plate sheet metal at equation $P_{sn}=f(x_1-x_2)$ is taken as an analytical expression taking into account metal variable resistance shear depending on plastic characteristics of cutting metal and cutting punch penetration depth.

Figure 1 illustrates diagrams of deformation resistance at cutting / punching and roll stamping. The part of diagram P_{sn} at material chips during cutting is described by the following equation:

$$P_{sn}^{max}=P_0+a[(h_M-(x_1-x_2))/(h_M-h_K)]^\wedge[(1-K_2)/K_2] \qquad (2)$$

where P_0 – pushing force of punch after cutting,

P_{sn}^{max} – maximum resistance of deformation of cutting material;

$a=P_{sn}^{max}-P_0$;

K_2 – coefficient of diagram loading factor of working loading in segment of chip (find by empirical method), $0< K_2<1$.

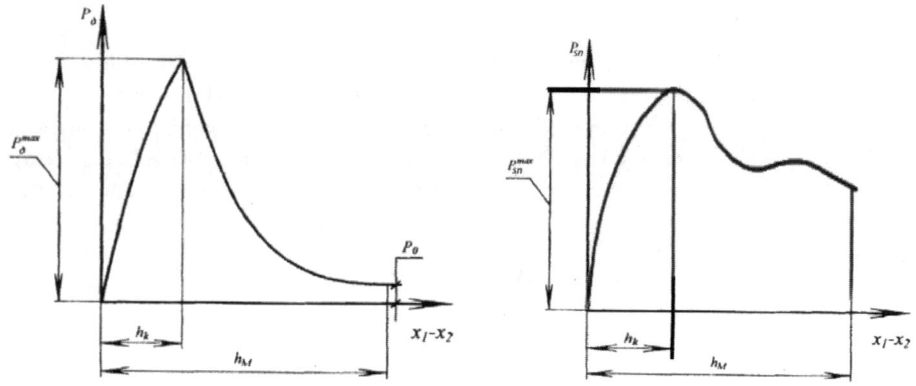

(a) (b)

Figure 1 – Diagrams of deformation resistance (a) at cutting / punching and (b) at roll stamping: h_M – full way of sliding bearing through the material, h_K – way of sliding bearing from the contact with slug till the beginning of chip

The contraction of area at fracture as a rule is equal to percent reduction of area, hence $Psn=Psn^{max}$.

Performed experimental researches allowed getting diagrams of resistance deformation at different technological processes. In a mathematical model (1) deformation resistance *Psn* is presented as curve equations such as equation (3) for technological process of roll stamping and it also helps us to improve accuracy of calculations.

$$(x):=(253.362x-86.3863x^2-14.6373x^3+10.9418x^4-1.81233x^5+0.141945x^6-0.00434566x^7)/(2.82589+11.199x-6.45885x^2+x3) \qquad (3)$$

Mathematical model (1) allows:

Studying of influence of hydraulic press parameters on dynamic processes at shearing operations;

Considering hydraulic press as a universal system where movement of its working parts, tool and processed slug that give an opportunity to design processes with optimal parameters and optimal control.

To realize computer modeling was chosen graphical programming environment LabVIEW. The advantage of graphical programming environment use of LabVIEW in modeling includes functions (virtual instruments) allowing to solve differential equations at the time of close to real by Runge-Kutta method of the second order.

The programs executing computer modeling and data visualization were created (Figure 2).

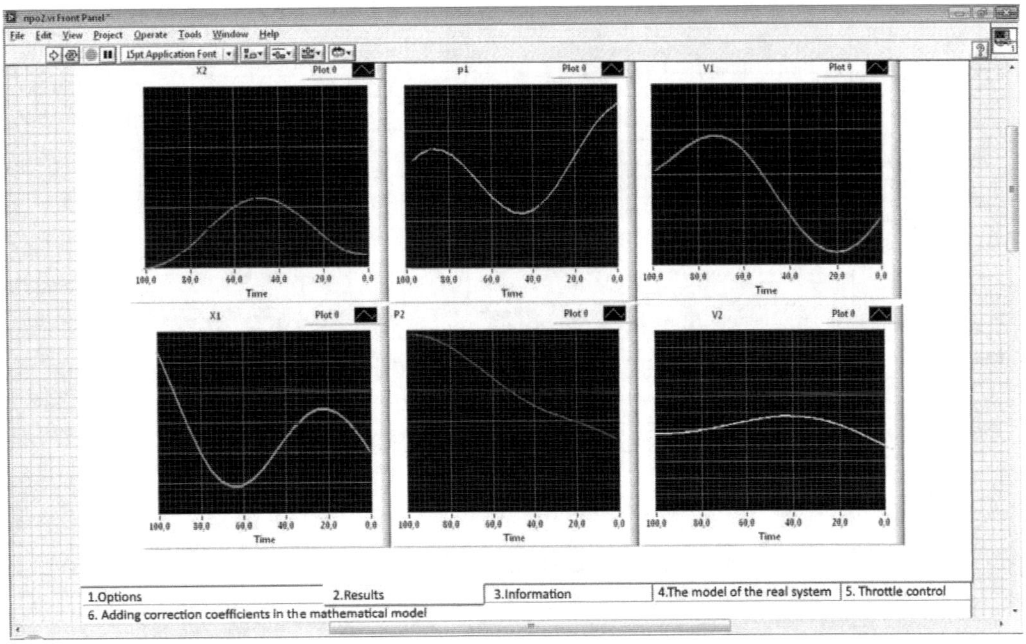

Figure 2 – Picture of computer modeling program window

Computer modeling system of hydraulic press work at technological processes of metal working process allowed to define dependences of equipment work parameters and acknowledge that control system of hydraulic equipment registering calibrated values deviations lets us to define technical malfunctions or equipment deterioration. Study of the results of computer modeling shows parameters demanding control and management.

Developed the structural scheme (figure 3) of automated control system of technological form changing process that is realized on the equipment including connections between the basic system parameters.

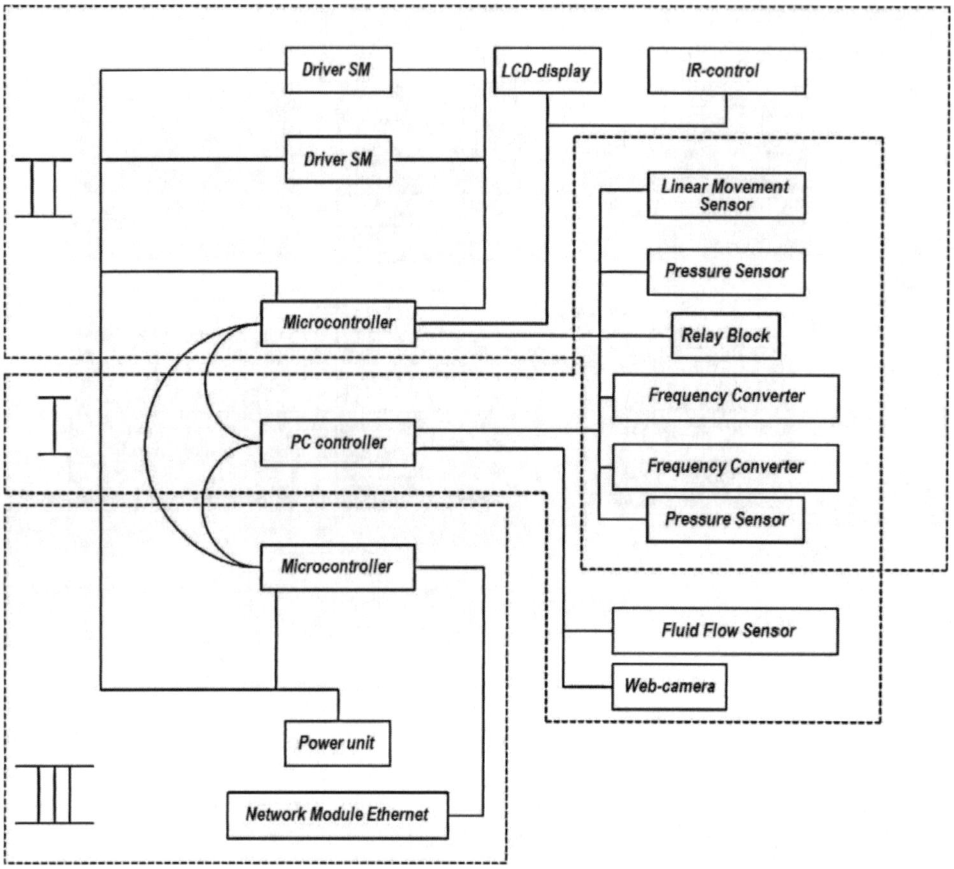

Figure 3 – Structural scheme of automated control system

The system in real-time mode receives information from RT pressure sensors situated on sliding bearing and frame; information through ATS is being delivered to the computer that is operator desk; production control is implemented with the help of computer vision, i.e. camera transmits a signal that is processed in LabVIEW by intellectual algorithms; electronic computing machine processes information and compares values with the values got by computer modeling of technological process. In case of system detecting deviation of one of the parameters of technological process, it gives the operator recommendations how to optimize the work. If the operator is out of his working place, the switch unit turns on the automation mode where automatic control system controls parameters of technological process with the help of step motors, frequency converters and relay blocks.

The common structure of control system is given in the figure 4. In automatic mode of work PC-controller presets initial values $x(t)=p_0(t),x_0(t),v_0(t)$, influencing on engine rotational speed of hydraulic pump ω and fluid consumption flowing through throttle q. Sensors located on hydraulic press deliver calculated parameters to the self-tuning block comparing data received by experimental way and computer modeling method defines correctional coefficient k_{cor}, using logical control law: if $|p_{exp}|<0,2*|p_{math}|$, then $k_{cor}=p_{exp}*k_1$; otherwise $k_{cor}=p_{exp}*k_2$; where $k_1<k_2$. Regulator forms control impact $U(t)$ and delivers it to the operating controller.

IOP Conf. Series: Materials Science and Engineering **127** (2016) 012048 doi:10.1088/1757-899X/127/1/012048

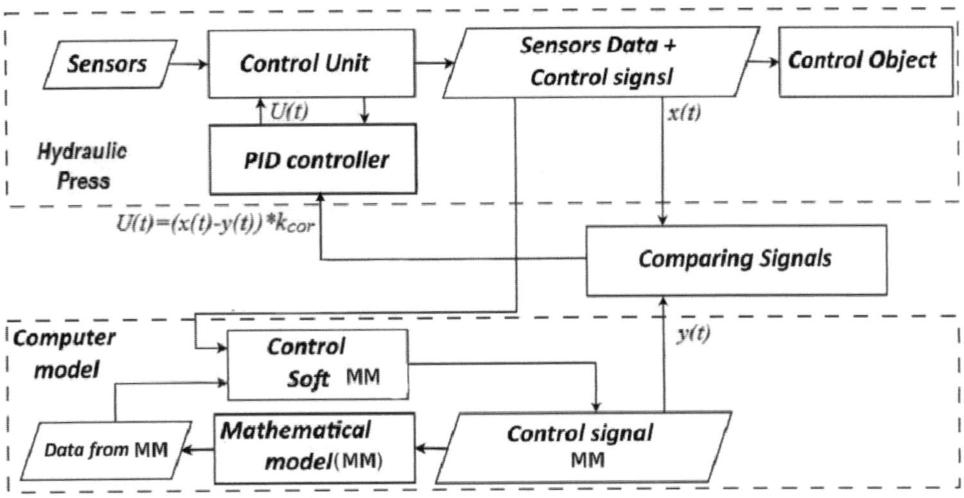

Figure 4 – Common structure of control system

Control system consists of two microcontrollers[5] and one PC-controller[6]. Algorithm of the basic control system for PC-controller is given in the figure 5. The following consequences of actions are carried out: 1 – Start; 2 – Input experiment name, comment and N – quantity of demanded cycles; 3 – Choice of technological process; 4 – Technological process of cutting/perforating; 5 – Technological process of roll blanking; 6 – Turn on frequency converter hydraulic pump and transfer beam in the maximum position; 7 – Login procedure of data acquisition device; 8 – Diagnostic routine of microcontroller modules including into control system; 9 – Mode selection: Automatic control; 10 – Demand assignment at microcontroller 1; 11 – Switch to microcontroller 1; 12 – Switch the basic system into sleep mode; 13 – Insert data about material: L, d, h; 14 –Selection of material and its parameters; 15 – 19 – Material type, density, strain resistance diagram and so on; 20 – Selection of working conditions for equipment; 21-25 – Parameters of regulation, pressure, speed and so on; 26 - Data acquisition from ATS; 27 – Data cleaning; 28 – Data processing and conversion into C; 29 – Solution of differential equations by second order Runge-Kutta method; 30 – Screen display of diagrams and data; 31 – Program of data comparison; 32 – $i=N$?; 33 – Do you wan to change N cycles and continue working; 34 – Do you finish your work or choose other parameters? 35 – End of program.

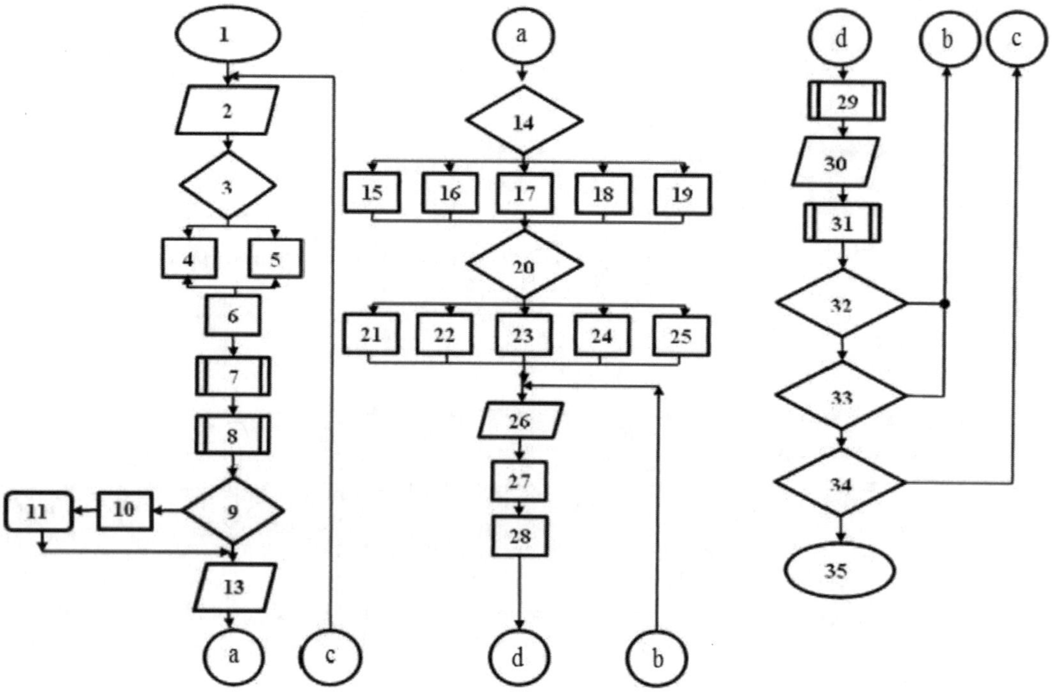

Figure 5 – Algorithm of Automated Control System

Working principle of PID controller, i.e. the difference between current pressure and pressure received due to the results of computer modeling is multiplied by configurable factor and we get capacity which is necessary to show at the present moment in output control device. Proportional term works at the moment of occurring mismatching. When cross beam starts moving, capacity begins decreasing, and when it achieves minimum mark, device is turned off. The effect from the impact is shown with a delay and environment also affects the object, so, cross beam mass, fluid and pipe line compressibility. To compensate "external" influences integral term was added in the circuit. As a result of such an approach integral becomes stable that is why value of output capacity turns out permanent. Moreover, as there is a required pressure then mismatching is absent, proportional term doesn't work at all. To compensate impact of delays between the influence and reaction of the system in the system was added a differential term.

Simply proportional regulator gives capacity all the time until pressure reaches required point; proportional-differential one starts decreasing supplied capacity before cross beam attains required point. As mismatching is reducing, there is a negative derivative decreasing the influence. That allows minimizing the pressure at higher scale transitions.

We made a plan of the experiment including 13 tests in each of six experiments; sample size was calculated and justified. And hereby as a variable parameter was chosen a controlled parameter of sectional area for throttle shelter q. Analyzed parameters were admitted the following parameters – pressure at pressure main p_l, cross beam speed v_l, cross beam movement x_l, frame movement x_2.

Correlation dependences of cross beam movement speed from throttle shelter at roll stamping procedure, shown in the figure 6 (a) and pressure dependences in the pressure main from throttle shelter area in the figure 6 (b), allows acknowledging adequacy of the developed mathematical model. Errors are situated within permitted values; hence, computer model designed on the basis of the mathematical one works correctly and can be used at the development of control systems of hydraulic equipment.

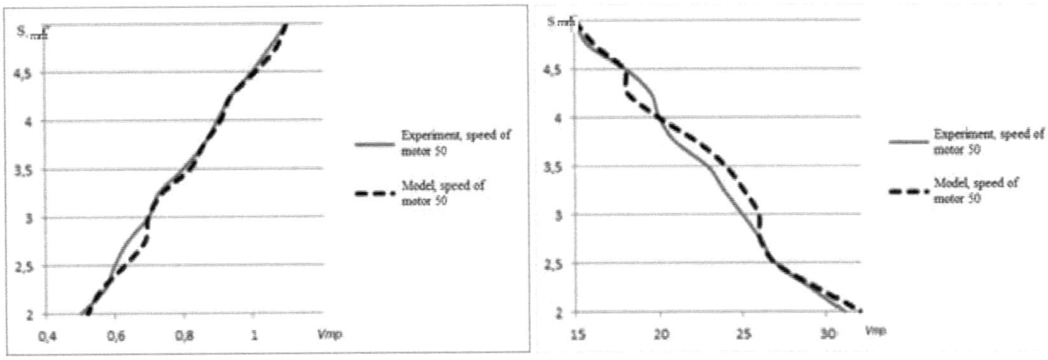

(a) (b)

Figure 6 – Diagrams of dependences at roll stamping procedures – speed of cross beam movement, b) pressures in the pressure main from throttle shelter area

Conclusion

Basing on the results of the theoretical and experimental researches got task solution at the development of automated control system for hydraulic equipment on the basis of adaptive algorithms built on mathematical models, being important to industrial enterprises or research societies.

Developed mathematical model allows us researching the influence of hydraulic unit parameters on dynamical processes at shearing operations; moreover, hydraulic press is considered as a single system where there is a movement of its working objects, tools and processed work piece that gives an opportunity to design machines with optimal parameters and with optimal control, and to serve as a basis for calculations of processes in presses at operations with long working cycle. Researches passed approbation in real production enterprise that is affirmed by adoption deed. Automated control of hydraulic working parameters lets to optimize energycosts for technological process, to reduce consumption and to increase working tool wear resistance.

References

[1] Vorontsov A L, Sultan-Zade N M, Albagachiev A Y 2008 Development of a new theory of cutting. *Russian Engineering Research.* 28(5) 461-471

[2] Morse A S 1992 High-order parameter tuners for the adaptive control of linear and nonlinear systems. *Systems, models and feedback: Theory and Applications.* 339-364.

[3] Thomas M B 2003 Advanced servo control of a pneumatic actuator. *Doctoral dissertation, The Ohio State University*

[4] Cao Z, Yue Z, Pender G 2011 Landslide dam failure and flood hydraulics Part II: coupled mathematical modelling. *Natural hazards.* 59 1021-1045

[5] Hiller B, Tuchscherer A 2008 Real-time destination-call elevator group control on embedded microcontrollers. *Springer Berlin Heidelberg.* 357-362

[6] Limroth J 2005 LabVIEW Real-Time for Networked Embedded Control. *Handbook of networked and embedded control systems.* 447-470

[7] Ni G J, Wang P, Liu L PID 2012 Optimization with Regulation-Based Formulas and Improved Differential Evolution Algorithm. *Advanced Engineering Forum.* 6 261-266

[8] Mean V, Abdullah N S, M Dawal S Z, Aoyama H 2013 Investigation on Musculoskeletal Symptoms and Ergonomic Risk Factors at Metal Stamping Industry. *Advanced Engineering Forum.* 10 293-299

Mechanical Properties of the Assembly Welded Joint of the Oil Transportation Tank After a Long-Term Service

A L Pimnev[1], M Yu Zemenkova[2], Yu D Zemenkov[3], D P Iljyashchenko[4]

[1,2,3] Tyumen State Oil and Gas University, 38 Volodarskogo st., Tyumen, 625000, Russia
[4] 652050, Kemerovo region, Yurga, Leningradskaya str.26 Yurga Technological Institute branch of Tomsk Polytechnic University

E-mail: pimnev@mail.ru

Abstract. The paper provides results of studying a change in mechanical properties of the design elements metal and assembly welded joints of a vertical steel tank after a long-term service in oil transport facilities.

Introduction

The reliability of pipeline transport systems depends on the reliability of all the facilities involved in a technological process [1,2]. Tank batteries play an important role in maintaining uninterrupted oil transmission.

Existing tank batteries are significantly worn out. Recently, the amount of repair, reconstruction and construction of new tanks has increased dramatically. This has resulted in a sharp rise in technical diagnosing of vertical steel tanks (VST) using new technical means and techniques. To solve the problem of evaluating the technical state of vertical steel tanks a two-level technical diagnosis system has been implemented according to which the frequency and type of inspection depends on the operation life. However, the existing technical diagnosis system does not fully meet current requirements.

In such conditions, it is necessary to take actions to increase the efficiency of tank use by prolonging a safe tank operation life in production. As vertical steel tanks are elements of complex engineering systems of oil and oil products production, transportation and storage, providing safety and evaluating a safe operation life of in-service tanks are related to increasing the operation life of the tank battery and the entire pipeline transportation system. Moreover, increasing the tank reliability results in saving the main funds.

The problem of the tank safety is connected to the issues of reducing oil and oil product losses caused by spills and accidents, as well as a poor technical condition of the tank shell and bottom [2]. Financial losses caused by tank accidents are great and tend to become even greater as the tank capacity increases. Because of numerous related expenses, the real loss exceeds that of the reclamation and product cost. Expenditures on contamination control alone caused by spills are 4 times greater than direct losses.

When diagnosing the technical state of VSTs after a long service it is necessary to take into account the industry expertise in the complex tank inspection during which special attention must be paid to changes in the geometric shape that can occur under conditions of the trunk line tank battery operation.

Defects influencing the reliability of vertical steel tanks can be brought in to their construction at any stage of their life cycle: from design to operation. Conditionally, these defects of tank design elements can be divided into five groups depending on the period of their formation:

- metallurgical — brought in to the design during rolled product manufacturing (folds, delamination, non-uniform alloying, scuffing, micro cracks, distortion of the rolled product geometry, etc.);
- manufacturing — occurred during fabricating work pieces (defects of assembling and welding);
- transport — originated during transportation of metal structure elements to the assembly site (dents, scuffing, tearing, crumpling, bending, corrugation, etc.);
- assembly —imperfections during tank assembling (defects of bases and substructures, defects of assembling and welding of metal structures, angularity of welds, remains of fixtures, etc.);
- operational — occurred during tank operation (deposits, stability loss, corrosion, buckling, etc.).

As a rule, these defect groups are detected during technical diagnosis of tanks by means of non-destructive testing: visual, ultrasonic, radiographic, etc. Conducting inspection of VSTs used at oil pump stations of the oil trunk lines allows us to conclude that basic defects occurring practically in all VSTs and having an essential impact on their operational reliability are corrosion defects and tank geometry irregularities. It is due to the fact that these particular defects have a strongly pronounced operational character of development.

It is known that in the course of long-term service of metal tank structures there can be a change of properties of parent metal and weld metal of weld seams. It can lead to strength decrease, impact strength change and plasticity increase.

In [3] results of an experimental evaluation of properties of parent metal and tank welds metal after various service lifespans up to 25 years in the area of the Tuapse city are given. It has been established that a long-term service under the influence of media leads to strain ageing of the metal, i.e. decrease in resistance to brittle fracture. The coefficient of strain ageing for the investigated steels after 25 years of service equals 1.1÷1.34. Tests have shown that strength characteristics have small deviations from standard values, however, a considerable decrease in plasticity characteristics is observed. It has been found in this work that changes in mechanical characteristics of various zones of a welded joint during service of steel 09G2S showed the most intensive degradation processes occurring in a heat-affected zone: σ_y rises by 12-13 % after 20 years of service; plasticity characteristics decrease (ε by 25-30 %).

Thus, a change of mechanical properties of a metal can be referred to as an operational defect of vertical steel tanks. A current problem is researching into changes of mechanical properties of a welded assembly joint of an oil transportation tank after a long-term service, as a considerable impact of welding on metal properties is known [4].

The subject of research is a vertical steel tank RVS-20000 (a type design № 704-1-60, CNIIPSK of N.P. Melnikov, Moscow). The tank has been assembled and put into operation in 1988 at the oil pumping station. The construction region falls within the ID climatic region, the ambient temperature on the coldest days being -48 °C, the absolute minimum ambient temperature— -55 °C.

After the tank has been routinely taken out of service, a due full technical diagnosis has been performed to determine its condition, evaluate the possibility of further operation and necessity of repair. Having undergone the repair work, the tank has been tested by water loading and put into operation once again. After one year of service, on the sheet of the second belt of the VST wall a crack was formed.

Modern studies have shown that fatigue destruction occurs in structures with a large number of technological and operational defects, such as quenching cracks, welding defects, non-metallic inclusions, pores, pitting corrosion, nicks, undercuts, scratches, machine marks and others. VST cyclic loading leads to accumulation of damages starting from local plastic deformation causing shear bands and micro cracks, one of which and, in some cases, several of them evolve into a main crack resulting in the limit state — tank destruction.

According to the regulations on tank service at oil pumping stations the presence of cracks of any size in VST bearing elements is prohibited. It is connected with the possibility of brittle fracture of vertical steel tanks [5,6].

2. Methods and results

For the research, two fragments 750×750 mm in size have been cut out from the tank wall: one fragment with a crack – from the second belt, and another one, defect-free – from the first belt. Sampling of such fragments is due to the following reasons. The first and second belts of the wall are the most loaded elements of the tank design and they experience the greatest radial motion. Internal stresses in the metal of the first belt are greater than in the second one, as the height of the tank wall and the liquid acting on the first belt is greater than the height of the same parameters acting on the second belt. However, despite this fact, the crack was formed not in the first belt, but in the second one. Therefore, it has been decided to compare metal properties of the first and second belts.

As an assumption concerning the possible reasons of the crack origin in the second belt of the tank a hypothesis has been put forward that the material of the wall doesn't conform to the designed material. To prove this assumption the analysis of chemical composition has been carried out which showed that the material of the first and second belts by its elements content matches the steel grade 09G2S. The results of chemical analysis of the fragments material and the requirements of the standard documentation are given in table 1.

Table 1. Chemical composition of investigated steels.

Mass fraction of elements	Standard value for steel 09G2S (GOST 19281-89*)	Value for fragment of first tank ring	Value for fragment of second tank ring
Carbon [%]	Not greater than 0.12 ±0.02	0.11	0.14
Silicon [%]	(0.5... 0.8) ±0.005	0.56	0.50
Manganese [%]	(1.3...1.7) ±0.1	1.46	1.57
Chromium [%]	Not greater than 0.3 ±0.005	0.05	0.15
Copper [%]	Not greater than 0.3 ±0.005	0.13	0.28
Nickel [%]	Not greater than 0.3 ±0.005	0.24	0.24

In order to determine the conformity of the investigated fragments mechanical properties to the requirements of the standard documentation, tension and impact strength tests were carried out. Mechanical properties – ultimate strength, yield strength and elongation – were determined on flat XV-type samples in accordance with GOST 6996-84. Impact strength was determined on standard VI-type (Figure 1a) samples 10×10×55 mm in size with U-shaped (Mesnager piece) at temperatures +20C °and -40C°. Tests were conducted according to GOST 1497-84 and GOST 7564-84.

a b

Figure 1. Impact strength type samples
(a – U-shaped (Mesnager piece); b – V-shaped (Charpy specimens))

Figure 2 provides stress-strain diagrams of cross-section specimens cut out from the first and second tank rings: the former has a clearly pronounced yield plateau, the latter has almost none. The results of processing the machine stress-strain diagrams are given in table 2.

Impact strength testing of Charpy specimens is harsher, as their notch radius is 0.25 mm. These tests allow us to reveal more precisely the susceptibility of steels to brittle fracture [5-9]. IX-type specimens (Charpy specimens) were tested at temperatures +20°C, -5°C and -40°C. Temperatures +20°C and -5°C were taken in order to determine the conformity of the investigated steels to SNiP 2.05.06-85 requirements. Temperature -40°C was taken considering the region of tank construction. Tests were conducted on an impact pendulum-type testing machine. Negative temperatures were gained by means of liquid nitrogen. Aviation kerosene was used as a coolant. Temperature was measured with a low-temperature thermometer graduated in 0.1 °C. Tests were conducted according to GOST 9454-84.

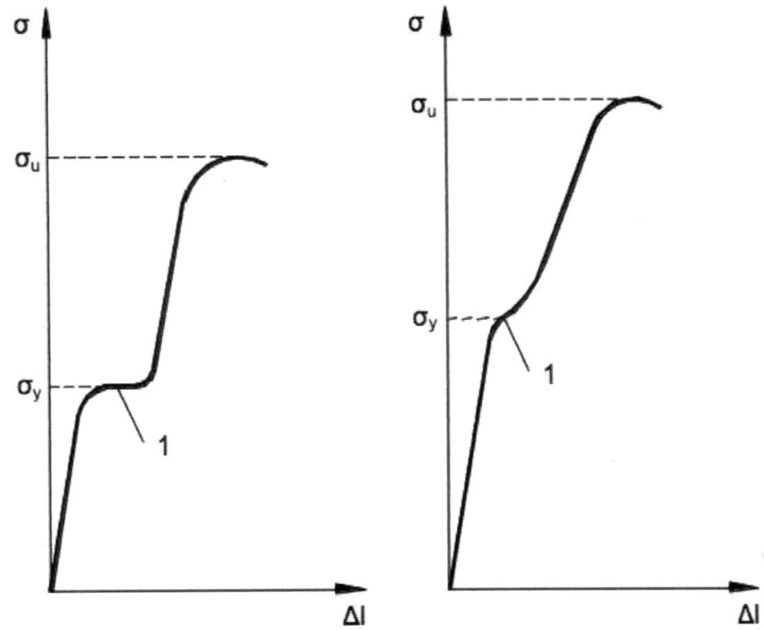

Figure 2. Stress-strain diagrams of specimens from the first and second tank rings (1 – yield plateau)

Table 2. Mechanical properties of investigated steels.

Physical property	Standard value for steel 09G2S (GOST 19281-89*)	Value for fragment of first tank ring	Value for fragment of second tank ring
Ultimate strength σ_u [MPa]	470	480	540
Yield strength σ_y [MPa]	325	330	353
Elongation ε [%]	21	24	18
Impact strength KCU [J/cm^2]	> 59	100 – 105	57 – 60
Impact strength KCU^{-40} [J/cm^2]	> 39	65 – 74	32 – 35

As is clear from tables 2 and 3, the values of impact strength for the first ring exceed the values specified by regulatory documents. The results of testing a fragment from the second ring are different. At room temperature the impact strength matches the requirements. However, at temperature -40°C the impact strength is below standard values. Besides, when testing Mesnager pieces the difference is small, and on Charpy specimens the values differ very strongly – almost twofold.

Table 3. Impact strength when testing Charpy specimens.

Impact strength	Standard value for steel 09G2S (SNiP 2.05.06-85)	Value for fragment of first tank ring	Value for fragment of second tank ring
KCV [J/cm^2]	Not regulated	83 – 86	46 – 51
KCV^{-5} [J/cm^2]	Not regulated	Not determined	27 – 31
KCV^{-40} [J/cm^2]	> 29.4	35 – 41	14 – 18

Thus, it demonstrates that the rings made of the same grade steel have different properties. To find out what is the difference in steel properties caused by, hardness measurement has been performed. The results of hardness measurements are given in table 4. Hardness measurements were conducted with a TSh-2M device at ball diameter D=10 mm and load P=30 kN. Tests were carried out according to GOST 9012-59.

Table 4. Results of measuring hardness of investigated steels.

Fragment	Value of parent metal hardness [HB]	Value of assembly weld hardness [HB]	Value of hardness in heat-affected zone of assembly weld [HB]
The first ring	148	155	-
The second ring	166	176	170

As it can be seen from the data given in table 4, the value of hardness for the second ring is greater than that for the first one. The value of hardness depends on where it was measured.

3. Discussion

Design structural stresses in vertical steel tanks reach up to 80 % of the yield point. Under such conditions, local defects that are stress concentrators cause the formation of zones where the value of yield strain is so great that micro cracks occur at an early stage of loading.

There are cases when structures, for example, gasholders, that passed acceptance tests at loads 1.4 times exceeding the design ones, either were destroyed after 5 years of operation or previously unregistered cracks were discovered upon inspection.

Analysis of test results allows us to deduce an inference that referring to changes in mechanical properties of a material of vertical steel tank design elements as operational defects is justified (table 5).

Table 5. Changes in mechanical properties of investigated steels.

Physical property	Change for fragment of first tank ring	Change for fragment of second tank ring
Ultimate strength σ_u	2.1 %	14.9 %
Yield strength σ_y	1.5 %	8.6 %
Elongation ε	14.3 %	-14.3 %

It is obvious that the presence of assembly welding fixtures speeds up degradation processes in steel. So, in the zone of the assembly weld changes in mechanical properties are considerably greater: ultimate strength is increased by 15 %, yield strength – by 9 %. A hardness increase in this zone is thus observed.

Almost all large-size tanks, operating in the oil transportation system, were assembled by the scrolling production technique. Some disadvantages of this technique are multiple imperfections of the geometric shape and assembly welded joints of metal structures. At present, as the operation life of most tanks exceeds the designed one, there arose a problem of evaluating the stress-strain state of such VSTs.

The place of crack formation in the investigated tank coincides with the largest wall deformations under the operational load. This can indicate a more intensive development of fatigue damage in metal tank structures with imperfections of their geometric shape.

Timely and high-quality evaluation of the technical state and elimination of defects of structures improves their service reliability. Such evaluation can be obtained only based on a complex inspection, including non-destructive testing of weld joints, metal quality testing, well thickness control, geometric shape control, etc. The required extent of such inspection considering the operation life: for new tanks — total control, for in-service tanks — partial control, for tanks at the end of operation life — inspection according to an individually developed program and investigation of changes in metal properties of tank structures.

4. Results and conclusion

The results of studying the RVS-20000 tank metal after long-term service show a change of mechanical properties, with the greatest changes occurring in a place of the assembly welded joint at the second ring level. At operation life of more than 10 years there occurred an increase in ultimate strength by 2 % and yield strength by 1.5 % for a defect-free design element – the first belt of the tank wall; and approximately by 15 % and 9 %, correspondingly, for the second ring having an assembly welded joint. This confirms the necessity to record changes in mechanical properties of metal as an operational defect of vertical steel tanks when determining the residual operation life during technical diagnosing. It is necessary to pay special attention to places of welding the assembly welded fixtures.

References

[1] V Kurushina, Y Zemenkov Innovative cyclical development of the Russian pipeline system WIT Transactions on Ecology and the Environment (2014)

[2] Yu D Zemenkov, V V Shalay, M Yu Zemenkova Expert Systems of Multivariable Predictive Control of Oil and Gas Facilities Reliability J Procedia Engineering **113** (2015) 312-315

[3] A V Votinov Ocenka strukturnyh parametrov stalej i resursnyh harakteristik rezervuarov dlja hranenija nefteproduktov v uslovijah dlitelnoj jekspluatacii (2006) 24

[4] R A Mamadaliev, V N Kuskov, U D Zemenkov, A A Popova Influence of High-Concentrated Heat Sources on Alloying Elements Transition into the Weld Metal J Applied Mechanics and Materials **770** (2015) 19-22

[5] I M Rozenshtein Brittle fracture of steel weldments J Zavodskaja laboratorija. Diagnostika materialov Factory Laboratory Diagnosis materials (2008) 48-51

[6] I M Rozenshtein On the issue of the oil product storage tank brittle fracture hazard analysis J Territorija «NEFTEGAZ» Oil and Gas Territory **10** (2015) 110-114

[7] P P Lazarevsky R A Gizatulin Y Romanenko D V Valuev A V Valueva A Serikbol Extraction of chromium from carbon ferrochromium residual wastes, J. IOP Conference Series: Materials Science and Engineering. **91** (2015)

[8] R A Gizatulin D V Valuev, V A Trifonov A V Valueva A Serikbol Oxidation processes in blowing steel with inert gas into the ladle J. IOP Conference Series: Materials Science and Engineering **91** (2015) p. 1-7

[9] G V Galevsky V V Rudneva A K Garbuzova D V Valuev Titanium carbide: nanotechnology, properties, application J IOP Conference Series: Materials Science and Engineering **91** (2015)

IOP Conf. Series: Materials Science and Engineering **127** (2016) 012050 doi:10.1088/1757-899X/127/1/012050

Metallurgical Properties and Phase Transformations of Barium-Strontium Modifier

M A Platonov[1, a], I S Sulimova[2, b], I D Rozhikhina[3,c], V I Dmitrienko[4,c], G V Horoshun[5]

[1,2,5]National Research Tomsk Polytechnic University, Yurga Technological Institute branch of Tomsk Polytechnic University 652050, Kemerovo region
[3,4]Siberian State Industrial University, Novokuznetsk, Russia

E-mail: [a] maxrauber@mail.ru, [b] siris78@mail.ru, [c] kafamsf@sibsiu.ru

Abstract. Metallurgical properties and phase transformations of barium-strontium modifier were tested in laboratory conditions resembling steel processing in furnace and ladle. When heating barium-strontium modifier start of melting, kinetics of decomposition, phase and structure transformation were studied. The concentrate under consideration has been revealed to be a complex mineral compound containing barytocalcite, calcite, calciostrontianite, dolomite and siderite. The reaction kinetics of decomposing mineral components of barium-strontium modifier to oxides does not considerably affect slag formation in conditions of out-of-furnace steel processing.

Introduction

Barium-strontium modifier BSC-2 is used in metallurgical production. To develop efficient technologies of its application detailed investigation into the processes taking place when introducing modifier into steel is required.

Chemical composition of barium-strontium modifier BSC-2, %: 16.0 BaO; 5.5 SrO; 21.5 CaO; 24.8 SiO_2; 0.9 MgO; 3.0 K_2O; 1.5 Na_2O; 4.0 Fe_2O_3; 0.2 MnO; 2.9 Al_2O_3; 0.9 TiO_2; 18.0 CO_2 (Technical Conditions 1717-001-75073896).

The research into mineral composition of barium-strontium modifier and integrated investigations of phase transformations provide means of obtaining data on nature and special features of modifier mineral formation, as well as change in the structure of raw materials under heating.

X-ray phase analysis at temperature T = 293 K (figure 1) has revealed core compounds barium-strontium modifier consists of are barytocalcite $BaCa(CO_3)_2$, calcite $CaCO_3$, calciostrontianite $CaSr(CO_3)_2$, dolomite $MgCO_3$ and siderite $FeCO_3$ impurities.

Differential thermal analysis implemented by means of thermal gravimetrical method was applied to study phase and structure transformations occurring in ore mineral components of barium-strontium modifier under heating.

[1,a] To whom any correspondence should be addressed.

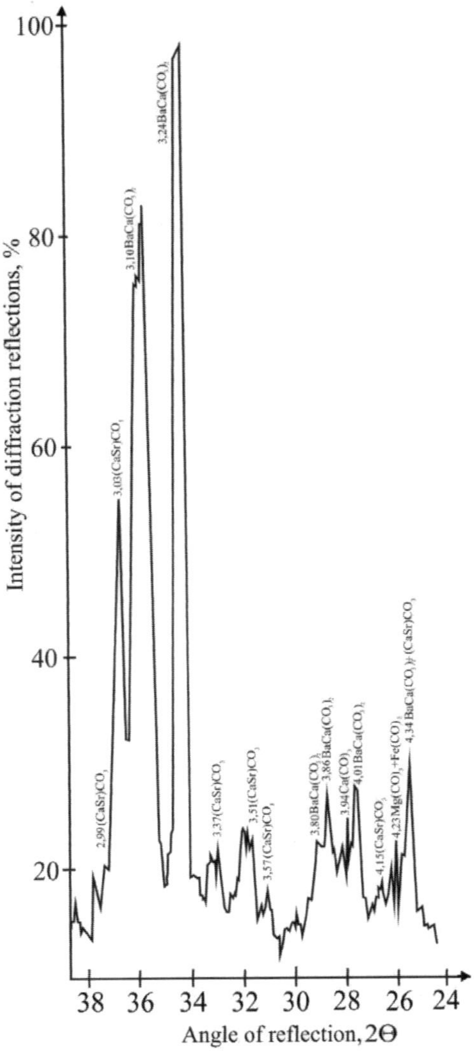

Figure 1. Fragment of barium-strontium modifier X-ray diffraction pattern.

The sample was heated up to 1223 K at rate 10 degree a minute. Survey mode: DTA = 1/5; DTG = 1/10, Pmass = 500 mg. Table 1 presents the results of differential thermal analysis of barium-strontium modifier.

Table 1. Results of differential thermal analysis of barium-strontium modifier

T, K	Phase composition
381	Phase transformations in impurities
568	Phase transformations in impurities
743	Phase transformations in impurities
913	Phase transformations in impurities
1018	Dolomite dissociation
1173	Calcite dissociation
1223	Barytocalcite and calciostrontianite dissociation

Figure 2 displays results obtained in the course of differential thermal analysis of barium-strontium modifier thermogram.

Thermo gravimetrical analysis of the modifier identified adsorbed water removal when heating up to 381 K. Further heating up to 381 – 913 K conditions phase transformations in impurities. Within temperature interval 913 – 1018 K dolomite dissociation resulting in MgO formation takes place. Further heating up to 1018 – 1223 K causes dissociation of calcite, barytocalcite, and calciostrontianite, alongside with mass loss 11.57 %.

Carried out investigations show barium-strontium modifier is a complex mineral compound containing barytocalcite, calcite, calciostrontianite, dolomite and siderite. Differential thermal analysis determined the sequence of phase transformations, arising in ore mineral components of barium-strontium modifier under heating.

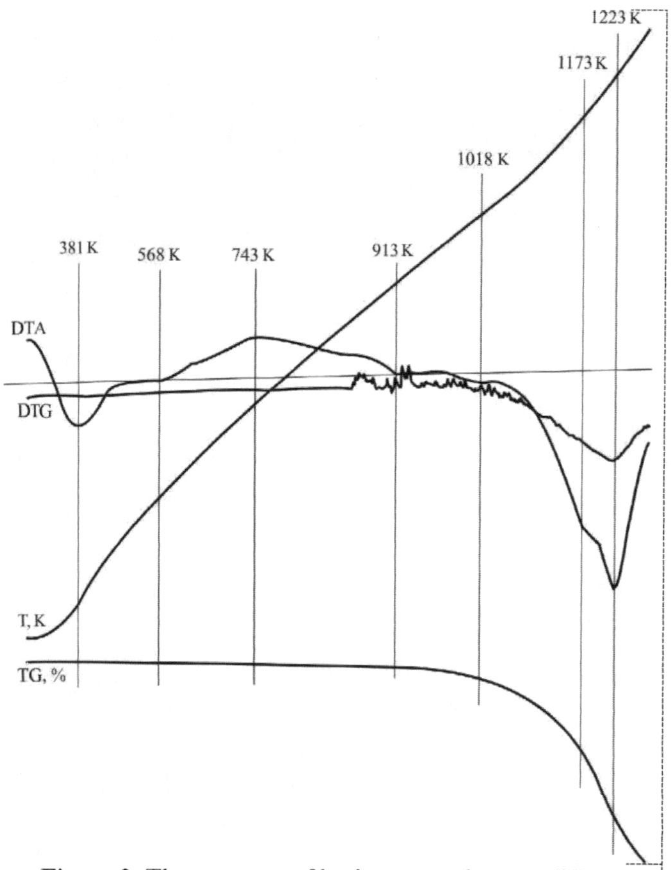

Figure 2. Thermogram of barium-strontium modifier subjected to differential thermal analysis.

The kinetics of modifier decomposition was tested to determine the influence of barium-strontium modifier on properties of forming slag. The kinetics of decomposition was studied gravimetrically at temperatures 1673 and 1773 K according to mass decrease of 20 g modifier sample, particle dimensions 0.01 – 0.1 mm. Mass decrease was registered every minute. The layout of the appliance is shown in figure 3.

Differential thermal analysis and investigations into kinetics of barium-strontium modifier decomposition (figure 4) pointed at successive reactions when heating barium-strontium modifier:

$$MgCO_3 \rightarrow MgO + CO_2\uparrow,$$
$$CaCO_3 \rightarrow CaO + CO_2\uparrow,$$

$$BaCa(CO_3)_2 \rightarrow BaCO_3 + CaO + CO_2\uparrow,$$
$$CaSr(CO_3)_2 \rightarrow SrCO_3 + CaO + CO_2\uparrow,$$
$$BaCO_3 \xrightarrow{>1273K} BaO + CO_2\uparrow,$$
$$SrCO_3 \xrightarrow{1373K} SrO + CO_2\uparrow.$$

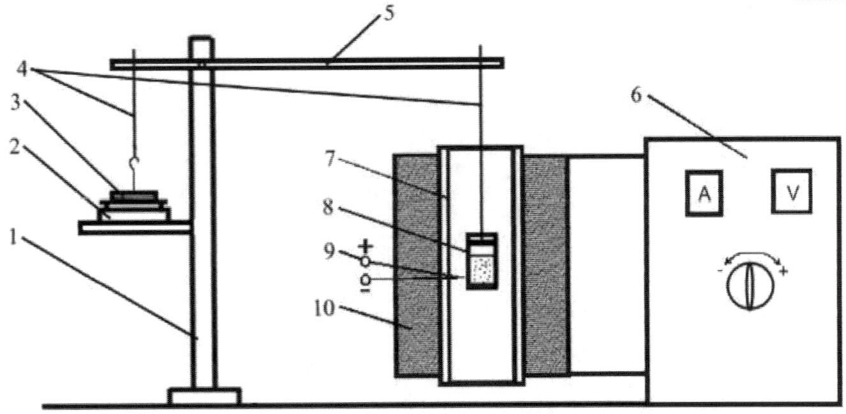

1 – support; 2 – electronic balance; 3 – counterweight; 4 – hanger; 5 – rocker; 6 – power transformer with furnace control unit;7 – heater; 8 – alundum crucible in graphitic cup containing barium-strontium modifier sample; 9 – thermocouple; 10 – resistance furnace with coal heater

Figure 3. Layout of experimental appliance.

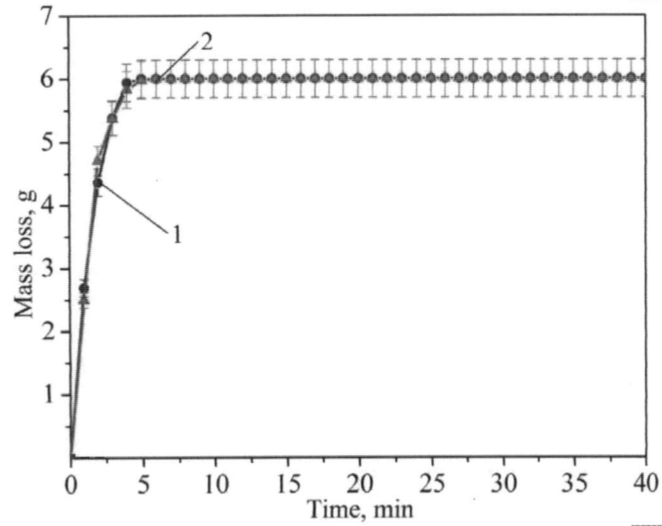

Figure 4. Dissociation kinetics of barium-strontium modifier at temperatures 1673 K (1) and 1773 K (2).

Therefore, temperature increase up to 1673 and 1773 K, reaction rate of decomposing mineral components of barium-strontium modifier to oxides do not affect slag formation in conditions of out-of-furnace steel processing.

Resistance furnace with platinum heater was used to specify the start of barium-strontium modifier melting. The layout of the furnace is displayed in figure 5.

When heating a platinum plate behavior of barium-strontium modifier placed on a hot junction was observed visually by a binocular microscope with 30 – 50 fold magnification in the reflected light.

To specify the start of melting a sample of barium-strontium modifier was grinded in the mortar so grains 0.01 – 0.1 mm were obtained, then 5 – 7 grains of matter under consideration were placed under a hot junction. The furnace heating rate up to temperature 1473 – 1573 K was 100 – 150 degree a second, then 20 – 30 degree a second, about the temperature of melting – 3 – 5 degree a second. The first attempt to determine melting temperature of barium-strontium modifier is an approximate one, used to define temperature boundaries of the experiment.

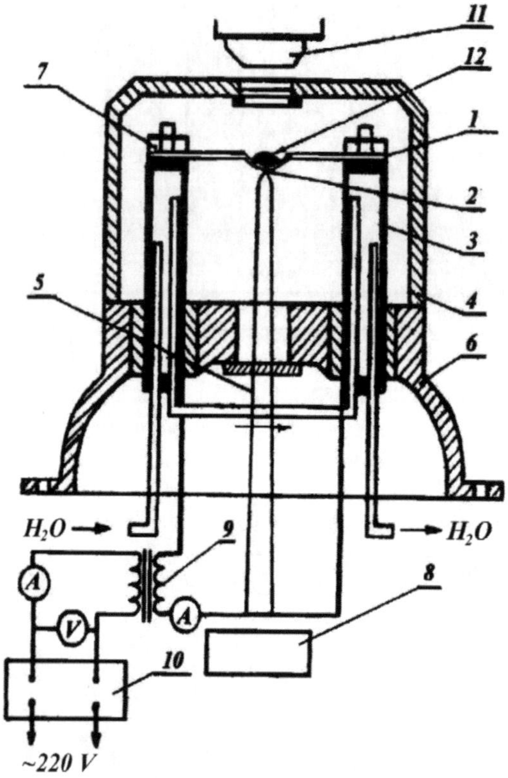

1 – platinum plate (heater); 2 – hot junction;
3 – shunt; 4 – furnace cover; 5 – thermocouple; 6 – case;
7 – heater fasteners; 8 – potentiometer; 9 – power transformer;
10 – adjusting transformer; 11 – microscope; 12 – furnace charge

Figure 5. Micro-furnace layout.

The temperature when sharp edges of barium-strontium modifier particles fuse was considered the start of melting. Starts of barium-strontium modifier melting are 1641 – 1655 K according to the results of carried out experiments represented in table 2.

Table 2. Starts of melting of barium-strontium modifier

№	Start of melting, K	
experiment	Barium-strontium modifier	Slag mixture
1	1653	1640
2	1641	1645
3	1655	1651
4	1653	1653
5	1645	1644
6	1653	1655
7	1653	1648
8	1644	1643
9	1653	1655
10	1653	1642

Mixture containing modifier (15 %) and lime (85 %) was melted at temperature 1873 K to find out the influence of barium-strontium modifier on properties of slag formed when processing steel in resistance furnace with coal tube heater. After the sample had been cooled down, melting temperatures of a slag mixture were tested according to the described above procedure. The results of carried out experiments demonstrated (table 2) starts of melting of a slag mixture are similar to those of barium-strontium modifier and are equal to 1640 – 1655 K.

Conclusions

The tests of material composition revealed the concentrate under consideration to be a complex mineral compound containing barytocalcite, calcite, calciostrontianite, dolomite and siderite. The reaction kinetics of decomposing mineral components of barium-strontium modifier to oxides does not considerably affect slag formation in conditions of out-of-furnace steel processing. Starts of melting are 1641 – 1655 K for barium-strontium modifier; 1640 – 1654 K for slag mixture containing the modifier (15 %) and lime (85 %).

References

[1] S S Chernyak High-Manganese Steel for Dredge in the construction State University Publishing Irkutsk (1996)
[2] V P Ivanova Thermal analysis of minerals and rocks Depths (1974) 399
[3] D M Heyker X-ray difractometry State Publishing of Physical and Mathematical Literature (1963) 380
[4] L M Kovba X-ray phase analysis Moscow University Press (1976) 232
[5] Ya S Umansky Roentgenography of metals Metallurgy (1976) 235
[6] P Paulik Certificate of derivatographic system Theoretical fundamentals Budapest: Hungarian optical factory (1974) 146
[7] L G Berg Introduction into thermography Science (1969) 395
[8] R A Gizatulin, D V Valuev, A V Valueva, C V Edesheva Melting of corrosion-resisting steels using air in bath agitation at the end of oxygen blowing J IOP Conference Series Materials Science and Engineering **66** (2014) 1-4
[9] G V Galevsky Rudneva, V V Garbuzova, A K Valuev, D V Titanium carbide: nanotechnology, properties, application J IOP Conference Series: Materials Science and Engineering **91** (2015) 1-7
[10] R A Gizatulin, D V Valuev, A V Valueva, A Serikbol, I F Borovikov Optimizing technological parameters of the reduction processes in treating steels in a ladle furnace J IOP Conference Series: Materials Science and Engineering **91** (2015) 1-7

The Stress-Strain Condition Estimation of Detail in Crack Tip by Integral Strain Gauges

V Syzrantsev[1] **and K Syzrantseva**[2]

[1,2]Tyumen State Oil and Gas University, 38 Volodarskogo st., Tyumen, 625000, Russia

E-mail: v_syzrantsev@mail.ru, kv.syzr@gmail.com.

Abstract. The paper considers the task of stress-strain condition calculation of experimental sample in fatigue crack tip on weld boundary at its cyclic deforming. For this task decision authors use the information obtained by original means of cyclic strains measurement: Integral Strain Gauges. The results of carried experimental researches are compared with data of stress-strain condition estimation of detail in crack tip calculated by Finish Element Method.

Introduction

There are presented in works [1, 2, 3] the methods of experimental estimation of strains and stresses, appearing on details surfaces at cyclic loading. These methods are based on new means of measurement – Integral Strain Gauges (ISG). It considers the gauge manufacture technology, the ways of obtaining information by gauges and data processing methods, allowing to estimate the loading, to calculate the longevity and to forecast the resource of machine parts and metalware at block and casual loading modes in conditions of stand and operational tests. The monograph [3] describes the methods of Integral Strain Gauges applications for experimental research of parts of transmissions and carrying systems of transport machines.

Accumulation of fatigue damages in process of cyclic deforming of detail leads to crack initiation, appearing as usual in stress concentration zone. At further cyclic deforming of detail the microcrack grows to macrocrack, which causes detail destruction [4, 5]. For estimation of fatigue crack growth rate it is necessary to determine the parameters of fracture mechanics, which can be calculated by computer simulation. Now for the solution of this task the Finish Element Method [6, 7, 8] is widely used. At the same time for correct use of simulation results it is necessary to have the data of its comparison with actual stress distribution law in crack tip. ISG pasted to the detail surface with crack allows to obtain the demanded data. The goal of described research was determination with the help an ISG the stress-strain condition of detail in the fatigue crack tip on weld boundary at its cyclic deforming. For this task decision authors carried out both experimental research using ISG and Finish Element Analysis by software ANSYS [8, 9, 10, 11] for the determination of strain distribution law in crack tip.

1. Experimental researches

The welded samples (Figure 1) made of a welded pipe were object of the experiment. Pipe material: 14HGS. Weld was placed on a middle part of a sample perpendicular to its longitudinal axis. ISG being the aluminum foil made on special technology [1, 3] was pasted to the sample face. The sample on a pulsating stand was subjected to deforming in the conditions of stretching at constant amplitude (F_{min} = 6 000 N, F_{max} = 35 000 N) during 80 000 cycles. During last 20 000 cycles the sample had crack. By photo equipment the digital pictures of ISG reaction on a face surface of welded sample (Figure 2), on the top surface of weld and on the lower surface of weld (Figure 3) were obtained. The analysis of the achieved information provides that intensity of ISG reaction in various zones of a weld significantly differs. The greatest ISG reaction is registered on a surface of the lower weld on weld boundary (Figure 3a). The maximum ISG reaction testifies to stress (strains) concentration in the researched place and, accordingly, the greatest accumulated fatigue damage. Confirmation of described fact is fatigue crack initiation on a welded sample in a zone of weld boundary of the lower weld with the main metal (Figure 3a). The obtained photos of ISG reaction (similar to data of work [12] show that stress distribution (accumulated during cyclic deforming of a sample) on a surface of a welded sample in its different zones (weld, zone of thermal influence, main metal, weld boundary) is very heterogeneous. The area in crack tip is characterized by still grated heterogeneity of stress distribution. Figure 4 illustrates the reaction of ISG placed on a sample face (Figure 1) in a location of the crack initiation (on the weld boundary).

a)

b)

Figure 1. a) The welded sample with ISG; b) face-grind of welded sample

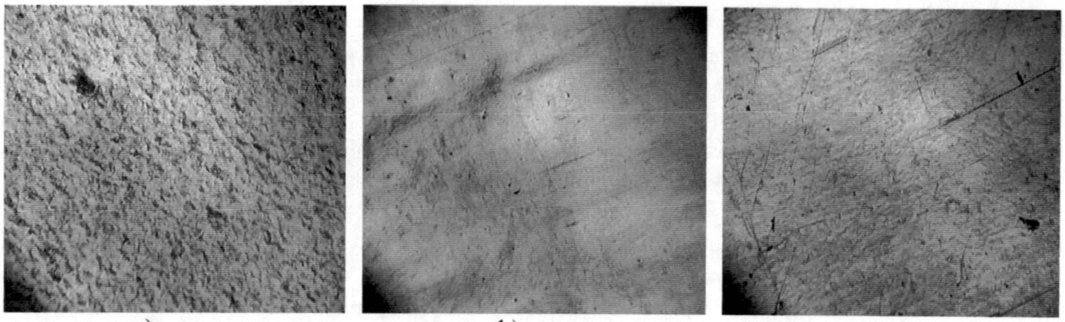

a) b) c)

Figure 2. The ISG reaction on different face sites of welded sample:
a) the top weld in a weld zone with the main metal; b) the lower weld in a zone of thermal influence; c) the lower weld in a weld zone with the main metal

a) b) c)

Figure 3. An ISG reaction after weld sample deforming during 60000 cycles:
a) on a surface of lower weld in zone of weld boundary; b) on a surface of main metal (top surface of a sample); c) on a surface of main metal (lower surface of a sample)

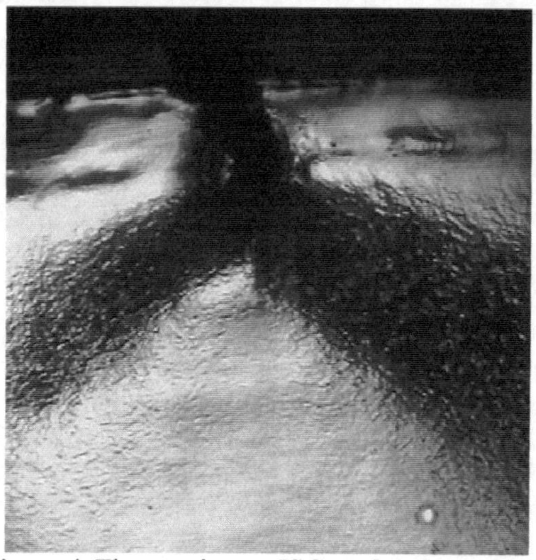

Figure 4. The reaction on ISG surface in crack tip

For an explanation of the obtained picture it is necessary to determine what deformations correspond to borders of ISG reaction. In the book [3] at research of load distribution in Novikov's gearing with the help an ISG it was established that ISG reaction border corresponds to compression strains of a constant value. For the purpose of verification of this assumption the stress-strain condition calculation of sample with crack was executed for identification of a picture of Figure 4.

2. Theoretical researches by the Finish Element Analysis

For the solution of a task the sample (Figure 5a) was used. Because of symmetry of a sample about two planes, only ¼ part of sample was subjected to the analysis. The calculation model of sample loading is submitted on Figure 5b. The task was solved by program complex ANSYS [9, 12].

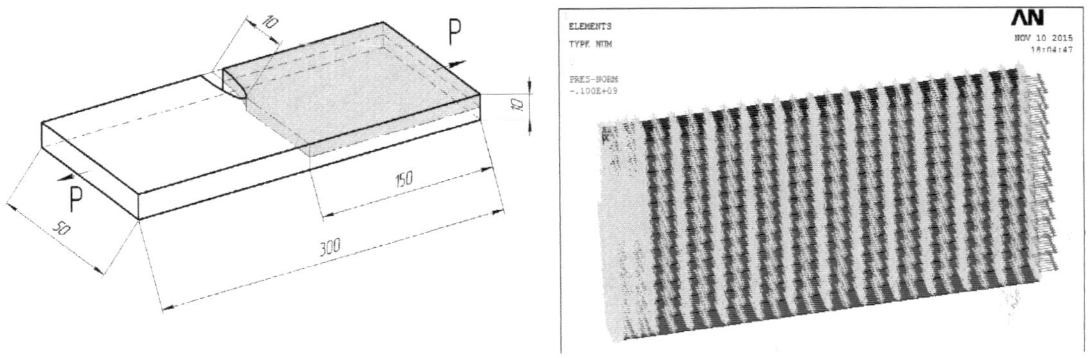

a) the geometrical description b) finish element model

Figure 5. The sample and its loading calculation model

The geometrical model was constructed taking into account possibility of calculation of destruction mechanics parameters: near crack tip field the auxiliary areas outlined by arches for generating in crack tip the singular elements were created. The areas defining the top side of a sample were meshed by two-dimensional quadratic finish elements. Around the crack tip the triangular degenerate singular elements with the intermediate nodes shifted on ¼ towards tip were created. Such elements allow to describe correctly stress surge in crack tip [8]. Other elements are set by the square quadrangular. As a result of extrude operation the three-dimensional quadratic elements in whole sample volume were generated.

On the section planes of sample the symmetry boundary conditions were defined (nodes with zero degrees of freedom are shown in Figure 5b by labels). Pressure 100 MPas modeling stretching of a sample and similar to the carried-out experimental researches was applied to a side surface of model. For the solution of this task the Sparse Solver was chosen. This solver is used in ANSYS by default.

After the solution is done it is possible in ANSYS to plot the results - standard physical quantities, such as displacements, strains, stresses and others. However for interpretation of ISG reaction it is necessary to have pictures of compression strains which aren't present in the ANSYS results list. The special macro [8] in the ANSYS Parametric Design Language realizing operations with arrays of element tables [13] was developed for creation of a compression strains distribution picture in this paper.

As a result of the executed computer simulation the pictures of compression strains in crack tip of a sample (Figure 6) are obtained which form is sufficiently close to the registered ISG reaction (Figure 4).

For comparison, Figure 7 illustrates the picture of distribution of equivalent stresses von Mises [14] near crack tip (standard result in ANSYS) which shows very high stress concentration in crack tip. The deformed shape of models shown on figure 6 and figure 7 is exaggerated by 50 times for visualization.

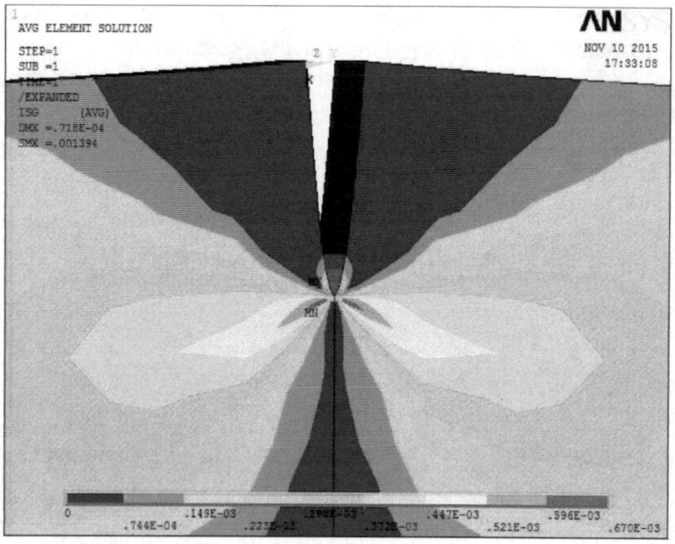

Figure 6. The results of calculation of compression strains near crack tip.

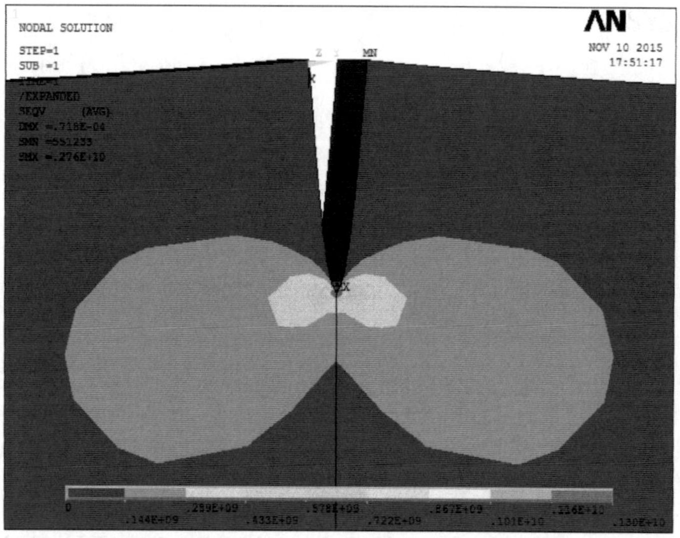

Figure 7. Equivalent stresses distribution in sample.

Two parameters are necessary for the analysis of crack growth: coefficient of stress intensity K and invariant J-integral [5, 15]. As shown in [8, 15], crack opening arises if K exceeds maximum limit. In case of the solution of an elastic task the value of J-integral is necessary only for more precise definition of K.

In this analysis for determination of the K_I – coefficient of stress intensity [15, 16] for a separation crack the KCALC command of an ANSYS was used, and for calculation of J-integral the special macro in the APDL-language was used [8]. As a result of calculations the following values were obtained:

$J = 5621.12$;

$K_I = 34.752 MPa \cdot \sqrt{m}$ in accordance with the method of displacements;

$K_I = 36.05 MPa \cdot \sqrt{m}$ in accordance with the method of J-integral.

The divergence in 3.6% in results of calculations of K_I by two different methods is very small and quite allowable.

Conclusion

The proposed computational-experimental method allows with sufficient degree of reliability to estimate the values of parameters characterizing the process of crack growth and appearing in equations for durability estimation and lifetime calculation of metal manufactures.

References

[1] Patent 2209412 C2 (RU). A method of manufacture of gauges for control of cyclic strains / V.N.Syzrantsev; Published in B.i.-2003. - No. 21. – 4 pages.

[2] Syzrantsev V., Golofast S., Syzrantseva K., Malenkov A. New means and methods of experimental research on cylindrical gears. Mechanics in Design MID-98.- Nottingham: Trend University, 1998. – P. 553-562.

[3] Syzrantsev V.N., Golofast S.L., Syzrantseva K.V. (2004) Diagnostics of loading and resource of parts of machines transmissions and carrying systems according to integral strain gauges indications. Novosibirsk: «Nauka», 2004, 188p. ISBN 5-02-031898-1.

[4] J. Toribio, J.C. Matos, B. González, J. Escuadra. Numerical modelling of crack shape evolution for surface flaws in round bars under tensile loading. Engineering Failure Analysis. Volume 16, Issue 2, March 2009, Pages 618–630. DOI:10.1016/j.engfailanal.2008.02.014

[5] S. Sankararaman, Y. Ling, C. Shantz, S. Mahadevan. Uncertainty quantification in fatigue crack growth prognosis. International Journal of Prognostics and Health Management. Vol.2 (2011) Available online: http://www.phmsociety.org/node/287

[6] J.Tinsley Oden (2010) Finite element method. Scholarpedia, 5(5):9836. DOI: 10.4249/scholarpedia. 9836.

[7] Wittbrodt E, Szczotka M, Maczy'nski A, Wojciech S (2013) Rigid finite element method in analysis of dynamics of offshore structures. Springer, Berlin. 252 p. 150 illus.— ISBN 978-3-642-29885-1 e-ISBN 978-3-642-29886-8, DOI 10.1007/978-3-642-29886-8.

[8] K.V. Syzrantseva. Computer aided engineering of load ability and deformability of oil and gas equipment units. Monograph. – Tyumen, TSOGU, 2009. – 124p.

[9] A.V. Oshibkov, D.D. Vodorezov, K.V. Syzrantseva, S.A. Frolov, M.V. Dvoynikov, V.N. Svetashov, D.L. Bakirov. (2015) The stress-strain condition estimation of construction of multilateral well junction. Oil Industry, 2015, no.6, p. 83-85.

[10] S. Nadimi, R.J. Khoushehmehr, B. Rohani, A. Mostafapour. Investigation and Analysis of Weld Induced Residual Stresses in Two Dissimilar Pipes by Finite Element Modeling. Journal of Applied Sciences, 8 (2008): 1014-1020. DOI: 10.3923/jas.2008.1014.1020

[11] V. M. Joy Varghese, M.R. Sumanth, M.R. Suresh. Numerical Simulation of Residual Stress in a Spot Welded Low Carbon Steel Plate. Procedia Engineering. Volume 38, 2012, Pages 2913–2921. DOI:10.1016/j.proeng.2012.06.340

[12] Mohammed Asif. M, Kulkarni Anup Shrikrishana, P. Sathiya. Finite element modelling and characterization of friction welding on UNS S31803 duplex stainless steel joints. Engineering Science and Technology, an International Journal. Volume 18, Issue 4, December 2015, Pages 704–712. DOI:10.1016/j.jestch.2015.05.002

[13] M.G. Gökçen, V. Kaleneroğlu. Finite element analysis of plasticity-induced fatigue crack closure using contact elements. High Performance Structures and Materials V. WIT Press (2010), 587p.

[14] I Salam, M A Malik, M Abid, M Farooque. Experimental and numerical study of fatigue crack propagation in a thick-walled cylinder under cyclic hoop stress. International Symposium on

Advanced Materials (ISAM 2013) IOP Publishing IOP Conf. Series: Materials Science and Engineering 60 (2014) 012026. DOI:10.1088/1757-899X/60/1/012026
[15] Manuel da Fonte, Manuel de Freitas. Stress Intensity Factors for semi-elliptical surface cracks in round bars under bending and torsion. International Journal of Fatigue. Volume 21, Issue 5, May 1999, Pages 457–463. DOI:10.1016/S0142-1123(98)00090-5
[16] Khoo Sze Wei, Saravanan Karuppanan. Stress Intensity Factor for Cracks Emanating from a Shaft. Journal of Applied Sciences, 11 (2011) 1839-1844. DOI: 10.3923/jas.2011.1839.1844

Improving Efficiency of Gear Shaping of Wheels with Internal Non-involute Gears

A Tarapanov[1], R Anisimov[2], N Kanatnikov[3], A Pilipenko[4]

[1,2,3,4]State University– Education-Science-Production Complex, Orel, Naugorskoe shosse 29 Russia 302020

E-mail: NVKanatnikov1989@gmail.com

Abstract. In the article questions of improving efficiency of gear shaping of wheels with internal non-involute profile gears. The technique of the analysis of kinematics parameters of gear shaping process is presented. Researches devoted to the influence of root profile form of processed wheel with interna. In the article questions of improving efficiency of gear shaping of wheels with internal non-involute profile gears. The technique of the analysis of kinematics parameters of gear shaping process is presented. Researches devoted to the influence of root profile form of processed wheel with internal non-involute gears on the thickness of the cutting-off metal layer, size of components of cutting forces and a roughness of the processed surface are given.

Introduction

In designs of modern cars and mechanisms details with internal periodic non-involute profiles are actively applied. These details are components of different gears and connections, such as cycloidal drives, epicyclic gearings, gear pumps, ratchet mechanisms, splines, etc., which are used in construction and mining machinery, agricultural machinery, an aircraft industry, the hoisting-and-transport equipment, railway transport and etc [1, 2].

One of the main problems of receiving internal gears of items is the operation determining the accuracy of gear wheel. More or less the only one operation allowing to receive internal gear in a mass production and providing the necessary productivity and accuracy at the minimum prime cost is gear shaping [3]. Modern designs of gear shaping machines include a large number of the axes operated from CNC for ensuring efficient work, reducing the time for readjustment, and solutions of various technological problems of a gear treatment, for example, cutting of two and more mutually oriented wheels. The similar equipment is created by the principles of modular configuration on the basis of one basic model, and in a modification design the concept of the main and support functions automation is realized. Advanced technologies of gear wheels production use machines with the CNC and the tool with high resistance that provides considerable speeds of processing (up to 320 m/min at gear shaping).

It should be noted that the technology of non-involute gear processing and tool profiling are developed insufficiently. In scientific and technical literature data on parameters and the processing modes is not provided.

Launching of gear wheels with internal non-involute profile gears, emergence of the new tool and processed materials forces to make new labor-consuming and expensive experimental investigations on the definition of the cutting modes and the extent on its influence on surface accuracy. As a result of it, it is necessary to conduct theoretical researches of the impact of method, processing mode, and a gear shape on a kinematic, power characteristics of process, accuracy and a roughness [4] of the received surface, tool wear [5] and process productivity.

Receiving the desired quality indicators and the maximum process productivity at a stage of preliminary design makes actual a problem of technique development of the complex analysis of a tooth profile parameters, physicomechanical properties of tool material and material of preparation, parameters of tool and preparation geometry, kinematic and dynamic characteristics of processing of gear wheel with internal gears.

1. The Complex Analysis of Shaping Process

The analysis of gear shaping processing parameters is based on mathematical display of the cutting scheme. Processing of wheel with internal non-involute gear profiles is characterized by a mutual generating cutting process of the cutting tool edges (figure 1) [3], which remain within hypocycloidal trajectory (extended and truncated) relative to the processing billet [6]. While developing this representation the following assumptions were accepted:

- the technological system is accepted as absolutely rigid;
- tool edges represent geometrical lines.

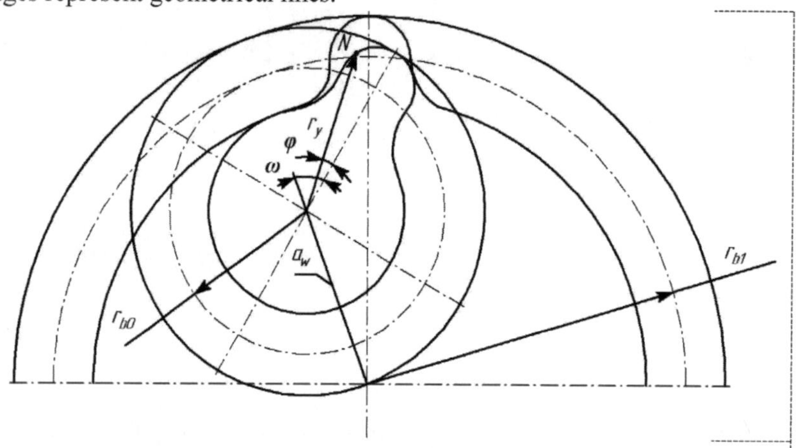

Figure 3. Scheme of shaping cutter position definition concerning a gear wheel in the complex analysis of shaping process on the example of a cycloidal profile.

The equations of points movement of the shaping cutter cutting edges and its center relative to the detail is presented in a parametrical form:

$$X = a_w \sin\left[U_0\left(\omega + \frac{S}{r_0}\left(1+\frac{l}{2L}\right)\right)\right] + \sqrt{x_{0i}^2 + y_{0i}^2}\,\sin\left[\left(1-U_0\right)\left(\omega+\frac{S}{r_0}\left(1+\frac{l}{2L}\right)\right) + \arcsin\frac{x_{0i}}{\sqrt{x_{0i}^2+y_{0i}^2}}\right],$$

$$X = a_w \cos\left[U_0\left(\omega + \frac{S}{r_0}\left(1+\frac{l}{2L}\right)\right)\right] + \sqrt{x_{0i}^2 + y_{0i}^2}\,\cos\left[\left(1-U_0\right)\left(\omega+\frac{S}{r_0}\left(1+\frac{l}{2L}\right)\right) + \arcsin\frac{x_{0i}}{\sqrt{x_{0i}^2+y_{0i}^2}}\right], \quad (1)$$

$$Z = l.$$

where a_w – is interaxal distance of a shaping cutter and the processed wheel, U_0 – is the relation of number of shaping cutter gear to the number of wheel gear, ω – is the corner defining position of shaping cutter tooth relative to the interaxal distance, S – is district giving, r_0 – is the radius of a dividing circle of a shaping cutter, l – is a width of a gear wheels, L – is stroke of shaping cutter, x_{0i} and y_{0i} are positioning coordinate of shaping cutter gear profile, received after profiling.

The analysis of the received representation of the cutting scheme allows to determine kinematic changes of corners by perimeter of an edge and thickness of the cut-off material layer for any point of a profile and at any time. These data are a basis of forecasting of cutting force components, and subsequently – of roughnesses of a gear wheel surface of non-involute wheels with internals gears and the accuracy of their processing.

Definition of kinematic changes of geometrical parameters of internal non-involute gear wheels processing (thickness of the cut-off layer, kinematic change of a forward corner and kinematic change of a back corner) was carried out with use of differential geometry methods, according to the formulas and the scheme that is presented in figure 2:

$$a = \frac{a_X V_{Xs} + a_Y V_{Ys} + a_Z V_{Zs}}{\sqrt{a_X^2 + a_Y^2 + a_Z^2}},$$
$$\Delta\gamma = \arctan \frac{a}{\sqrt{V_{Xv}^2 + V_{Yv}^2 + V_{Zv}^2}}, \qquad (2)$$
$$\Delta\alpha = \arctan \frac{\sqrt{V_{Xs}^2 + V_{Ys}^2 + V_{Zs}^2}}{\sqrt{V_{Xv}^2 + V_{Yv}^2 + V_{Zv}^2}}.$$

where a_X, a_Y, a_Z – are vector a coordinates, V_{Xs}, V_{Ys}, V_{Zs} – are vector V_S coordinates, V_{Xv}, V_{Yv}, V_{Zv} – are vector V_V coordinates.

Theoretical researches of processing by a wheels gear shaping with internal triangular, cycloidal gears and gears of ratchet gearing were carried out at the following general parameters: height h of gears is 5 mm, wheel number z_1 is 19, gear number of shaping cutter z_0 is 14, thickness of gear wheel b is 90 mm, gear wheel material is steel 45 (1045, C 45), district giving S is 0.44 mm/double stroke, cutting speed V is 20 m/min.

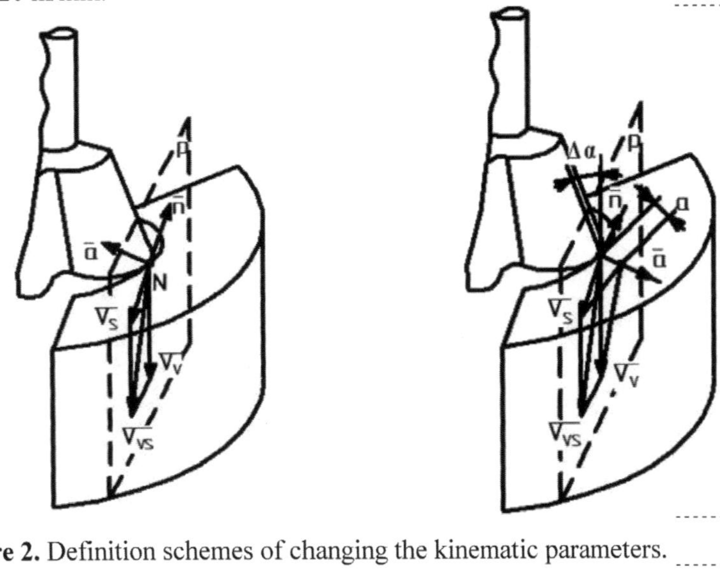

Figure 2. Definition schemes of changing the kinematic parameters.

It is established that at gear shaping the instrument gear has essential impact on change of thickness of the cut-off layers. S_o, at the identical modes of processing and heights of a profile the ratio between loading of an input and output edge is 1.2 for an evolvent root; 1.3 for a triangular and ratchet root; 2 for the cycloidal root.

Cutting forces are the key parameter of gear surface processing which defines quality of the cutting gear wheels and firmness of the cutting tool. Knowing cutting forces it is possible to calculate or reasonably choose the cutting tool, devices, and also the power spent for cutting and a rational operating machine mode.

Optimization of operational opportunities of gear shaping machines on its productivity and quality of the cut wheels with internal gears is, first of all, connected with determination of the maximum amplitude of cutting forces and their change during gear processing. Calculation of components of cutting force of wheels gear shaping process with internal gears consists in determination of specific forces per unit of length of the shaping edge cutting edge and their subsequent summation on all cutting an edge, participating in cutting. For determination of specific cutting force ΔP experimentally receiving dependences of cutting force for the range of thickness of the cut-off layer from 0.01 to 0.5 mm, from -0.0175 to 0.0349 radians for forward corners and from 0.87 310- 2 to 0. 0524 radian for back corners and also the range of the cut-off thickness from 0 to 0.01 mm. Such division is necessary for the accounting of cutting process and a for metal layer crushing that is deleted by various sites of the cutting edge. Values of components of cutting force for all cutting edges are received by summation of the specific forces arising on an elementary part of the cutting edge:

$$\begin{aligned} P_Y &= \sum \Delta P_y \Delta b \cos(\Phi + \psi_{y0})K, \\ P_X &= \sum \Delta P_y \Delta b \sin(\Phi + \psi_{y0})K, \\ P_Z &= \sum \Delta P_Z \Delta b K. \end{aligned} \tag{3}$$

where ΔP_Y, ΔP_Z are specific cutting forces, Δb – an elementary part of the cutting edge, F is a corner defining the wheel root positioning regard to interaxial distance, Δy_0 – is a corner between the point radius vector of the cutting edge and the line of gear symmetry, K is complexity coefficient of shaving formation.

In figure 3 schedules of components change of cutting force at processing by one gear of shaping cutter with internal gears of an evolvent, triangular, cycloidal profile and a profile of ratchet gearing from steel 45 (1045, C 45) are submitted with a height of tooth of h=5 of mm, S is 0.44 mm / double stroke, at the standard modes of cutting. Figures shows that the profile of the processed root has direct impact on the making cutting forces that is connected with change of length of the active cutting edge of an edge of shaping cutter gear and the total area of the cut-off material layer during processing.

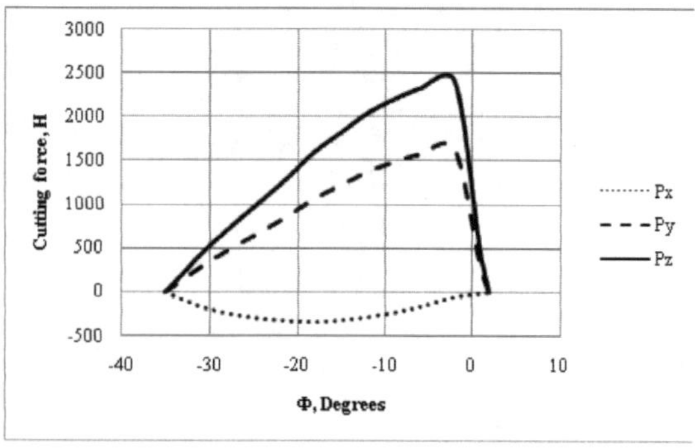

a

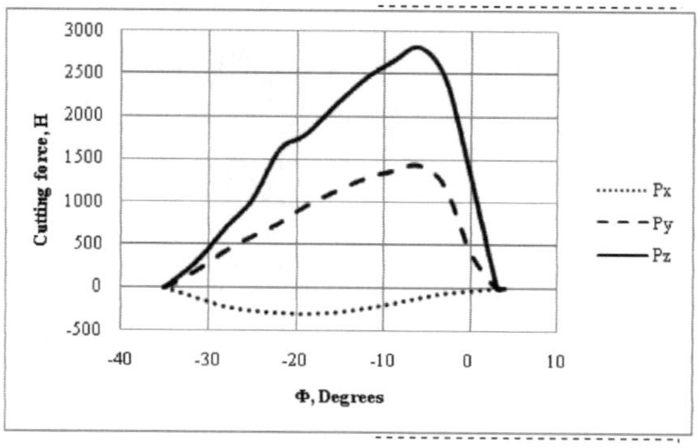

b

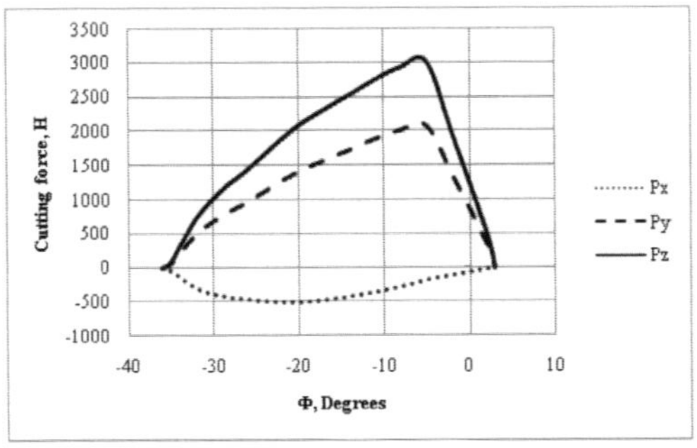

c

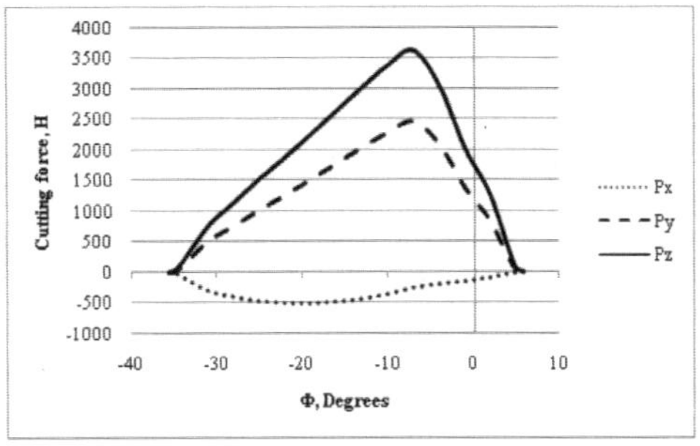

d

Figure 3. Cutting force during the work of one gear of shaping cutter at a gear shaping of wheels with internal gears: a – for evolvent profile, b – for triangular profile, c – for cycloidal profile, d – for a profile of ratchet gearing.

In Fig. 4 change of cutting force components in operating time of all gears which are in cutting when processing wheels with internal evolvent, triangular, cycloidal gears and gears of ratchet gearing is presented. Researches of cutting force components showed that the gear shaping of wheels with internal gears of non-involute profile from steel 45 (1045, C 45) with a gear height is h=5 mm, S is 0.44 mm / double stroke, V = 20 m/min is characterized by the following ratios: processing of a triangular and evolvent profile – P_{YT}/P_{YE} = 0.81, P_{ZT}/P_{ZE} = 1.09, P_{ZT}/P_{ZE} = 1.28, a cycloidal and evolvent profile P_{YC}/P_{YE} = 1.38, P_{ZC}/P_{ZE} = 1.38, a ratchet and evolvent profile – P_{YR}/Y_E = 1.27, P_{ZR}/P_{ZE} = 1.28.

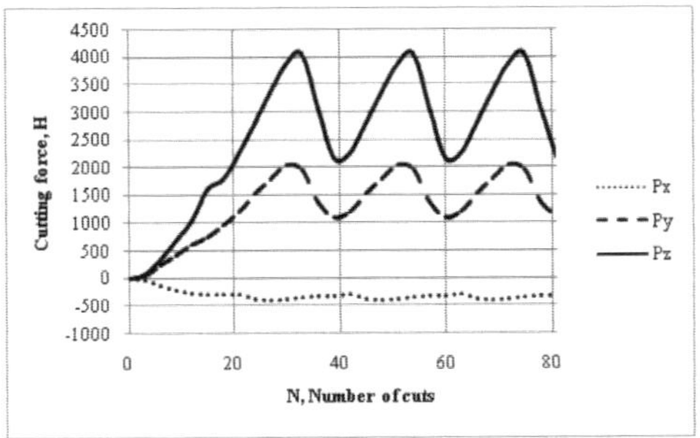

a

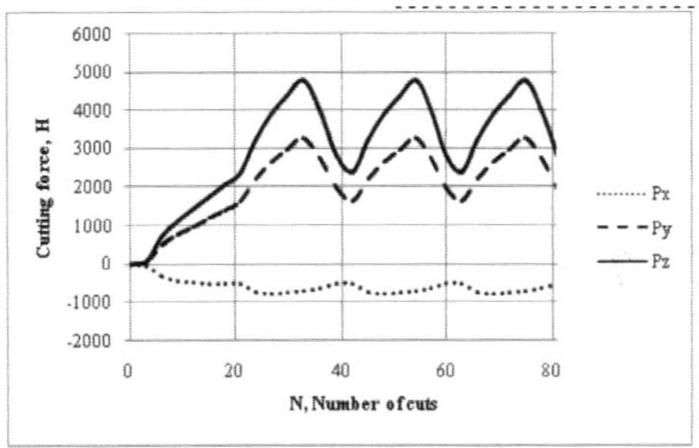

b

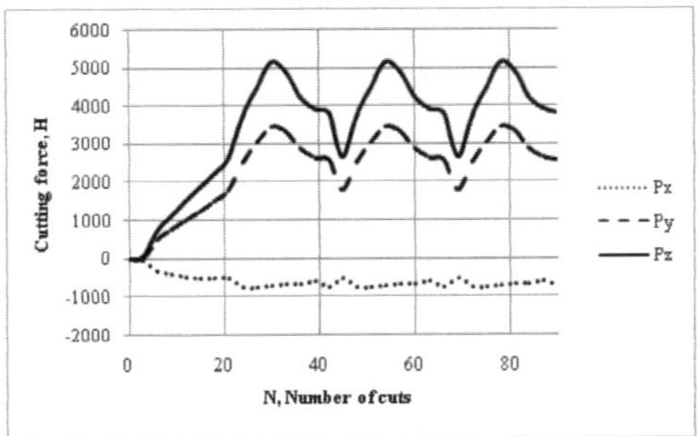

c

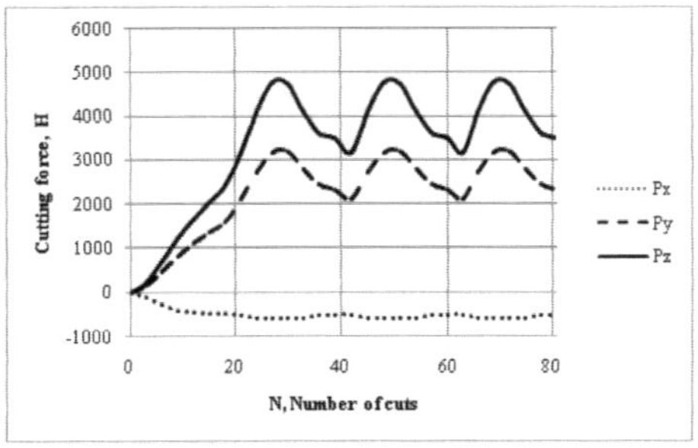

d

Figure 4. Cutting forces at a gear shaping of wheels with internal gears: a – of evolvent profile, b – of triangular profile, c – of cycloidal profile, d – for a profile of ratchet gearing.

2. Experimental Research

Experimental research of cutting force components at a gear shaping of wheels with internal non-involute gears was made by means of a universal dynamometer of UDM-600, the personal computer and the LabView environment. Works were performed on the gear shaping machine 5140. The program of pilot studies provided measurement of three mutually perpendicular components of cutting force at processing a wheel with internal gears of a triangular profile. The universal dynamometer UDM-600 was applied to the definition of three components of cutting force.

The system works as follows. Signals from the strain gages pasted on elastic elements of a dynamometer arrive on a payment of balancing resistance, then on amplifiers of signals of SG - 3016 which are working from a voltage of 15V. After that signals from amplifiers are given on a payment of ADC of the personal computer which directly transmits signals to the LabView software. By means of a program application on the monitor of personal computer curves that are equivalent to cutting forces are displayed.

When carrying out experiment specially developed adaptation serving for fixing of billet of a wheel with internal gears of non-involute profile in UDM (figure 5) was used. The adaptation is attached on UDM-600 by means of 6 openings. Fixing of billet is carried out by holding straps on openings. Experiment duration on the developed installation takes few minutes. Before measurement of cutting forces the dynamometer calibration in a static state by loading of cutting (serial) in the direction of force components action is carried out and the indication of a virtual oscillograph in the LabView environment on this component is fixed.

Because of casual errors of separate measurements cannot be excluded completely, for reduction of value of measure of inaccuracy some parallel experiments were made. At measurement of cutting force their number equaled 6, 7 or 8.

In the process of experimental researches establishment of cutting force components dependence on the processing modes (speed of cutting or giving), parameters of the processed billet (a profile of gear or the module) was planned. Preparations for research were selected from one party of delivery and were checked for hardness before researches, for more complete elimination of influence of fluctuations of hardness and a chemical composition of preparations on an error of measurement of cutting force. To exclude influence of scratches of the cutting edge on an error of measurements, before experiences the cutting tool is examining under a microscope. Wear on a back lateral surface of gears is allowed no more than 0.03 mm.

Figure 5. Appearance of experimental installation for measurement cutting forces

The obtained data when carrying out experiment are given in Fig. 6, 7, 8, 9. The difference in values of cutting force between theoretical data and received experimentally makes 4-11%. Besides, P_X differs up to 4%, components P_Y and P_Z – to 11%. Also it was established that at increase in circular giving the making forces of cutting and amplitude of their fluctuation grow not in proportion to giving. This results from the fact that at increase in district giving thickness of the cut-off layer increase and the specific force that contains cutting force growth decreases.

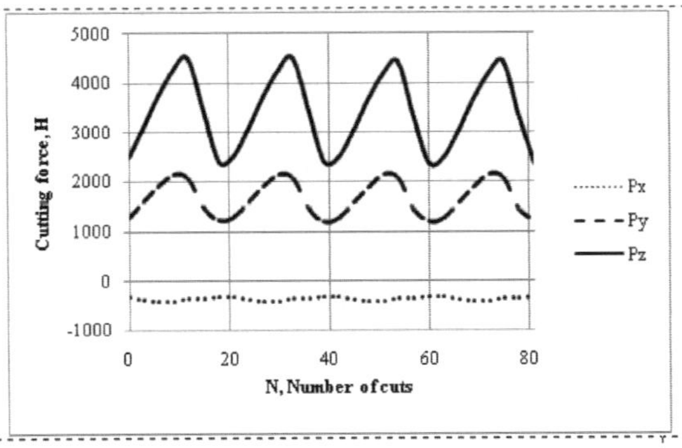

Figure 6. Change of cutting force components at a gear shaping of wheels with internal gears of the triangular profile.

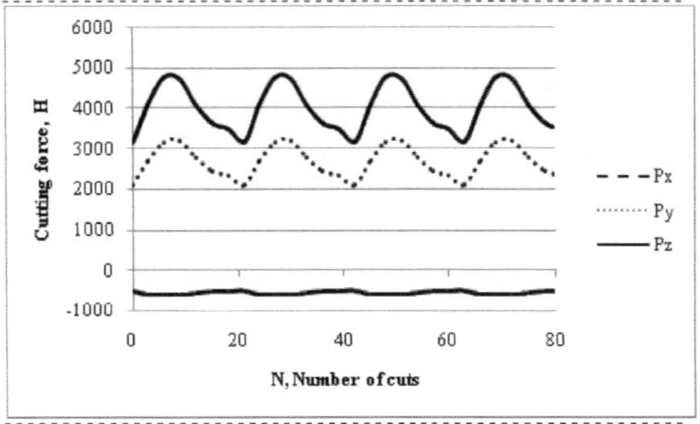

Figure 7. Change of cutting force components at a gear shaping of wheels with internal gears of the ratchet profile.

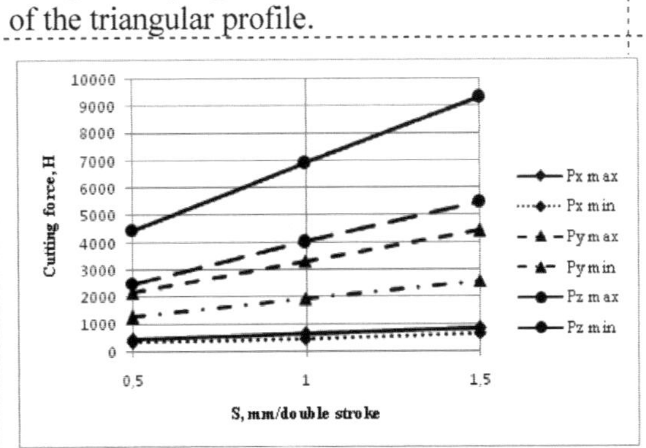

Figure 8. Cutting forces at a gear shaping of wheels with internal gears of the ratchet profile.

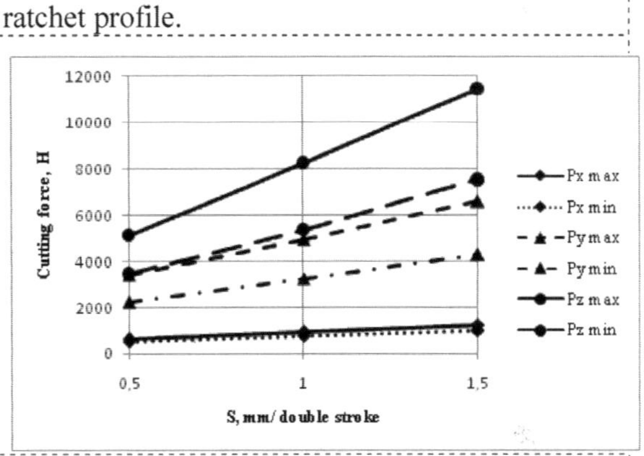

Figure 9. Cutting forces at a gear shaping of wheels with internal gears of the triangular profile.

3. Conclusions

The complex of universal mathematical software of display of the cutting scheme of a gear shaping of wheels with internal non-involute gears including the description of the tool movements and billet which analysis by means of provisions of differential geometry allowed to establish analytical dependences for definition of working corners of the cutting edge, thickness of the cut-off layer on a profile of the cutting edges, is developed. This further allows to define fluctuations of components of cutting force which have solving impacts on values of a roughness and accuracy of a product with internal non-involute gears. Dependences of ratios between loadings entrance and the output cutting edge are established at a gear shaping of wheels with internal gears with an identical height of gear which make: 1.2 for evolvent gears; 1.3 for triangular gears and gears of ratchet gearing; 2 for cycloidal gears. Numerical dependences between sizes of components of cutting force and a gear profile of shaping cutter are established.

IOP Conf. Series: Materials Science and Engineering **127** (2016) 012052 doi:10.1088/1757-899X/127/1/012052

References

[1] K Nagamura, K Ikejo, F G Tutulan Design and performance of gear pumps with a non-involute tooth profile Proceedings of the Institution of Mechanical Engineers, Part B: J of Engineering Manufacture. **218** (2004) 699-711

[2] F L Litvin, I Gonzalez-Perez, A Fuentes, K Hayasaka Design and investigation of gear drives with non-circular gears applied for speed variation and generation of functions Computer methods in applied mechanics and engineering **45** (2008) 3783-3802

[3] K D Bouzakis Manufacturing of cylindrical gears by generating cutting processes: A critical synthesis of analysis methods CIRP Annals-Manufacturing Technology **57** (2008) 676-696

[4] A Gok new approach to minimization of the surface roughness and cutting force via fuzzy TOPSIS multi-objective grey design and RSA Measurement **70** (2015) 100-109

[5] J Gerth, On the wear of PVD-coated HSS hobs in dry gear cutting Wear **266** (2009) 444-452

[6] G Figliolini, P Rea Effects of the design parameters of involute gears generated by rack-cutters International Gear Conference (2014) 294

IOP Conf. Series: Materials Science and Engineering **127** (2016) 012053 doi:10.1088/1757-899X/127/1/012053

Manufacturing Component Parts of Mining Equipment With Application of Hardening Technologies

D V Valuev[1,a], N N Malushin[2,b], A V Valueva[1,c], R S Dariev[1,d], R A Mamadaliev[3,f]

[1] 652050, Kemerovo region, Yurga, Leningradskaya str.26 Yurga Technological Institute branch of Tomsk Polytechnic University
[2] 654042 Kemerovo region, Novokuznetsk, Cosmic Highway 16, OJSC Evraz WSMK
[3] 625000, Tyumen, 38 Volodarskogo st.,Tyumen State Oil and Gas University

[a] email: valuev@tpu.ru

Abstract. To ensure highest hardness and wear resistance the authors develop an aggregate technology of manufacturing mining equipment component parts. The aggregate technology of manufacturing faced component parts includes the following operations: plasma-jet hard-facing with high-speed steels; high-temperature tempering after the facing, ultrasonic surface strengthening treatment, additional tempering, reconstructive facing.

Introduction

The materials developed on the base of chromium-tungsten hot-working and high-speed steels are the ones that most completely correspond to the requirements for the added metal applied for strengthening mining equipment component parts. Hot-working steels like R18, R6M5, R9, R2M8, 3H2V8 and other combine high working properties with insufficient weld ability. Normally, to prevent cold cracking the traditional facing technology involves mandatory application of high-temperature preliminary and additional heating ($T_{heating} = 400 - 700^{\circ}C$) and retarded cooling of the work piece. This process is accompanied by the formation of plastic products of austenite degradation characterized by low hardness and low wear resistance which, in its turn requires complicated thermal processing. In the process of heat treatment of a bimetallic work piece it is not always possible to take advantage of the properties of high-alloyed metal and provide its highest hardness [1-4].

Objective of the work – development of aggregate strengthening technology for manufacturing mining equipment component parts on the base of plasma-jet hard-facing with hard hot-working steels.

The peculiarity of hard-facing methods developed in Siberian State Industrial University is application of low-temperature preliminary and additional heating ($T_{heating} = 230\text{-}280\ ^{\circ}C$). To obtain the added metal with low crack formation capability we regulate the level of temporary stress during the hard-facing process by their partial relaxation due to kinetic plasticity effect which accompanies martensitic or bainitic transformations. The suggested thermal cycle of hard-facing consists of three stages. The first one provides limited time of heating and increased cooling rate in the region of high temperatures, prevents grain growth and austenite degradation with formation of equilibrium but low-strength structures. This stage can be implemented with application of high concentrated heating sources (for example, constricted or plasma arc) and cooling. The second stage of the thermal cycle provides the austenite state of the added metal when depositing all layers in the process of hard-facing.

It is achieved by application of heating with $T_{heating} = M_н + (50 – 100^{\circ}C)$. To obtain added metal with low capability of fracture formation we regulate the level of temporary stresses in the process of hard-facing at the third stage of the thermal cycle by temporary reduction of $T_{heating}$ below $M_н$. The temporary stresses reduce due to partial relaxation in the moment of martensitic or bainitic transformations. It allows obtaining added hard surfaced metal with low level of residual stresses. This facilitates the process of hard-facing in comparison to traditional technology, productivity of hard-facing increases, the properties of the high-alloyed added layer are implemented at their most [2 - 4].

The methods of multilayer hard-facing with high-speed steels developed and studied by us were taken as a basis for the technology of manufacturing bimetallic cold rolling work rolls [2, p.124]. The technological process of deposited work rolls manufacturing comprises the following stages: mechanical processing of the workpiece for hard-facing, plasma-jet hard-facing, facing of the deposited work roll, finishing mechanical processing, quality control of the active layer.

We recommend to apply 30HGSA steel as the base metal due to its high mechanical properties. When we complete facing according to the recommended thermal cycle the cooling rates in the region of minimal stability of 30HGSNA steel austenite are lower than the tolerance ones equaling 2.5–6.0°C/s. That prevents formation of cold cracks in the base metal when facing according to the developed method.

Facing of the workpiece was completed with application of the plant for plasma-jet facing of rotation bodies. The rollers were faced with plasma arc with feeding non-current-carrying welding powder wire PP-R181YuN into the weld pool. Argon was used as the plasma-supporting gas, nitrogen was used as the shielding one. Plasma-jet facing was completed according to the thermal cycle presented in Figure 1[3].

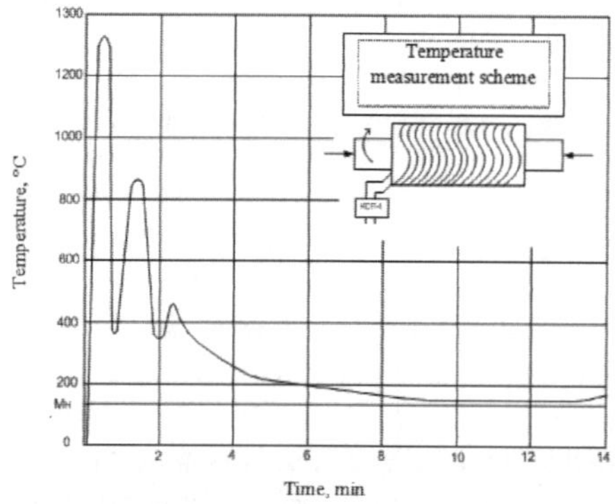

Figure 1. Thermal cycle under plasma-jet facing of the first layer of the roll 10 mm in diameter

The workpiece with the allowances for facing of 10-12 mm for one side is placed in the centers of the facing plant, previously heated up to 230 °C, and the roll neck cooling with water flow up to 2 l/min was applied. Cooling of the rolls with internal opening was completed with the help of a special device. After the preparation 5-6 layer facing was completed. Time duration of facing is 3 hours, when facing finished the roll is cooled in the air.

Pilot lots of rolls were faced for the rolling mill 6/100×315, rolling mill 425 and for ceramic sheet rolling according to the given technology. Visible inspection did not reveal any defects of facing. The quality of the faced surface is satisfactory. Hardness of the faced metal is HRC 54–58. Four-time one-hour tempering under 580 °C was completed to increase hardness and wear resistance. Influence of the number of tempering times upon chromium-tungsten metal is shown in Figure 2. As we can see from Figure 2 hardness of the active layer of the faced rolls answers the requirements imposed for the surface layer of the work rolls for cold rolling. After the tempering finish machining was completed.

It has been established that the work rolls of the cold mill produced with the application of plasma-jet facing of the active layer by hot-working high-hardness steels demonstrate increased wear resistance (by 1.5 … 2.0) in comparison to batch rolls.

Increased wear resistance of the faced rolls can be explained by fine carbides M_6C и MC in their structure. Compressive stresses in the surface layer of the faced rolls and absence of abrupt transition of compressive stresses into the stretching ones may be one of the factors contributing to the wear resistance increase. A typical structure of the faced metal (surface layer) is shown in Figure 2 (microscope OLYPUSGX– 51, pickling solution–HNO_3, 3% alcohol solvent).

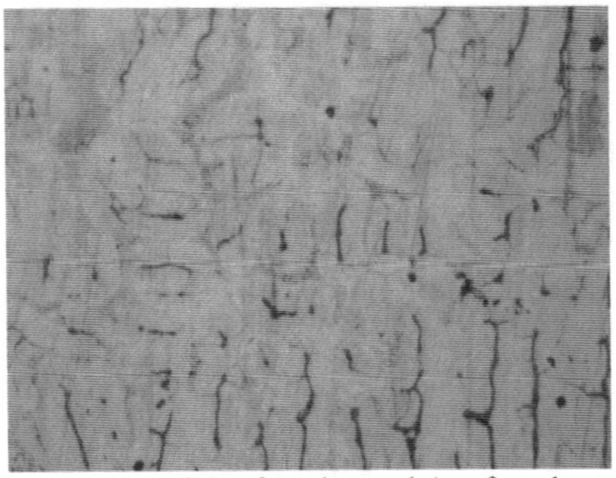

Figure 2. Structure of the faced metal (surface layer) × 1000

The main causes of failures of the faced rolls as well as those of rolls produced according to traditional technology include natural wear, delaminating and spalls. In [2, p.102] the authors note that delaminating is related to formation of fatigue failure sites inside the roll. Fatigue failure sites are situated in the weakened zone of the weld joint – the fusion area where the base metal was welded with the deposit one and where the metal properties change abruptly.

To improve the properties of the faced highly alloyed metal as well as to ensure favorable stressed condition we can additionally apply high temperature tempering after the facing. So, metal hardness after facing is HRC 52–57 and after tempering for secondary hardness it reaches HRC 62–64 which corresponds to 95–102 HSh (see Figure 3).

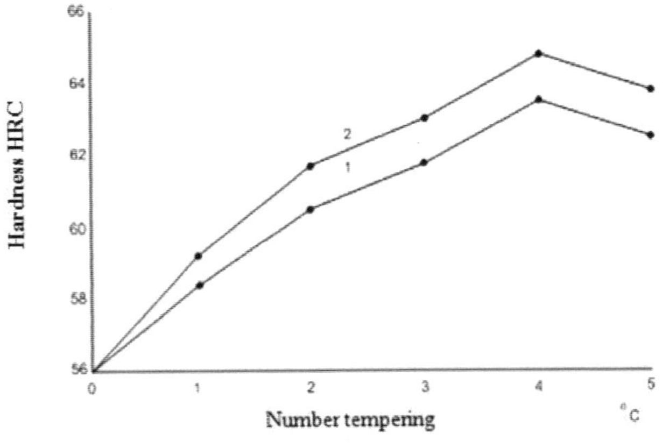

Figure 3. Influence of the number of tempers upon the hardness of the faced rolls

Additional application of ultrasonic surface hardening treatment after the facing allows increasing hardness up to HRC 64–66. Under the certain values of ultrasonic treatment intensity some characteristics of the hardened layer may deteriorate. At the same time there is the area of optimal values (P ≈ 10 daN, A ≈ 20 um, V ≈ 20 m/min) of the technology factors of ultrasonic surface treatment where we observe abrupt increase of all characteristics of the hardened layer of the faced metal [6].

Hardness of the surface of the faced disks is up to HRC – 65 and microhardness H_{100} is up to 12200 Mpa, hardening depth reaching 4 mm. In the work we reveal the optimal area of ultrasonic treatment parameters ($P_{cr} \approx 10$ daN, $A \approx 20$ um, $V \approx 20$ m/min) which, after certain adjustage and correction, can be used for surface strengthening of the faced rolls resulting in increase of the wear resistance of the rolls [3-8].

Additional application of nitrogen hardening of the faced high-speed metal is not recommended as the hardening effect is insignificant as it is shown in the work. Attempts of application of the faced metal for further nitrogen hardening showed that with high-alloyed steels of R18 type we cannot produce layer with the depth over 0.20-0.25 mm and hardness of 1100-1240 HV after nitrogen hardening. That is why introduction of additional hardening technology, such as nitrogen hardening, is inappropriate [9].

For laser-jet facing of the surface layer of metallurgic equipment component we developed low-alloyed flux-cored wire of steel 38HMYuAT type which allows increasing wear and corrosion resistance of the surface layer. Faced and nitrogen hardened components (flattering rolls) showed high wear and corrosion resistance which proves efficiency of complex application of plasma-jet facing with low-alloyed wire with further nitrogen hardening. The reserves for increasing the quality of the components faced by hot-working high hardness steels are included into the process of their exploitation by application further high-temperature tempering [10].

Application of high-temperature tempering in the process of exploitation of faced rolls for cold rolling brings the plastically deformed metal into more stable structural condition and the recrystallizing process removes strain hardness of the base and deposit metal fusion area and restores fatigue resistance of the base metal up to its initial values. Removal of the strain hardness prevents formation of the fatigue failure sites and significantly decreases the possibility of flaking. Thus, the main cause of rolls failure – strain hardness of the transition area – is eliminated and their resistance improves. The aggregate technology flow scheme of working efficiency increase of the faced components at all stages of their production and exploitation is provided in Figure 4.

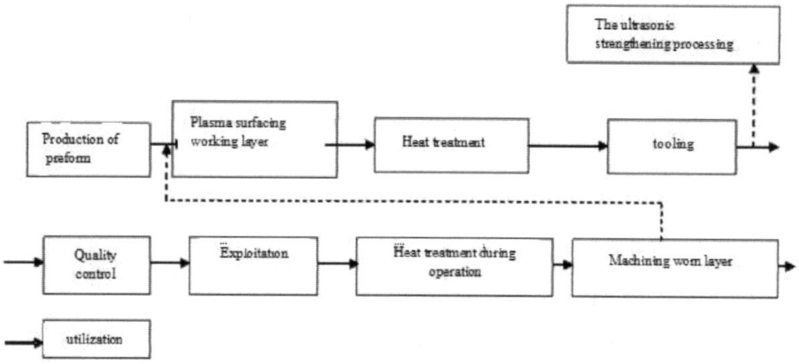

Figure 4. Aggregate technology of working efficiency increase of the faced components

We suggest applying active working layer of the roll by means of plasma-jet facing with high-speed high harness steels. To improve the properties of the faced high-alloy metal and to ensure favorable strained condition we can apply additional high-temperature tempering after the facing. Application of high-temperature tempering in the process of exploitation of the faced rolls results in additional increase of their resistance. In case of roll failure it is possible to complete its reconstructive facing.

Industrial tests showed that the suggested aggregate technology can be successfully used for producing faced metal with high hardness (≥HRC 64) and avoiding cold cracking.

IOP Conf. Series: Materials Science and Engineering **127** (2016) 012053 doi:10.1088/1757-899X/127/1/012053

The developed technology with application of the suggested facing methods, facing materials and other strengthening technologies was used for manufacturing a number of components for mining and metallurgical, mining equipment and agricultural machines.

Conclusion

1. Basing upon the completed research we developed an aggregate technology of manufacturing faced component parts based on the application of plasma-jet facing with non-current-carrying welding powder wires in the shielding and alloying atmosphere of nitrogen.

2. We developed an aggregate technology of faced component parts production which includes the following operations: 1) the basic technology – plasma-jet facing with high-speed steels; 2) additional operations – high-temperature tempering after the facing, ultrasonic hardening surface treatment after the finish-grinding, additional tempering in the process of exploitation, if further exploitation of the worn layer restoring facing is completed.

References

[1] Geller Yu A Tool steels, Metallurgiya 1975

[2] Malushin N N Valuev D V Quality control of metallurgical equipment component parts at all stages of their life cycle by means of plasma-jet facing with hot-working steels Tomsk Polytechnic University Press 2013

[3] Il'yaschenko D P Chinakhov D A Gotovschik Y M Influence of power supply energy characteristics upon the stability of MMA process J *Applied Mechanics and Materials* 756 (2015) 97-100

[4] Valuev D V Serikbol A Stress state and its definition in hardfacing heat-resistant steel cold rolling mill rollers J *Advanced Materials Research* 1040 (2014) 703-707

[5] Malushin N N Vostretsova T G Regulated thermal cycle for plasma-jet facing forming rolls with hot-working steels J *Blank productions in engineering* 3 (2014) 14 – 16

[6] Malushin N N Osetkovsky V L Facing component parts of metallurgical equipment with hot-working high-hardness steels with low-temperature heating. Technologies of strengthening, coating and repairing: theory and practice: in 2 parts Part 1 Works of 16th International scientific and practical conference Saint-Petersburg Polytechnic University Press (2014) pp 104-110

[7] Chinakhov D A Gas dynamic control of properties of welded joints from high strength Alloyed steels J *China Welding* 3 (2014) 27-31

[8] Saprykina N A Saprykin A A Matrunchik M S Formation of surface layer of cobalt chrome molybdenum powder products with differentiation of laser sintering modes J *Applied Mechanics and Materials* 682 (2014) 294-298

[9] Malushin N N Valuev D V Valueva A V Serikbol A Testing of metallurgical equipment on contact strength and wear resistance J *Applied Mechanics and Materials* 682 (2014) 53-57

[10] Paton B E Technology of electric welding of metals and alloys by fusion Mashinostroyeniye 1974

IOP Conf. Series: Materials Science and Engineering **127** (2016) 012054 doi:10.1088/1757-899X/127/1/012054

Traditional Technology of Chromium-Tungsten Steels Facing, its Disadvantages and Suggestions for their Eliminations

D V Valuev[1,a], N N Malushin[2,b], A V Valueva[1,c], R S Dariev[1,d], R A Mamadaliev[3,f]

[1] 652050, Kemerovo region, Yurga, Leningradskaya str.26 Yurga Technological Institute branch of Tomsk Polytechnic University
[2] 654042 Kemerovo region, Novokuznetsk, Cosmic Highway 16, OJSC Evraz WSMK
[3] 625000, Tyumen, 38 Volodarskogo st.,Tyumen State Oil and Gas University

[a] email: valuev@tpu.ru

Abstract. To reveal the disadvantages of the traditional technology of facing with chromium-tungsten steels analysis of the given technology was completed. The analysis showed that the main disadvantages of the technology are high-temperature heating and underutilization of high-alloyed metal properties. To eliminate the disadvantages we developed the methods of facing allowing obtaining faced metal which state is close to that of the hardened one without cracks.

Introduction

Facing materials developed on the base of chromium-tungsten heat-resisting high-speed steels are applied to strengthen the components of mining and metallurgical equipment. A characteristic feature of the given group of steels is the structure of the deposited metal consisting of martensite, retained austenite and carbides. Basically all variety of faced metal composition is further development and modification of traditional well-known and widely used tool steels of 3H2V8 and R18 types. The basic alloying elements in chromium-tungsten facing materials are carbon (0.2–1.5 %), chromium (1–6 %) and tungsten (1–18 %) [1]. Chromium-tundsten electrode materials are widely used when facing hot rolls, press matrices and hollow bars, grippers of crab cranes, blank dies, blades, scissors, fillerings, drift pins, cutters, rolls of the roller-tables, various axles, reels of the rolling mills, components of hammer breakers and other parts [2, 4].

Work objective – analysis of traditional technology of facing hardening chromium-tungsten steels and development of suggestions to eliminate its disadvantages.

The main difficulty which appears when facing with chromium-tungsten steels is formation of cold (hardening strain) cracks during the process and after the facing. Crack formation is impeded by temperature increase of martensite transformation and slowing down the cooling rate. The most efficient and the easiest method of preventing cracks is regulation of the thermal cycle by means of choosing corresponding methods and facing conditions as well as application preliminary heating if necessary. "Ideal" thermal cycle ensuring highest resistance against cold cracks formation is shown in Figure 1. Such thermal cycle allows avoiding overheating due to high-speed heating and cooling of metal under the temperatures over point A_1. Slowed down cooling under the temperatures below point A_1 decreases the possibility of hardening structures, thus, improving steel resistance to static fatigue and cold cracks formation as slow cooling contributes to development of pearlite and intermediate transformations of overcooled austenite in the faced metal and eliminates or dislocates $\gamma \rightarrow \alpha$ – martensite transformation into the high temperature region. It is rather actively promoted by slowed down cooling of the faced component in the interval of martensite transformation. As a result of martensite self-tempering under the temperatures below 350 C the crystalline structure of metal in the areas adjacent to the grain boundaries becomes more regular, general metal ductility increases and formation and development of cracks impedes. Real thermal cycles close to the "ideal" ones can be obtained under electron-beam welding with preliminary heating or under multilayer arc welding of thick metal with preliminary heating. These are the basic theorizes which underlie the traditional technology of facing hardening chromium-tungsten steels [2].

To obtain faced metal free of cracks and ensure required hardness and wear resistance when facing mining and metallurgic equipment components major attention is paid to the thermal conditions of facing: temperature of preliminary heating, thermal condition of the component in the process of facing and in the process of cooling after facing. Special attention is paid to more uniform and slow heating as too fast heating of certain places up to 350–450 C and cold neighbouring areas may lead to crack formation.

Temperature of component heating for facing depends upon many factors: composition of base and faced metal, form of the component, facing conditions and required properties of the faced metal. The more the faced metal is liable to crack formation the higher the temperature of preliminary heating must be as it is considered that otherwise crack prevention is impossible. To prevent cracks, reduce internal stresses and produce quite plastic structure of the faced metal we apply obligatory preliminary heating of components up to 350–400 C when facing with highly alloyed electrode wire PP-3H2V8 or whole-drawn EI-701. To eliminate cracks when facing high-speed steels we need preliminary and concurrent heating of the workpieces up to 500–700 C.

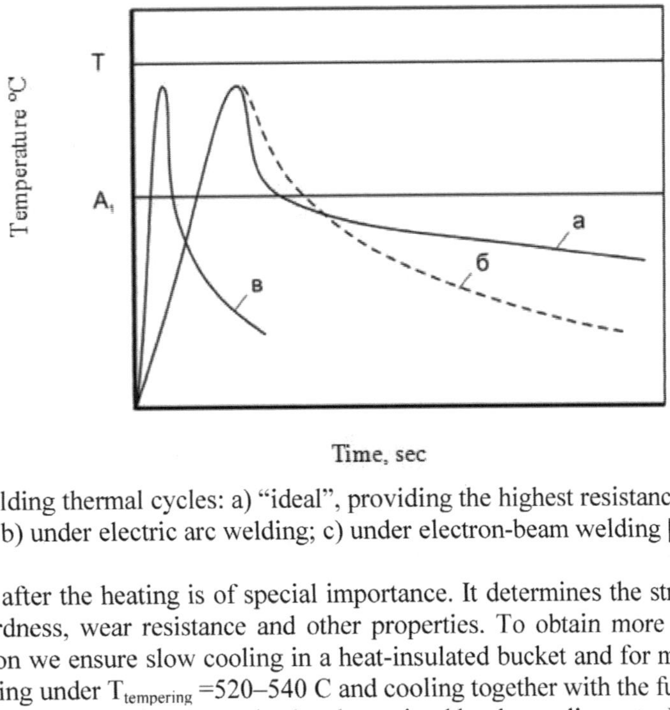

Figure 1. Welding thermal cycles: a) "ideal", providing the highest resistance to cold cracks formation; b) under electric arc welding; c) under electron-beam welding [4, p. 719]

The cooling rate after the heating is of special importance. It determines the structure of the faced metal and, thus, hardness, wear resistance and other properties. To obtain more plastic products of austenite deterioration we ensure slow cooling in a heat-insulated bucket and for massive components we complete tempering under $T_{tempering}$ =520–540 C and cooling together with the furnace [2].

Structure of the faced metal of R18 type is also determined by the cooling rate. Under the moderate cooling rate (facing with heating up to 500–600 C and further cooling with the furnace) crystals of δ–ferrite (contained in the solid solution of W and Cr) with reduced carbon content are the first to be formed. The melt is carburized with peritectic reaction taking place and δ–ferrite partly transforming into austenite. At the end of the hardening process ledeburite eutectic is exuded in the form of a net along the boundaries of dendrite crystals [1, 2].

The faced components are usually machined with preliminary annealing. When facing with steel of R18 type we recommend the following rather complicated cycle of the faced workpieces treatment [9]. Heating up to 660 C and holding for 4 hours, heating up to 880 C and holding for 9 hours, cooling with the furnace up to 750 C during 3.5-4 hours and holding under 750 C for 6 hours, cooling with the furnace up to 400 C and bloom discharge. After annealing hardness of the faced metal is 220-270 HB. Isothermal annealing of the faced metal of 3 H2V8 type is completed with holding under 750 C during 6 hours. Hardness after the annealing is 190-220 HB.

Application of slow cooling rates of the faced metal recommended for prevention of cold cracks formation results in the low hardness of the faced layer and, thus, in low wear resistance. Current technology of facing hardening chromium- tungsten steels does not allow producing faced metal without cold cracks and with high hardness right after the facing. To improve the service properties of the faced metal we complete hardening and tempering of the bimetallic item. For chromium-tungsten steels after hardening we recommend heat treatment – 3-4-time 1-hour tempering under 560–580 °C. After that the given steels have optimal structure and high hardness HRS 60-62 and red hardness.

As we can see traditional technology of bimetallic items production (facing with heating – annealing – hardening – tempering) is unreasonably time and labor consuming. A significant disadvantage of the technology is mandatory application of high-temperature heating ($T_{heating}$=350–700°C), slow cooling and, as a result, further hardening of the faced component while high working properties of hardening steels are not made maximal use of. Application of high-temperature heating when facing chromium-tungsten steels significantly complicates the technological process of producing a bimetallic faced product. The costs for the facing increase due to the necessity of heating and high temperature maintenance. The process of facing with application of high-temperature preliminary and additional heating and slow cooling significantly prolongs. Slow cooling of the component recommended to prevent cold cracks generates a need for conducting further hardening of the bimetallic product which, by itself, is a difficult task due to different properties of faced high-alloyed layer and low-alloyed core. Hardening of some large-sized components is practically unrealizable. Current technological processes of tool steels facing do not allow complete enough application of the faced layer metal properties [2, 4].

That is why the problem of developing new methods of chromium-tungsten steels facing allowing elimination of the mentioned above disadvantages of traditional technology and complete use of high hardness and wear resistance designed by metallurgists and metallographers. It is rational to combine the facing and hardening processes which allows producing a faced layer in the condition close to the hardened one. It is also necessary to avoid heating or, at least, decrease its temperature applying new methods of cold cracks elimination.

We can produce faced metal demonstrating high resistance to cold cracking and high mechanical properties right after the facing by regulating the thermal cycle of facing. The thermal cycle for the multi-layer facing chromium-tungsten hardening steels is shown in Figure 2 [2-3].

The specific feature of the suggested thermal cycle is in three stages of heating. The first provides limited time of heating and increased cooling rate in the high-temperature region which prevents growth of grains and austenite deterioration with formation of equilibrium low-strength structures.

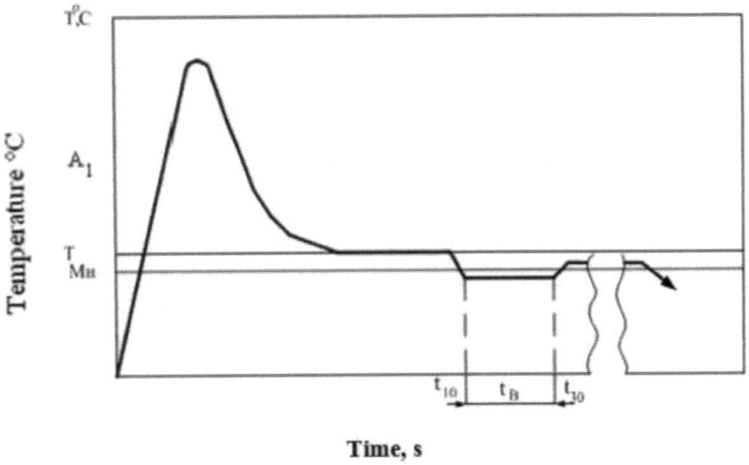

Figure 2. The diagram of the thermal cycle when facing with heat-resisting steels

The first stage of the thermal cycle is provided by application of highly-concentrated heating sources (constricted or plasma arc) and additional heating of the faced component. The second stage of the thermal cycle ensures faced metal remaining in the austenite condition when completing all layers in the process of facing. It is achieved by application of preliminary heating by 50–100 C over the temperature of abnormal plasticity occurrence. Cooling of the faced component at the third stage of the thermal cycle after the facing is finished is caused by phase transitions in the chromium-tungsten faced metal accompanied by abnormal plasticity (superplasticity) occurrence. The internal stresses partly relax which allows producing faced metal with low level of retained stresses and, thus, with low liability of cold cracks formation.

Cooling of the faced metal after the process is over leads to formation of martensite. In this case there is no necessity of hardening the faced component to improve its strength, hardness and wear resistance. There is also no necessity in completing further complicated thermal treatment of the faced component (annealing – hardening – tempering). According to the suggested method it is only necessary to conduct one additional operation – tempering – to improve hardness and wear resistance.

To produce faced metal with low capability of forming cracks we suggest regulating the level of temporary stresses in the process of facing. Such regulation may be implemented by lowering the temperature of heating by 20–100 C below the temperature where phase transformations start. This allows decreasing the temporary stresses due to their partial relaxation under phase transformations accompanied by abnormal plasticity. The degree to which the heating temperature is lowered down depends upon the required level of temporary stresses decrease to produce faced metal structure with the certain amount of austenite and martensite and phase transformation interval. Formation of insignificant amount of martensite in the structure is aimed. The heating temperature should be increased after partial relaxation of temporary stresses. After the facing process is finished the component must be cooled down to the temperature where phase transitions stop (the third stage of the thermal cycle) [4-6].

The suggested method is efficient when facing high-speed chromium-tungsten steels with low temperatures of phase transformations start (M_s=180° – 380 C). It allows significantly decrease the temperature of preliminary heating for R18 steel from 500–600°C to 230–280°C and even to 80–160°C in the suggested method which simplifies the technology of chromium-tungsten steels facing. The suggested method of hardening steels facing ensures high resistance of the faced metal to formation of hardening cracks and allows obtaining high hardness and wear resistance of the layer hardened in the process of facing. The structure of the faced metal in the area of fusion with the base metal is shown in Figure 3.

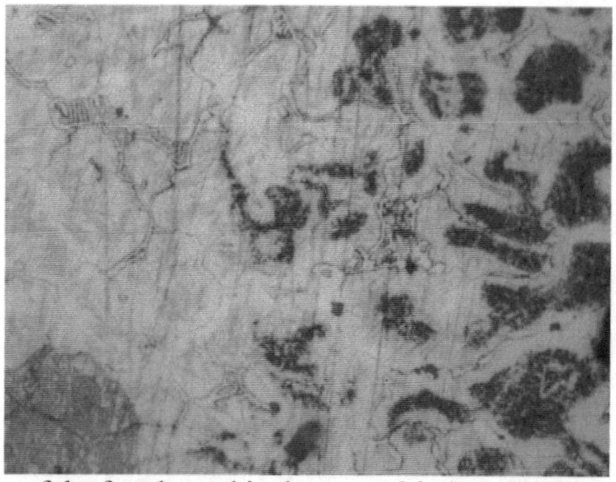

Figure 3. Structure of the faced metal in the area of fusion with the base metal, × 1000

Metal hardness after the facing according to the given method approaches hardness of hardened steels (HRC 52-58). Due to high tempering hardness can be increased up to HRC 62-66. Facing completed according to the suggested method allows full use of the properties of the faced metal [7-8].

Conclusion

1. The traditional technology of facing chromium-tungsten steels with high-temperature heating and slow cooling is labor-consuming and does not allow full use of service properties of the faced metal.

2. It is possible to produce faced metal, hardened, without cracks and with low level of residual stress, through application of new facing methods according to the thermal cycle of facing which specific features include three stages of thermal heating and application of low-temperature heating.

3. It was suggested to regulate the stressed state in the faced components with application of kinetic plasticity effect due to short-time lowering of the heating temperature and time-wise holding the component under the given temperature conditions.

References

[1] Geller Yu A Tool steels Mettallurgiya 1975

[2] Malushin N N Providing the quality of metallurgic equipment components at all stages of their life cycle through application plasma-jet facing with hot-working steels Tomsk TPU Press 2013 358

[3] Buyalich G D Anuchin A V Dronov A A 2015 The Numerical Analysis of Accuracy of Hydraulic Leg Cylinder in Modeling Using Solid Works Simulation J. Applied Mechanics and Materials 770 pp 456-460

[4] Saprykina N A, Saprykin A A, Borovikov I F, Sharkeev Y P Influence of layer-by-layer laser sintering conditions on the quality of sintered surface layer of products IOP Conference Series: Materials Science and Engineering 91 2015

[5] Babakova E V Gradoboev A V Saprykin A A Ibragimov E A Yakovlev V I Sobachkin A V Comparison of activation technologies powder ECP-1 for the synthesis of products using SLS J Applied Mechanics and Materials 756 (2015) 220-224

[6] Malushin N N Vostretsova T G The regulated thermal cycle for plasma-jet facing of the rolls with hot-working steels J Blank production in engineering 3 (2014) 14 – 16

[7] Paton B E The technology of electric welding of metals and alloys by fusion Mashinostroyeniye 1974

[8] Kryukov R E Kozyrev N A Galevsky G V Goryushkin V F Valuev D V Some aspects of oxidation-reduction reactions under carbon-bearing flux weldin IOP Conference Series: Materials Science and Engineering 91 2015

IOP Conf. Series: Materials Science and Engineering **127** (2016) 012055 doi:10.1088/1757-899X/127/1/012055

System of Controlling the Reliability of Hydraulic Machinery in Oil and Gas Facilities

M Yu Zemenkova[1], Yu D Zemenkov[2], A L Pimnev[3], E V Kurushina[4]

[1,2,3,4]Tyumen State Oil and Gas University, 38 Volodarskogo st., Tyumen, 625000, Russia

E-mail: muzemenkova@mail.ru

Abstract. Under current conditions of operating hazardous production facilities and the level of environmental regulations, a crucial issue is evaluation and control of technological machinery reliability in hydrocarbon transportation. The authors have developed methodologies and a complex of models and algorithms aimed at creating a support system for control decision-making. It is based on the concepts of the reliability theory and regulatory requirements on technological machinery operation. A mathematical complex has been developed for predicting technical state changes and reliability in order to improve the decision-making efficiency concerning the necessity of preventive maintenance. The example is fundamental statements on reliability evaluation using fluctuation analysis.

Introduction

Today, the Russian Federation is implementing a complex of government programs intended to provide a reliable and safe operation of pipeline systems. Modernisation of the technical regulation system in 2009-2014, stiffening of requirements on production facility safety and corporate responsibility raised the significance of advancements in the field of equipment reliability monitoring and planning of measures for accident and incident prevention. The trunk line system is an elaborate complex of engineering structures with a system of providing safety and environmental compatibility. Analysis of operation practices on trunk line transportation facilities has demonstrated that around 95 % of incidents occur due to equipment failures and damage.

1. Control system requirements

In existing economic and ecological situation, a highly important task is determining the optimal inter-repair period considering its actual technical condition of equipment.

Current regulating industry standards [1,2] oblige hazardous production companies to operate their facilities under control of monitoring systems with quite precisely specified requirements and functions. According to government standards, a modern reliability monitoring service must perform the following functions:

- develop recommendations on equipment operation to increase its inter-repair period;
- develop recommendations on time schedules of taking equipment down to repairs;
- collect data on equipment reliability and actual state based on repair and operation results;
- analyse data from diagnostic network server in order to control the work of services on maintenance and repair of technological equipment, diagnostic network maintenance, etc.

In conditions of toughening the requirements it is necessary to develop decision-making support systems based on equipment reliability monitoring. Decision-making support systems [1,2,3] include properties of diagnostic systems and must provide instructions for personnel to prevent hazardous state of the facility and bring it to its normal state.

Generally accepted standards on inter-repair periods have been established for equipment with a set operation resource and are still meant for stabilised operation periods (figure 1). In some cases, for low-cost components with simple wear-out reliability models it is rather effective. However, under modern operation conditions and the complexity of technological equipment with practices in prolonging statutory service life based on life ratio parameters, a most relevant problem is monitoring the reliability of hydraulic machinery with estimated parameters. Most reliability parameter evaluation models are meant for the normal operation period (t_n-t_{cr}), however, in conditions of technical control of the system, even in the simplest case, the equipment reliability model changes (figure 2). The interval length increases (t_n-t_{cr} *), i.e. the normal operation period, and the starting point of the wear-out period (point t_{cr} *) and the function of the failure rate variable curve are unknown and depend on time and a number of factors x_n:

$$\lambda_{i(t)}=F(x_1,x_2...x_n;\ t) \tag{1}$$

Determining by methods of mathematical analysis with the help of the expert system the predicted formula and value of function $\lambda_{i(t)}=F(x_1,x_2...x_n;\ t)$, with probability Pi and the reliability parameter value, we can legitimately evaluate the machine reliability in the period (t_{cr}-t_{cr}*) and make decisions on the possibility of its operation or control decision-making upon reaching maximum admissible rate parameter λ_{adm}.

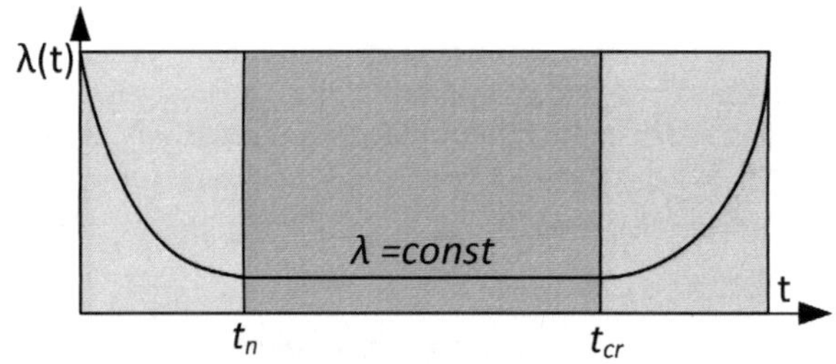

Figure 1. A typical curve of failure rate.

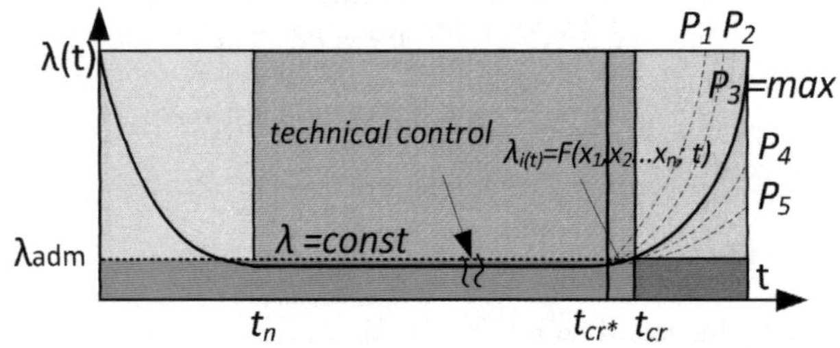

Figure 2. A failure rate curve of a complex machine in result of technical control.

The designed expert system is based on the multivariable predictive control (MPC) technology and the technology of real-time optimisation according to actual operation data. Such systems are designed individually taking into account the features and structure of the production facility using the methodological evaluation complex and the reliability parameters predicting complex. They enable transitioning from a set system of scheduled preventive repair and maintenance by operating time to servicing by actual and predicted state.

So, the primary goal of operational services is maintaining serviceable and fail-safe operation of the hydraulic machinery. The solution of this hard problem is based on strict observance of requirements of the specifications and technical documentation on the organisation of equipment maintenance and repair. These requirements, as a rule, come down to the annual time plan of scheduled preventive maintenance (SPM), according to which it is necessary to qualitatively and in a timely manner perform planned repairs on pumping power units. However, this situation does not eliminate premature equipment breakdown.

In a number of cases, forecasting system malfunction is possible when using the models of parametric failures that can be predicted by state parameters behaviour. Using mathematical methods determines the dependence of state parameters on other physical parameters («perturbations»), which accidental fluctuations are the primary cause of malfunctions later developing into failure.

2. Methods, results and discussion

The authors have developed innovative design procedures and mathematical models of reliability parameters [3-14] compatible with:

- the modern production system of maintenance;
- the system of recording operational supervisory data;
- the system of non-destructive control diagnosis in discrete measurement;
- the automated systems of technological process control.

The developed models of parameters of oil and gas facilities reliability provide the possibility for:

- forecasting the technical facilities parameters in real-time or over a fixed time span;
- the structural and factor system analysis for optimum service.

The methodology of determining reliability on the basis of studying the parameters characterising the technical state of the system, involves identifying the regularities of degradation processes leading to failures and finding analytical connection of these regularities with the reliability parameters. Depending on the character of the test parameter implementation various mathematical models are applied.

In Tyumen State Oil and Gas University a complex of mathematical models and algorithms for reliability prediction on parameter implementation has been developed.

The problem solution is based on the mathematical model of facility functioning as a dynamic system. A deterministic functioning model is taken as a basis. The model represents transformations of random input parameters ("perturbations") into target parameters of the dynamic system ("states"). The main forecasting stages are given in figure 3.

In a number of parametric reliability calculation problems it is impossible to neglect the state variables variation in time. The random process is identified by means of both theoretical and experimental research methods.

Studying the behaviour of parameters is conducted based on the sampled data and divided into three basic stages: 1) recording the parameter in a number of sections; 2) analysing each interval; 3) constructing a smoothed average, identifying the reference value and analysing the parameter trend.

In this case, the calculation method is based on the theory of random processes overshoot. The random process overshot $\zeta(t)$ over the Xcr level means the presence of such a continuous section within which $\zeta(t) \geq X_{кр}$, and before and after which $\zeta(t) < X_{кр}$.

A stream of overshoots is characterised by the operating time before the first overshoot (crossing), the operating time between overshoots, the duration or area of staying outside the set limits, etc. The parameters of the stream of overshoots are: mathematical expectation of the operating time before or between overshoots, rate of the first overshoots or overshoots in general, probability of an overshoot at a given operating time, mathematical expectation of the number of overshoots within the given interval, relative duration or area of staying outside the set limits, etc. The simplest and most frequently occurring are models in which failure is represented as the first crossing by each implementation of a random process $\zeta\ (t)$ with the critical level Xcr (figure 3).

Any of the basic technical parameters at the set modes and service conditions, as a rule, can be presented as a compound random process (2):

$$\zeta(t) = Q + \eta(t) + \chi(t), \tag{2}$$

where Q - the parameter reference value; $\eta(t)$ – the parameter trend; $\chi(t)$ – parameter fluctuations (harmonious changes of the kinematic or dynamic origin).

The parameter of the *i-th* system is presented in the form of a certain function ${}^{i}\zeta(t)$ with structure (3) matching (2):

$${}^{i}\zeta(t) = {}^{i}Q + {}^{i}\eta(t) + {}^{i}\chi(t), \tag{3}$$

where ${}^{i}Q$ - the parameter reference value in the *i*-th implementation; ${}^{i}\eta(t)$ – the parameter trend of the *i*-th implementation; ${}^{i}\chi(t)$ – parameter fluctuations of the *i*-th implementation (harmonious changes of the kinematic or dynamic origin).

For example, operation of gas compressor units (GCU) and pumps is characterised by a complex interaction of its components. Depending on the unit type (brand, model, drive, purpose), both a complex of measurements and a technique of the analysis of all parameters and their interconnections will be different.

Condition monitoring of gas compressor units can be carried out on the basis of key parameters defined during operation with a high precision, such as: effective power, fuel consumption per hour, rotational speed of rotors, temperature characteristics (for example, temperature of gases before the turbine, input and output values of gas temperature and pressure), lube oil pressure and temperature, gas composition, etc. Based on the dynamics of test parameters it is possible to carry out the analysis of malfunctions according to design plans and regulations. However, for preventive control of unit reliability, its structural or functional subsystem, forecasting of parameters applying the developed mathematical apparatus for the analysis of registration parameters is necessary.

In tests, mathematical analysis of observation results, as a rule, is based on the statistical analysis of each successive recorded implementation. The reference value and trend are then calculated on the basis of a smoothed average (4)

$${}^{(i)}\overline{\zeta}(t) = {}^{i}Q + {}^{(i)}\eta(t) \tag{4}$$

A smoothed average is the reference value and slow changes of the parameter on operating time t.

In case the process can be described as a deterministic periodic process, the periodic component is calculated as harmonious elements using the Fourier transformation and peak analysis (5), then:

$${}^{i}Q = {}^{i}z_k(t) = \sum_{v}\left[A_k \cdot \sin\omega t + B_k \cdot \cos\omega t\right] \tag{5}$$

where ${}^{i}z_k(t)$ - the deterministic periodic component of the process parameter; v - significant peaks of the Fourier transformation; A_k, B_k, ω, - harmonic function parameters; k - ζ(t)=xnumber of the observation interval.

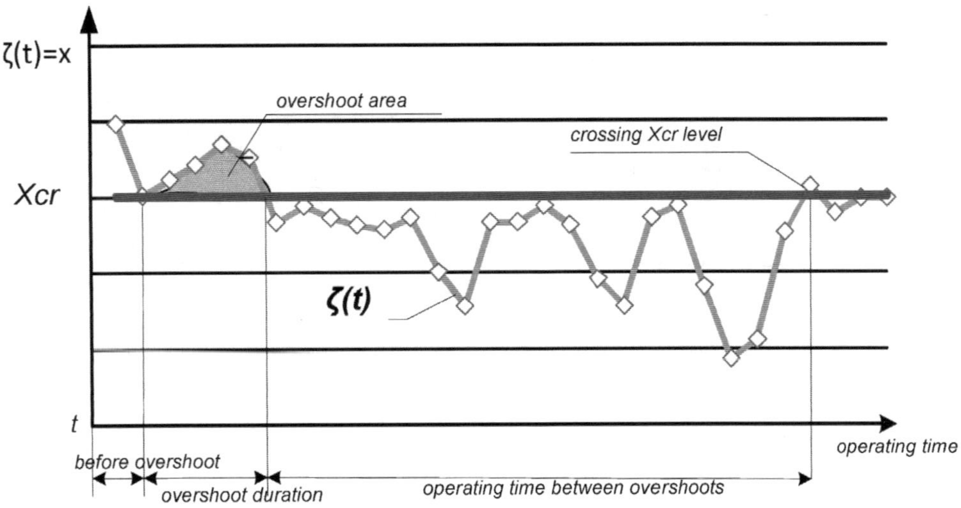

Figure 3. Random process overshoots above the admissible level (the X_{cr} level).

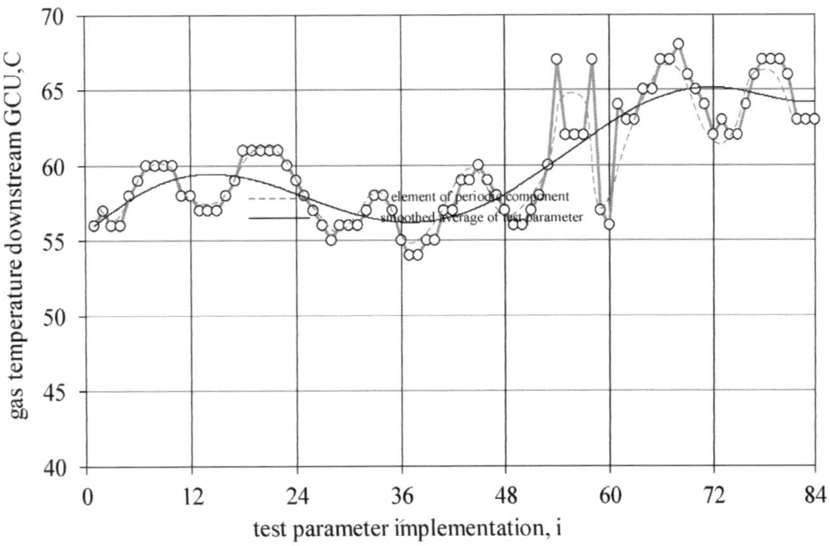

Figure 4. An example of test parameter implementation on the discrete model of recording GCU parameters.

Basing on mathematical analysis, the law of random parameter component distribution is determined. Unlike known models of reliability, the suggested technique and algorithms (figure 5) allows us to use the most adequate types of distributions from a complex of typical distributions both at a constant and variable failure rate.

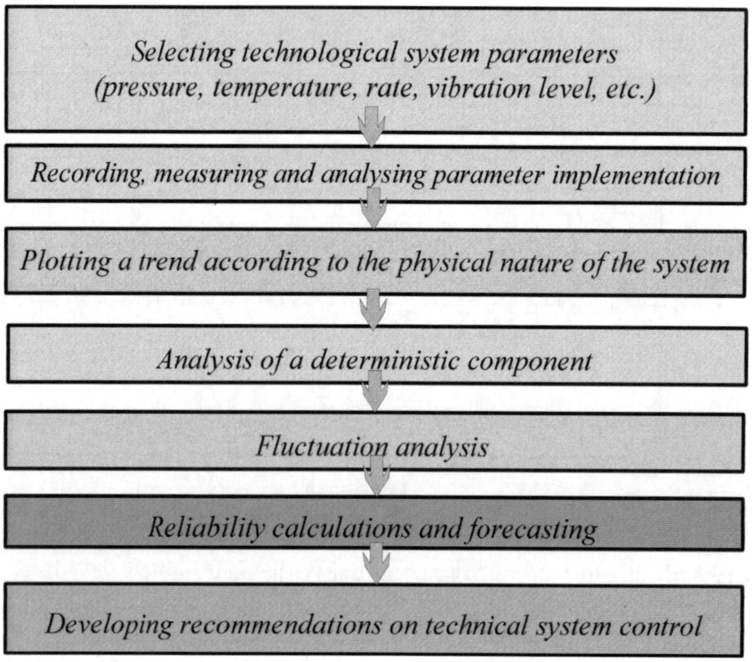

Figure 5. An algorithm of forecasting on the test parameter dynamics using fluctuation analysis.

Results can be gained both in the form of the expected parameter value and the probability of parameter staying within the admissible area. Calculation techniques are an extensive and elaborate mathematical complex, so calculations are done using relative computer software.

In practice, these methods can be used together with diagnostic software to improve the reliability and efficiency of equipment operation, as well as to prevent failures using the acquired information on system conditions.

Conducting repairs based on the actual technical state reduces the expenses for a part of the works performed during planned repairs, and maintenance in an optimum mode leads to fuel saving and prolonged inter-repair periods in particular.

3. Summary

Practical value of this work is seen in developing a complex of mathematical models and methods for the on-line reliability parameters monitoring system facilitating the transition from the "after-failure" service and maintenance system to "preventive" based on the predicted reliability parameters.

Therefore, solving the problem of predicting changes in the technical state of a complex system on the test parameter dynamics, it is possible to ensure safe operation by means of adjusting modes, as well as creating a system of early detection of negative changes. The proposed method of parameter processing and forecasting makes it possible to produce well-timed recommendations on the facility state in order to plan justified periods of taking the equipment down for repairs, the extent of control and material and labour supply.

References

[1] Condition monitoring and diagnostics of machines Hazardous equipment monitoring. Requirements for monitoring systems GOST R 53564 (2009-2010)

[2] Condition monitoring and diagnostics of machines Hazardous equipment monitoring. Organizational procedures GOST R 53563 (2009-2010)

IOP Conf. Series: Materials Science and Engineering **127** (2016) 012055 doi:10.1088/1757-899X/127/1/012055

[3] Yu D Zemenkov, V V Shalay, M Yu Zemenkova Immediate analyses and calculation of saturated steam pressure of gas condensates for transportation conditions Procedia Engineering **113** (2015) 254-258

[4] Yu D Zemenkov, V V Shalay, M Yu Zemenkova Expert systems of multivariable predictive control of oil and gas facilities reliability Procedia Engineering **113** (2015) 312-315

[5] Y N Antip'ev, A P Nevolin, Yu D Zemenkov Operation of intermediate pumping stations on transport of gas saturated oils 10 (1981) 46

[6] P Iljyashchenko D A Chinakhov Y M Gotovshchik MAW productivity development and reduction of its harmful effect on human organisms J Applied Mechanics and Materials **682** (2014) 122-126

[7] D V Valuev V I Danilov Reasons for negative formation of structures in carbon steel processing of pressure J. Proceedings 2012 7th International Forum on Strategic Technology IFOST 2012. 2 (2012) 151-154

[8] D P Iljyashchenko D A Chinakhov Y M Gotovshchik Investigating the influence of the power supply type upon the properties of the weld joints under manual arc welding J Advanced Materials Research 1040 (2014) 837-844

[9] R A Gizatulin D V Valuev A V Valueva A Serikbol I F Borovikov Optimizing technological parameters of the reduction processes in treating steels in a ladle furnace J IOP Conference Series: Materials Science and Engineering 91 (2015)

[10] Stepanov S., Cheremisin N., Sokolov S., Altunin A., Shabarov A. Computational method for relative permeability curves based on the bernoulli generalized porous media equations ECMOR 2012 - 13th European Conference on the Mathematics of Oil Recovery (2012)

Mathematic Modeling of Complex Hydraulic Machinery Systems When Evaluating Reliability Using Graph Theory

M Yu Zemenkova[1], A N Shipovalov[2] and Yu D Zemenkov[3]

[1,3]Tyumen State Oil and Gas University, 38 Volodarskogo st., Tyumen, 625000, Russia
[2]Long-Yugansky LPU of MG, JSC Gazprom Transgaz Yugorsk, Yugorsk, Mira St., 15, Khanty-Mansi Autonomous Area, Tyumen region, 628260, Russia

E-mail: muzemenkova@mail.ru

Abstract. The main technological equipment of pipeline transport of hydrocarbons are hydraulic machines. During transportation of oil mainly used of centrifugal pumps, designed to work in the "pumping station-pipeline" system. Composition of a standard pumping station consists of several pumps, complex hydraulic piping. The authors have developed a set of models and algorithms for calculating system reliability of pumps. It is based on the theory of reliability. As an example, considered one of the estimation methods with the application of graph theory.

Introduction

The system of trunk pipeline transport is a complex set of engineering structures with the system of security and environmental protection. The analysis of the experience of exploitation of main facilities of transport showed that most of them happen because of failures and damage to the hardware. Using the developed methods and apparatus of semi-Markov models and the theory of construction of graph States, let us analyze the algorithm for calculating the persistence on a difficult technical systems of pipeline transport of oil.

1. Problem definition

Monitoring the reliability of hydraulic machines is a difficult task. In the practice of their operation in the oil and gas sector, such monitoring is carried out using a complex of diagnostic parameters such as bearing temperature, vibration, pressure, performance. In addition, the monitoring of leaks and the parameters determining the possibility of occurrence of cavitation regime. One such operational parameter is the coefficient of hydraulic reliability.

For the implementation of the proactive functions the evaluation of reliability is reduced to the problem of determining the probabilistic characteristics of the life of the system and determining availability according to the criterion of performance persistence, for example, hydraulic reliability.

Consider a typical scheme of a pumping station, a pump with a design capacity Q_1 и Q_2, the power N_1=N_2. The reliability function of the pumps has the form $F_1{}^*(t) = c_1 e^{-\lambda_1 t}$, $F_2{}^*(t) = c_2 e^{-\lambda_2 t}$, the recovery time of each machine has an exponential distribution $F_{1rep}(t) = e^{-\mu_1 t}$, $F_{2rep}(t) = e^{-\mu_2 t}$, with parameters μ_1 and μ_2, the required minimum acceptable level of reliability J_{ac} , schedule actual hydraulic reliability J_{act} defined in real time. On site there is a tank farm with the volume V, the time for park empty t_r, assessment of the temporary reserve $D(t)$.

It is required to determine: probabilities of transitions to all possible conditions and their moments, an availability quotient of system in case of the set working schedule; a shelf life of system, the predicted maintenance time by criterion of hydraulic reliability of system[1-3].

2. Methods, results and discussion

Let at the time of the beginning of the analysis the system be in operating state. In the line No. 1 transfer of a product with a productivity is performed Q_1 , on the second – with a productivity Q_2. Two options of functioning of system are possible. According to the first option in operation there are both pumps. Then let t_1 — time, during which both machines at the same time are in work according to the plan of transfer for 1 calendar year t_y, i.e. both machines have no structural allowance. Then provided that reliability of one of objects decreases, reliability of system in general decreases, we accept probabilistic parameter α, characterizing probability of decrease in reliability of system is lower than reliability of one, admissible owing to loss, of elements, equal $\alpha_1 = t_1 / t_Y$. By the second option of operation in work there is only one hydraulic machine. Thus the second carries out function of not loaded allowance. Then

$$\alpha_2 = t_2 / t_y ,$$

where t_2 - is time during which the element of system has a structural allowance.

It is necessary to construct a state graph for the considered system, accepting that E_0 hydraulic reliability is able $Js_{act} \geq Js_{al.}$ (figure 1).

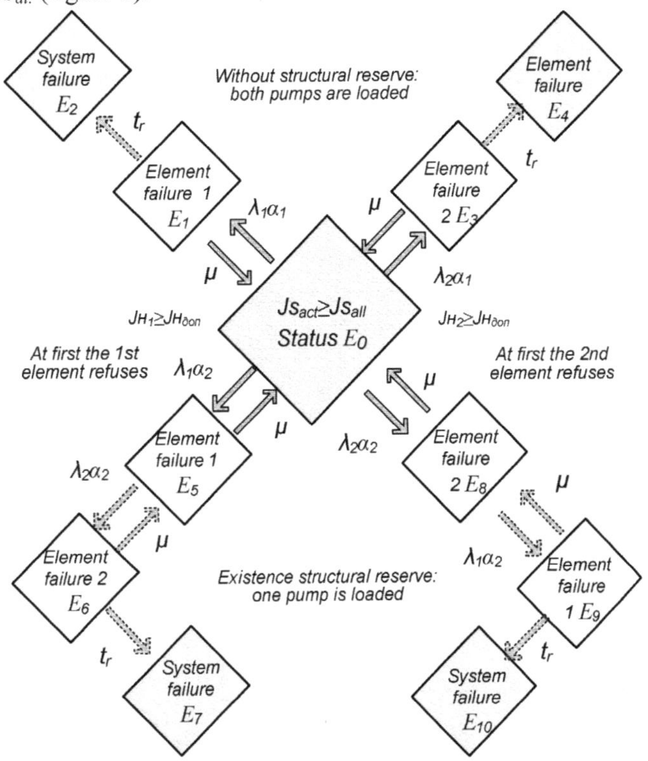

Figure1. State graph example for an assessment of reliability of system.

In case of decrease in hydraulic reliability of an element No.1 or No.2 to level $J_{act}<J_{al}$ there is a transition of system to conditions E_1 and E_3 respectively. Decrease in reliability $J_{act}<J_{al}$ in this case means that the system in general doesn't conform to the requirement of reliability and technical intervention is necessary. The system is in conditions E_1 and E_3 also technical intervention with intensity is required μ_1 and μ_2. In case of recovery of an element in time *tr* the system passes into a reliable condition E_0, otherwise – in refusal E_2 and E_4. During operation when there is a structural allowance, the system with intensity of $\lambda_1\alpha_2$ and $\lambda_1\alpha_2$ passes into the conditions of E_5 and E_8, similar to E_1 and E_3, however transfer can be performed by means of machines, carrying out function reserve while the system recovery is made [3-11].

In case of decrease in reliability of a reserve thread is down to one level lower admissible after float time the system passes into conditions of E_7 and E_{10} with the level of reliability is lower than the admissible.

The shelf life of system is determined by the time spent in conditions E_0-E_1, E_0-E_3, E_0-E_6, E_0-E_9.

System of the equations in an integrated form for the count:

$$\begin{cases} Q_0(t,t_r) = \int_0^t Q_1(t-x,t_r)dP_{0,1}(x) + \int_0^t Q_3(t-x,t_r)dP_{0,3}(x) + \\ + \int_0^t Q_5(t-x,t_r)dP_{0,5}(x) + \int_0^t Q_8(t-x,t_r)dP_{0,8}(x); \\ Q_1(t,t_r) = \int_0^t Q_0(t-x,t_r)dP_{1,0}(x) + P_{1,2}(t); \\ Q_3(t,t_r) = \int_0^t Q_0(t-x,t_r)dP_{3,0}(x) + P_{3,4}(t); \\ Q_5(t,t_r) = \int_0^t Q_0(t-x,t_r)dP_{5,0}(x) + \int_0^t Q_6(t-x,t_r)\cdot dP_{5,6}(x); \\ Q_6(t,t_r) = \int_0^t Q_6(t-x,t_r)dP_{6,5}(x) + P_{6,7}(t); \\ Q_8(t,t_r) = \int_0^t Q_0(t-x,t_r)dP_{8,0}(x) + \int_0^t Q_9(t-x,t_r)\cdot dP_{8,9}(x); \\ Q_9(t,t_r) = \int_0^t Q_9(t-x,t_r)dP_{9,8}(x) + P_{9,10}(t) \end{cases} \quad (1)$$

System of the equations in a probabilistic form:

$$\begin{cases} P_0(t,t_r) = 1 - P_{01}(t) - P_{03}(t) + P_{05}(t) + P_{08}(t) - \int_0^t P_1(t-x,t_r)dP_{0,1}(x) + \\ + \int_0^t P_3(t-x,t_r)dP_{0,3}(x) + + \int_0^t P_5(t-x,t_r)dP_{0,5}(x) + \int_0^t P_8(t-x,t_r)dP_{0,8}(x); \\ P_1(t,t_r) = 1 - P_{1,0}(t) - P_{1,2}(t) + \int_0^t P_0(t-x,t_r)dP_{1,0}(x); \\ P_3(t,t_r) = 1 - P_{3,0}(t) - P_{3,4}(t) + \int_0^t P_0(t-x,t_r)dP_{3,0}(x); \\ P_5(t,t_r) = 1 - P_{5,0}(t) - P_{5,6}(t) + \int_0^t P_0(t-x,t_r)dP_{5,0}(x) + \int_0^t P_6(t-x,t_r)\cdot d_{5,6}(x); \\ P_6(t,t_r) = 1 - P_{6,5}(t) - P_{6,7}(t) + \int_0^t P_6(t-x,t_r)dP_{6,5}(x); \\ P_8(t,t_r) = 1 - P_{8,0}(t) - P_{8,9}(t) + \int_0^t P_0(t-x,t_r)dP_{8,0}(x) + \int_0^t P_9(t-x,t_r)\cdot d_{8,9}(x); \\ P_9(t,t_r) = 1 - P_{9,8}(t) - P_{9,10}(t) + \int_0^t P_8(t-x,t_r)dP_{9,8}(x) \end{cases} \quad (2)$$

System of the equations in Laplace-Carson's forms:

$$\begin{cases} Q^*_0(s,t_r) = Q^*_3(s)P^*_{0,3}(s) + Q^*_1(s,t_r)P^*_{0,1}(x) \\ + Q^*_5(s,t)P^*_{0,5}(s) + Q^*_8(s,t_r)P^*_{0,8}(s); \\ Q^*_1(s,t_r) = Q^*_0(s,t_r)\cdot P^*_{1,2}(s); \\ Q^*_3(s,t_r) = Q^*_0(s,t_r)P^*_{3,4}(t); \\ Q^*_5(s,t_r) = Q^*_0(s,t_r)P^*_{5,0}(x) + Q^*_6(s,t_r)P^*_{5,6}(s); \\ Q^*_6(s,t_r) = Q^*_6(s,t_r)P^*_{6,7}(s); \\ Q^*_8(s,t_r) = Q^*_0(s,t_r)P^*_{8,0}(s) + Q^*_9(s,t_r)P^*_{8,9}(s); \\ Q^*_9(t,t_r) = Q^*_9(s,t_r)P^*_{9,10}(s) \end{cases} \quad (3)$$

Using this model we set semi-Markov process in a general view as $P(t) = \|P_{ij(}(t)\|$. Then, if the system is in a status *i*, the transition probabilities between statuses of system will have an appearance:

$$P_{ij}(t) = \int_0^t \prod_{k=1}^n (1 - F_{ik}(x))dF_{ij}(x) \quad , \quad (4)$$

where $F_{ij}(x)$ - the distribution function of time of stay of an element is able i upon the following transition to a status *j*;

$k \in N, k \in (1;\mathrm{n})$ - numbers of statuses except a status of j in which one-step transition from a status is possible *i*;

$F_{ik}(x)$ - the distribution function of time of stay of an element is able i upon the following transition to a status *k*.

For determination of a distribution function of transitions from a status to a status it is necessary to determine statistically function of change of a keeping of system by criterion of hydraulic reliability F_{ij}

, in particular *F(Js,t)*. For example, for the transmission system the function of distribution of hydraulic reliability *Js(t)* will be constructed in real time with use of methods of statistical modeling and the least squares on the base to operation properties of the equipment and the dispatcher data since the beginning of the analyzable period (figure 4).

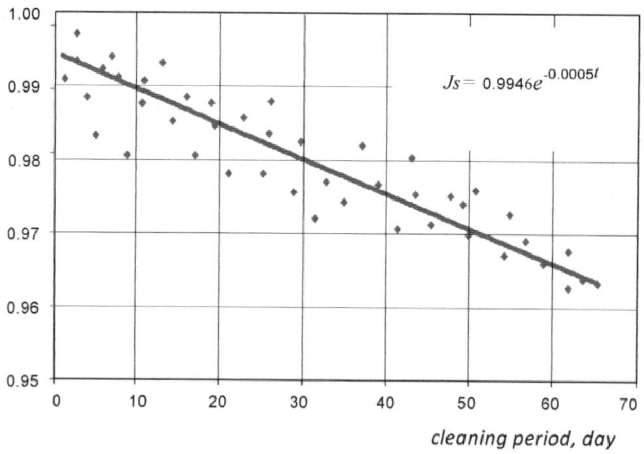

Figure 2. Distribution function of $F_1(t)$ for the oil pipeline

Further we will determine distribution parameters of a keeping of F_{ij} by the diagram of change of hydraulic reliability under the law of change of coefficient of hydraulic reliability for the interclearing period of 65 days from conditions:

$$Js_{al} = Js(t^*);\quad t^* = \frac{\ln(Js_{al})}{C};\quad \lambda = 1/t^*, \tag{5}$$

where t^* - time to failure; Js_{al} - critical level of an index of hydraulic reliability.

Thus, based on empirical function of change of coefficient of hydraulic reliability, we will receive a distribution function of failures (a distribution function of frequencies of achievement by system of critical values of coefficient Js and failure density by criterion of Js).

For determination of the transition probability of system from a status 0 in a status 1 we will use a graph of fig. 1, supposing that adjacent statuses are No. 3,5,8. Without the need for the accounting of a temporal reserve the formula (3) will assume an air:

$$P_{01}(t) = \int_0^t (1 - F_{03}(x)) \cdot (1 - F_{05}(x)) \cdot (1 - F_{08}(x)) dF_{01}(x) \tag{6}$$

Time distribution functions for statuses can be defined as

$$\int_0^t f(u)du, \quad u = F(x), \tag{7}$$

where u - density function.

- in particular:

$$F_{01}(x) = 1 - e^{-\lambda_1 \alpha_1(x)},$$
$$F_{08}(x) = 1 - e^{-\lambda_2 \alpha_2(x)},$$
$$F_{03}(x) = 1 - e^{-\lambda_2 (1-\alpha_2)(x)},$$
$$F_{05}(x) = 1 - e^{-\lambda_1 (1-\alpha_1)(x)}, \quad \alpha_1 = 1 - \alpha_2. \tag{8}$$

The transition probability of P_{01} is calculated as follows:

$$P_{01}(t)=\int_0^t(1-F_{08}(x))\cdot(1-F_{05}(x))\cdot(1-F_{0,3}(x))dF_{01}(x)=$$

$$=\int_0^t(1-(1-e^{-\lambda_2 q_2(x)}))\cdot(1-(1-e^{-\lambda_2(1-q_2)(x)}))\cdot(1-1-e^{-\lambda_1(1-q_1)(x)}))dF_{01}(x)=$$

$$=\int_0^t(e^{-\lambda_2 q_2(x)})\cdot(e^{-\lambda_2(1-q_2)(x)})\cdot(e^{-\lambda_1(1-q_1)(x)})dF_{01}(x)=$$

$$=\int_0^t(e^{-\lambda_2 q_2 x-\lambda_2(1-q_2)x-\lambda_1(1-q_1)x})dF_{01}(x)=\int_0^t(e^{-\lambda_2 q_2 x-\lambda_2 x+\lambda_2 q_2 x-\lambda_1 x+\lambda_1 x q_1})dF_{01}(x)=$$

$$\int_0^t(e^{-\lambda_2 x-\lambda_1 x+\lambda_1 x q_1})dF_{01}(x)=\int_0^t(e^{-\lambda_2 x-\lambda_1 x+\lambda_1 x q_1})dF_{01}(x)=\int_0^t(e^{-\lambda_1 x q_1(\frac{\lambda_2}{\lambda_1 q_1}+\frac{1}{q_1})+\lambda_1 x q_1})dF_{01}(x)=$$

$$=\int_0^t(e^{-\lambda_1 x q_1(\frac{\lambda_2+\lambda_1}{\lambda_1 q_1})+\lambda_1 x q_1})dF_{01}(x)=\frac{\lambda_1 q_1}{\lambda_2+\lambda_1}\left(1-e^{-(\lambda_1+\lambda_2)t}\right) \tag{9}$$

Remaining transition probabilities are calculated similarly.

For determination of the transition probability of system from a status 1 in a status 0 we will receive:

$$P_{10}(t)=\int_0^t(1-F_{12}(x))dF_{10}(x) \quad , \tag{10}$$

where $F_{10}(x)=1-e^{-\mu_1(x)}$ - repairing function.

For determination of the transition probability of system from a status 1 in a status 2 taking into account a temporal reserve the formula will assume an air:

$$P_{12}(t)=\int_0^t(1-F_{10}(x))\cdot(1-D(x)))dF_{12}(x) \quad , \tag{11}$$

where *D(x)* – distribution function of a temporal reserve, *D(x)=1 npu* $t_r<t$; *D(x)=0 npu* $t_r\geq t$;

$F_{12}(x)=1-e^{-\mu_2(x)}$ - repairing function.

By results of calculations, conversions and integration the following dependences for determination of the transition probabilities are received:

- from a status E_0:

$$P_{01}(x)=\frac{\lambda_1\cdot\alpha_1}{\lambda_1+\lambda_2}(1-e^{-(\lambda_2+\lambda_1)t}),\quad P_{03}(x)=\frac{\lambda_2\cdot\alpha_2}{\lambda_1+\lambda_2}(1-e^{-(\lambda_2+\lambda_1)t}),$$

$$P_{05}(x)=\frac{\lambda_1\cdot\alpha_2}{\lambda_1+\lambda_2}(1-e^{-(\lambda_2+\lambda_1)t}),\quad P_{0,8}(x)=\frac{\lambda_2\cdot\alpha_2}{\lambda_1+\lambda_2}(1-e^{-(\lambda_2+\lambda_1)t}) \quad . \tag{12}$$

- from statuses E_5, E_8:

$$P_{56}(x)=\frac{\lambda_2\cdot\alpha_2}{\lambda_2+\mu_1}(1-e^{-(\lambda_2+\mu_1)t}),\qquad P_{89}(x)=\frac{\lambda_1\cdot\alpha_2}{\lambda_1+\mu_2}(1-e^{-(\lambda_1+\mu_2)t}),\tag{13}$$

$$P_{50}(x)=\frac{\mu_1}{\lambda_2\alpha_2+\mu_1}(1-e^{-(\lambda_2\alpha_2+\mu_1)t}),\ P_{80}(x)=\frac{\mu_2}{\lambda_1\alpha_2+\mu_2}(1-e^{(\lambda_1\alpha_2+\mu_2)t}) .\tag{14}$$

- from statuses E_1, E_3, E_6, E_9:

$$P_{3,0}(t)=e^{-\mu_2 tr},\ t\geq tr\ ,\ P_{3,0}(t)=e^{-\mu_2 t},\ t<tr\ ,\tag{15}$$

$$P_{1,0}(t)=e^{-\mu_1 tr},\ t\geq tr\ ,\ P_{1,0}(t)=e^{-\mu_1 t},\ t<tr\ ,\tag{16}$$

$$P_{6,5}(t)=e^{-\mu_2 tr},\ t\geq tr\ ,\ P_{6,5}(t)=e^{-\mu_2 t},\ t<tr\ ,\tag{17}$$

$$P_{9,8}(t)=e^{-\mu_1 tr},\ t\geq tr\ ,\ P_{9,8}(t)=e^{-\mu_1 t},\ t<tr\ .\tag{18}$$

- from statuses E_1, E_3, E_6, E_9:

in case $t<tr$, that probabilities $P_{1,2}$, $P_{3,4}$, $P_{6,7}$, $P_{9,10}$ are equal 0, and $t\geq tr$ - equal 1.

By results of the carried-out transformations and integration, the following dependences for definition of time of stay in statuses (T_i) according to the count are received:

- for status E_0:

$$T_0=\int_0^\infty\prod_j(1-F_{ij}(t))dt=\frac{1}{\Lambda_1}=\frac{1}{\sum_{i=1}^{m}\lambda_i\alpha_m}=\frac{1}{\lambda_1\alpha_1+\lambda_2\alpha_2+\lambda_1\alpha_2+\lambda_2\alpha_1}=\frac{1}{\lambda_1+\lambda_2}\ ,\tag{19}$$

$$\Lambda_1=\sum_{i=1}^{m}\lambda_i\cdot\alpha_m\ ,\qquad T_0=\frac{1}{\lambda_1+\lambda_2}\ .\tag{20}$$

- for statuses E_1, E_5, E_3, E_8 :

$$T_i=\frac{1}{\mu_i}(1-e^{-\mu_i t_r});\tag{21}$$

$$T_1=\frac{1}{\mu_1}(1-e^{-\mu_1 t_r});\ T_3=\frac{1}{\mu_2}(1-e^{-\mu_2 t_r});\ T_5=\frac{1}{\mu_1}(1-e^{-\mu_1 t_r});\ T_8=\frac{1}{\mu_2}(1-e^{-\mu_2 t_r})\ ;\tag{4.21}$$

- for statuses E_2, E_4, E_{10}, E_7 :

$$T_i=\frac{1}{\mu_i}\ ,\tag{22}$$

$$T_2=\frac{1}{\mu_1};\ T_4=\frac{1}{\mu_2};\ T_7=\frac{1}{\mu_2};\ T_{10}=\frac{1}{\mu_1};\tag{23}$$

- for statuses E_5, E_8:

$$T_5=\frac{1}{\lambda_2+\mu_1}\ ,\qquad T_8=\frac{1}{\lambda_1+\mu_2}\ .\tag{24}$$

On a formula of conversion of Laplace-Carson for function $F(S)$:

$$F^*(S)=S\int_0^\infty e^{-st}\cdot F(t)\cdot dt\tag{25}$$

we will receive formulas for determination of images for system of equations of an analysable graph.

The readiness quotient determined by criterion of a keeping from dependences and the equations for nested Markov «chains»:

$$K_R(J_S,t)=\sum_{n\in E(Js\geq Js_{al})} P_n(t)=\frac{\sum_{i\in E_+} T_i\pi_i}{\sum_{i\in E} T_i\pi_i};\quad \sum\pi_i=1;\ \pi_i=\sum\pi_i P_{ij}. \tag{26}$$

For $\mu_1=\mu_2$, $\lambda_1=\lambda_2$ we will receive dependences for determination of an availability quotient:

$$K_R(J_s,t)=\sum_{n\in E(J_S\geq Js_{al})} P_n(t)=\frac{\sum_{i\in E_+} T_i\pi_i}{\sum_{i\in E} T_i\pi_i}=\frac{\frac{1}{\lambda}+\frac{2\alpha_2}{\mu}(1-e^{-\mu t_r})+\frac{\alpha_1}{\mu_1}+\frac{2\lambda\alpha_1^2}{\mu^2}(1-e^{-\mu tr})}{\frac{1}{\lambda}+\frac{2\alpha_2}{\mu}+\frac{\alpha_1}{\mu}+\frac{\lambda\alpha_1^2}{\mu^2}},$$

$$K_R{}^*(Js_i,T_{pr})=\frac{1}{T_{pr}}\int_0^{Tpr} K_R(Js_i;t) \tag{27}$$

The keeping index - a shelf life of system is defined by the time spent in statuses E_0-E_1, E_0-E_3, E_0-E_6, E_0-E_9 inclusive:

$$T_{0-1*}=\frac{1}{\lambda_1}+\frac{1}{\mu_1}(1-e^{-\mu_1 t_r});\ T_{0-3*}=\frac{1}{\lambda_2}+\frac{1}{\mu_2}(1-e^{-\mu_2 t_r});$$

$$T_{0-6*}=\frac{1}{\lambda_1+\lambda_2}+\frac{1}{\mu_1}(1-e^{-\mu_1 t_r});\ T_{0-9*}=\frac{1}{\lambda_1+\lambda_2}+\frac{1}{\mu_2}(1-e^{-\mu_2 t_r}).$$

$$K_R(J_s,t)=\sum_{n\in E(Js\geq Js_{al})} P_n(t)=\frac{\sum_{i\in E_+} T_i\pi_i}{\sum_{i\in E} T_i\pi_i}=\frac{\frac{1}{\lambda_1+\lambda_2}+\frac{\alpha_1}{\mu_1}+\frac{\alpha_2}{\mu_2}+\frac{\alpha_2(\lambda_1+\lambda_2)}{\mu_2^2}\left(1-e^{-\mu_2 t_r}\right)+\frac{(\lambda_1+\lambda_2)\alpha_1\cdot}{\mu_1^2}\left(1-e^{-\mu_1 t_r}\right)}{\frac{1}{\lambda_1+\lambda_2}+\frac{\alpha_1}{\mu_1}+\frac{\alpha_2}{\mu_2}+\frac{\alpha_2(\lambda_1+\lambda_2)}{\mu_2^2}+\frac{(\lambda_1+\lambda_2)\alpha_1}{\mu_1^2}}. \tag{28}$$

Thus, with use of the graph theory analytical dependences for the description of a set of statuses of system of oil pipelines and determination of probable characteristics are received.

The developed system of equations of Laplace-Carson allows to determine stationary probabilities by numerical methods using packets of mathematical computation.

3. Summary

The received dependences allow to characterize a set of statuses of systems of hydraulic machines which in a general view consist of the elements described by different communications, connections, state graphs and intensity of failures and restoration. In general the technique can be applied by preparation of information to acceptance of administrative decisions in tasks of monitoring of reliability, faultlessness, a keeping and safety in case of reliability assessment of hydraulic machines.

References

[1] V Kurushina, Y Zemenkov Innovative cyclical development of the Russian pipeline system WIT Transactions on Ecology and the Environment-**2** (2014) 881-888

[2] Yu D Zemenkov, V V Shalay, M Yu Zemenkova Expert systems of multivariable predictive control of oil and gas facilities reliability Procedia Engineering 113 (2015) 312-315

[3] R A Mamadaliev, V N Kuskov, U D Zemenkov, A A Popova Influence of high-concentrated heat sources on alloying elements transition into the weld metal Applied Mechanics and Materials **770** (2015) 19-22

[4] D P Iljyashchenko D A Chinakhov Y M Gotovshchik Physical nature of the processes in forming structures, phase and chemical compositions of medium-carbon steel welds J IOP Conf. Series: Materials Science and Engineering **91** (2015)

[5] N N Malushin D V Valuev A V Valueva A Serikbol Kinetic study of the effect of plasticity and its role in stress relaxation in the weld speed steel during the martensitic transformation J Applied Mechanics and Materials 682 (2014) p. 58-63

[6] D P Iljyashchenko D A Chinakhov Y M Gotovshchik Increasing strength and operational reliability of fixed joints of tubes by MMA welding J IOP Conf. Series: Materials Science and Engineering **91** (2015)

[7] R A Gizatulin D V Valuev A V Valueva A Serikbol I F Borovikov Optimizing technological parameters of the reduction processes in treating steels in a ladle furnace J IOP Conference Series: Materials Science and Engineering **91** (2015)

[8] V N Kuskov, R A Mamadaliev, A A Galinskiy The influence of a source of heating for alloying the weld metal upon receipt of the connection of high-alloyed steels Fundamental research **3** (2014) 515-518

[9] V N Kuskov, R A Mamadaliev, A G Obukhov The transition of the alloying elements in the weld metal when welding steel 12X18H10T Fundamental research **11** (2013) 1794-1797

[10] Stepanov S., Cheremisin N., Sokolov S., Altunin A., Shabarov A. Computational method for relative permeability curves based on the bernoulli generalized porous media equations ECMOR 2012 - 13th European Conference on the Mathematics of Oil Recovery (2012)

YIT-UPMME 2015 IOP Publishing
IOP Conf. Series: Materials Science and Engineering **127** (2016) 012057 doi:10.1088/1757-899X/127/1/012057

Problems of Solid-Phase Synthesis in Cylindrical Ampoules under Explosive Loading

S A Zelepugin[1,2], O V Ivanova[2], A S Yunoshev[3], A S Zelepugin[1,2]

[1] National Research Tomsk State University, 36 Lenin Ave., 634050, Tomsk, Russia
[2] Tomsk Scientific Center of SB RAS, 10/3 Akademichesky Ave., 634055, Tomsk, Russia
[3] Lavrentyev Institute of Hydrodynamics of SB RAS, 15 Lavrentyev Ave., 630090, Novosibirsk, Russia

E-mail: bliz3@yandex.ru

Abstract. The peculiarities of solid-phase synthesis are studied experimentally and numerically in the aluminum-fluoroplastic and aluminum-sulfur mixtures in cylindrical ampoules under explosive loading. The experimental results show that the use of a mixture capable of ultrafast exothermic reactions leads to the destruction of a cylindrical ampoule under explosive loading. When the transient shock wave is reflected from the bottom lid of the ampoule as a compression wave, there is a sharp increase in pressure in the lower part of the ampoule, which is accompanied by the increase in rate of the chemical reaction. The high rate of heat release during the chemical reaction in the lower part of the ampoule causes the formation of a gas phase, which leads to a further increase in pressure and destruction of the ampoule.

Introduction

Explosive technologies are commonly used in metal processing, including the technologies of shaping, welding, cutting, hardening, and compaction. At present, the prospective studies are the studies in the area of explosive pressing ceramic powders, which are represented mainly by experimental works [1]. A lot of experiments are conducted to obtain new materials by combining the various physical processes. For example, in [2] the technology is developed for producing composites based on titanium and chromium borides using a one-stage method that combines the self-propagating high-temperature synthesis (SHS) and pressing of hot products. The work [3] demonstrates the fact that there is a possibility to combine welding and explosive compaction with SHS for production of layered ceramic-metal materials. Many of these methods are already used in production, while the possibilities and prospects for the solid-phase synthesis of materials under explosive loading are not sufficiently investigated and have not yet reached the level of technology due to the lack of experimental data and numerical techniques correctly describing the process.

Synthesis of new materials according to equilibrium phase diagrams is exhausted. The perspectives are connected with producing metastable combinations in nonequilibrium conditions. Explosive loading gives big opportunities to create and control such conditions. High pressure and speed of substances create extreme conditions to obtain materials with unique properties. But the substance reaction to such extreme effect can be changed depending on the substance characteristics. It is necessary to take into account short duration of the explosive loading process ($\sim 10^{-6}$ s). Since various processes and their stages require different time intervals for the development, not all the processes have enough time to develop within a short period of time which explosive loading is characterized [4].

The developed numerical models in combination with the available experimental data will provide the possibility to receive reliable information on the behavior of reacting media, including the mechanisms and kinetics of physical and chemical transformations and ways to form new states of substances.

The aim of this work is to study the synthesis of materials in the solid-phase mode and find the reasons for destruction of cylindrical ampoules subjected to explosive loading. A characteristic peculiarity of the studies conducted is an integrated approach combining the experiments and numerical simulation.

1. Formulation of the problem

The system of equations describing the nonstationary adiabatic motion of each component in a solid compressible mixture comprises the equations of continuity, momentum, and energy [5, 6]:

$$\frac{\partial}{\partial t}(\alpha_i\rho_i)+\nabla\alpha_i\rho_i\upsilon_i=0,\ (i=1,2,\ldots,N),$$

$$\alpha_i\rho_i\frac{d_i\upsilon_i}{dt}=\nabla\sigma_i+\alpha_i\sum_{j=1}^{N}\alpha_j\mathbf{R}_{ji},\ (i=1,2,\ldots,N),$$

$$\alpha_i\rho_i\frac{d_iE_i}{dt}=\sigma_i\varepsilon_i+\alpha_i\sum_{j=1}^{N}\alpha_j\Phi_{ji},\ (i=1,2,\ldots,N),$$

$$\text{where } \frac{d_i}{dt}\equiv\frac{\partial}{\partial t}+\upsilon_i^k\frac{\partial}{\partial x^k}.$$

Here t is the time, ρ_i is the density of the i-th component equal to the mass of i-th component per unit volume of the i-th component, υ_i is the velocity vector, E_i is the internal specific energy, ε_i is the strain rate tensor, $\sigma_i=-P_i\delta_i+S_i$ is the stress tensor, P_i is pressure, S_i is the stress deviator, R_{ji} is the intensity of the momentum exchange between the j-th and i-th components, Φ_{ji} is the intensity of the energy exchange between the j-th and i-th components, N is the number of components.

Volume fractions of the mixture occupied by each component [5] are given by:

$$\alpha_1+\alpha_2+\ldots+\alpha_N=1,\ (\alpha_i\ge 0),\ \alpha_i=\rho_i^*/\rho_i,$$

where ρ_i^* is the reduced density (mass of the i-th component per unit volume).

To solve the shock wave loading problems of porous reacting mixtures, the finite element method is used [7, 8]. On the basis of this method every component of a mixture simultaneously occupies the same volume as the mixture and consists of a set of the final elements connected with the node points. Inside every element, components interact with each other, exchanging momentum, energy, and mass (in the presence of chemical reactions) within the framework of the multicomponent medium model. After interaction of components and their summary contribution to node forces of an element, the components in a mixture obtain the velocity of the corresponding element.

In modeling chemical reactions under shock wave loading, we use the zeroth-order kinetic relation [9, 10] characterized by a constant rate of chemical transformations:

$$J_{ji}=\frac{d\eta}{dt}=\begin{cases}0, & \text{if } \eta=1 \text{ or } (T_i<T_\eta \text{ and } P<P_\eta)\\ f(P_\eta), & \text{if } \eta<1 \text{ and } (T_i\ge T_\eta \text{ or } P\ge P_\eta)\end{cases},$$

$$f(P_\eta)=\begin{cases}K_0, & \text{if } P<P_\eta\\ K_pK_0, & \text{if } P\ge P_\eta,\end{cases}$$

here T_i is the temperature, P is the matched pressure of components, and T_η, P_η, K_p, K_0 are the constants, η is the conversion degree.

Evolution of porosity in the material (compression and growth of pores) is simulated using a kinetic model of the active type, which determines changes in specific volume of pores influencing on the material properties and causing stress relaxation [11]:

$$\frac{dV_{fi}}{dt} = \begin{cases} 0, \ if \ |P_{si}| \le P_i^* \ or \ (P_{si} > P_i^* \ and \ V_{fi} = 0) \\ -\operatorname{sign}(P_{si}) K_{fi}(|P_{si}| - P_i^*)(V_{2i} + V_{fi}), \\ if \ P_{si} < -P_i^* \ or \ (P_{si} > P_i^* and \ V_{fi} > 0), \end{cases}$$

where $P_i^* = P_{ki}V_{1i}/(V_{fi} + V_{1i})$, P_{si} is the pressure in the solid (undamaged) part of the i-th component in the mixture, V_{1i}, V_{2i}, P_{ki}, K_{fi} are the experimentally determined constants of the material.

Pressure in the undamaged component of the mixture is a function of specific volume and internal energy, and over the entire range of loading conditions, it is determined by the Mie-Grüneisen equation of state according to the formula [10]:

$$P_i = \rho_{0_i} u_{ci}^2 \mu_i + \rho_{0_i} u_{ci}^2 [1 - \gamma_{0_i}/2 + 2(b_i - 1)]\mu_i^2 + \rho_{0_i} u_{ci}^2 [2(1 - \gamma_{0_i}/2)(b_i - 1) + 3(b_i - 1)^2]\mu_i^3 + \gamma_{0_i} \rho_{0_i} E_i ,$$

Here $\mu_i = V_{0_i}/(V_i - V_{f_i}) - 1$, γ_{0_i} is the Grüneisen coefficient, V_{0_i} and V_i are the initial and current specific volumes, respectively, and u_{c_i} and b_i are the constants of the Hugoniot shock adiabat described by the relation:

$$u_{s_i} = u_{c_i} + b_i u_{p_i} ,$$

here u_{s_i} is the shock wave velocity and u_{p_i} is the particle velocity of a component behind the shock wave front.

Studying the deformation of multicomponent media, it is necessary to take into account the state and response of each component, as well as, in contrast to a homogeneous mixture, not only the displacement of the external boundaries of the selected volume, but also the displacement of components in the selected volume of the mixture. Here, we consider the equal pressures of the components to be a condition for joint deformation of components during the interaction in the mixture, which determines volume concentrations of the components [9]:

$$P = P_i(V_i, E_i) = P_j(V_j, E_j) = \ldots = P_N(V_N, E_N).$$

The change in porosity is assumed to be influenced only by the spherical stress component or pressure, whereas the components of the stress deviator are bounded by the independent deviatoric yield function [12]:

$$2G_i\left(\varepsilon_{i_{kj}} - \frac{1}{3}\varepsilon_{i_{qq}}\delta_{i_{kj}}\right) = \frac{dS^0_{i_{kj}}}{dt} + \lambda_i S_{i_{kj}},$$

where $dS^0_{i_{kj}}/dt$ - is the Jaumann derivative given by:

$$\frac{dS^0_{i_{kj}}}{dt} = \frac{dS_{i_{kj}}}{dt} - S_{i_{kq}} W_{i_{jq}} - S_{i_{jq}} W_{i_{kq}},$$

and $2W_{i_{kj}} = \partial \upsilon_{i_i} / \partial x_j - \partial \upsilon_{i_j} / \partial x_i$. The parameter λ is zero for elastic deformation, and for plastic deformation is determined from the Mises yield criterion [12]:

$$S_{i_{kj}} S_{i_{kj}} = \frac{2}{3}\sigma_i^2 .$$

Here G_i is the shear modulus and σ_i is the dynamic yield point:

$$G_i = G_{0_i} K_{T_i}\left(1 + \frac{c_i P_i}{(1+\mu_i)^{1/3}}\right)\frac{V_{3_i}}{(V_{fi} + V_{3_i})},$$

$$\sigma_i = \begin{cases} \sigma_{0_i} K_{T_i}\left(1 + \dfrac{c_i P_i}{(1+\mu_i)^{1/3}}\right)\left(1 - \dfrac{V_{f_i}}{V_{4_i}}\right), \ if \ V_{f_i} \le V_{4_i}, \\ 0, \ if \ V_{f_i} > V_{4_i} \end{cases},$$

$$K_{T_i} = \begin{cases} 1, & if \ T_{0_i} \le T_i \le T_{1_i} \\ \dfrac{T_{m_i} - T_i}{T_{i_m} - T_{i_1}}, & if \ T_{1_i} < T_i < T_{m_i} \\ 0, & if \ T_i \ge T_{m_i} \end{cases}$$

In the computations, the function $K_{T_i}(T_i)$ was chosen to model the nonthermal character of plastic deformation and dynamic strength of solids at high strain rates (10^4 s^{-1} or higher).

The temperature was calculated using the following ratio:

$$dT_i = \begin{cases} d(E_i - E_{0xi})/c_{pi}, & if \ T_i < T_{mi} \\ 0, & if \ T_i = T_{mi}, \\ d(E_i - E_{0xi} - \Delta H_{mi})/c_{pi}, & if \ T_i > T_{mi} \end{cases}$$

where the specific heat capacity c_{pi} increases linearly with increasing the temperature up to the melting point of a substance, E_{0xi} is the "cold" component of the specific internal energy, T_{mi} is the melting temperature, ΔH_{mi} is the specific melting heat of the *i*-th component.

2. Experimental results

In the experiments the powder mixture of aluminum and fluoroplastic (Al/Tf) was used as a loaded material that was placed in a cylindrical ampoule. The powders were mixed in an AGO-2U planetary mill. The mass fractions of components were 30/70 (Al/Tf), which was close to the stoichiometry of the aluminum fluoride (AlF_3) formation. The initial porosity of the loaded mixture components was varied in the experiments. In the first case, the initial porosity of the mixture components was 0.5 (ratio between the volume of pores and the total volume) and in the second case, the initial porosity was 0.045. The height of the cylindrical ampoule was 95 mm, the external diameter was 20 mm. The thickness of the lateral wall of the ampoule was 3 mm. The ampoule was closed with lids 10 mm in sickness. The ampoule was loaded by an explosive (ammonite 6GV). The external diameter of the explosive was 45 mm. The density of the explosive was 1.0 g/cm^3, and the detonation velocity was 3.85 km/s.

The first ampoule was broken into large fragments (Fig. 1).

Figure 1. Fragments of the ampoule after explosive loading.

The second ampoule was cracked along the whole length of the material (Fig. 2).

Only the upper layer of the sample 5 mm in thickness was saved after explosive loading (Fig. 2). There are no any traces of the reaction in this part of the sample according to X-ray diffraction analysis.

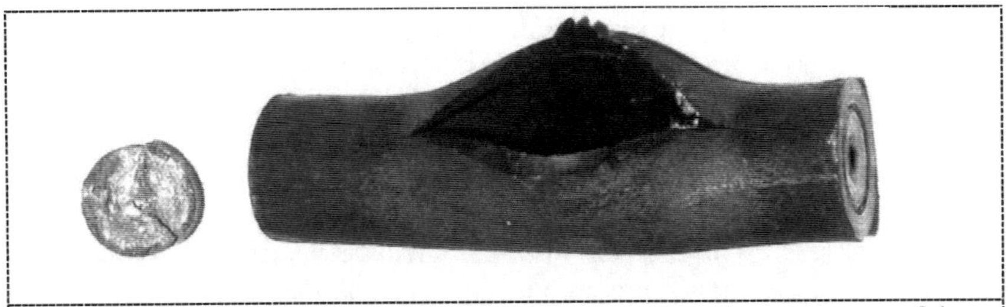

Figure 2. Ampoule and layer of the sample from the upper part of the ampoule after explosive loading.

3. Numerical results

To identify the reasons for destruction of the ampoules, the numerical computations were carried out by using the model of a multicomponent medium [7]. We numerically solved an axisymmetric problem of a cylindrical steel ampoule containing a porous Al–S mixture under explosive loading. The mass fractions of the components were as follows: Al – 0.35; S – 0.65. After initiation of a chemical reaction, aluminum sulfide Al_2S_3 was formed as a reaction product. We chose the reaction initiation criteria in terms of temperature, T_η = 933 K (aluminum melting point), and pressure, P_η= 0.9 GPa [7-10]. The reaction rate K_0 was taken to be 260.8 GJ/(kg s), and K_p = 20.0. Numerical computations have been carried out by using a research software package on the basis of the finite element method modified to simulate the high-speed loading.

Fig. 3 shows the distribution of pressure profiles in the Al–S mixture placed in a cylindrical ampoule at different moments of time, describing the dynamics of the shock wave process during the explosive synthesis of aluminum sulfide.

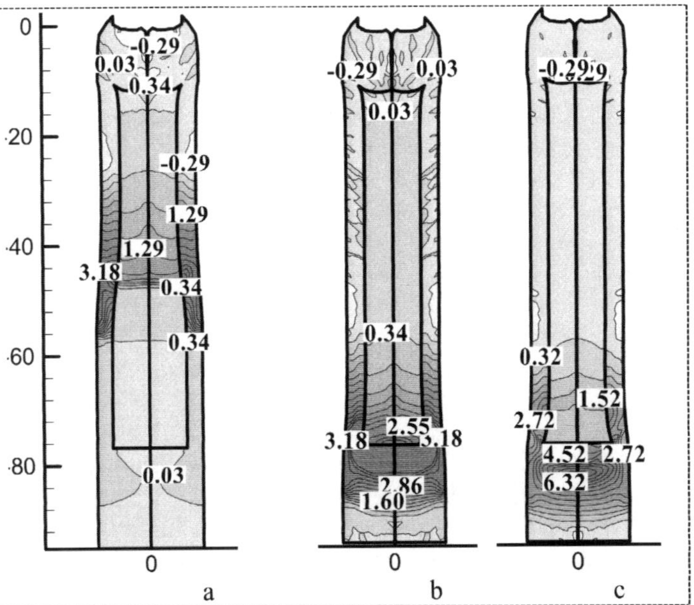

Figure 3. Distribution of pressure profiles (GPa) in a mixture of reacting porous components Al/S in an axial section of the ampoule at different moments of time: (a) 20 μs, (b) 31 μs, (c) 32 μs.

A shock wave propagates along the sample subjected to the action of explosion products. At the same time, the wave front in the mixture is located behind the wave front in the ampoule walls due to the collapse of pores in the mixture. After the reflection of the shock wave from the bottom of the ampoule as a compression wave, the pressure in the lower part of the sample increases considerably (Fig. 3c) and is higher than 6 GPa.

Analysis of the numerical and experimental results suggests the following dynamics for the development of chemical transformations in exothermic mixtures placed in the cylindrical ampoule under explosive loading. In the upper and central parts of the ampoule, the reaction is initiated in the transient shock wave. After reflection of the transient shock wave from the bottom lid of the ampoule as a compression wave, a sharp increase is observed in the bottom of the ampoule, which leads to the increase in rate of chemical transformations in the lower part of the mixture. The high heat release rate during the chemical reaction in the bottom of the ampoule causes the formation of a gas phase, which, in turn, leads to the increase in pressure in this area and the damage of the ampoule, and the damage process is initiated in the bottom of the ampoule.

Conclusion

The peculiarities of solid-phase synthesis were studied experimentally and numerically in the aluminum-fluoroplastic and aluminum-sulfur mixtures placed in cylindrical ampoules under explosive loading. The experimental results have shown that the use of a mixture capable of ultrafast exothermic reactions leads to destruction of a cylindrical ampoule under explosive loading.

The numerical results have shown that after reflection of the transient shock wave from the bottom lid of the ampoule as a compression wave, there is a sharp increase in pressure, which is accompanied by the increase in rate of chemical reactions. The high rate of heat release during the chemical reaction in the bottom of the ampoule causes the formation of a gas phase, which leads to a further increase in pressure and the destruction of the ampoule.

Acknowledgments

This work was supported by the Russian Foundation for Basic Research (project no. 14-03-00666a).

References

[1] L B Pervukhin, M I Alymov, I V Saikov, R D Kapustin, O L Pervukhina, E V Petrov, Explosive Compaction of Ceramic Powders, Letters on materials **5** (2015) 57-60

[2] V A Shcherbakov, A N Gryadunov, N V Sachkova, A V Samokhin, Combustion Synthesis of Composites Based on Titanium and Chromium Borides, Letters on materials **5** (2015) 20-23

[3] M I Alymov, L B Pervukhin, A S Rogachev, O L Pervukhina, I V Saikov, Combination of SHS and Shock Wave Compacting for Production of Composite Materials, Letters on materials **4** (2014) 153-158

[4] A Yu Dolgoborodov, M N Makhov, A N Streletsky, I V Kolbanev, M F Gogulya, V E Fortov, The Possible Presence of Detonation in a Mechanically Activated Composite of Aluminum – Teflon, .Khimicheskaya Fizika **23** (2004) 85-88.

[5] R I Nigmatulin, Dynamics of Multiphase Media. Vol. 1. New York: Hemisphere, (1991) 507.

[6] V F Kuropatenko, Momentum and Energy Exchange in Nonequilibrium Multicomponent Media, J Applied Mechanics and Technical Physics **46** (2005) 1-8.

[7] O Ivanova, S Zelepugin, A Yunoshev, V Silvestrov, A Multicomponent Medium Model for Reacting Porous Mixtures under Shock Wave Loading, J. Energ. Mater. **28** (2010) 303–317

[8] O V Ivanova, S A Zelepugin, Condition for Joint Deformation of Components in Mixtures under Shock Wave Compaction, J Tomsk State University Journal of Mathematics and Mechanics **1** (2009) 54-61

[9] O V Ivanova, S A Zelepugin, A S Yunoshev, V V Sil'vestrov, Experimental and Numerical Research in Explosive Loading of Two- and Three-Component Solid Mixtures, Eurasian Chemico-Technological Journal **16** (2014) 3-9

[10] S A Zelepugin, O V Ivanova, A S Yunoshev, V V Sil'vestrov, The Development of the Aluminum Sulfide Synthesis Reaction on Explosive Loading of a Cylindrical Ampoule, Doklady Physical Chemistry **434** (2010) 172–176

[11] G I Kanel, V E Fortov, S V Razorenov, Shock Waves in Condensed-State Physics, Physics-Uspekhi, **50** (2007) 771-791.

[12] M L Wilkins, Computation of Elastic-Plastic Flow, in Methods of Computational Physics. Edited by B. Alder, S. Fernbah and M. Rotenberg. Academic Press, New York, **3** (1964)